International Conference on Green Agro-Industry and Bioeconomy (ICGAB 2017)

IOP Conference Series: Earth and Environmental Science
Volume 131

Batu, Indonesia
24 - 25 October 2017

Editor:

Erryana Martati

ISBN: 978-1-5108-6010-0
ISSN: 1755-1307

Printed from e-media with permission by:

Curran Associates, Inc.
57 Morehouse Lane
Red Hook, NY 12571

Some format issues inherent in the e-media version may also appear in this print version.

Printed by Curran Associates, Inc. (2018)

For permission requests, please contact the Institute of Physics at the address below.

Institute of Physics
Dirac House, Temple Back
Bristol BS1 6BE UK

Phone: 44 1 17 929 7481
Fax: 44 1 17 920 0979

techtracking@iop.org

Additional copies of this publication are available from:

Curran Associates, Inc.
57 Morehouse Lane
Red Hook, NY 12571 USA
Phone: 845-758-0400
Fax: 845-758-2633
Email: curran@proceedings.com
Web: www.proceedings.com

Table of contents

Volume 131

International Conference on Green Agro-industry and Bioeconomy (ICGAB 2017)

24–25 October 2017, Batu City, East Java, Indonesia

Accepted papers received: 1 March 2018
Published online: 22 March 2018

Preface

Papers

ON FARM - FOOD PRODUCTION SYSTEM

OFF FARM - FOOD PRODUCTION SYSTEM

PRODUCTION SYSTEM MANAGEMENT AND REGULATION

BIOTECHNOLOGY AND BIOPROCESSING

RENEWABLE ENERGY

WASTE AND ENVIRONMENTAL MANAGEMENT

IOP Conf. Series: Earth and Environmental Science **131** (2018) 011001 doi:10.1088/1755-1315/131/1/011001

Preface

On behalf of the Committee, we would like to thank you to all participants from seven countries (the Netherland, Japan, Turkey, India, Thailand, Vietnam, and Indonesia) for their great supports and contribution to the International Conference on Green Agro-Industry and Bioeconomy (ICGAB) 2017 held on October 24-25, 2017 at Universitas Brawijaya Malang Indonesia. The conference is intended to provide a platform for academia, researchers, practitioners and policy makers to share their latest achievements and future challenges in terms of food, feed, bio-based products and bioenergy.

ICGAB 2017 was well attended with six topics covered, where all presenters did an outstanding job of sharing their expertise and experiences, beneficial to support the creation of circular economy in Indonesia. This year, we collaborated with PATPI (Indonesian Association of Food Technologist), PERMI (Indonesian Society for Microbiology), KBI (Indonesian Biotechnology Consortium), IATPI (Indonesian Society of Sanitary and Environmental Engineers),PERTETA (Indonesian Society of Agricultural Engineering), APTA Indonesia (Indonesian Society of Agricultural Technologist), and IPSBI (Indonesian Association of Biotechnology Program), in promoting and diseminating the event. We also collaborated with IOP Conference Series: Earth and Environmental Science to publish the conference proceedings.

This volume records the selected manuscripts submitted to the ICGAB 2017. All the manuscripts were peer-reviewed arranged by the editorial committee.

The Editors would like to thank you to the reviewers who provided their valuable comments and suggestions, to all the authors for their trust and contribution, and also to IOP publisher for their support for publication of the proceedings.

Editor-in-Chief

IOP Conf. Series: Earth and Environmental Science **131** (2018) 011002 doi:10.1088/1755-1315/131/1/011002

CONFERENCE COMMITTEE

Scientific Committee

Prof. Hiroyuki Sakakibara (Food Science and Nutrition - University of Miyazaki - Japan)
Prof. Jenshin Linn (Department of Food Science, National Pingtung University of Science and Technology-NPUST, Taiwan)
Prof. Jong Hoon Chung (Seoul National University, South Korea) - Biomaterial
Prof. Tsair Wang Chung (Chung Yuan Christian University, Taiwan) - Bioprocess Engineering
Prof. Irwandi Jaswir (International Islamic University Malaysia)
Assoc. Prof. Dogan Karagan (Yilditz Technical University, Turkey)
Assoc. Prof. Lynsey Melville (Birmingham City University, UK)
Dr. M. Roil Bilad (Universiti Teknologi Petronas, Malaysia) - Membrane Technology
Dr. Syazwani Idrus (University of Putra Malaysia, Putrajaya Malaysia)
Dr. Chan Weng Howe (Department of Software Engineering, Faculty of Computing - Universiti Teknologi Malaysia)
Dr. Mohd. Saberi Mohamad (Department of Software Engineering, Faculty of Computing - Universiti Teknologi Malaysia)
Prof. Rindit Pambayun [Indonesian Food Technologist Association (PATPI)]
Prof. Misri Gozan [Indonesian Biotechnology Consortium (KBI)]
Dr. Siswa Setyahadi [Indonesian Society for Microbiology (PERMI)]
Dr. Adi Joko Guritno [Indonesian Association of Agro-industry Technologist (APTA)]
Dr.rer.nat. Maria Goretti M. Purwanto (Technobiology Faculty, Universitas Surabaya - IPSBI)
Dr. Eng. Ir. Priana Sudjono [Institut Teknologi Bandung - Indonesian Society of Sanitary and Environmental Engineers (IATPI)]
Dr. M. Sintorini [Universitas Trisakti - Indonesian Society of Sanitary and Environmental Engineers (IATPI)]
Tunjung Mahatmanto, Ph.D [Faculty of Agricultural Technology Universitas Brawijaya-Indonesia/Arizona State University-USA]
Mochammad A. Junaidi, MP [Faculty of Agriculture Universitas Brawijaya – Indonesia]
Dr. Imam Santoso [Faculty of Agricultural Technology Universitas Brawijaya-Indonesia]

Steering Committee

Dr. Sudarminto Setyo Yuwono (Dean - Faculty of Agricultural Technology)
Prof. Sri Kumalaningsih (Agroindustrial Technology - Faculty of Agricultural Technology, Universitas Brawijaya)
Prof. Simon Bambang Widjanarko (Agricultural Product Technology - Faculty of Agricultural Technology, Universitas Brawijaya)
Prof. Bambang Suharto (Agricultural Engineering - Faculty of Agricultural Technology, Universitas Brawijaya)
Prof. Harijono (Agricultural Product Technology - Faculty of Agricultural Technology, Universitas Brawijaya)
Prof. Sumardi HS. (Agricultural Engineering - Faculty of Agricultural Technology, Universitas Brawijaya)
Prof. Wignyanto (Agroindustrial Technology - Faculty of Agricultural Technology, Universitas Brawijaya)

IOP Conf. Series: Earth and Environmental Science **131** (2018) 011002 doi:10.1088/1755-1315/131/1/011002

Prof. Yunianta (Agricultural Product Technology - Faculty of Agricultural Technology, Universitas Brawijaya)
Prof. Teti Estiasih (Agricultural Product Technology - Faculty of Agricultural Technology, Universitas Brawijaya)
Prof. Sutiman (Deapartment of Biology - Faculty of Mathematics and Natural Sciences – Universitas Brawijaya)

Editors

Dr. Erryana Martati (Editor in Chief)
Dr. Siti Narsito Wulan
Sri Suhartini, Ph.D
Dr. Eng. Akhmad Adi Sulianto
Yusuf Hendrawan, Ph.D
Yusuf Wibisono, Ph.D
Wenny Bekti S, Ph.D
Fajri Anugroho, Ph.D
Ade Gafar Abdullah, Ph.D
Hera Sisca Prasmita, M.Sc
Rosalina Ariesta L., M.Si
Retno Damayanti, MP
Shinta Rosalia Dewi, M.Sc
Nur Lailatul Rahmah, M.Si
Claudia Gadizza Perdani, M.Si
Dego Yusa Ali, M.Sc.
Wendra G. Rohmah, MP
Riska Septifani, MP

IOP Conf. Series: Earth and Environmental Science **131** (2018) 011002 doi:10.1088/1755-1315/131/1/011002

ORGANIZER LOGO

SPONSOR LOGO

Patpi
INDONESIAN FOOD TECHNOLOGIST ASSOCIATION

PERHIMPUNAN MIKROBIOLOGI INDONESIA
INDONESIAN SOCIETY FOR MICROBIOLOGY

IATPI
INDONESIAN SOCIETY OF SANITARY AND ENVIRONMENTAL ENGINEERS

IPSBI
INDONESIAN SOCIETY OF BIOTECHNOLOGY STUDY PROGRAMME

SEAMEO SEARCA

APTA

KBI Indonesian Biotechnology Consortium

IOP Conf. Series: Earth and Environmental Science **131** (2018) 011003 doi:10.1088/1755-1315/131/1/011003

Peer review statement

All papers published in this volume of *IOP Conference Series: Earth and Environmental Science* have been peer reviewed through processes administered by the proceedings Editors. Reviews were conducted by expert referees to the professional and scientific standards expected of a proceedings journal published by IOP Publishing.

IOP Conf. Series: Earth and Environmental Science **131** (2018) 012001 doi:10.1088/1755-1315/131/1/012001

Application of neem (*Azadirachta indica*) as biological pesticides in cocoa seed (*Theobroma cacao*) storage using various local adsorbent media

S A Mardiyani, Sunawan and A E Pawestri

Department of Agrotechnology, Faculty of Agriculture, University of Islam Malang, Indonesia

Email: asmadwip69@gmail.com; asmadwip69@unisma.ac.id

Abstract. Cocoa seeds are recalcitrant (the water content is more than 40%) that require special handling. The use of adsorbent media to reduce the decrease in the quality of cocoa seeds and extend their shelf life in this storage has not been widely done. Local adsorbent media such as sawdust, sand and ash have the potential to maintain the viability of cocoa seeds. The objective of this research was to determine the interaction of the application of neem (*Azadirachta indica*) as biological pesticides and the use of various natural adsorbent media in the storage of cocoa seeds (Theobroma cacao). It was an experimental study with a factorial design composed of three factors. The first factor was the medium adsorbent type for the storage of cocoa seed, which consists of three levels (river sand, ash, and sawdust). The second factor was the concentration of neem leaves for pre-storage treatment with three levels (10, 20, and 30%). The third factor was the storage time (10 and 20 days). The results of the study indicated that the combination of the three factors showed a significant interaction in the height of the plant and the diameter of the stem of the seedling at 28 days after sowing. The fresh weight of the seedlings of the seeds that were stored in ash media gave a better result than the seedlings of seeds that had been stored in the river sand and the sawdust as adsorbent media. The application of 20% extract of neem leaves gave the best influence for the seeds that were stored for 20 days.

1. Introduction

Cocoa beans (*Theobroma cacao*) is one of the most important agricultural export products in Indonesia. Along with the growing demand for the product, the production of cocoa beans has increased enormously due to the rapid expansion of the participation of small farmers. Indonesia is the third world exporter of cocoa beans [1,2]. The production of cocoa beans depends on many factors, including the quality of the seeds. It is one of the important production factors that determine the success of the expansion of the plantation area. The problem is that cocoa seeds are recalcitrant (they have more than 40% water content) that require special handling during storage [3]. Calistru et al. [4] said that recalcitrant hydrated seeds can be stored only for days or months. However, the condition of recalcitrant seeds (high in humidity) makes them easier to deteriorate and also lead to the proliferation of fungi and bacteria. Storing recalcitrant seeds in moist media will cause them to weaken rapidly, losing vigor and viability sooner. At that time, the seeds become susceptible to attack by microorganisms, especially fungi. Therefore, it needs special treatments to reduce the decrease in the quality of the cocoa seed during the entire storage time. The combination of the appropriate insecticide and media storage will maintain the moisture content of the seeds and reduce the deterioration of the seeds throughout the storage time.

The use of natural adsorbent media in the storage of cocoa seeds to extend shelf life is still limited. Natural adsorbent media such as sawdust, sand and ash have the potential to maintain the viability of cocoa seeds. It will delay germination of the seed and reduce the attack of the disease. The combination of natural adsorbent media with the application of neem as natural antifungal at several doses will provide a significant benefit to maintain the quality of the cocoa seed in a period of storage time. This research has carried out a combination of several natural local adsorbent media combined with the biological treatment of pesticides using neem leaves extract to maintain the quality of the cocoa seeds during storage. This study was aimed to determine the interaction of the application of neem leaves (*Azadirachta indica*) in several doses as organic pesticides and the use of various local adsorbent media in the storage of cocoa to increase performance of cocoa seedlings.

2. Material and Methods

2.1.Material

The research was carried out in the Agricultural Laboratory of UNISMA and in the greenhouse of the Faculty of Agriculture, University of Islam Malang. The materials used were cocoa sp seeds Creole f2. The seeds were selected based on the uniform size and weight. They must be free from disease and free from mechanical damage. And for the germination and seedling process, we use the mixture of soil, sand, and manure with a ratio of 2:1:1 as a medium. The tools used in this research were polyethylene bags, a digital scale, beaker glass, and polybags.

The experimental design used was a random block design, organized in factorial composed of three factors. The first factor was the use of natural adsorbent media (river sand, ash, and sawdust). The second factor was a concentration of neem leaves extract that consisted of 3 levels: 10%, 20%, and 30%. The third factor was days of seeds storage in the media (10 days and 20 days).

2.2. Research Implementation

The seeds were selected, washed and dried under the sun. The dried seeds were then dissolved in distilled water at a concentration of 10%, 20% and 30% neem extract for 10 minutes. The storage media used in this investigation (river sand, ash, and sawdust) were sifted with the same particle size and then pasteurized at a temperature of 70°C. The seeds were stored in each medium according to the treatments for 10 and 20 days. Then, the seeds were planted to be seedlings.

The variables measured in this study were seed percentage weight loss, moisture content and seed germination after according to the storage time; height, stem diameter, total leaves, the total length of the root and total dry weight of the seedlings. The seedling growth was observed at 7, 14, 21, 28 days after sowing. The data from three time replications for each treatment were analyzed using Analysis of Variance continued wit with Duncan Multiple Range Test (DMRT) 5% to see the difference between the treatments for the significant ones.

3. Results and Discussion

The result of the research showed that the seeds that were stored in sawdust media had the lowest weight decrease which was significantly different with the seeds that were kept in river sand and ash media (Table 1). Karthikeyan *et al.* [5] stated that ash is capable adsorbent media in keeping the seed moisture content and prolonging the storage time.

The result also showed that seeds stored for ten days had better germination percentage than seeds stored for 20 days. It showed that the quality of the seeds stored for 20 days had decreased. Table 2 shows that the combination of 20% leaves of neem extract and 10 days of storage gave the best germination compared to the other combination. Bewley *et al.* [6] stated that the viability of the seeds showed a close relationship with the storage period.

Table 1. Percentage weight loss of cocoa seed based on media treatment.

Media Treatment	Percentage weight loss (%)
River Sand	24.32±1.28 a
Ash	36.42±0.82 b
Sawdust	18.80±1.04 a

significantly different in ANOVA Test (5%)

Table 2. Seed germination after 7 days planting based on neem extract treatment and time storage.

Treatments		Germination (%)
% neem extract	Storage time	
10 %	10 days	84.44±2.68 bc
	20 days	70.00±2.64 ab
20 %	10 days	92.22±3.85 c
	20 days	60.00±2.29 a
30%	10 days	82.22±2.54 bc
	20 days	74.44±2.21 b
	Sig (P:0.05)	

The percentage of germination of the seeds treated with 20% extract of neem leaves and kept for 10 days has the highest value. It showed that the application of 20% neem leaves extract was more effective in preventing fungal attack during storage compared to the other concentrations. According to Bansod and Rai [7] neem leaves extract has an excellent potential to produce antifungal compounds against the fungus Candida albicans and Aspergillus niger. It was also affirmed that the extracts of neem leaves can be used as a alternative to natural insecticides to control insect pests [8].

Table 3. Seedling height and fresh weight after 28 days of planting.

Treatments			Height (cm)	Fresh Weight (g)
Media	% of neem extraxt	Time storage		
	10% (P1)	10 days (H1)	18.52±1.36 cd	4,87±0,78 c
		20 days (H2)	12.58±0.52 bcd	3,69±0,56 bc
	20% (P2)	10 days(H1)	18.26±0,65 cd	4,81±0,73 c
River Sand		20 days(H2)	6.65±0,51 bc	2,28±0,62 bc
	30% (P3)	10 days (H1)	18.42±0,52 cd	4,80±0,72 c
		20 days (H2)	9.77±0,67 bc	3,74±0,65 c
Ash	10% (P1)	10 days(H1)	18,99±1,31 d	5,15±0,74 c
		20 days(H2)	14.43±1.07 bcd	4,39±0,35 c
	20% (P2)	10 days (H1)	17,09±0,80 cd	4,77±0,68 c
		20 days (H2)	15.66±0,77 cd	4,43±0,51 c
	30% (P3)	10 days(H1)	18.25±0,75 cd	4,54±0,50 c
		20 days(H2)	14.67±1.52 cd	3,97±0,50 c
Sawdust	10% (P1)	10 days (H1)	18.56±0,51 cd	4,71±0,62 c
		20 days (H2)	0.00±0.00 a	0,00±0,00 c
	20% (P2)	10 days(H1)	16.12±0,78 c	4,26±0,25 c
		20 days(H2)	9,06±0,41 bc	3,08±0,39 bc
	30% (P3)	10 days(H1)	15.36±0,36 c	4,77±0,86 c
		20 days(H2)	8,03±0,96 c	3,89±0,84 c
			Sig(P:0.05)	Sig(P:0.05)

The statistical analysis showed an interaction between the 3 factors for the height of the stem and the fresh weight of the seedling 28 days after sowing. Table 3 shows that seeds that had been stored in ash media with 10% treatment with neem extract for ten days resulted in the highest and significantly different seedling with some other treatments. The seedlings of the seeds that had been stored in ash media with 10% neem extract for ten days had the highest result. It was the best combination for the height of the seedlings compared to the other combinations. It also showed that seeds stored in ash media treated with 10% biological pesticides for ten days had the highest fresh weight. Karthikeyan *et al.*,[3] affirmed that the use of ash and neem oil in the storage of seeds has been widely used to prevent the deterioration of seeds caused by insects and attacks of microorganisms. The application of ashes to the seed storage media can control weight seed losses in sorghum, considerably up to 80%. The use of neem oil is known as popular post-harvest procedure in India. Neem oil acted as the repellent of several insects. The result also showed that the seeds that had been stored for 20 days had a worse vigor and viability compared to the seeds that had been stored for 10 days. Olaiya (2016) [9] said that the longer retention period of seed cocoa after harvest will result in worse performance of the seedlings.

4. Conclusion

The combination of the media, the concentration of the neem leaf extract and the storage time showed a significant interaction in the height of the seedlings and the fresh weight of the seedlings after 28 days after sowing. The yield of seedlings from seed that had been stored in ash media was better than seedlings stored in sand and sawdust. The application of neem extract had a positive effect on the performance of the seedlings.

References

[1] Effendy E H, Setiawan B, & Muhaimin A W 2013 *J Agric Sci.* **5** 12 72
[2] Junaid, M. 2015 *Manufacturing Confectioner*, **47**
[3] Bewley J D and M Black 1994 *Seeds ; Physiology of Development and Germinaiton* Plenum Press, New York and London. p 367
[4] Calistru C., McLean M., Pammenter N W & Berjak P 2000 *Seed Sci.Res.* **10** 3 341-353
[5] Karthikeyan C D Veeraragavathatham D Karpagam and A Firdouse 2009 *Indian Journal of Traditional Knowledge* **8** 4 564-568
[6] Bewley J D Bradford K J Hilhorst H W & Nonogaki H 2013 *Seeds Physiology of Development, Germination and Dormancy*, 3rd Edition Springer New York p392
[7] Bansod S & Rai M. 2008 *World Journal of Medical Sciences* **3** 2 81-88
[8] Girish K & Shankara Bhat S 2008 *EJBio.* **4** 3 201-111
[9] Olaiya A O 2016 *Nigerian Journal of Ecology* **15** 1 19-23

Influence of Paclobutrazol (PP_{333}) and Sridiamin (Human hair-derived aminoacid mixture) on growth and quality of Tomato PKM-1

S Suja[1] and N Anusuya[2]

[1]Associate professor and Head of Biochemistry Department, Bharathiar University, Coimbatore, India
[2]Research Scholar, Biochemistry Department, Bharathiar University, India

Email: sujaramalingam08@gmail.com

Abstract. Tomato is one of the most popular vegetable in subtropics and tropics. Plant growth regulators have potential for manipulating growth of many agricultural crops. Among the plant growth retardants, paclobutrazol (PP333) has been reported to exert profound effects on improving the yield of certain vegetables. Aminoacids are essential prerequisite for plant growth. Sridiamin a natural blend of 17 essential L-aminoacids, fortified with vitamins ensuring better crop growth and higher productivity. Therefore the present study was designed with 5mg and 10 mg concentration of PP333 as soil drench and a foliar spray of sridiamin of 0.5% and 1% concentration as individual and as combined treatment improved the yield and quality of tomato PKM1.Various biometric parameters, along with chlorophyll, starch, aminoacid and protein content were analysed in the leaves. In fruit analysis like titrable acidity, total soluble solids, ascorbic acid, lycopene, total sugars, macronutrients and micronutrients were analysed.

1. Introduction

One of the most popular Solanaceae families of self-pollinated crop is Tomato (*Lycopersicone sculentum* Mill.). It cultivated in all countries from temperate to tropical region. It is a functional food, plays a major role in human nutrition due to its good levels of vitamins, minerals and excellent antioxidant qualities [1,2]. Tomato requires a day temperature of 21–28°C and moderately cool night temperature of 15–20°C for proper fruit setting. Its production in the world in 2013 reached to 163.9 million tons and average yield of 34.7 tons per hectare [3]. The utility of tomato vary based on the type; cherry tomato for salads, big slicers for sandwiches and firm fleshed paste tomatoes for sauces, freezing and canning [4]. In 1978 sour type PKM-1 tomatoes were released and cultivated in Tamil Nadu Agricultural University and the result was extremely high yield and good quality of tomatoes.

The beneficial effects of growth regulators may vary based upon the type, concentration, and method of application [5]. For instance plant growth regulators of auxin and gibberellins are effective in increasing both yield and quality of tomato reported one by [6]. Abscisic acid is a direct mediator for inhibition of lateral root development, photosynthesis and respiration in leaves [7]. Salicylic acid plays a vital role in plant defense response against pathogen attack. Therefore the use of fertilizers and synthetic growth regulants not only increase fruit set and fruit ripening but also alters the quality of tomato fruit [8].

The anti-gibberellin a potent growth retardant and triazole group of fungicide is "Paclobutrazol" (PP_{333}), used for flower induction, fruit setting and reducing the extension of shoot growth for many crops. In addition, it is more stable in xylem and sub-xylem vessels [9,10,11]. Soil drenches and foliar application is the two main methods of application.

Another essential and required material for plant growth is the amino acids. Their importance has been accepted worldwide as a means to increase the yield and overall quality of crops. Sridiamin is a mixture of 17 natural and essential amino acids derived from natural protein (Keratin). It is non-poisonous, non-hazardous and a liquid amino acid product (protein hydrolysate). This formulated product was developed by Srinivasa Cystine Ltd., Hyderabad, for the first time in India. And it was fortified with vitamins ensuring better crop growth and higher productivity. Foliar nutrition in the form of protein hydrolysates (liquid amino acid) and foliar spray provide readymade building blocks for protein synthesis [12].

The present investigation was conducted to evaluate the effect of Paclobutrazol (PP_{333}), and Sridiamin (Human hair derived amino acid) on the growth and quality of Tomato cv.PKM-1.

2. Material and Methods

The experiment was carried out using randomized block design with four replications for each treatment.

Crop Husbandry- Initially the breeder's quality seeds of tomato cv.PKM-1 were sown in raised beds. Then the seedlings were 12-15cm tall, they were transplanted on the sides of ridges spaced at 60cm in the row and 60cm between rows.

Treatment Schedule- Paclobutrazol (PP_{333}) was applied as soil drench one month after transplanting. Sridiamin, exogenous mixture was applied on fortnightly basis as a foliar spray upto a period of 2 months. The T_1-Control, T_2- 0.5% Sridiamin, T_3-1.0% Sridiamin, T_4- 5mg Paclobutrazol, T_5- 10mg Paclobutrazol, T_6- 0.5% Sridiamin + 5mg Paclobutrazol, T_7- 1.0% Sridiamin + 5mg Paclobutrazol, T_8- 0.5% Sridiamin + 10mg Paclobutrazol, T_9- 1.0% Sridiamin + 10mg Paclobutrazol.

Plant Biometric Parameters- The plant height and spread, number of laterals per plant, flower clusters and flowers per cluster, days to first flowering were counted in the experiment part.

Fruit Characters- The number of fruits and yield per plant were counted and expressed in grams. Mean fruit weight was also obtained by,

Mean fruit weight = Total yield per plant (g) / Total number of fruits per plant (g).

Biochemical Parameters

Estimation of total leaf chlorophyll- The total chlorophyll estimation in leaf was followed by the method of [13].

Estimation of starch in leaves- Starch estimation in leaves was determined by the method of [14].

Total protein in leaves- The estimation of total protein in fruits and leaves were identified by the method [15].

Total free aminoacids in leaves- The total free amino acids in the fruit and leaves were estimated by [16] method.

Titrable acidity in fruits- The titrable acidity in fruits was estimation was followed by the method [17].

Total soluble solids (TSS)- The TSS of fruit pulp was determined using ERMA Hand Refractometer and expressed as degree brix.

Ascorbic acid in fruits- The ascorbic acid in fruits was estimated by the [18] method.

Lycopene estimation in fruits- The estimation of lycopene content in fruits was followed by the method of [19].

Determination of total sugars in fruits- The total sugars in fruits were determined by [20] method.

Determination of Total Nitrogen- The total nitrogen content determination was followed by the method of [21].
Estimation of Phosphorous- The estimation of phosphorus was determined by [22] method.
Estimation of Potassium- The estimation of potassium content in the sample was measured by [23] method.
Estimation of Manganese, Zinc, Iron and Copper- The estimation of micronutrients was followed by the method of [24].

3. Result and Discussion

3.1 Plant biometric parameters

a. Plant height and spread

In tomato plant biometric parameters to the various treatments are given Table 1. As the foliar spray of Sridiamin individually influenced the plant height and spread to a greater extent than the other treatments. This might be due to enhancement of cell division and cell elongation. Paclobutrazol has reduced the plant height about 16.1% and spread about 13.1% at 10mg per plant (T_5) compare to control. The reduction percentage in T_5 was found to be on par with T_4. However, in the combined treatment T_7 showed increase in plant height and spread when compared to the other treatments.

b. Number of laterals per plant

Application of Sridiamin at T_2 and T_3 levels compare to control increased the laterals to about 13.6% and 27.2%. This was lesser when compared to that found in T_4. The increment in T_4 and T_5 was found to be 95.4% and 72.7% respectively. In combined treatments, T_7showed best result. Thus reduction in number of laterals due to combined treatment compare to individual treatment of PP333 was observed. This might be due to reduced apical dominance. The similar result was also reported in *Brassica junecea* by [25].

c. Number of days to first flowering

PP_{333} alone or combination with Sridiamin reduces the days to first flowering significantly compare to control. PP_{333} owing to its antigibberellin activity could induce or intensify flowering by blocking the conversion of kaurene to kaurenoic acid a precursor of gibberellins [26]. The similar result obtained by [27].

Table 1. Biometric parameters.

Treatments	Plant Height (in cm)	Plant Spread (in cm)	Laterals per plant (in number)	Days to first flowering (in number)	Flower clusters per plant	Flower per cluster (in number)
T_1	65.6±1.02	67.7±1.02	5.5±0.04	69.3±1.02	5.5±0.05	4.6±0.05
T_2	79.8±1.01	69.6±1.01	6.3±0.03	63.5±1.04	6.8±0.03	5.6±0.02
T_3	82.8±1.01	73.0±1.01	7.0±0.02	64.3±1.03	8.3±0.02	5.9±0.01
T_4	58.5±1.03	52.0±1.03	10.8±0.01	57.0±1.05	6.3±0.04	5.3±0.04
T_5	55.0±1.05	46.6±1.04	9.5±0.02	55.5±1.07	5.8±0.04	5.4±0.03
T_6	64.9±1.02	54.3±1.02	8.3±0.05	54.3±1.08	7.0±0.01	5.5±0.02
T_7	65.9±1.01	54.9±1.02	8.5±0.04	57.8±1.06	7.3±0.01	5.4±0.04
T_8	53.8±1.4	43.6±1.04	7.0±0.01	58.5±1.06	6.5±0.05	5.3±0.03
T_9	50.4±1.5	44.3±1.01	7.5±0.03	56.5±1.05	6.3±0.03	5.2±0.01
SED	2.521	2.289	0.918	2.475	0.413	0.154
CD (0.05%)	5.202	4.725	1.895	5.108	0.851	0.318

Values are mean± SD of four replications

d. Number of flower clusters and flowers per cluster per plant

In T_3 and T_2 (8.3 and 5.9) increased number of flower clusters and flowers per cluster were seen among all the treatments compared with control but less than T_3. PP_{333} in T_4 and T_5 had more flower cluster (6.3 and 5.8) than control. However in T_6 and T_7 had (7 and 7.3) clusters, therefore a similar result was found.

e. Fruit Characters

Among all the treatments mean fruit weight, number of fruits per plant and yield per plant were found to be higher in $T_7$and T_6compared to the other treated group.

Table 2. Fruit Characteristics

Treatments	Mean fruit weight (g)	Number of fruits per plant	Yield per plant (kg)
T_1	39.7 ±0.02	41.0 ±1.03	1.6 ±0.04
T_2	42.6 ±0.03	46..0 ±0.02	2.4 ±0.07
T_3	45.1 ±0.02	47.0 ±0.01	2.0 ±0.05
T_4	30.6 ±0.04	46.0 ±0.03	1.4 ±0.03
T_5	23.4 ±0.01	42.0 ±0.04	1.0 ±0.11
T_6	55.2 ±0.01	48.0 ±0.02	2.6 ±0.01
T_7	46.5 ±0.02	51.0 ±0.01	2.0 ±0.03
T_8	38.3 ±0.04	46.0 ±0.04	1.8 ±0.04
T_9	33.3 ±0.01	41.0 ±0.05	1.7 ±0.06
SED	1.058	1.036	0.059
CD (0.05%)	2.184	2.198	0.124

Values are mean ± SD of four replications

3.2 Biochemical parameters

a. Total leaf chlorophyll and starch

Among all the treatments T_9 resulted in highest content of chlorophyll and all other treatment results found to be on par with each other. In addition the high level of starch observed (1.95g/100g) in T_5 followed by T_8. Application of both PP_{333} and Sridiamin individually and in combination has influenced the chlorophyll as well as starch content of the leaves. Paclobutrazol increases the chlorophyll content when compared to control in all the sampling of barley seedlings and tomato [28].

Table 3- Total chlorophyll and starch content in leaves.

Treatments	Total chlorophyll (mg/100g)	Starch content (g/100g)
T_1	0.313±0.08	0.85± 0.02
T_2	0.338±0.05	1.07±0.01
T_3	0.339±0.07	1.17±0.03
T_4	0.346±0.03	1.64±0.01
T_5	0.349±0.01	1.96±0.02
T_6	0.342±0.01	1.35±0.01
T_7	0.346±0.04	1.46±0.07
T_8	0.352±0.02	1.80±0.06
T_9	0.372±0.01	1.51±0.08
SED	0.004	0.043
CD (0.05%)	0.009	0.088

Values are mean ± SD of four replications

b. Total protein and free amino acids in leaves

In leaves highest protein content was noticed in T_3 and T_9 followed by T_7 and T_8. There was no significant difference noticed between T_2 and T_6 (11.9mg/g). Thus in leaves PP_{333} alone and in combination not showed any increased protein content but in fruits the level of protein is high when compared with Sridiamin alone and in combination treatments. Therefore combined application of Sridiamin and PP_{333} in either concentration has increased the quantity of protein in leaves and fruits.

The total free aminoacids content was found to be higher in fruits than leaves. Therefore Sridiamin shown highest amount of aminoacids individually and in combination. The results were depicted in Table 4. A positive correlation (r=+0.3866) and (r=+0.2791) were observed between total protein and free amino acids in leaves and fruit. [29] Has also shown that PP333 increased the aminoacid content of olive plants.

Table 4. Total protein and total free amino acids in leaves.

Treatments	Total protein (mg/100g)	Total free aminoacids (g/100g)
T_1	10.4±0.04	3.5±0.31
T_2	11.9±0.01	3.9±0.11
T_3	13.4±0.07	3.4±0.07
T_4	10.9±0.05	4.1±0.03
T_5	11.4±0.02	4.5±0.06
T_6	11.9±0.04	4.2±0.01
T_7	12.9±0.01	5.0±0.02
T_8	12.4±0.03	5.3±1.04
T_9	13.4±0.08	5.6±0.10
SED	0.085	0.141
CD (0.05%)	0.179	0.291

Values are mean ± SD of four replications

c. Titrable acidity, total soluble solids, ascorbic acid and lycopene in fruits

Titrable acidity levels higher in T_2 and T_3 about 19.6% and 37.2% compared to control. However plants treated with PP_{333} both individually and in combination resulted higher titrable acidity than control. The Total soluble solids content of the fruit varied between 4.03-5.48 degree brix.

Ascorbic acid content of the treatment was higher than the control. In T_5 followed by T_4 the highest ascorbic acid content was noticed. All other treatments did not differ significantly in the ascorbic acid content of fruits. PP_{333} had increased ascorbic acid content in *Actinidia arguta* [30] and in Tomato cv. Pusa ruby [31]. Lycopene is responsible for the red colour in tomato, though it has no nutritional value, its contribution to the colour of tomato has a great role in consumer acceptability [32]. P_{333} at T_4 and T_5 caused 60.9% and 85.85% increased lycopene content compared to control were observed. About 94.5% of highest lycopene content was noted in the T_9 plant followed by T_7. Therefore titrable acidity, total soluble solids, ascorbic acid and lycopene contents in fruits were higher in plants treated with PP_{333} individually or in combination were showed in Table 5.

Table 5. Titrable acidity, total soluble solids, ascorbic acid, and lycopene content in fruits.

Treatments	Titrable acidity (g/100g)	Total Soluble Solids (Degree birx)	Ascorbic acid (mg/100g)	Lycopene (mg/100g)
T_1	0.51±0.21	4.03±0.17	25.6±0.14	1.28±0.10
T_2	0.61±0.11	5.00±0.19	28.4±0.22	1.54±0.21
T_3	0.70±0.45	5.10±0.01	26.4±0.35	1.62±0.18
T_4	0.86±0.31	5.23±0.22	30.0±0.12	2.06±0.29
T_5	0.86±0.19	5.48±0.51	31.6±0.27	2.37±0.17
T_6	0.80±0.08	5.25±0.45	28.6±0.01	1.87±0.38
T_7	0.83±0.20	5.38±0.36	27.2±1.01	2.25±0.28
T_8	0.86±0.11	5.10±0.50	26.4±0.25	1.96±0.05
T_9	0.83±0.23	5.45±0.34	29.2±0.11	2.49±0.13
SED	0.013	0.043	0.272	0.076
CD (0.05%)	0.026	0.088	0.562	0.158

Values are mean ± SD of four replications

d. Total sugars in fruits

The content of total sugars in the fruits are shown in Table 6. The total sugars were found to be more in T4 and T5 followed by T7 where the sugar content was found to be higher in the combined treatments. [33] Showed that application of paclobutrazol reduced stem length, number of tubers per plant and sugar content of potato.

Table 6. Total, reducing and non-reducing sugar content of fruits

Treatments	Total sugar (g/100g)
T_1	9.6±0.21
T_2	12.4±0.23
T_3	13.2±0.56
T_4	15.2±0.11
T_5	14.4±0.38
T_6	13.0±0.22
T_7	14.8±0.31
T_8	13.6±0.11
T_9	13.2±0.29
SED	0.049
CD (0.05%)	1.101

Values are mean ± SD of four replications

3.3 Determination of macronutrients in fruits and leaves

The plants up taking macronutrients of nitrogen, phosphorus, potassium were determined in the fruits are described in Table 7.

a. **Nitrogen-** Among all the treatments nitrogen content of the fruit T_9 was the highest followed by T_8. Individual application of Sridiamin and PP_{333} not showed any significant influence on the nitrogen content of the fruit.
b. **Phosphorus-** The phosphorus content was also high in T_9 fruits as well as leaves in both stages. The individual treatments, both Sridiamin and PP_{333} increased level of phosphorus but that was not effective.
c. **Potassium-** Sridiamin and PP_{333} as individual spray at both concentration not effective levels of potassium found in both leaves and fruits, compare to combined treatment. All other treatments did not show significant difference in the quantity of potassium. Among the treatments the macronutrient content was higher in the combined treatment (T_9) fruits and leaves were observed. PP_{333} inhibited growth and increased most of the macronutrients in the leaves of sweet peppers [34]. The results obtained are agreement with the findings.

Table 7. Macronutrients- N, P and K in fruits.

Treatments	Nitrogen (g %)	Phosphorus (g %)	Potassium (g %)
T_1	1.30±0.34	0.68±0.22	2.70±0.034
T_2	1.58±0.12	0.68±0.18	3.01±0.22
T_3	1.81±0.16	0.70±0.36	3.20±0.15
T_4	1.89±0.45	0.72±0.11	3.05±0.28
T_5	2.00±0.23	0.84±0.25	3.20±0.24
T_6	1.76±0.31	0.88±0.16	3.18±0.19
T_7	1.65±0.12	0.97±0.35	3.18±0.10
T_8	1.94±0.04	1.20±0.17	3.70±0.23
T_9	2.06±0.01	1.40±0.21	3.98±0.12
SED	0.019	0.030	0.060
CD (0.05%)	0.040	0.060	0.130

Values are mean ± SD of four replications

3.4 Determination of micronutrients in fruits

Micronutrient of Mn, Zn, Fe and Cu contents were shown in Table 8.

a. Manganese- The highest manganese content was also observed in combined treatment (T_9) followed by T_5 (fruits) and T_7 and T_8 (leaves) when compare to control.

b. Zinc-A similar amount of zinc content was found in fruits except T_5, T_8 and T_9 which contained 40ppm. During the harvest stage like other nutrients zinc content was low.

c. Iron-Iron content remained high in the combined treatment than in the individual treatment, which also showed a significant difference when compared to the control.

d. Copper- The copper content in fruits ranged between 16-22ppm. The highest copper content in fruits was noted in T_3 followed by T_7. [35] Have reported that PP_{333} had enhanced the growth parameters as well as Ca, Mg, Fe and Zn in the leaves of Mango.

Table 8. Micronutrients- Mn, Zn, Fe and Cu in fruits.

Treatments	Manganese(ppm)	Zinc(ppm)	Iron(ppm)	Copper(ppm)
T_1	31±0.12	30±0.32	160±0.19	16±1.01
T_2	40±1.01	30±0.21	198±0.76	18±0.12
T_3	45±1.12	30±0.34	213±0.26	22±0.56
T_4	45±0.12	30±0.45	199±0.67	19±0.72
T_5	47±0.45	40±0.49	206±0.45	19±0.56
T_6	32±0.52	30±0.19	218±0.35	17±0.57
T_7	36±1.17	30±0.19	267±0.11	21±0.01
T_8	46±0.24	40±0.50	252±0.21	18±0.14
T_9	50±1.33	40±0.15	279±0.17	20±0.18
SED	1.223	0.408	0.544	0.817
CD (0.05%)	2.596	0.866	1.154	1.731

Values are mean± SD of four replications

4. Conclusion

Therefore, the results revealed that PP_{333} individually reduces the plant height but also the yield of fruit, whereas application of 1% Sridiamin enhances the effect. From the result it was concluded that application of PP_{333} at 5mg concentration as soil drench and Sridiamin at 1% concentration as foliar spray provided a good yield and enhanced in nutrition value when compared to the other treatments.

References

[1] Varela AM, Seif A and Lohr B 2003 *A Guide to IPM in Tomato Production* (The International Centre of Insect Physiology and Ecology No.14)

[2] Naika S, Van Lidt de Jeude V, De Goffau M, Hilmi M and Van Dam B 2005 Cultivation of tomato: production processing and marketing. *CTA*. Series no 17

[3] Araujo JC and Telhado SF 2015 Organic food: a comparative study of the effect of tomato cultivars and cultivation conditions on the physico-chemical properties *Foods* **3** 263-270.

[4] Nick J, Bradley FM and Atthowe H, 1994 *Growing fruits & vegetables organically* Distributed in the book trade by St. Martin's Press

[5] Tiwari AK and Singh DK 2014 Use of plant growth regulators in tomato (Solanum lycopersicum L) under tarai conditions of Uttarkhand *Indian J Hill Fmg.* **2**

[6] Gemici M, Bengu T and Kit T 2006 Effects of 2 4-D and 4-CPA on yield and quality of the tomato (Lycopersicon esculentum Miller) *J. Fac. Sci.* **29** 24-32

7] Zhou L, Jang JC, Jones TL and Sheen J 1998 Glucose and ethylene signal transduction crosstalk revealed by an Arabidopsis glucose-insensitive mutant. *Proceedings of the National Academy of Sciences* **17** 10294-10299

[8] Baruah GKS, Arora SK and Pandita ML 1994 Effect of paclobutrazol (PP_{333}) on fruit set and fruit ripening of tomato *Veg. Sci.* **21** 106-110

[9] Webster AD and Quinlan JD 1984 Chemical control of tree growth of plum (Prunus domestica L.)I preliminary studies with the growth retardant paclobutrazol (PP333) *J. Hor. Sci.* **3** 367-375
[10] Richardson PJ, Webster AD and Quinlan JD 1986 The effect of paclobutrazol sprays with or without the addition of surfactants on the shoot growth, yield and fruit quality of the apple cultivars Cox and Suntan *J. Hor. Sci.* **4** 439-446
[11] Dheim MA and Browning G 1988 The mechanism of the effect of (2RS, 3RS) paclobutrazol on flower initiation of pear cv. Doyenne du Comice and Conference *J. Hort. Sci.* **63** 393-405.
[12] Beevers L and Hageman RH 1983 Uptake and reduction of nitrate bacteria and higher plants In Inorganic plant nutrition Springer Berlin Heidelberg. 351-375
[13] Arnon DI 1949 Copper enzymes in isolated chloroplasts Polyphenoloxidase in Beta vulgaris *Plant Physiol.***1** 1
[14] Hedge IE and Hofreiter BT 1962 Methods in carbohydrate chemistry (Eds Whistler RL, and Be Miller JN,) Academic Press New York
[15] Lowry OH, Rosebrough NJ, Farr AL and Randall RJ 1951 Protein measurement with the Folin phenol reagent. *J.Bio.Chem.***1** 265-275
[16] Moore S and Stein WH 1948 In methods in enzymol (Eds Colowick SP and Kalpan ND) Academic Press New York **3** 468
[17] Official AOAC 1960 Tentative methods of analysis (Association of Official Agricultural Chemists) Washington DC 494.
[18] Harris L and Ray SN 1935 Diagnosis of vitamin-C subnutrition by urine analysis with a note on the antiscorbutic value of human milk *The Lancet* **5811** 71-77.
[19] Ranganna S 1976 In manual analysis of fruits and vegetable product McGraw Hill New Delhi 77.
[20] Hedge IE and Hofreiter BT 1962 Methods in carbohydrate chemistry (Eds Whistler RL and Be Miller JN) Academic Press New York.
[21] APHA 1976 American Public Health Association, American water works association and water pollution control federation. Standard methods for the examination of water and waste water. (14th Edition, American Public Health Association Inc) New York **2** 35
[22] Jackson ML 1973 Soil chemical analysis (Prentice Hall of India Private Ltd) New Delhi
[23] Piper CS 1966 Soil and plant analysis (Hans Publications) Bombay 368.
[24] Lindsay WL and Norvell WA 1978 Development of a DTPA soil test for zinc, iron, manganese, and copper *Soil Sci Soc Am J.* **3** 421-428.
[25] Setia N, Setia RC and Kaur P 1997 Morphological development and anatomical features of *Brassica juncea* as influenced by paclobutrazol, *Phytomorphology* **47** 427-436.
[26] Voon CH, Pitakpaivan C and Tan SJ 1989 Mango cropping manipulation with Cultar. *III International Mango Symposium* **291** 219-228.
[27] Kalpana S & Saroja S 1999 Effect of paclobutrazol (pp333) and human hair-derived amino acids mixture (sridiamin) on growth yield and quality of tomato cv. PKM-1 (Lycopersicon esculentum Mill) South Indian Hortculture **1/6** 262-266.
[28] Muthukumarasamy M and Panneerselvam R 1997 Amelioration of NaCl stress by triadimefon in peanut seedlings. *Plant growth regulation* **3** 157-162.
[29] Thakur A, Thakur PS and Singh RP 1998 Influence of paclobutrazol and triacontanol on growth and water relations in olive varieties under water stress *Indian J Plant Physiol* **3** 116-120.
[30] Yulan , Yuxiang and Yueshen M 1997 Effect of PP_{333} on the growth and development of Actinida Arguta [J]. *Scientia Agricultura Sinica* **30** 65-70.
[31] Baruah GKS, Arora SK and Pandita ML 1994 Effect of Paclobutrazol (PP_{333}) on fruit set and fruit ripening of tomato *Veg. Sci.* **21** 106-110.
[32] Sadasivam S and Manickam A 1996 Biochemical Methods (II Edition, New Age International Publishers) 188-189.

[33] Mabvongwe O, Manenji BT, Gwazane M and Chandiposha M 2016 The effect of Paclobutrazol application time and variety on growth, yield, and quality of potato (Solanum tuberosum L.) *Advances in Agriculture*.

[34] Borowski E, Blamowski ZK and Kozlowska L 1998 Growth and physiological reactions to cold in weet peppers exposed to Paclobutrazol *Horticultura* **6** 129-136.

[35] Helail BM and Eissa MA 1997 Effect of some cultural practices and growth regulator treatments on growth of mango seedlings *Annals of Agricultural Science* Moshtohor (Egypt) **35** 883-894.

Prebiotic attributes of inulin enriched biscuits: sensory and nutritional aspects

C Raihing [1] and S U Mageshwari [2]

[1] Research scholar, Department of Food service Management and Dietetics Avinashilingam Institute for Home science and Higher education for Women, Coimbatore -43, Tamil Nadu, India
[2] Associate Professor, Department of Food service Management and Dietetics Avinashilingam Institute for Home science and Higher education for Women, Coimbatore -43, Tamil Nadu, India

Email : chingriyo@gmail.com, uma.adu3@ gmail.com

Abstract. The study aims to develop an enriched functional biscuit with prebiotic inulin and to understand the impact of addition of different levels of inulin on sensory and nutritional properties of the biscuits. A standard biscuit and three variations namely P1, P2 and P3 were formulated and standardized through test and trial method to get consistent final products. Inulin was added replacing fat in the standard-0%, P1- 100% P2-62% and P3-37%. The biscuits were evaluated by 15 panel members for sensory attributes i.e. appearance, colour, flavour, taste, texture and overall acceptability. Nutrients analysis for energy, carbohydrate, protein, fat, fibre, free fatty acid and total antioxidant activity were done using standard procedures. Addition of inulin can give rise to products with different rheological behaviour and sensory characteristics. P2 (62% fat replaced) was acceptable with a high score for overall acceptability (7.45±0.39). Fat content was found to be highest in P3 (25.88gm). The inulin content in P1, P2 and P3 was 25.7gm, 34.5gm and 19.1 gm respectively. Antioxidant activity was also observed in all the biscuits and activity was highest in P2 (229.54μg). The partial replacement of fat by inulin in the production of biscuits was effective in reducing the fat content in the final product. A significant difference in taste ($p=0.004$) and texture ($p= 0.048$) and was observed among the biscuit variations. The formulation P2, with 62 % fat replacement by inulin, presented similar results to the standard product, being the formulation with the greatest sensory acceptance and physical characteristics.

1. Introduction

The changing trends in eating habits have made food production to take a pivotal role in promoting health. Increased energy intake, unbalanced diets and highly processed foods are huge problems of concern as these are contributory factors causing lifestyle diseases. The trend in the demand of functional food providing both taste and health claim is increasing and attention is focused on development of enriched staple food or commonly consumed snacks which is widely accepted. Prebiotics is used as a functional food ingredient to enhance health and wellbeing [1]. An option to improve the intake of prebiotics is to fortify commonly consumed foods with prebiotics. Prebiotic ingredients such as inulin have the advantage that they can be added to a wide range of commonly consumed foods because of their technological properties [2]. Many studies have proved its role as a good fat replacer for developing products with healthier properties with desired sensory characteristics. Excessive intake fast foods and processed foods which are loaded with dietary fats such as saturated fats and trans fats used in fast food outlets have a deteriorating effect on health. The possible health benefits of prebiotic inulin facilitated by its nutritional and technological properties can be help in developing appropriate dietary interventions to stem the tide of increasing lifestyle diseases linked to modern society. The study aims to develop an enriched functional biscuit with prebiotic inulin and understand the impact of addition of different levels of inulin on sensory and nutritional properties of the biscuits.

2. Materials and Methods

2.1 Selection of prebiotic foods

A baked food product was formulated using inulin as an adjunct to incorporate the prebiotic functionality into the product. Inulin extracted from the chicory root (*Cichoriumintybus*) was used as it offers dual benefit of improving the organoleptic quality and offer functional benefit [1]. Fructafit® HD inulin is a natural powdered ingredient extracted from chicory roots. It is classified as a dietary fibre and used in food industry for its remarkable functional properties such as the ability to act as a fat or sugar replacer [3].

2.2 Standardisation of prebiotic food

Biscuits were formulated with three variations using inulin as fat and sugar replacer. A standard biscuit and three variations namely P1 (Fat and sugar removed), P2 (62% fat replaced and 100% sugar replaced) and P3 (37% fat replaced). The biscuits were standardised through test and trial method for all the variations to get consistent final products.

The biscuits were evaluated for sensory attributes and were compared with standard. Fifteen judges evaluated the biscuits using a 9 point Hedonic Scale. Each attribute was assigned a numerical value for a subjective factor i.e. appearance, colour, flavour, taste, texture, and overall acceptability. The biscuits were evaluated and the results were recorded. The analysis of variance (ANOVA) test was performed in order to find out the significant differences between the variations for different sensory characters by using SPSS software

2.3 Nutrient Analysis

Analysis for the three variations P1, P2 and P3 was conducted for nutrients such as energy, protein, carbohydrate, fat, fibre, free fatty acid and total antioxidant activity using standard procedures. Carbohydrate was analyzed using Anthrone method. Kjeldhal method of protein estimation was used to determine the protein conent , fat was estimated using soxhlet method, free fatty acid content was estimated using AOAC: 2012 method and total antioxidant activity in the biscuits were analyzed using FRAP method. Inulin content was estimated using DGHS method. The nutrient content of inulin was analyzed separately for carbohydrate, fat, fibre and total antioxidant activity.

3. Results

It was noted that addition of inulin changes the physical characteristics of the biscuits. Similar findings by Kuntz *et al* 2013 demonstrated that addition of inulin alters the physiochemical and sensory characteristics in foods [4]. Incorporation of inulin changes the dimensions and aeration of the biscuit. On baking, increase in fat replacement resulted in reduction in the size of the biscuits compared to the standard formulation and consequently giving a more firm and crispy final product. The physical change can be associated with the ability of elastic recoil presented by inulin [5].

3.1 Sensory evaluation of the prebiotic biscuits
It can be noted that there was no significant difference in sensory attributes such as appearance (p =0.15), colour (p= 0.69), flavour (p=0.10) and overall acceptability (p= 0.26) among three variation of biscuits prepared with inulin and the standard. A significant difference in taste (p=0.004) and texture (p= 0.048) and was observed among the biscuit variations. The mean acceptability scores for appearance, taste and texture attributes among the three variations were highest in P2 (7.45±0.69; 7.80±0.60; 7.45±0.44). Maximum scores for colour attribute were noted in P3 (7.75±0.44). The mean score for flavour was observed to be similar in all the variations. It can be observed that among the fat replaced P1, P2 and P3, P2 was acceptable with a high score for overall acceptability (7.45±0.39) and hence P2 presented better results among the three variations and was accepted. Partial replacement of fat with inulin in the development of biscuits resulted in formulations of biscuits with better sensory acceptance and physical characteristics [6].

Table 1. Mean acceptability scores of standard and inulin biscuits.

Biscuit	Appearance	Colour	Taste	Texture	Flavour	Overall acceptability
Standard	7.50±0.51	7.35±0.59	7.85±0.67	7.60±0.51	7.40±0.60	7.41±0.14
P1 n=15	7.20±0.62	7.60±0.75	7.45±0.51	7.25±0.64	7.30±0.83	7.40±0.17
P2 n=15	7.45±0.69	7.65±0.51	7.80±0.60	7.45±0.44	7.30±0.70	7.45±0.0.39
P3 n=15	6.75±0.44	7.75±0.44	5.75±0.44	6.75±0.44	7.25±0.44	7.18±0.60
F-value	1.821^{NS}	0.485	4.839*	2.70*	2.153^{NS}	1.343^{NS}

Values are mean± SD of the scores obtained
*- Significant at 5 per cent level ($p<0.05$)
NS- The difference between the mean values within the column is not significant.
P1 –Fat and sugar removed; P2- Fat replaced and sugar removed; P3-Fat replaced

3.2 Nutrients content of the prebiotic biscuits
Among the three variations, fat content was found to be highest in P3 (25.88 g) which can be attributed to the fat added in the formulation since only 32 per cent of fat used in the formulation was replaced with inulin hence higher fat content compared to P1 and P2. The protein content of standard biscuits (6.06 g) was found lower than that of the three biscuit variations (11 g, 7.08 g and 6.26 g for the variations P1, P2 and P3 respectively.) Prebiotic ingredient may have contributed to the protein content of the biscuits hence higher values were found in the variations compared to the standard. Crude fiber content of standard biscuits (0.45 g) was observed to be lowest.

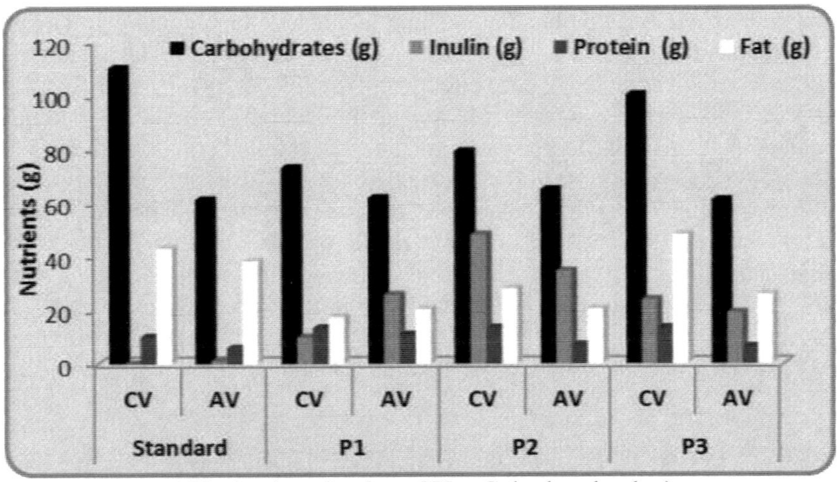

AV- Analysed value; CV – Calculated value)

Figure 1. Nutrient composition

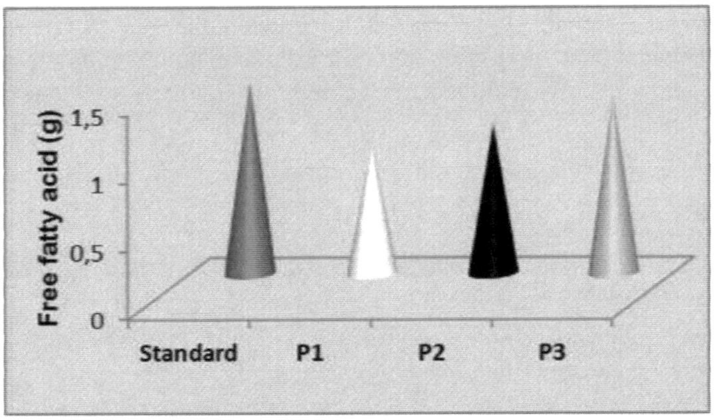

Figure 2. Free fatty acid content

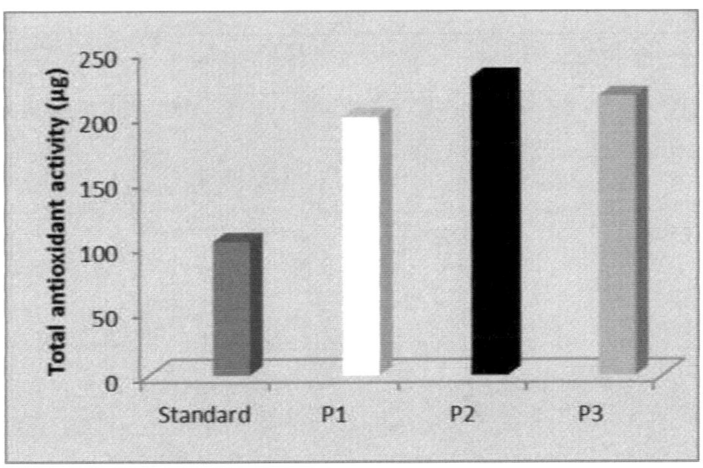

Figure 3. Total antioxidant activity

The inulin content in P1, P2 and P3 was 25.7 g, 34.5 g and 19.1 g respectively. The difference in the values may be attributed to the amount of inulin used in the formulation. Antioxidant activity was also observed in all the biscuits and activity was highest in P2 (229.54 µg). P1 and P3 showed an activity of 198.6 µg and 214.96 µg respectively. However standard biscuit showed the least activity (102.8 µg). Hence it can be inferred that incorporation of inulin in biscuits enhances the antioxidant activity. This observation is in line with a study by Pasqualetti *et al.,* 2014 which showed that antioxidant activity of inulin was significantly higher compared to simple sugars. The study concluded that the effects of inulin on the health of the host are attributable, not only to its bifidogenic [6]. Fats and oils are key ingredients in baked that confer desirable characteristics such as soft texture, aeration and flavour [8]. With the replacement of baking fat with inulin the free fatty acid content of the variations also decreased. Hence P3 (37% fat replaced) having the lowest replacement showed higher content of free fatty acid of 1.30 mg compared to P1 (0.93 g) and P2 (1.09 g).

4. Conclusion

The partial replacement of fat by inulin in the production of biscuits was effective in reducing the fat content in the final product. A significant difference in taste (p=0.004) and texture (p= 0.048) and was observed among the biscuit variations. Sensory attributes such as appearance, colour, flavour and overall acceptability was not significantly different among the variations and the standard. The formulation of P2, with 62 % fat replacement by inulin, presented similar results to the standard product, being the formulation with the greatest sensory acceptance and physical characteristics. Inulin presents as a versatile functional food ingredient with diverse applications which can be widely used in food industries. Fat and carbohydrate replacement with inulin can provide the advantage of delivering a nutritionally enhanced food product with health promoting benefits without compromising on taste and texture.

References

[1] Roberfriod M 2002 Functional food concept and its application to prebiotics *Digestive and Liver disease* **35** 2 S105- S110
[2] González S 2015 Inulin in food products: prebiotic and functional ingredient *British Food Journal* **117** 1 371-387
[3] Zahn S, Pepke F and Rohm H 2010 Effect of inulin as a fat replacer on texture and sensory properties of muffins *International Journal of Food Science and Technology* **45** 12 2531-2537
[4[Kuntz M G F, Fiates G M R, and Teixeira E 2013 Characteristics of prebiotic food products containing inulin *British Food Journal* **115** 2 235-251
[5] Garcia J R *et al* 2013 Effect of Fat Replacement by Inulin on Textural and Structural Properties of Short Dough Biscuits *Food and Bioprocess Technology* **6** 10 2739–2750
[6] Lourencetti R E *et al* 2013 Development of biscuit type cookie with partial replacement of fat by inulin *International Journal of Nutrition and Food Sciences* **2** 5 261-265
[7] Pasqualetti V *et al* 2014 Antioxidant Activity of Inulin and Its Role in the Prevention of Human Colonic Muscle Cell Impairment Induced by Lipopolysaccharide Mucosal Exposure PLoS ONE **9** 5 p e98031
[8] Rios R V *et al* 2014 Application of fats in some food products *Food Sci. Technol Campinas* **34** 13-15

Detecting aroma changes of local flavored green tea (*Camellia sinensis*) using electronic nose

D Ralisnawati[1], A C Sukartiko[1], A Suryandono[1] and K Triyana[2]

[1]Department of Agro-industrial Technology, Universitas Gadjah Mada, Jl. Flora 1 Bulaksumur Yogyakarta 55281, Indonesia
[2]Department of Physics, Faculty of Mathematics and Natural Sciences, Universitas Gadjah Mada, Yogyakarta 55281, Indonesia

E-mail: daniralisnawati@gmail.com

Abstract. Indonesia is currently the sixth largest tea producer in the world. However, consumption of the product in the country was considered low. Besides tea, the country also has various local flavor ingredients that are potential to be developed. The addition of local flavored ingredients such as ginger, lemon grass, and lime leaves on green tea products is gaining acceptance from consumers and producers. The aroma of local flavored green tea was suspected to changes during storage, while its sensory testing has some limitations. Therefore, the study aimed to detect aroma changes of local flavors added in green tea using electronic nose (e-nose), an instrument developed to mimic the function of the human nose. The test was performed on a four-gram sample. The data was collected with 120 seconds of sensing time and 60 seconds of blowing time. Principal Component Analysis (PCA) was used to find out the aroma changes of local flavored green tea during storage. We observed that electronic nose could detect aroma changes of ginger flavored green tea from day 0 to day 6 with variance percentage 99.6%. Variance proportion of aroma changes of lemon grass flavored green tea from day 0 to day 6 was 99.3%. Variance proportion of aroma changes of lime leaves flavored green tea from day 0 to day 6 was 99.4%.

1. Introduction

Tea has an important role in Indonesia's economic activity. The country is currently the sixth largest tea producer in the world [1]. However, despite having relatively high production level, local consumption of tea drinks is considered relatively low, only 0.2 kg/capita/year, compared to some countries, such as Ireland, UK, Pakistan, and India, respectively with a consumption rate of 3.5 kg/capita/year, 2.5 kg/capita/year, 1 kg/capita/year, 0.6 kg/capita/year [2]. Both condition, high in production level and low in local consumption, indicates the enormous economic potential that can be developed from the product.

Although Indonesia has a variety of local flavor sources, the potential has not been optimally utilized yet, including as local flavor that can be added to tea products. The addition of local flavor, besides increasing the consumption of tea, can also improve the utilization of local Indonesian flavor, so that more familiar not only to people in this country but also people around the world. Flavors that can be added to tea drinks can be sourced from various parts of the plant, i.e., leaves, fruit, seeds, stems or skin, or roots. Based on the criteria of consumer acceptance, availability, and processing yield, eight local flavors meet the criteria [3]. Three of these flavors, namely ginger, lemongrass and lime leaves were then used in this study. Green tea was used in this study because it contains more polyphenols than black tea or oolong tea due to differences in the processing of tea leaves [4].

The addition of local flavors to green tea products has highly accepted by tea consumers as well as sensory experts of tea producers. From several assessed aspects, the aroma of tea that has been added with local flavors, such as ginger, lemongrass, jasmine and lime leaves, becomes the most important aspect of consumer acceptance [5]. The flavor of tea can be divided into two categories: taste (non-volatile compounds) and aroma (volatile compounds) [6]. Aroma was an important indicator in determining the quality of a product [7]. The local aroma that added in green tea was suspected to change during storage. Meanwhile, according to [8], the sensory test has limitations because it is influenced by subjective factors that can cause errors. Therefore, this study was conducted to detect aroma changes of local flavors that added to tea products using electronic nose (e-nose).

2. Materials and Methods

A dry green Peko tea obtained from PT Pagilaran was used as green tea sample, while ginger, lemongrass and lime leaves were used as local flavor. The extraction was done using steam distillation method. About 1 kg of local flavor sample was extracted for 6 hours to obtain essential oil to be applied on dried tea leaf samples. The essential oil obtained was then sprayed on a dried tea leaves sample using a nebulizer (NE-C28 Omron, China). Nebulizer is a breathing apparatus with the working principle of breaking the liquid particles into smaller particles making it easier to evaporate. The tea and oil composition used was 0.5 ml of essential oil for 25 g of dried tea. Tea that has been added with local flavor was stored in room temperature inside an aluminum foil with the weight of each packing material 4 g.

The testing process was done by electronic nose which was developed by Laboratory of Material Physics and Instrumentation Universitas Gadjah Mada. This series of electronic nose was connected to a computer that already installed with Electronic nose G2 Delphi software. The result of the sensor output then transmitted serially to the data recording system on the computer. The analog data from the sensor will be converted into digital data by Analog to Digital Converter (ADC) to be saved to the computer and analyzed further.

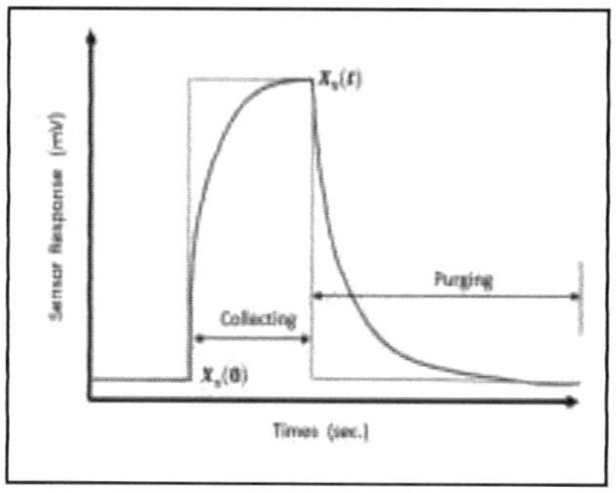

Figure 1. Sensor response with maximum (Xs (t)) and minimum response (Xs(0)).

As shown in Figure 1, the typical of sensor response consists of baseline, collecting (sensing), and purging samples aroma. The type of signal manipulation is based on a fractional method. It means that the dimensionless sensor response ($y_s(t)$) can be formulated as Eq. 1 [9]

$$Y_s(t) = (x_s(t) - x_s(0))/x_s(0) \quad (1)$$

where ($x_s(t)$) is the maximum of sensor response, while ($x_s(0)$) is the baseline of sensor response.

The sensor response data then performed by signal preprocessing and feature extraction. Feature extraction aims to extract or retrieve the important values of a digital tracker signal. Taking the feature of a data was necessary to obtain main information from the data to be analyzed for the next process [10]. In this research, we used eight variables that represent the number of sensors in electronic nose, namely TGS813, TGS822, TGS2600, TGS826, TGS2611, TGS2620, TGS2612, and TGS2602. These eight variables will be reduced into two dimensions that consist the first principle component or PC_1 and the second principle component or PC_2. These components can represent a significant percentage the variance value of the total data variance and was used two dimensional data visualization graph for qualitative analysis and information interpretation [11]. To perform PCA analysis, in this research we used MINITAB 17.

3. Results and Discussion

Based on the observation of electronic nose sensor response, it can be seen that electronic nose produced different response signal characteristics in each type of tea sample with flavor of ginger, lemongrass and lime leaves.

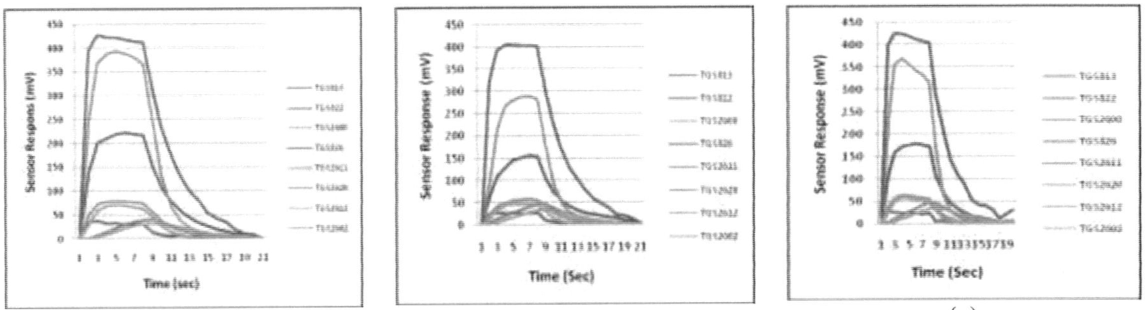

(a) (b) (c)

Figure 2. Typical response of sensor array to flavored tea samples (a) ginger, (b) lemongrass, (c) lime leaves

The aroma of the samples was converted into electrical signals (voltage) by the gas sensors. The electrical signals represent gas sensor response. The sensor response was a description of each of the gas sensor which response to the aroma samples with its characteristics [12]. This gas concentration from electrical signal was not for one type of gas, but for a group of gases. That was because the TGS gas sensor was non-selective gas sensor. The sensors in electronic nose are only capable of responding to classify different samples, not to test the content in the sample under test [13].

In general, the sensor response (Figure 2) for the three samples of flavored green tea (ginger, lemongrass, and lime leaves) have similarities, both speed, and amplitude response. But if we further analyze, TGS 826, TGS 2600 and TGS 822 sensors have large or strong amplitude compared to TGS 813, TGS 2611, TGS 2620, TGS 2612, and TGS 2602 sensors. Both TGS 826 and TGS 2600 sensors have the amplitude signal over 350 mV on ginger and lime leaves flavor, and then the signal amplitude beneath them are owned by TGS 822 sensors. The difference between the three samples of flavored green tea lies in TGS 826, TGS 2600, and TGS 822. While TGS 826 and TGS 2600 sensor gives strong respond in ginger (427 mV; 392 mV) and lime leaves (426 mV; 357 mV), but it gives a weak response in lemongrass (405 mV; 289 mV). Furthermore, TGS 822 sensors give strong respond in ginger (222 mV), but give a weak response in lemongrass (157 mV) and lime leaves (180 mV).

To make it easier to see the pattern of data generated from feature extraction, the data can be displayed in graphical form as shown in Figure 3 (a-c).

The pattern from the three graphs was almost the same that was a very drastic decline in the third day test on certain sensors. On average, there was three sensors that show a very high response on day 0 test. The decline data after third day was more invisible. In some sensors also seen flat pattern, did not show a significant decrease in response.

In the test of ginger flavored tea has a greater response value compared with the sensor response value for lemongrass and lime leaves flavored tea. This suggests that ginger flavored tea samples have a stronger aroma than the tea samples of lemongrass and lime leaves.

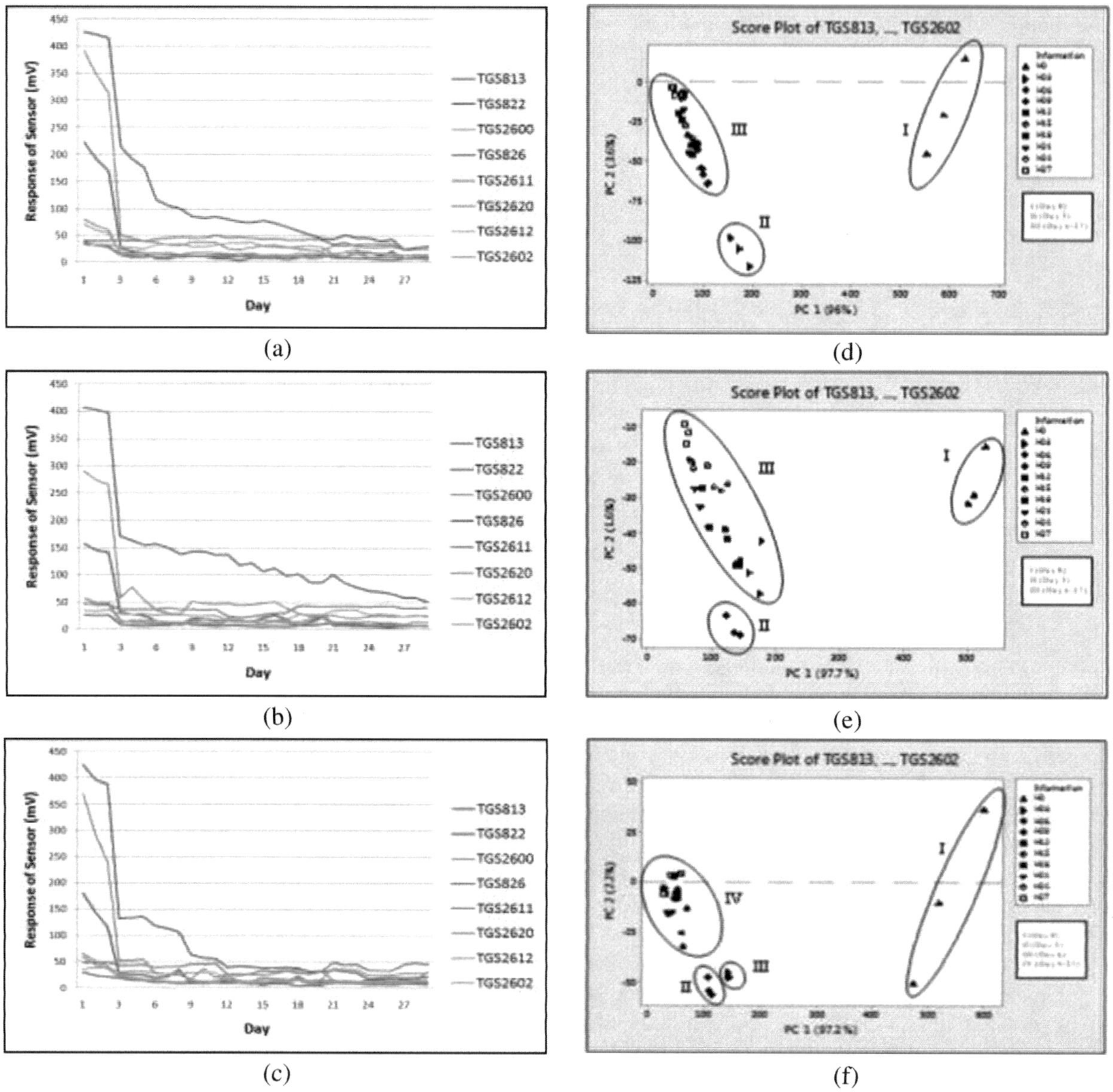

Figure 3. Feature extraction result of sensor response (a-c) and PCA score plot (d-f) of flavored tea samples, respectively from top to bottom, for ginger; lemongrass; and lime leaves. In score plot (d-e), the data were grouped into three, ie I (day 0); II (day 3); and III (day 6-27), whereas score plot (f), the data were grouped into four, ie I (day 0); II (day 3); III (day 6); and IV (day 9-27).

Figure 3 (a-c) displays the data in the form of graphic patterns from each sensor so that it can be compared the response of each sensor. However, the graph has not been able to analyze the aroma changes by combining the entire sensor response as input. Therefore, further data processing was done by Principal Component Analysis (PCA) that used eight electronic nose sensors as variables. PCA was an unsupervised pattern recognition methods used for multivariate analysis [14].

The analysis using PCA aims to reduce the dimensions of mutually correlated variables into linearly uncorrelated variables called the principle components to explain the maximum possible variance with the minimum number of major components [15].

To obtain PCA, in this research using software MINITAB 17. The number of dimensions in PCA processing can be determined by creating a score plot. Based on the score plot, only one dominant factor and on the second factor and so on has a low eigenvalue or it can be said to be zero. Based on these results, the eight variables will be reduced to two dimensions consisting of the first principle component or PC_1 and the second principle component or PC_2.

In general, the number of PCs that should be used as much as two or three PCs. Then the selected PC has a percentage of variance that indicates the PC's ability to describe the entire information of all previous initial variables [16].

The result of two-dimensional graphical visualization of PCA processor can be seen in Figure 3 (d-f). In the PCA processing of tea samples of ginger and lemongrass flavor, it can be seen that the collection of data points on days zero (group I) and third day (group II) can be completely separated from other points. As for testing day 6 to 27 cannot be separated perfectly and still overlap in one area (group III). It shows that electronic nose was able to distinguish the aroma of ginger tea and lemongrass tea sample on the day zero and the third day but not yet able to group the test results on day 6 to day 27 (group III). This was because on the day zero to the third day and on the third day until the sixth day clearly shows the decline in the aroma of the sample while on day 6 to day 27 (group III) there was a very small decline so it cannot be read clearly. The small response of sensors to fragrance on days 6 to day 27 was due to the volatile oil so it is likely that the test most of the added essential oil has released [17].

In the PCA processing results of lime leaves flavored tea samples can be seen that the collection of data points on days zero, third day, and sixth day (group I, II, and III) can be completely separated. As for the test day to 9 to 27 cannot be separated perfectly and still collected in one place (group IV). It shows that the electronic nose was able to distinguish lime leaves flavored tea samples on day zero, third day, and sixth day but not yet able to classify the test result on the 9th day until the 27th day (group IV). The result of the score plot for lime leaves flavored tea sample was different to the score plot of ginger flavored tea samples and lemongrass because in this score plot can already classify the test until the sixth day. It may because citrus essential oil has high rendement so it has a strong aroma on the sample [18].

PCA was employed to reduce the dimensionality and visualization of the datasets while retaining as much as possible of the variation present in the original dataset [19]. The eigenvalues are related to the variance (percentage variance) that can be explained by the main components [20]. On ginger flavored tea sample, PC_1 explained 96% variation while PC_2 explain 3.6% of the variation so that the accumulation of two principal components obtained 99.6% of variance data. For lemongrass flavored tea samples, the percentage value of variance for PC_1 was 97.7% and for PC_2 was 1.6% so that the cumulative variance percentage value of the two main components was 99.3% of variance data. The percentage value of lime leaves flavored tea for PC_1 was 97.2% and for PC_2 was 2.2% so that the cumulative variance percentage value of the two main components was 99.4% of variance data.

It means if the eight gas sensors are used as variables to classify three flavored tea samples, then the gas sensors are reduced to two variables. Hence, the two variables are able to explain respectively 99.6%, 97.7%, and 99.4% for ginger, lemongrass, and lime leaves flavored tea of the total variability of the eight variables. Based on [21] that studied about Japanese green tea flavor using electronic nose and used three principal component with the total variance percentage are 95.7%, it can be said that it has a good variance percentage. So, this reasearch has a good percentage of variation that can provide the enough information for the discrimination of the flavored tea samples. It means that electronic nose with eight sensors can only distinguish aroma ginger and lemongrass flavored tea until the sixth day, and for lime leaves flavored tea until the ninth day.

4. Conclusion

Changes in flavored tea aroma, added with ginger and lemongrass flavor, until the sixth day, can be detected by electronic nose using eight sensors with percentage of variation of, respectively, 99.6% and 99.3%. Meanwhile in lime leaves flavor, until ninth day, can be detected with percentage of variation 99.4%.

References

[1] C B of Statistics 2013 Statistik Teh Indonesia

[2] C B of Statistics 2015 Statistik Teh Indonesia

[3] Febrianti F 2016 Formulasi Pembuatan Tablet Effervescent Teh Hijau (*Camellia sinensis*) dengan Penambahan Citarasa Tesis Universitas Gadjah Mada Yogyakarta

[4] Senanayake S P J N 2013 Green tea extract: Chemistry, antioxidant properties and food applications - A review *J. Funct. Foods* **5** 4 1529–1541

[5] Supartono W, Sukartiko A C, Yuliando H, Kristanti N E 2015 Possibility of some indigenous spices as flavor agent of green tea *Agric. Agric. Sci. Procedia* **3** 62–66

[6] Ho C T, Zheng X, Li S 2015 Tea aroma formation *Food Sci. Hum Wellness* **4** 1 9–27

[7] Baietto M and Wilson A D 2015 Electronic-nose applications for fruit identification, ripeness and quality grading *Sensors* **15** 1 899–931

[8] Yu H, Wang Y, Wang J 2009 Identification of tea storage times by linear discrimination analysis and back-propagation neural network techniques based on the eigenvalues of principal components analysis of E-nose sensor signals *Sensors* **9** 10 8073–8082

[9] Chen Q, Zhao J, Chen Z, Lin H, Zhao D A 2011 Discrimination of green tea quality using the electronic nose technique and the human panel test, comparison of linear and nonlinear classification tools *Sensors and actuators B: Chem.* **159** 1 294–300

[10] Jia P, Tian F, He Q, Fan S, Liu J, Yang S X 2014 Feature extraction of wound infection data for electronic nose based on a novel weighted KPCA *Sensors and Actuators B: Chem.* **201** 555–556

[11] Godoy J L, Vega J R, Marchett J L 2014 Relationships between PCA and PLS-regression *Chemom Intell Lab Syst.* **130** 182–191

[12] Lelono D, Triyana K, Hartati S, Istiyanto J E 2016 Principal component analysis classification of Indonesia black Teas Based on Quality by Using Electronic Nose and Principal Component Analysis *AIP Conf Proc.* 1755 pp 20003–1

[13] Hidayat S N and Triyana K 2016 Optimized back-propagation combined with radial basic neural network for improving performance of the electronic nose: case study on the fermentation process of tempeh *AIP Conference Proceedings* 1755

[14] Nimsuk N 2014 Neural Network Based Electronic Nose for Discrimination of Fish Sauces. Electrical Engineering/ Electronics, Computer, Telecommunications and Information Technology (ECTI-CON), 11th International Conference, IEEE Conference Publications pp 1–4

[15] Abdurachman S H, Komalig H, Nainggolan N 2014 Penggunaan analisis komponen utama dalam penggabungan data peubah ganda pada kasus produksi pertanian dan perkebunan di wilayah Bolaang Mongondow tahun 2008 *JdC* **3** 2 1–7

[16] James D B 2009 Choosing the right number of components or factors in PCA and EFA, *JALT Test Eval. SIG.* **13** 2 19–23

[17] Singh G, Kapoor I P S, Singh P, de Heluani C S, de Lampasona M P, C A N 2008 Catalan chemistry, antioxidant and antimicrobial investigations on essential oil and oleoresins of Zingiber officinale, *Food Chem. Toxicol.* **46** 10 3295–3302

[18] Kasuan N, Yusoff Z M, Muhammad Z, Nordin M N N, Rahiman M H F, Taib M N 2012 Essential oil extraction with automated steam distillation: FMRLC for steam temperature regulation. Proceedings of IEEE International Conference on Control System, Computing and Engineering, ICCSCE pp 198–203

[19] Wang B, Xu S, Sun D W 2010 Application of the electronic nose to the identification of different milk flavorings *Food Res Int.* **43** 1 255–262

[20] Liang Z and Lee Y 2013 Eigen-analysis of nonlinear PCA with polynomial kernels, *Stat Anal Data Min.* **6** 6 529–544

[21] Yang Z, Dong F, Shimizu K, Kinoshita T, Kanamori M, Morita A, Watanabe N 2009 Identification of coumarin-enriched Japanese green teas and their particular flavor using electronic nose *J. Food Eng.* **92** 3 312–316

The mineral composition and the effect of particle size of carbonized rice straw as colorant of a traditional cake *kue jongkong Surabaya*

E S Murtini, S S Yuwono and H Y Setyawan

Department of Agricultural Product Technology, Faculty of Agricultural Technology, Universitas Brawijaya, Malang, Indonesia

E-mail: erni.murtini@ub.ac.id

Abstract. Carbonized rice straw (CRS) is a term defined for the residue of incomplete combustion of rice straw. Utilization of CRS as a natural food coloring agent has been the local Indonesian wisdom. However, study of this local food coloring agent is rare in the literature. This study was aimed to determine the mineral composition of the CRS, and to investigate the effect of particle size of the CRS to the black color intensity of a traditional Indonesian cake called *kue jongkong Surabaya*. The mineral content of the CRS was analyzed using X-ray fluorescence (XRF). The CRS was grounded and sieved passing through different screen sizes (40, 80, 100, 120 and 200 mesh).The particle size distribution was measured using particle size analyzer. The CRS with different particle sizes were then applied as a natural coloring agent of the *kue jongkong*, from which the intensity of black color was determined using a color reader. It was found that the dominant minerals of the CRS were SiO_2, carbon, and K_2O. Other trace elements found were Cl, CaO, Na_2O, MgO, P, S, Fe, Al_2O_3 and Mn. The CRS which passed to the sieve of 40 mesh has particle size distribution of 28µm, 115µm, and 348µm for a standard of D10, D50, and D90, respectively. However, CRS that passing through the sieve of 60-200 mesh have similar particle sizes (D10: 12-14µm, D50: 49-60µm, and D90: 114-145 µm). The smaller of CRS particle size produced a darker color of the *kue jongkong* due to better molecule dispersion and wider surface area.

1. Introduction

The common harvesting technique of rice is cutting the top part of rice straw, then put it in the rice mill top separate the rice seed. This technique left the top part of the rice straw as a waste. The abundant quantity of this waste is expected due to the high rice production, particularly in Indonesia which among the top three of largest world rice producers after China and India[1]. A range of 0.7-1.4 rice straw waste is produced for every 1 kg of harvested rice [2]. In 2015, the production of rice in Indonesia reached 75 million tons [3] , leaving ≈52-105 million tons of rice straw in the process. However, the utilization of this large number of rice straw is limited.

One of the potential utilizations of rice straw is as a natural colorant, particularly black color. Popularity of natural colorants are increasing, since people pay more attention to the health safety issues of synthetic food colorants [4]. In Indonesia, rice straw has been used as a natural colorant for traditional food, such as *kue jongkong Surabaya*. *Kue jongkong* is a wet and gel-like cake, and has black color. This cake is made from a mixture of rice flour, coconut milk, sugar, salt, and black colorant, all of which the batter is placed in a container then steamed.

To use the rice straw as a colorant, it requires carbonization process. Thus, the colorant product from rice straw is called carbonized rice straw (CRS). The products of carbonization are carbon and un-vaporized minerals. Thus, the benefit of using natural colorant is renewable and contains useful minerals. However, investigation of CRS is limited and a better understanding of this natural colorant particularly on its mineral properties is required. Therefore, in this research, the mineral components were studied using XRF.

Further investigation on its particle size is also required to study the effective mixture that can be formed by this natural colorant in a traditional food. Investigation of the particle size was aimed to meet the requirement of the Panel on Food Additives and Nutrient Sources added to Food [5]. Based on this standard, the smallest particle size allowed is no less than 275nm. In addition, this research studied the effect of particle size on the intensity of the black color which can provide a recommended particle size of CRS for the making of food colorant.

2. Method

Rice straw was burnt in a ceramic container, from which the CRS was collected. The sample was then analyzed using XRF to determine the mineral composition based on modified method of Hazra et al [6]. In brief, the rice straw was burn at a set furnace of 110 and 1000°C, and the loss on Ignition (LOI1000) was determined using a robotic TGA system. The sample was then investigated using XRF to determine its mineral properties.

For particle size analysis, the CRS collected from the ceramic container was then grinded and sieved at 40, 80, 100, 120 and 200 mesh. The size variation of the particles was then measured by using particle size analyzer (Cilas 1090). This method provided a standard of particle size analysis described in D-Values (D10, D50 and D90), which are the intercepts for 10%, 50% and 90% of the cumulative mass.

Application of different particle sizes which pass through the 40-200 sieves was then used as a colorant of *kue jongkong*. A 20g CRS was diluted with 50ml for 24h. The suspended solid was separated and the final CRS solution obtained. The *kue jongkong* was made from 230g rice flour, 20g tapioca starch, 800ml coconut milk, 125g sugar, 1.6g salt, 50ml CRS solutions. All of the ingredients were mixed and the batter then is placed on a 20x10x3 cm tray then steamed at 90±2°C for 30 minutes. Five different samples were created and triplicate based on the particle size of the CRS which through the sieves. Measurement of the color of the kue jongkong was done by using a color reader (Minolta CR-100). The colors of *kue jongkong* were measured based on the lab space color (L) where L=0 is the darkest and L=100 is the lightest. Data of L color of *kue jongkong* was statistically analized using Anova single factor and post hoc using Tukey (α=5%).

3. Results and Discussion

3.1. Mineral composition of CRS

Table 1 shows XRF result for the mineral composition of CRS. Silica is the highest component in CRS, followed by carbon, K_2O, and Cl. A few components such as CaO, Na_2O, MgO was found at concentration of <0.1%. There were no toxic chemicals, such as Pb and Co found in the CRS, suggesting that CRS suitable for food additive in low concentration. It has been reported that similar commercial product of bamboo charcoal powder (BCP) is declared safe to be consumed and no-observed-adverse-effect level (NOAEL) at daily consumption of 11.24 g/kg BW/day [7]. However, further research on toxicology of CRS is still required.

By nature, silica is one of most abundance components on earth as it is easily found in leafy green plants, which is the characteristics of rice straw. Silica may safely used directly in food according to FDA on Code of Federal Regulations Title 21. However, the recommended usage of silica is not more than 2wt% of the food as an anti-caking agent.

Table 1. Mineral composition of CRS.

Mineral	Unit	Content	Mineral	Unit	Content
SiO_2	%	42.51±0.06	As	%	0.00±0
C	%	36.92±0.09	TiO_2	%	0.01±0
K_2O	%	17.16±0.01	Pb	%	0.00±0
Cl	%	1.79±0	Cu	%	0.00±0
CaO	%	0.80±0	Ba	%	0.00±0
Na_2O	%	0.71±0	V	%	0.00±0
MgO	%	0.62±0.01	Cr	%	0.00±0
P XRF	%	0.32±0	Ni	%	0.00±0
S XRF	%	0.32±0	Co	%	0.00±0
Al_2O_3	%	0.19±0.01	Sn	%	0.00±0
Fe	%	0.12±0.01	Sr	%	0.00±0
Mn	%	0.09±0	Zr	%	0.00±0
Zn	%	0.02±0			

As the highest proportion of CRS, silica has the greatest effects on the properties of CRS. Silica has high surface area and easily adsorbs gas and oil [8], which suitable as food additive such as anti-caking agent [9] and bacterial anti-adhesion food-contact surface [10].

Carbon is the second largest proportion which determine the black color of the CRS as food additive. Purified carbon from plants combustion has been commonly used as a black colorant for food [7]. Table 3 shows that over 30wt% of CRS consist of carbon makes it potential source of food colorant.

3.2. Particle size of CRS

The CRS was grinded and sieved with 40-200 mesh size. Table 2 shows the particle size distribution of CRS. Overall, the particle that pass through of the 40mesh-sieve is larger compare to that of 80-200mesh-sieve. It is notable that the particle size decrease with increasing mesh value.

Table 2. Particle size of CRS sieved at various mesh sizes.

Sieve (mesh)	Particle size distribution (μm)		
	D10	D50	D90
40	28.32	115.37	348.91
80	10.31	69.36	145.41
100	11.78	66.83	143.96
120	14.83	62.2	135.1
200	12.76	49.13	114.31

The particle size of CRS from 40-200 sieve sizes is larger than the minimum EFSA requirement for vegetable carbon (E 153), which is not under 275 nm (0.275 μm), showing that CRS particle size complied with the EFSA requirement. However, compared to commercial black colorant, the E 153, the CRS particle was larger. The current standard, according to NATCOL (2011) in EFSA Panel on Food Additives and Nutrient Sources added to Food [5], the particle size distribution of the commercial available preparations is characteristically 10% < 2 μm, 50% < 5 μm, and 90% < 55 μm or 10% < 1 μm, 50% < 4 μm, and 90% < 10 μm for the finer product, with an absence of particles below 275 nm. Therefore, further research is required to reduce the particle size of CRS. Smaller particle size will effect on surface area. The smaller the particle, the wider the surface area and the particle will easily to disperse in a solution at more stable state. The presence of silica in the CRS also assist the stability of the solution created.

3.3 The effect of particle size to kue jongkong Surabaya color

Table 3 shows the lab space color test (L) which show the darkness color of a material. Overall, higher sieve size, which produce smaller particle size of CRS, correspond to the darkness of *kue jongkong*. The CRS particles from 40mesh-sieve resulting on L value of 34 and the CRS particles obtained from 200mesh-sieve produce L value of 27 in the production of *kue jongkong*. This shows that smaller particles produced darker color of the *kue jongkong*.

Table 3. Effect of sieve size to L value of *kue Jongkong Surabaya*.

Sieve size (mesh)	L value	Tukey-test (α=5%)
40	34.43±2.12	a
80	32.93±0.26	ab
100	30.60±0.86	bc
120	28.63±0.70	cd
200	27.23±0.31	d
Kue jongkong Surabaya: commercial sample	30.53±0.14	

The commercial *kue jongkong* L value was 30. Therefore, larger CRS particles collected from 40 and 80 mesh-sieve were not met the darkness standard commercial of *kue jongkong*. The CRS particle which pass through 100-200 mesh-sieve shows L color value that similar and lower than that of the commercial one when applied to *kue jongkong*, suggesting that 100 mesh-sieve is the minimum sieve size to be used for CRS sieving.

4. Conclusion

Incomplete combustion of rice straw produced carbonized rice straw (CRS) that can be used as food colorant. The CRS contained various minerals (63%) and carbon (37%). The CRS particles sieved with 40-200mesh–sieve produced particle size that comply with the standard for commercial vegetable carbon. When the CRS particle was applied to *kue jongkong*, a minimum of 100 mesh-sieve size was required for producing CRS particle which can match the standard of darkness of *kue jongkong* commercial.

Acknowledgement

This paper is part of INSINASriset pratama project funded by Ministry of Research, Technology and Higher Education of the Republic of Indonesia.

References

[1] FAOSTAT 2017 Rice production quantity 2014. Retrived from http://www.fao.org/faostat/en/#data/QC.

[2] IRRI 2017 Rice Straw. Retrived from http://www.knowledgebank.irri.org/step-by-step-production/postharvest/rice-by-products/rice-straw

[3] BPS 2017 Produksi Padi Menurut Provinsi (ton) 1993-2015 (Rice Production based on Province (ton) 2013-2015. Retrived from https://www.bps.go.id/linkTableDinamis/view/id/865. (In Indonesian)

[4] Amchova P, Kotolova H and Ruda-Kucerova J 2015 Health safety issues of synthetic food colorants Regul. *Toxicol. Pharmacol.* **73** 3 914-22.

[5] EFSA 2012 Scientific Opinion on the re-evaluation of vegetable carbon (E 153) as a food additive *EFSA Journal* **10** 4 2592-n/a.

[6] Hazra B, Varma AK, Bandopadhyay AK, et al 2016 FTIR, XRF, XRD and SEM characteristics of Permian shales, India *J Nat Gas Sci Eng* **32** (Supplement C) 239-255.
[7] Zhenchao J, Yuting Z, Jiuming Y, Yedan L, Yang S, Jinyao C and Lishi Z et al 2015 Safety assessment of dietary bamboo charcoal powder: A 90-day subchronic oral toxicity and mutagenicity studies *Food Chem Toxicol* **75** Supplement C 50-57.
[8] Morsy FA, El-Sheikh SM and Barhoum A 2014 Nano-silica and SiO2/CaCO3 nanocomposite prepared from semi-burned rice straw ash as modified papermaking fillers *Arab J Chem* Article in Press
[9] Contado C, Ravani L and Passarella M 2013 Size characterization by Sedimentation Field Flow Fractionation of silica particles used as food additives *Anal. Chim. Acta* **788** (Supplement C) 183-192.
[10] Oh J K, Perez K, Jayaraman A, Cisneros-Zevallos L et al 2015 Hydrophobically-modified silica aerogels: Novel food-contact surfaces with bacterial anti-adhesion properties. *Food Control* **52** (Supplement C) 132-141

IOP Conf. Series: Earth and Environmental Science **131** (2018) 012006 doi:10.1088/1755-1315/131/1/012006

Microwave-assisted extraction (MAE) of bioactive saponin from mahogany seed (*Swietenia mahogany* Jacq)

E Waziiroh, Harijono and K Kamilia

Department of Agricultural Product Technology, Faculty of Agricultural Technology, Universitas Brawijaya, Malang, Indonesia

E-mail: elokwz@yahoo.co.id

Abstract. Mahogany is frequently used for medicines for cancer, tumor, and diabetes, as it contains saponin and flavonoid. Saponin is a complex glycosydic compound consisted of triterpenoids or steroids. Saponin can be extracted from a plant by using a solvent extraction. Microwave Assisted Extraction (MAE) is a non-conventional extraction method that use micro waves in the process. This research was conducted by a Complete Random Design with two factors which were extraction time (120, 150, and 180 seconds) and solvent ratio (10:1, 15:1, and 20:1 v/w). The best treatment of MAE were the solvent ratio 15:1 (v/w) for 180 seconds. The best treatment resulting crude saponin extract yield of 41.46%, containing 11.53% total saponins, and 49.17% of antioxidant activity. Meanwhile, the treatment of maceration method were the solvent ratio 20:1 (v/w) for 48 hours resulting 39.86% yield of saponin crude extract, 9.26% total saponins and 56.23% of antioxidant activity. The results showed MAE was more efficient (less time of extraction and solvent amount) than maceration method.

1. Introduction

Mahogany (*Sweetenia mahogany Jacq*) is one of the widely grown plants in Indonesia and is widely used as medicine for several diseases. Habsah et al. [1] reported that mahogany seeds have anti-inflammatory, anti-cancer, anti-viral, anti-infective, anti-tumor and anti-diabetic effects. Mahogany seeds contain saponins and flavonoids, therefore often used to cure various diseases [2]. Saponin is a complex glycoside compound formed from triterpenoids or alkaloid steroids, non-volatile and amphiphilic [3]. Steroid saponin are common type of saponin playing a role as health-promoting properties.

Saponin can be separated from other components in a material by extraction using a solvent. Commonly, saponin is extracted by using a solvent like water, ethanol, and methanol. Conventional methods such as reflux and maceration are common method to extract saponin. The conventional methods have negative effect on saponin extract as time consuming tended to extract degradation. Recently, non-conventional methods are used to overcome the negative effect of conventional method, such as Microwave-assisted extraction [4].

Microwaves are non-ionizing electromagnetic waves with a frequency range from 0.3 to 300 GHz [4]. The use of MAE to extract saponin in green tea showed that the yield was increased by increasing the process length [5], while the yield of the extract was determined by the ratio of the material and the solvent [5]. Using low ratio of solvent will make the solvent quickly saturated by the extracted solid

[6]. Therefore, this research uses the Microwave-Assisted Extraction to extract of saponin of mahogany seeds.

2. Materials and Method

2.1. Materials

Mahogany seeds are obtained from mahogany plantations in Tuban area of East Java. The chemicals were vanillin (Sigma-Aldrich), 70-72% perchloric acid (Sigmaco), oleanolic acid (Sigma-Aldrich), 0.02 mM DPPH reagent having grade pa except 96% ethanol (Smart-Lab Factory) concentrated sulfate (Merck), petroleum ether (Smart-Lab Factory) and acetic acid (Smart-Lab Factory).

2.2 Apparatus

The equipments were cabinet dryer (OVG-12), blender (Miyako), 40 mesh sieve (ATE-126), analytical balance (Denver M-310), plastic basin, ultrasonic bath (Elmasonic S40) 45 kHz, rotary vacuum evaporator (Buchi B-490), spectrophotometer (Labomed Inc.), Oven (Memmert), UV-Vis spectrofotometer (Jenway 6305), and FTIR spectroscopy F108 wavelength 500-4000 cm-1 (Shimadzu).

2.3 Extraction by MAE [7]

Defatted mahogany seeds extracted with ethanol solvent (boiling point 78.37 ° C) using microwave with predetermined variables. The varibles were solvent ratio of 10: 1 (v/b), 15: 1 (v/b), and 20:1 (v/b) and time of 120s, 150s and 180s. The extracts were filtered and concentrated using rotary vacuum evaporator at 50 rpm and 50°C for 15 min.

2.4 Extraction by Maceration [8]

The maseration method was used as a comparison. The extraction of mahogany seed by maceration method was done by using solvent ratio of 10:1 (v/b), 15:1 (v/b), and 20:1 (v/b) with 48 hours extraction time. The extracts were filtered and concentrated using rotary vacuum evaporator at 50 rpm and 50°C for 15 min.

2.5 Chemical Analysis

Saponin extract from mahogany seeds was analyzed for saponin compound [9], type of saponin using Liebermann-Burchard method [10], saponin content [11], and antioxidant activity [12]. The extraction process yield was determined by [13]. The best treatment was analyzed by its functional group with FTIR (Fourier Transmitter Infra-Red).

3. Results and Discussion

3.1 Qualitative Properties of Mahogany Ethanol Extracts

3.1.1 Identification of Saponin Compound

The saponin compound can be seen from its ability to form foam when shaking. Both extract obtained by MAE and maceration can form a stable foam 1-2 cm for 30 minutes.

3.1.2 Determination of Saponin Types

The identification of the type of saponin whether terpenoid or steroidal agitone in saponin compounds can be done by Liebermann-Burchard method [14]. In this study, brown color was formed on Liebermann-Burchard test, which indicated the triterpenoid type saponins in mahogany seed extracts. The color changes occurred due to the oxidation of terpenoid compounds through the formation of the conjugated double bond.

3.1.3 Identification of Functional Groups of Mahogany Ethanol Extract with FTIR (Fourier Transmitter Infra-Red)

[15] reported that the C = H (Alkene) group that read at the peak of 1340-1470 (cm^{-1}), the C = O group was read at the peak of 1690-1760 (cm^{-1}), and the C = C group was read peak 2100-2260 (cm^{-1}), indicates the presence of a trisaccharide (sugar group) in the triterpene. OH group was read at the peak of 3200-3600 (cm^{-1}), C = H group was read at the peak of 675-995 (cm^{-1}), C = C group was read at peak 2100-2260 (cm^{-1}), cluster C = O which is read at the peak of 1690-1760 (cm^{-1}), and the CO group read at the peak of 1050-1300 (cm^{-1}), indicating the presence of triterpenoid oleanane which is read on mahogany seed extract.

The FTIR (Figure 1) showed O-C = O group indicating a carboxylic acid (oleanolic acid) group or an ester group [16]. Crude extract contains many impurities of protein (amine/amide) was read at the point of 1180-1360 cm-1 and other alkaloids indicated by the number of peaks was read on functional group reading with FTIR, but the saponins compound is still read according to the standard of quillaja saponin.

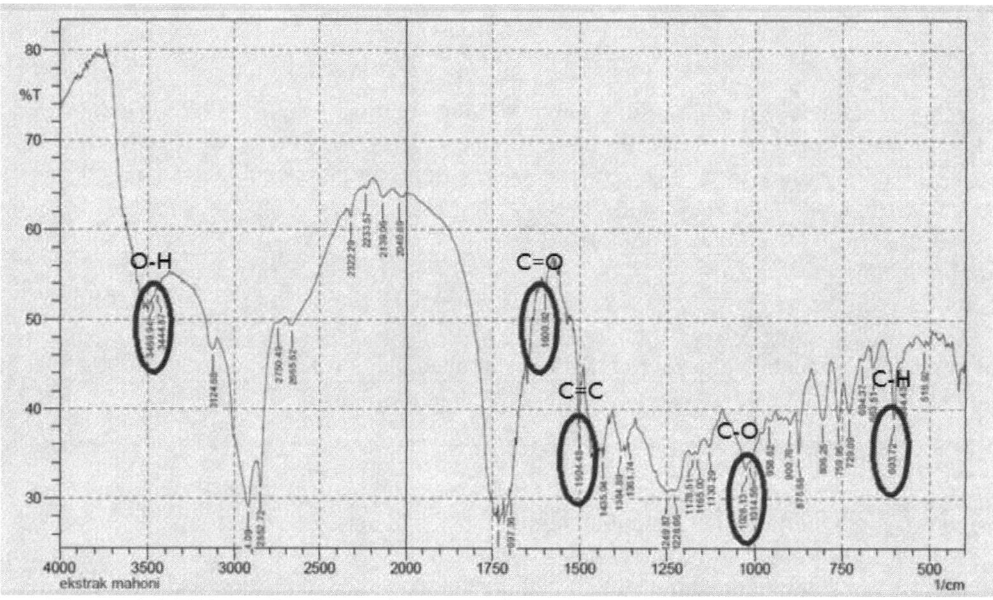

Figure 1. FTIR result of saponin extract from defatted mahogany seed with MAE

3.2 Quantitative Properties of Ethanol Extract of Mahogany Seed

3.2.1 Yield of extract

The yield of ethanol extract of mahogany seeds was 30.83% - 41.46%. Extending the extraction time resulted in greater yield of the extract, but the solvent ratio had a limit so that the increase in the solvent ratio leads to a decrease in the yield. The addition of solvent to MAE can lower the yield of the extracted compound [17]. This was due to a decrease in microwave radiation energy. The energy was absorbed by the solvent and decreased the energy received by the sample. Its energy capability to dissolve and pull the solute out of the sample was also smaller so that the solvent ratio needs to be limited.

In contrast to the yield obtained from the MAE method, the highest yield on the extraction by maceration was obtained on the use of solvent ratio of 20: 1 for 48 hours. The yield tended to increase with the addition of the ratio solvent: sample above 20: 1.

3.2.2 Saponin content

The saponin content of ethanolic extract from mahogany seeds was 8.69% -11.53% (Figure 2). Based on analysis of variance (α=0.05), the highest saponin content was obtained at solvent ratio 15: 1 for 180 seconds, while in the greater amount of solvent, the lower the saponin content. The addition of an excessive portion of the solvent may result in the reduction of the extracted compound [18]. More solvents will reduce the microwave energy received by the solute to be extracted so it will not dissolve. More time is needed for the extraction process when the likelihood of degradation of the compound being extracted is greater.

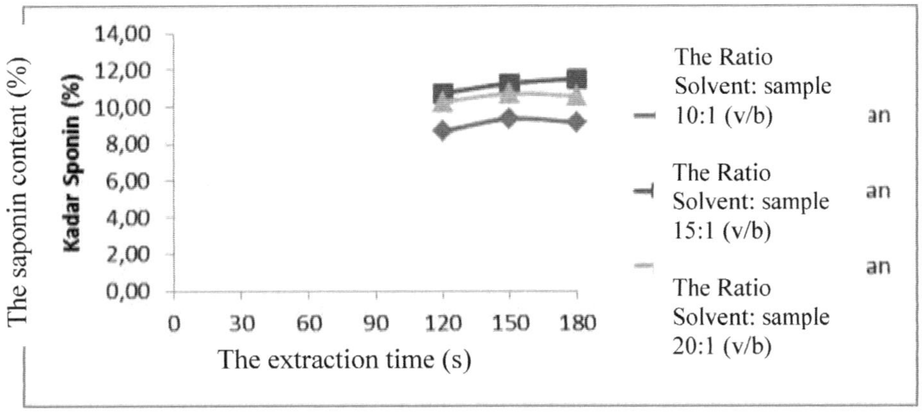

Figure 2. Saponin Content of extract from defatted mahogany seed with MAE

Based on Figure 2, the extraction with a longer time will result in the decrease of saponin content. Yulianti et al. [19] reported that at certain extraction times, saponins increased and then reached their peak point. After that stage, the addition of extraction time will not increase the saponin content produced.

Different results on saponin content were obtained by maceration for 48 hours. Based on analysis of variance (α=0.05), the highest saponin content yielded by the solvent : material ratio of 20: 1 (v/b) resulting in 9.26% of saponin. There was a tendency that saponin levels can still increase when it used a larger portion of solvent. In this study, maceration was done with a mild shaking, different with MAE. Microwave exposure helps the process of cell tissue destruction, as well as rotation of dipolar that occurs in the system will accelerate the occurrence of collisions between molecules of solute anf solvent causing solute expenditure occurs much faster [17].

3.2.3 Analysis of Antioxidant Activity

Antioxidant activity of ethanol extract of mahogany seed showed DPPH inhibition ability of 39.62% - 49.17% (Figure 3). The condition showed an indication that DPPH inhibitor compound was dominated by saponin compounds contained in the ethanol extract of mahogany seeds. The lower the saponin extract, the antioxidant ability to capture free radical DPPH was decreasing [6] (Figure 3). Based on analysis of variance (α=0.05), the highest antioxidant activity was obtained at solvent ratio 15: 1 for 150 seconds. Kusumawati [20] reported that antioxidant activity can be affected by saponin and also phenolic compounds or other flavonoid compounds.

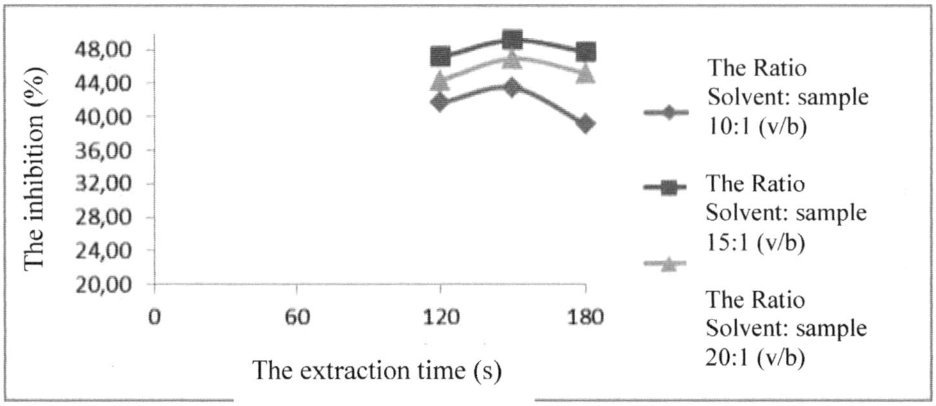

Figure 3. Antioxidant activity of ethanol extract from deffated mahogany seed

3.3 The Best Treatment and Comparative Effectiveness of MAE Extraction Process with Maseration

3.3.1 Best Treatment

The selection of the best treatment was done using Multiple Attribute. Assessment included yield of extract, saponin content and also antioxidant activity. The best treatment was obtained at solvent ratio of 15:1 (v/b) for 180 seconds. The best treatment parameter values of MAE can be seen on Table 1.

Table 1. Best treatment of saponin extraction from mahogany seeds by MAE method.

Ratio of Solvent:Sample (v/b)	Extraction Time	Average of Yield Extract (%)	Average of Saponin content (%)	Percentage of Inhibition (%)
15:1	180 s	41.46	11.3	49.17

3.3.2 The Effectiveness of Extraction

The total antioxidant capacity of ethanol extract of mahogany seeds can be expressed as the multiplication of the quantity value of the extracted compound with its antioxidant activity. In this study, the antioxidant capacity of the extract obtained was represented by the value of the effectiveness of the extraction process of antioxidant compounds from mahogany seeds with ethanol solvent.

Table 2. Comparison of best treatment of extraction with MAE and maceration method.

Methods	Ratio of Solvent:Sample(v/b)	Time	Yield (%) (=A)	Saponin Content (%)	Antioxidant activity (%) (= B)	Process effectiveness (A x B)
MAE	1:15	180 s	41.46	11.53	49.17	1940.07
Maceration	1:20	48 h	39.86	9.26	56.23	2241.14
T Test	-	-	0.024	0.0004	0.0034	0.018

Extraction of mahogany seeds using microwave was more effective in terms of the length of extraction time and the amount of solvent needed compared to maceration. Zhou [4] reported that the need for fewer solvents was caused by some energy generated in extraction with microwaves such as electrical energy and magnetic energy that can produce thermal energy causing the process of dissolving the cell and diffusion of compounds into the solvent are better with fewer solvents.

IOP Conf. Series: Earth and Environmental Science **131** (2018) 012006 doi:10.1088/1755-1315/131/1/012006

4. Conclusion

The best treatment of microwave-assisted saponin extraction using ratio of solvent:sample was 15: 1 (v/b) and extraction time of 180 seconds. The yield of extract was 41.46%, the inhibition percentage was 46.83%, the saponin content was 11, 53% and the effectiveness of extraction was 1940 units. The extraction results were significantly different compared to the maceration extraction method for 48 hours with the solvent: sample ratio of 20:1 (v/b). Thus it can be said that the microwave-assisted extraction method for mahogany saponin with 96% ethanol solvent was more efficiently used than the maceration method.

Acknowledgments

This work was financially supported by the Brawijaya University.

References

[1] Habsah A K 2013 Biological Activities and Phytochemicals of *Swietenia macrophylla King* Molecules Journal **18** 10465-10483

[2] Metty 2015 Antimicrobial activity of Bioactive Compound of Mahogany Seed (*Sweetenia Mahagoni (L.) Jacq.)* into Mycobacterium Tuberculosis Journal of Medika Respati **10** 2 68-77

[3] Sirohi, S K 2013 Utilization of Saponins, a Plant Secondary Metabolite in Enteric Methane Mitigation and Rumen Modulation Annual Research & Review in Biology **4** 1 1-19

[4] Zhou, T 2011 Evaluation Of Microwave-Assisted Extraction For Aristolochic Acid From Aristolochiae Fructus By Chromatographic Analysis Coupled With Nephrotoxicity Studies Biomed. Chromatogr. **26** 166–171

[5] Farida, R 2015 Antiosianin Extraction of Mangosteen Waste Using Microwave Assisted Extraction Journal of Pangan Dan Agroindustri **3** 2 362-373

[6] Putri, A R W dan Nisa, F C 2015 Antosianin Extraction of Red Rose (*Rosa Damascene Mill*) Using Microwave Assisted Extraction Journal of Pangan Dan Agroindustri **3** 2 701-712

[7] Chen, Y , Xie, M Y , dan Gong, X F 2007 Microwave-assisted extraction used for isolated of total triterpenoid saponins from *Ganoderma atrum Journal of Food Engineering* **81** 162–170

[8] Sianturi AHM 2001 Isolatioan and Fractination of Bioactive Compound of Mahogany Seed (*Swietenia mahagoni* Jacq) [Minor Thesis] Bogor: Institut Pertanian Bogor

[9] Harborne, J B 1987 Modern Method of Analysisi of Plant Bandung Institut Teknologi Bandung

[10] Edoga H O, Okwu, D E and Mbaebie, B O 2005 Phytochemical Constituents of Some Nigerian Medicinal Plants *African Journal of Biotechnology* **4** 7 685-688

[11] Xiang, Z B , Tang, C H, Chen, G, & Shi, Y S 2001 Studies on colorimetric determination of oleanolic acid in Chinese quince *Natural Product Research and Development* **13** 4 23–26

[12] Molyneax P 2004 The Use Of The Stable Free Radical Dyhenylpicrylhydrazil (DPPH) For Estimating Antioxidant Activity *Journal of Science Technology* **26** 2211-219

[13] Sudarmadji, S 1984 Analysis Procedure of Food dan Agricultural Material Third Edition Yogyakarta: Liberty

[14] Sangi, M , Runtuwene, M R J , Simbala, H E I dan Makang V M A 2008 Analysis of Phytochemistry of Medicine Herbs in North Minahasa [Minor Thesis]

[15] Kirmizigul S 2002 New Triterpenic Saponins from *Celphalaria transsylvanica* Turk Journal Chem **26** 947-954

[16] Handayani D 2014 Optimation Of Green Tea Waste Axtraction Using Microwave Assisted Extraction To Yield Green Tea Extract Traditional Medicine Journal **19** 1 29-35

[17] Dhobi M, Hemalatha S, and Mandal, V 2009 Optimization Of Microwave Assisted Extraction Of Bioactive Flavonolignan–silybinin *J Chem. Metrl.* **22** 13–23.

[18] Guo, Z. K. 2001. Microwave-Assisted Extraction of Effective Constituents from a Chinese Herbal Medicine *Radix puerariae*. Analytica Chimica Acta. 41–47.

[19] Yulianti, D., Susio, B., dan Yulianingsih, R. 2014. Effect of Time Extraction on Physic-Chemistry Charecteristic of Stevia Leaves (*Stevia Rebaudiana* Bertoni M.) Using *Microwave Assisted Extraction* (MAE) *Journal of Bioproses Komoditas Tropis* **2** 1 35-41, Juni 2014.

[20] Kusumowati, I. K. D. 2011. Correlation of Phenolic Compound and Antioxidant Activity of Cashew Leaves. Journal of Biomedika. **3**: 2.

IOP Conf. Series: Earth and Environmental Science **131** (2018) 012007 doi:10.1088/1755-1315/131/1/012007

Characteristics of okara color change during convective drying process

I Taruna and J Astuti

Department of Agricultural Engineering, Faculty of Agricultural Technology, University of Jember, Jember, Indonesia

E-mail: taruna@unej.ac.id

Abstract. Okara is a byproduct of the tofu factory that is perishable in nature and requires an immediate handling and treatment to preserve it. Drying has been a preferred method since it can increase the product compactness. However, the majority of okara drying studies done in the past are disregarding the influences of process condition on color change characteristic. Hence, this work aimed to investigate the characteristic of okara color changes during convective drying at temperatures of 85, 100, and 115°C. The okara color was expressed in terms of L a b values and its derivative attributes, i.e. whiteness, hue angle, chroma and the total color difference. The results showed that an increase in drying time and temperatures decreased the lightness, yellowness and the total color difference, but conversely increased the redness of okara. Fitting of two exponential models to the experimental data revealed that the Page's equation was better in predicting the okara color degradation than Newton's model.

1. Introduction

Manufactures of tofu or soymilk always generates okara, a byproduct that appears usually in a creamy white color and semi-solid form with 75-80% moisture on wet basis. Okara consists of remaining nutrients derived from soymilk such as protein, lipid, and fiber [1]. These nutrients quality are greatly high and therefore, it encourages people to utilize it as ingredient for human diets. However, okara must be processed rapidly through refrigeration or drying to prevent quick spoilage of this material.

Although refrigeration is a good method to preserve okara, but the people prefer mostly a drying operation because it can reduce the weight of final product. Several works have been reported on drying of okara for different methods, i.e. hot-air drying [2], spouted-bed drying [3], pneumatic tube and rotational drum-drying [4], flash drying [5], combination of vacuum-tray and microwave drying [6]. These studies are beneficial for them who will select drying methods for drying okara. Unluckily, there were no studies on okara drying that have been done in the past regarding to the okara color change during drying process. Hence, the aim of this study was to investigate the characteristic of okara color changes during convective drying for different temperatures between 85 and 115°C.

2. Materials and Methods

2.1. Raw okara

Raw okara was obtained from a small-scale tofu enterprise in Jember city, East Java, Indonesia. This was packaged in plastic bags and stored in a refrigerator (~4°C) for less than 4 days before the drying experiments. Initial moisture content of okara was determined using an AOAC standard method [7]. About 3 g of sample was weighed accurately and rapidly in the dish and then dried at 105°C in the oven for 3 h. The samples were cooled in a desiccator to the room temperature prior to weighing.

2.2. Experimental procedure

Three identical aluminum containers were filled with a thin layer of ±100 g okara and that of 1.5 cm thickness. They were placed horizontally on the same shelf in a hot-air oven (Memmert UNB400) and then dried at temperatures of 85, 100 and 115°C. Weight losses of samples were measured every hour till drying process was completed when the product moisture content reached 5% on wet basis. These data were then used to determine the change of okara moisture content during convective drying.

The color of okara was measured using a tristimulus colorimeter (Konica Minolta CR-10) in terms of Hunter L, a, and b values. The "L" or lightness value varies from 0 (black) to 100 (white). The "a" value represents the color range between green (-a) and red (+a), while the "b" value ranges from blue (-b) to yellow (+b). The change of okara color were measured periodically based on their L, a, and b values under similar interval time done for a sample weighing. Four readings of L, a and b values were taken from four different positions on sample surface. Derived color attributes such as whiteness (WI), chroma (Ch), hue angle (ϕ) and total color difference (ΔE) were calculated using Equation 1-4 [8, 9].

$$\text{Whiteness(WI)} = 100 - \left[(100 - L)^2 + a^2 + b^2\right]^{1/2} \quad (1)$$

$$\text{Chroma(Ch)} = \left(a^2 + b^2\right)^{1/2} \quad (2)$$

$$\text{Hue Angle}(\phi) = \text{ArcTan}\left(b/a\right) \quad (3)$$

$$\Delta E = \left[(L - L_o)^2 + (a - a_o)^2 + (b - b_o)^2\right]^{1/2} \quad (4)$$

where L_o, a_o and b_o represent the reading at time zero, and L, a and b represent the instant individual readings during drying process. All experiments were performed in three replicates

2.3. Data analysis

The relationships between the okara color attributes and drying time were developed for different drying temperatures. The change of okara color was assumed to follow first–order reaction kinetics and expressed by Equation 5 [10, 11]. The kinetic of change for selected okara color attributes during drying were analyzed using exponential models listed in Table 1.

$$CR = \frac{C}{Co} = \exp(-kt) \quad (5)$$

where C is a measured color attribute value based on L a b values (dimensionless) and Co is a measured color attribute value when drying did not begin yet (t=0).

Table 1. Exponential models selected for approximation.

No	Model	Name
1.	CR = exp(-kt)	Newton
2.	CR = exp(-kt^n)	Page

The constant value of k and n were determined by nonlinear regression analysis using SPSS 14 software. The accuracy of the models in estimating experimental data was evaluated using R^2, RMSE and P values as shown in Equation 6 to 8. The R^2 is a coefficient of determination that measures a goodness of fit, while a root mean square error (RMSE) shows a deviation between the experimental and estimated values. The P is the percentage error of estimated value by model in comparison to experimental value. The best fit of model should result in specifically the P value less than 10% [12].

$$R^2 = 1 - \left[\sum_{i=1}^{N} \left(CR_{est,i} - CR_{obs,i} \right)^2 \Big/ \sum_{i=1}^{N} \left(CR_{obs,i} - \overline{CR}_{obs} \right)^2 \right] \quad (6)$$

$$RMSE = \left[\frac{1}{N} \sum_{i=1}^{N} \left(CR_{est,i} - CR_{obs,i} \right)^2 \right]^{1/2} \quad (7)$$

$$P = \frac{100}{N} \sum_{i=1}^{N} \frac{\left| CR_{est,i} - CR_{obs,i} \right|}{CR_{obs,i}} \quad (8)$$

3. Results and Discussion

The present experimental setup could dry okara from 82.5-83.7% to 2.5-3.8% moisture on wet basis for about 8-10 h depending upon drying temperatures. In general, the okara dried entirely during the falling rate period. It might be due to the highly porous nature of okara that make easy moisture transfer rate. The change of okara color during convective drying are theoretically almost always three-dimensional, but not all three dimensions may be of practical importance [11]. Therefore, the color change analysis in the present study was based on the simplified methods proposed by former researchers, as discussed in the followings

3.1. Color change characteristics

The L a and b values of okara obtained during drying experiment were represented in the forms of L/L_o, a/a_o and b/b_o, where L_o, a_o and b_o indicated the values when the drying time was a zero. The relationship between these okara color parameters and drying time were shown in Figure 1to 3. It is shown in Figure 1 that an increase in drying time and temperature decreased the L/L_o values, causing a slight darker appearance on okara in comparison with the fresh conditions. The L values of dried okara degraded about 6.3-10.4% of its initial values. The highest change of L values was found for a sample dried at 115°C, while the lower changes were indicated by the samples dehydrated at 100 and 85°C, respectively. It was concluded that convective drying of okara at low drying temperature resulted in brighter color appearance, and thus produced the better product quality. This result was comparable to study that investigating the change of mango color due to drying process [13].

It is depicted in Figure 2 that the a/a_o values of okara increased when the drying temperature and time increased. The present drying experiment tended to produce a redder look of okara color at the final drying process. It was confirmed that the longest duration of drying performed at 85°C revealed a strongest effect in increasing redness of okara among the drying temperatures.

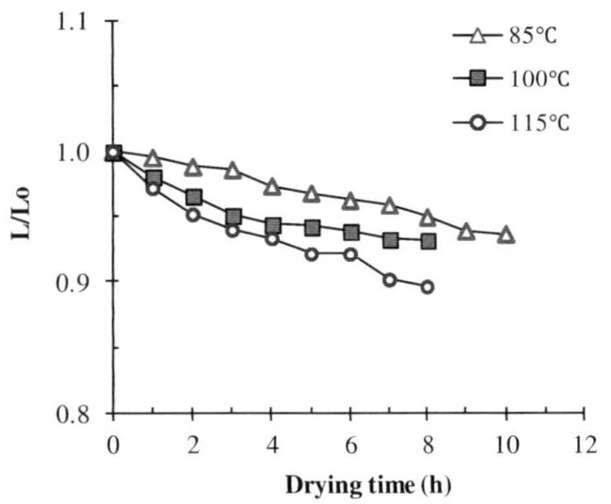

Figure 1. The lightness change of okara during drying at 85, 100, and 115°C

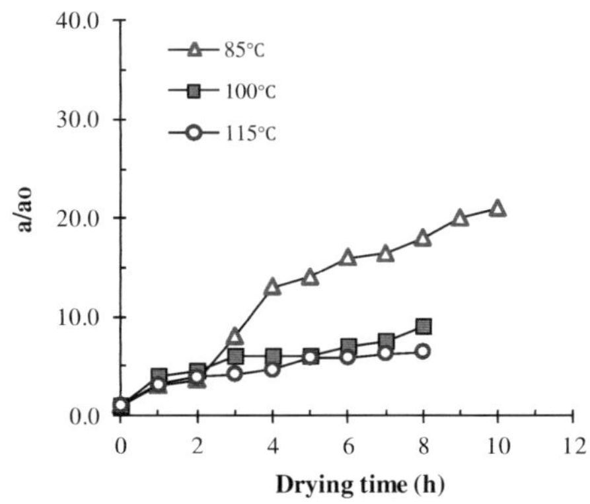

Figure 2. The redness change of okara during drying at 85, 100, and 115°C

The decreased relative b values (b/b_o) of okara were observed while drying time and temperatures increased from 85 to 115°C as shown in Figure 3. It seemed that okara was experiencing a yellowness loss during convective drying process. This phenomenon together with an increase in redness (a value) was claimed to cause the degradation of the okara lightness that had been found in this study. This was in agreement with the result obtained by Marcel et al [14].

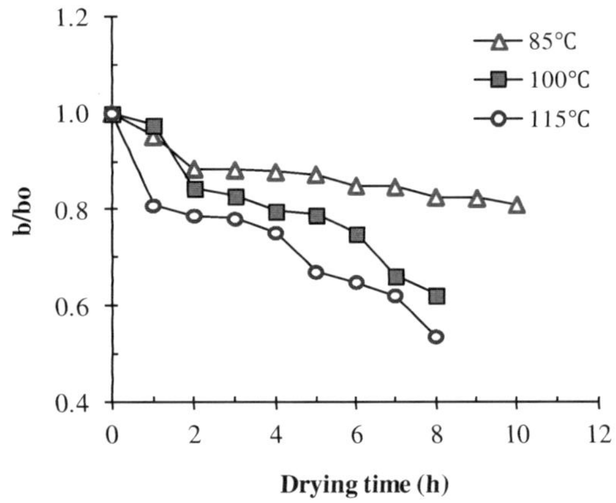

Figure 3. The yellowness change of okara during drying at 85, 100, and 115°C

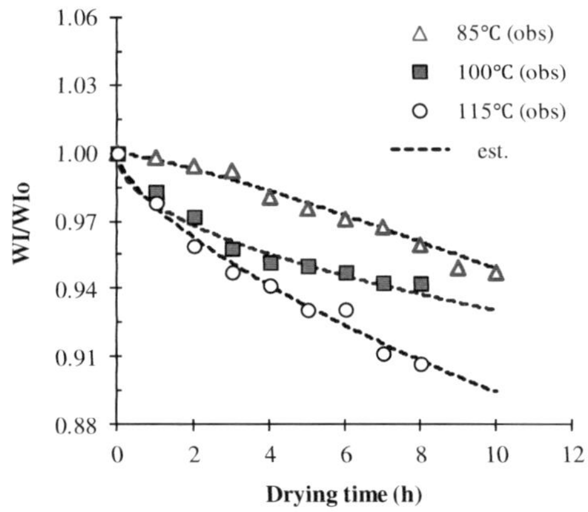

Figure 4. The change of whiteness of okara during drying at 85, 100, and 115°C

The characteristics of derivative color attributes, i.e. whiteness, chroma, hue angle and total color difference, during drying of okara are presented in Figure 4 to 7. The whiteness decreased throughout the drying as indicated by decreasing the WI/WI_o values and due to the drying time and temperatures. Figure 4 show that okara samples dried at drying temperatures from 85 to 115°C suffered the whiteness loss about 5.3-9.3% at the end of process. The relative chroma values (Ch/Ch_o) decreased during drying, suggesting the change of okara color intensity due to drying process as shown in Figure 5. An increase in temperature decreased significantly the chroma around 17.8-45.0% from an original condition. Changes in chroma values were determined predominantly by its redness and yellowness properties. The relative hue angle (ϕ/ϕ_o) was decreased due to an increase in drying time and temperatures. The use of higher drying temperature resulted in a lower ϕ value, indicating redder okara samples. Total color difference (ΔE) is parameter that frequently used to describe color variation in a test sample relative to a reference. The drying time and temperature used for drying okara was found significantly to affect the ΔE values. An increase in drying time and temperature from 0 to 8-10 h and from 85 to 115°C, respectively, increased the ΔE value. A higher ΔE values indicated a greater change on okara color [15]. Accordingly, it is shown in Figure 7 that the drying of okara at temperatures of 85°C produced the least discoloration in comparison with that of using 100 and 115°C.

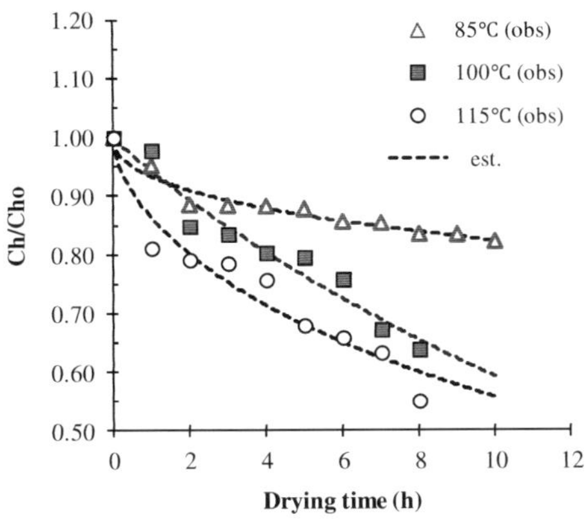

Figure 5. The change of chroma of okara during drying at 85, 100, and 115°C

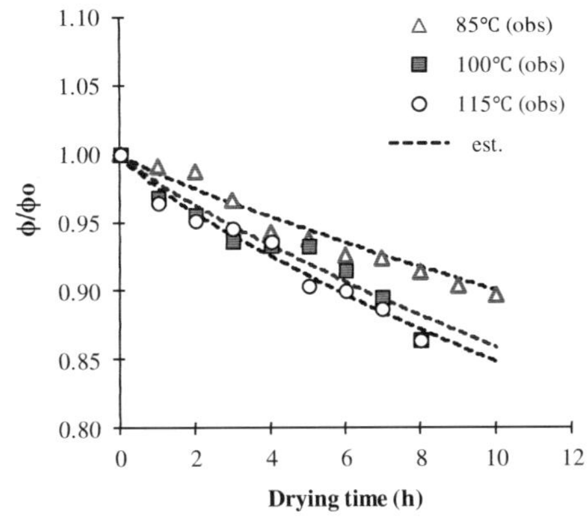

Figure 6. The change of hue angle of okara during drying at 85, 100, and 115°C

Table 2. Values of constants and statistical parameters for the models of okara color change.

Parameter	Model	T (°C)	Constants k (h^{-1})	Constants n	R^2	RMSE	P (%)
WI/WI_o	Newton	85	0.005	-	0.960	0.004	3.0
		100	0.009	-	0.761	0.009	7.1
		115	0.013	-	0.905	0.009	7.0
	Page	85	0.003	1.245	0.980	0.003	2.3
		100	0.023	0.496	0.969	0.003	2.6
		115	0.024	0.667	0.985	0.003	2.6
Ch/Ch_o	Newton	85	0.024	-	0.919	0.031	>10.0
		100	0.053	-	0.980	0.026	>10.0
		115	0.075	-	0.959	0.048	>10.0
	Page	85	0.069	0.451	0.990	0.011	9.2
		100	0.058	0.957	0.981	0.026	>10.0
		115	0.147	0.600	0.984	0.030	>10.0
ϕ/ϕ_o	Newton	85	0.011	-	0.979	0.007	5.6
		100	0.017	-	0.948	0.011	8.9
		115	0.018	-	0.961	0.009	6.3
	Page	85	0.014	0.874	0.979	0.007	6.6
		100	0.021	0.859	0.956	0.010	7.5
		115	0.025	0.818	0.976	0.007	5.7
1-ΔE	Newton	85	0.172	-	0.975	0.567	>10.0
		100	0.361	-	0.998	0.229	>10.0
		115	0.293	-	0.994	0.511	>10.0
	Page	85	0.056	1.637	0.992	0.318	>10.0
		100	0.308	1.134	0.999	0.187	>10.0
		115	0.289	1.010	0.994	0.511	>10.0

3.2. Effect of temperature on kinetics of color change

The kinetics of okara color degradation during drying process was evaluated in terms of whiteness (WI), chroma (Ch), hue angle (ϕ) and total color difference (ΔE). Nonlinear regression analysis of the models (Table 1) for selected color parameters in correlation with the drying time had resulted in the value of constants k and n as summarized in Table 2. The influence of drying temperatures on the constants value and accompanied by the respective R^2, RMSE and P values is shown in Table 2. It was confirmed that an increase in drying temperature increased generally the rate of okara color change as indicated by the k values for selected color parameters. The values of R^2 indicated a good fit for all models since they were more than 0.9. The Page's model proved generally as the most fitted equation to the experimental CR data, as indicated by the higher R^2 and the lower RMSE and P values in comparison to the Newton's model. However, it is not suggested definitely to use this model for estimating the change of total color difference (ΔE) since it revealed the P value greater than 10%.

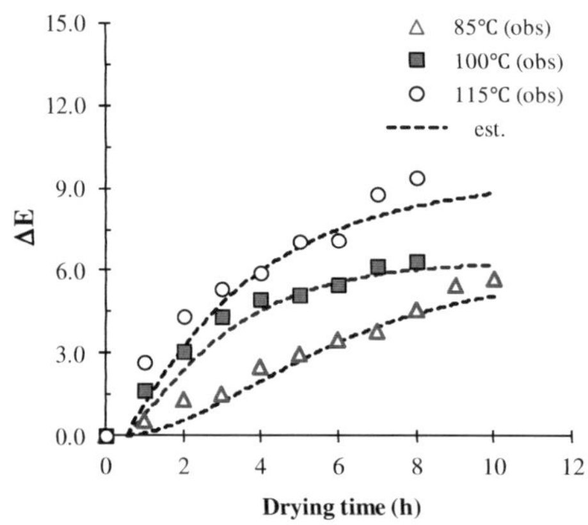

Figure 7. Total color difference of okara during drying at 85, 100, and 115°C

The plots between the observed and the estimated of okara color change based on the Page's model are presented in Figure 4 to7. It is verified that a Page's model were generally suitable in describing the okara color change during convective drying process. However, it is important to understand that application of such Page's model allow design calculations within the range of temperatures varied in this work and for similar experimental setup.

4. Conclusion

Drying of okara at temperature of 85, 100 and 115°C resulted in change of okara color. An increase in drying time and temperature decreased the lightness and yellowness but increased the redness appearance of okara. The Page's model was in general better than Newton's model in estimating experimental data of kinetics of color change. The information obtained in this works may be used as a rule for selecting the operating condition of drying to reduce the quality degradation of okara.

References

[1] Van der Riet W B., Wight A W, Cilliers J J L, Datel J M 1989 Food chemical investigation of tofu and its byproduct okara *Food Chemistry* **34** 193-202
[2] Lema, A A, Aguerre, R J, Suarez, C 1998 Drying kinetics of soybean draff (okara) Proceedings of the 11th International Drying Symposium, Halkidiki, Greece, August 19-22, 1998, pp. 2149-2155
[3] Taruna I, Jindal V K 2002 Drying of soy pulp (okara) in a bed of inert particles *Drying Technology* **20** 4–5 1035-1051
[4] Perussello C A, Amarante A C C, Mariani V C 2009 Convective drying kinetics and darkening of

okara *Drying Technology* **27** 1132–1141
[5] Grizottoi R K, Aguirre J M 2011 Study of the flash drying of the residue from soymilk processing- “okara” *Ciênc. Tecnol. Aliment., Campinas.* **31** 3 645-653
[6] Sengupta S, Chakraborty M, Bhowal J, Bhattacharya D K 2012 Study on the effects of drying process on the composition and quality of wet okara *Int'l J. of Sci. Envi. and Tech.* **1** 4 319 – 330
[7] AOAC 1999 Official Method of Analysis. Association of Official Analytical Chemist. Washington DC USA
[8] Bolin H R, Huxsoll C C 1991 Control of minimally processed carrot (*Daucus carota*) surface discoloration caused by abrasion peeling *J. Food Sci.* **56** 416–418
[9] Francis F J 2005 Colorimetric Properties of Foods in Engineering properties of foods (Edited by Rao M A, Rizvi S S H, Datta A K) 3rd Ed CRC Press Boca Raton FL USA 703-732
[10] Ahmed J, Shivhare U S, Raghavan G S V 2001 Color degradation kinetics and rheological characteristics of onion puree *Transactions of the ASAE* **44** 1 95–98
[11] Koca N, Burdurlu H S, Karadeniz F 2007 Kinetics of colour changes in dehydrated carrots *J. Food Eng.* **78** 2 449-455
[12] Mohapatra D, Rao P S 2005 A thin layer drying model of parboiled wheat *J. Food Eng.* **66** 513-518
[13] Akoy E O M, Hörsten D V, Luecke W 2008 Drying kinetics and colour change of mango slices as affected by drying temperature and time. Proceeding: TheTropentag International Conference on Competition for Resources in a Changing World: New Drive for Rural Development” Hohenheim, Germany 7-9 October 2008
[14] Marcel E, Alexis K, François G 2014 Effect of thermal process and drying principle on color loss of pineapple slices *Am. J. Food Sci. Technol.* **2** 1 17-20
[15] Maskan M 2001 Kinetics of colour change of kiwifruits during hot air and microwave drying *J. Food Eng.* **48** 169-175

IOP Conf. Series: Earth and Environmental Science **131** (2018) 012008 doi:10.1088/1755-1315/131/1/012008

Physical quality characteristics of the microwave-dried breadfruit powders due to different processing conditions

I Taruna, A L Hakim and Sutarsi

Department of Agricultural Engineering, Faculty of Agricultural Technology, University of Jember, Jember, Indonesia

E-mail: taruna@unej.ac.id

Abstract. Production of breadfruit powder has been an option to make easy its uses in various food processing. Accordingly, there is a need recently to apply advanced drying method, i.e. microwave drying, for improving quality since conventional methods produced highly variable product quality and required longer process. The present work was aimed to study the effect of microwave power and grinding time on physical quality of breadfruit powders. The experiment was done initially by drying breadfruit slices in a microwave dryer at power level of 420, 540, and 720 W and then grinding for 3, 5, and 7 min to get powdery product of less than 80 mesh. The physical quality of breadfruit powders were measured in terms of fineness modulus (FM), average particle size (D), whiteness (WI), total color difference (ΔE), water absorption (W_a), oil absorption (L_a), bulk density (ρ_b) and consistency gel (G_c). The results showed that physical quality of powders and its ranged-values included the FM (2.08-2.62), D (0.44-0.68 mm), WI (75.2-77.9), ΔE (7.4-10.5), W_a (5.5-6.2 ml/g), L_a (0.7-0.9 ml/g), ρ_b (0.62-0.70 g/cm^3) and G_c (41.3-46.8 mm). The experiment revealed that variation of microwave power and grinding time affected significantly the quality of the breadfruit powders. However, microwave power was more dominant factor to affect quality of breadfruit powder in comparison to the grinding time.

1. Introduction

Breadfruit (*Artocarpus altilis*) is a tropical fruit that consumed in Indonesia in various processed foods such as boiled, steamed, fried and grilled, due to its high carbohydrate content. Converting breadfruit into a powdery product is preferred recently to extend its shelf life and to ease its utilization. Previous authors used successfully breadfruit powder as main ingredient of cake, noodle and breadfruit bar [1, 2, 3]. Such powder may be made through drying, grinding and sieving process. Thus, insight of the processing effects on the powder quality can help to determine more effective production process.

Some studies on the application of different methods and various process conditions for breadfruit drying had been carried out previously to get data about its effect on final product and optimum operating condition [4, 5]. However, it was done using conventional drying methods that might cause a high variable product quality and a longer drying process. Accordingly, initiative to examine more advanced drying methods is still needed to get better quality products and to shorten drying time.

Application of microwave heating is becoming popular now in industrial food drying because it is faster, more uniform, more energy efficient and better final product quality when compared with other drying techniques [6]. The microwave drying influence on fruits and vegetables quality are reported in many literatures [7, 8], but very little of them concerning on physical quality properties of breadfruit powders due to different processing conditions. Therefore, the present study aimed to investigate the effects of microwave power and grinding time on physical quality characteristics of breadfruit powders.

2. Materials and Methods

2.1. Preparation of breadfruit powders

Fresh breadfruit was bought from the local market in Jember, East Java, Indonesia. This was initially peeled, washed and cut to form sample slices with ~5 mm thickness. Initial moisture of sample was determined according to standard method [9].

About 150 g breadfruit slices were put on a glass dish and dried intermittently in a microwave oven using the following steps: during the first 8 min it was off every 4 min, for the next 4 min, off every 2 min and for the rest drying time, off each min. This step was ended when sample moisture reached 7.0±1.0% w.b. that was then powdered through grinding and sieving process. The sample prepared for this experiment was based on a full 3x3 factorial design, randomized completely and replicated. The microwave power (420, 540 and 720 W) and grinding time (3, 5 and 7 min) were two factors used in this experimental design. A control sample was a powder prepared by soaking breadfruit in 0.02% Sodium Bisulfite solution for 45 min before hot-air drying at 60°C and then grinding for 5 min.

2.2. Evaluation of the powder properties

The physical properties of breadfruit powder was evaluated in terms of the fineness modulus, average particle size, color attributes, water and oil absorption capacity, bulk density, and gel consistency. Fineness modulus (FM) was measured by sieving about 150 g ground powder in a Retsch sieve's shakers containing specific sieves with mesh number of 4, 8, 16, 30, 50, 100 and a pan. The FM was a one hundredth of the sum of powder percentages retained on designated sieves. The average particle size (D) of powder (mm) was calculated using Equation 1 as suggested by Henderson and Perry [10].

$$D = 0.10414 * (2)^{FM} \tag{1}$$

Except for the FM and D measurements, each sample should be sieved through 80 mesh sieve prior to the physical properties analysis. Color attributes included whiteness (WI) and total color difference (ΔE) that were calculated using Equation 2 and 3, respectively based on L, a and b values recorded by a colorimeter (Konica CR-10). The ΔE value represents the total color difference between a control sample (subscript c) and tested sample.

$$WI = 100 - \left[(100 - L)^2 + a^2 + b^2\right]^{1/2} \tag{2}$$

$$\Delta E = \left[(L - L_C)^2 + (a - a_C)^2 + (b - b_C)^2\right]^{1/2} \tag{3}$$

Water absorption (W_a) and oil absorption capacity (La) were measured using minor modified method described by Traina and Breene [11]. About 1.05-1.36 g powder was mixed with 10 ml distilled water in a test tube. It was then shaken for 1 min and allowed to stand for 30 min at 24°C before it was centrifuged at 2000 RPM for 25 min. The retained water in a test tube after discarding the excess water was measured as W_a. The L_a value was obtained using similar procedure by replacing the distilled water with 3 ml palm oil. The bulk density (ρ_b) was measured by weighing powder that fully occupied a known-volume container. The gel consistency (G) of sample was quantified based on Cagampang et al. method [12]. About 100 mg powder was mixed with 0.2 ml 95% ethanol containing 0.025% thymol blue in test tubes and then added immediately about 2.0 ml of 0.2N KOH. Once the mixture has been dispersed, the tubes were covered with glass marbles and placed in a boiling water bath for 8 min. The samples released from a water bath were cooled at room temperature for 5 min and then followed using ice water for 15 min. The tubes were finally laid horizontally over a mm-paper for about 30-60 min. The

G_c value was the length of gel measured along test tube that indicated the following: hard (27-35 mm), medium hard (36-40 mm), medium (41-60 mm) and soft (61-100 mm).

3. Results and Discussion

Microwave drying was conveniently dried the sample from moisture content of 69.16% to 7.51% on wet basis. It appeared that an increase in microwave power from 420 to 720 W shortened the drying time from 24 to 16 min. Correlation analysis between physical properties of powders and processing condition i.e. microwave power and grinding time resulted in coefficients (r) as shown in Table 1. These r values labeled as significant by SPSS 14 software indicate the significant relationship between processing conditions and physical properties of powder as discussed in the following sections.

3.1 Fineness modulus and average size of powder

The uniformity of grind in the resultant product was expressed as FM, which a smaller value indicates a fine powder while a large value reveals a coarse powder. The FM value of breadfruit powder ranged between 2.08 and 2.62 depending on the processing conditions. It is shown in Table 1 that the grinding time (r = -0.937) and the microwave power (r = 0.075) showed respectively to be the most important factors affecting the FM values. Prolonged grinding time granted a longer phase of size reduction and thus produced a finer powder. However, the FM values of tested samples were significantly higher than a control sample, indicating a coarser powder than a control sample as shown in Figure 1. Variation of drying methods might cause changes in texture of dry breadfruit, which could affect the grinding performance in resulting in fine powder. Former study showed that the use of microwave drying tended to increase the hardness of vegetable material in comparison to sun drying [13].

Table 1. Correlation coefficients between powder properties and processing conditions.

Response Variables	Values		Correlation Coefficients (r)	
	Range	Mean	Microwave Power (W)	Grinding time (min)
FM	2.08-2.62	2.33	0.075	-0.937 **
D (mm)	0.44-0.68	0.53	0.133	-0.910 **
L	83.5-85.0	84.5	-0.479 *	-0.440
a	-2.7--0.5	-1.9	0.569 *	0.581 *
b	15.9-18.5	16.9	0.570 *	0.316
WI	75.2-77.9	76.9	-0.569 *	-0.377
ΔE	7.4-10.5	8.5	0.577 *	0.391
W_a (ml/g)	5.5-6.2	5.8	0.584 *	-0.611 **
L_a (ml/g)	0.7-0.9	0.8	-0.371	0.654 **
ρ_b (g/cm^3)	0.62-0.70	0.66	-0.026	0.722 **
G_c (mm)	41.3-46.8	43.9	0.378	0.041

Note: * Significant ($p<0.05$); ** Significant ($p<0.01$)

The average particle size (D) of powders was in the range of 0.44-0.68 mm depending on the processing conditions. A small D-value indicated a fine powder and thus resulted from a small FM value. Table 1 presents that the grinding time (r= -0.910) and microwave power (r= 0.133) affected significantly the D value. An increase in grinding time decreased the D values as shown in Figure 2.

3.2. Color attributes

The color attributes of the breadfruit powder was expressed in terms of whiteness (WI) and the total color difference (ΔE). It is presented in Figure 3 that WI values were in a very narrow range between 75.2 and 77.9 for the powders produced from all treatments. There was no significant difference on WI values among the tested samples, except for those that produced using 420 W with 3 min and 720 W with 7 min of microwave power and grinding time, respectively. Table 1 confirmed that the most significant factors affecting WI was a microwave power (r = -0.569). A control sample showed WI value higher than microwave-dried samples though exposing under longer heating. Soaking a control sample in Sodium Bisulfite solution might prevent darkening of this sample during convective drying.

Total color difference (ΔE) between a control sample and microwave-dried-breadfruit powder is presented in Figure 4. The ΔE values were in the range of 7.4-10.5 depending upon processing conditions. The lowest and the highest ΔE values were shown by the samples that dried using the microwave power of 420 W and 720 W, respectively. This is verified in Table 1 that the ΔE values correlated strongly with microwave power (r = 0.577). The increased color change in a microwave-dried-product was due to an increase in microwave power and exposure time [14].

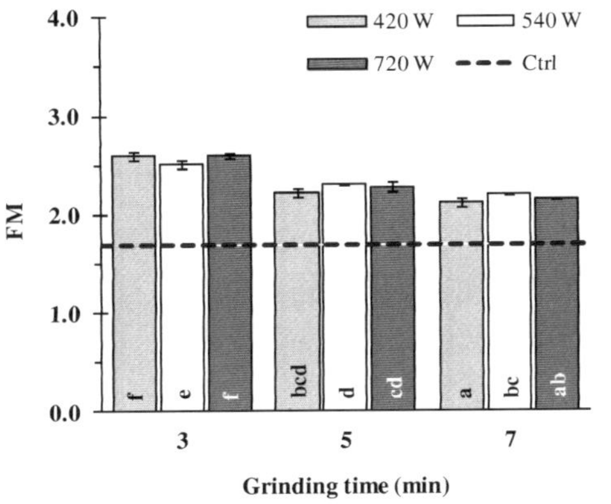

Figure 1. Fineness modulus (FM) of powder as affected by grinding time and MW power

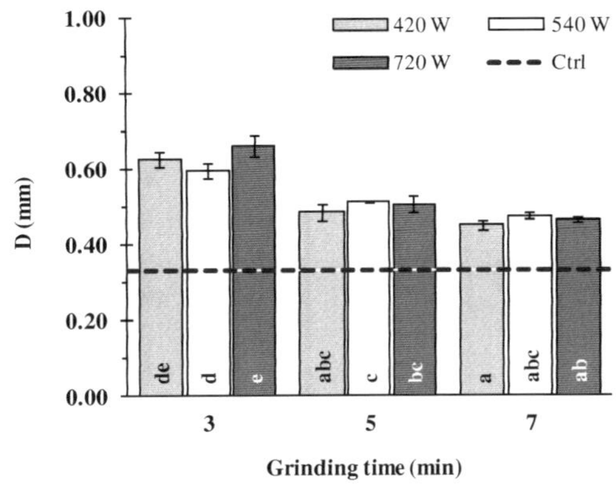

Figure 2. Average particle size (D) as affected by grinding time and MW power

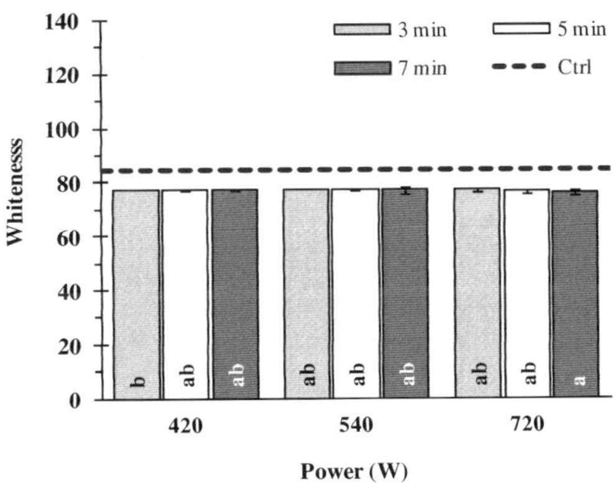

Figure 3. Whiteness of powder as affected by MW power and grinding time

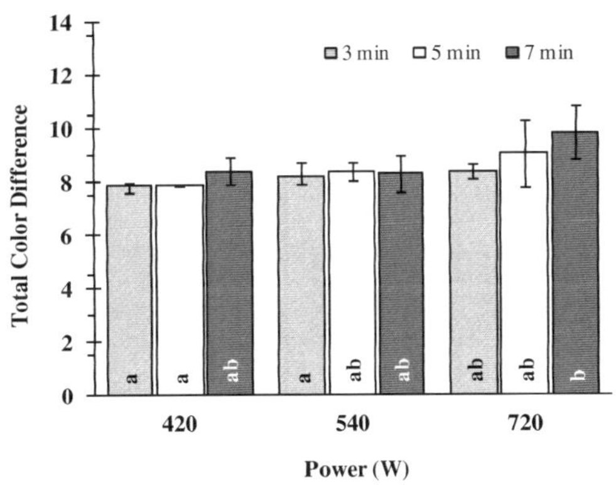

Figure 4. Total color difference between tested powder and a control sample

3.3. Water absorption capacity

Water absorption capacity (W_a) of powders obtained in this study ranged between 5.5 and 6.2 ml/g as presented in Figure 5. It is shown in Table 1 that the W_a values increased with microwave power (r = 0.584) and decreased with increasing of grinding time (r = -0.611). The increased grinding time was observed to produce a finer powder that possible to increase a particle surface area and facilitated more water absorption. It was comparable to the result found in palm kernel shell powders [15]. However, W_a value of a control sample was lower than the tested samples as shown in Figure 5. It was indicated in former study that such differences in W_a values might be due to the application of different drying methods [16]. Besides, application of microwave drying method showed a tendency to increase the product porosity and caused a more room space in powder to absorb water [17].

3.4. Oil absorption capacity

The effect of grinding time and microwave power on oil absorption capacity (L_a) of breadfruit powder is shown in Figure 6. The L_a values obtained from all treatments were in the ranges of 0.7- 0.9 ml/g, indicating a small variety among the tested samples. The grinding time (r= 0.654) was observed more controlling L_a values than microwave power (r=-0.371). The L_a values showed a tendency to increase with grinding time, while decreased with increasing microwave power. The longer grinding process produced a finer powder that might enlarge powder surface area for absorbing more oil. Comparable phenomenon on the effect of particle size on L_a value was found in cowpea flour [18]. The L_a of a control sample was about 1.1±0.1 ml/g and significantly higher than the tested powders. However, it was concluded that the breadfruit powder preferred to absorb more water than palm oil.

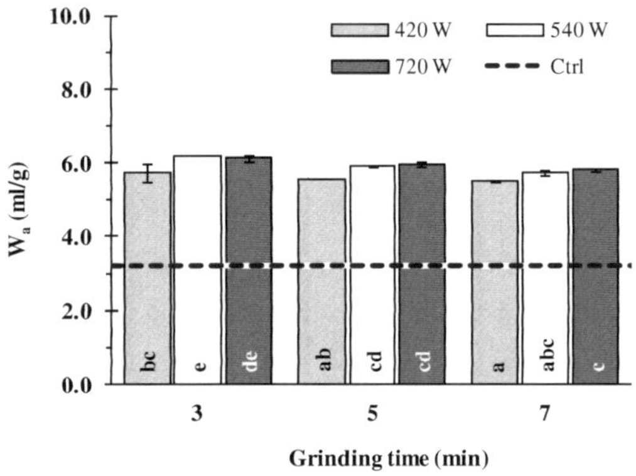

Figure 5. Water absorption of powder as affected by grinding time and MW power

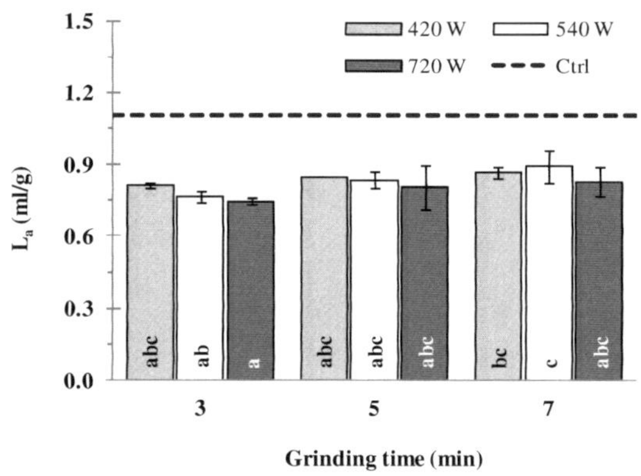

Figure 6. Oil absorption of powder as affected by grinding time and MW power

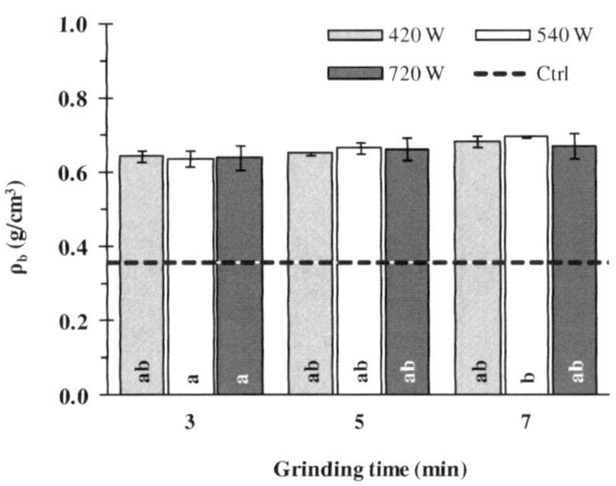

Figure 7. Bulk density of powder as affected by grinding time and MW power

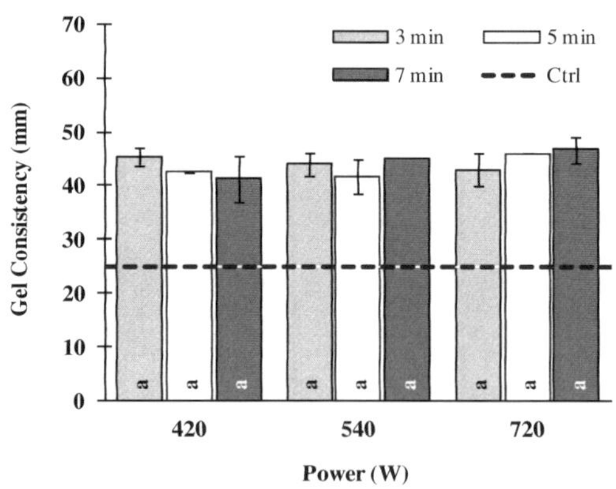

Figure 8. Gel consistency of powder as affected by MW power and grinding time

3.5. Bulk density and gel consistency

Bulk density (ρ_b) is useful for characterizing, handling, and processing of powder system in the food industry. The ρ_b values for microwave-dried breadfruit powder are presented in Figure 7 and in the range of 0.62-0.70 g/cm^3. Grinding time was found predictably to be the most factors affecting the ρ_b values because it correlated strongly to the particle size of powder. Longer grinding time resulted in more fine powder, which occupied less container volume and thus increased the bulk density.

Gel consistency (G_c) characterizes the tenderness or hardness of cooked breadfruit when cooling down. Figure 8 reveals that the G_c value of microwave-dried powders varied from 41.3 to 46.8 mm depending upon processing condition and classified as *medium* material. Table 1 shows that the effect of microwave power (r = 0.378) on G_c value was higher than grinding time (r = 0.041). The G_c value of control sample was lower than microwave-dried powder as illustrated in Figure 8. This suggested that the use of microwave drying method could increase the softness of cooked breadfruit powder.

4. Conclusion

Drying of breadfruit under exposure of microwave was successful to reduce the sample moisture content from 69.16% to 7.51% on wet basis. Variation of microwave power and grinding time proved to affect significantly the physical quality characteristics of breadfruit powders. The physical quality of microwave-dried breadfruit powders and its range values included the fineness modulus (2.08-2.62), average particle size (0.44-0.68 mm), whiteness (75.2-77.9), total color difference (7.4-10.5), water absorption capacity (5.5-6.2 ml/g), oil absorption capacity (0.7-0.9 ml/g), bulk density (0.62-0.70 g/cm3) and gel consistency (41.3-46.8 mm). It appeared that the breadfruit powder preferred to absorb more water than oil.

References

[1] Galappatti K M, Jayamanne V S 2010 Manufacture of cake using breadfruit flour (*Artocarpus altilis Fosb.*) and evaluation of its shelf-life, physicochemical and sensory properties. Proceedings of the 8th Academic Sessions University of Ruhuna 2010 **8** P7

[2] Akanbi T O, Nazamid S, Adebowale A A, Farooq A, Olaoye A O 2011 Breadfruit starch-wheat flour noodles: preparation, proximate compositions and culinary properties *Int'l Food Res. J.* **18** 1283-1287

[3] Nochera C L, Ragone D 2016 Preparation of a breadfruit flour bar *Foods* **5** 37 1-7

[4] Famurewa J A V, Esan Y O, Pele G I, Arewa O A 2015 Effect of maturity and drying methods on rheological and physico-chemical properties of reconstituted breadfruit (*Artocarpus altilis*) flour *IOSR J. Eng.* **5** 2 1-9

[5] George C, Mogil Q, Andrews M, Ewing G 2017 Thin layer drying curves for shredded breadfruit (*Artocarpus altilis*) *J.. Food Proces. Preserv.* **41** 5 1-8

[6] Orsat V, Changrue V, Raghavan G S V 2006 Microwave drying of fruits and vegetables *Stewart Postharvest Rev.* **6** 4 1-7

[7] Varith J, Dijkanarukkul P, Achariyaviriya A, Achariyaviriya S 2007 Combined microwave hot air drying of peeled longan *J. Food Eng.* **81** 459-468

[8] Sorour H, El-Mesery H 2014 Effect of microwave and infrared radiation on drying of onion slices *Int'l J. Res. Appl. Natur. Soc. Sci.* **2** 5 119-130

[9] AOAC 1999 Official Method of Analysis. AOAC Washington DC usa

[10] Henderson S M, Perry R L 1976 Agricultural process engineering 3rd Dd. The AVI Publ. Comp. Inc. Wesport Connecticut USA

[11] Traina M S, Breene W M 1994 Composition, functionality and some chemical and physical properties of eight commercial full-fat soy flour *J. of Food Proces. Preserv.* **8** 229-252

[12] Cagampang G B, Perez C M, Juliano B O 1973 A gel consistency test for eating quality in rice *J. Sci. Food Agr.* **24** 1589-1594

[13] Kermani A M, Khashehchi M, Kouravand S, Sadeghi A 2017 Effects of intermittent microwave drying on quality characteristics of pistachio nuts *Drying Technol.* **35** 9 1108-1116

[14] Das I, Shah N G, Kumar G 2014 Cashew nut quality as influenced by microwave heating used for stored grain insect control *Int'l J. of Food Sci.* **2014** 1-7 [http://dx.doi.org/10.1155/2014/516702]

[15] Dagwa I M, Builders P F, Achebo J 2012 Characterization of palm kernel shell powder for use in polymer matrix composites *Int'l J. of Mechanical and Mechatronics Eng.* **12** 4 88-93

[16] Chen J Y, Piva M, Labuza T P 1984 Evaluation of water binding capacity (wbc) of food fiber sources *J. Food Sci.* **49** 59-63

[17] Krokida M K, Maroulis Z B 1999 Effect of microwave drying on some quality properties of dehydrated products *Drying Technology* **17** 3 449-466

[18] Kerr W L, Ward C D W, McWatters K H, Resurreccion A V A 2000 Effect of milling and particle size on functionality and physicochemical properties of cowpea flour *Cereal Chem.* **77** 2 213–219

Effect of brewing technique and particle size of the ground coffee on sensory profiling of brewed Dampit robusta coffee

K Fibrianto , Y R Febryana and E S Wulandari

Food Sensory Research Group, Department of Agricultural Product Technology, Universitas Brawijaya, Malang, Indonesia

Email: kiki.fibrianto@ub.ac.id

Abstract. This study aimed to assess the effect of different brewing techniques with the use of appropriate particle size standard of Apresiocoffee cafe (Category 1) compared to the difference brewing techniques with the use of the same particle size (coarse) (Category 2) of the sensory attributes Dampit robusta coffee. Rate-All-That-Apply (RATA) method was applied in this study, and the data was analysed by ANOVA General Linier Model (GLM) on Minitab-16. The influence of brewing techniques (tubruk, French-press, drips, syphon) and type of particle size ground coffee (fine, medium, coarse) were sensorially observed. The result showed that only two attributes, including *bitter taste, and astringent/rough-mouth-feel* were affected by brewing techniques (p-value <0.05) as observed for brewed coarse coffee powder.

1. Introduction

Robusta coffee is one of the most widely cultivated types of coffee in Indonesia. According to [1] the southern part of Malang, precisely in Dampit area, is the largest producer of Robusta coffee in East Java with an area of ± 12,000 ha, with total production of ± 48,000 tons per year. To date, investigation on Dampit Robusta coffee is limited, in which they mostly focus on the comparison of composition [2]. Since the sensory profiling is imporant to identify consumer segmentation, the profiling of Dampit Robusta is required.

Meanwhile, it has been reported that coffee preparation techniques as one factor which influences the amount of some compounds in the beverages [3]. Similar study was also reported as they only focuses on aroma, flavor and aftertaste of Arabica coffee from Colombia with different method of electric brewing (coffee maker, infusion, espresso, and cupping) [4]. Different preparation techniques involve different polyphenols extraction, caffeine content, antioxidant activity level, total solid content and volatile profile [5]. Each brewing techniques applies different water and ground coffee ratio, water temperature and extraction time, and different water pressure and final volumes the beverage [6]. One of cafes in Malang region area that apply manual brewing techniques of Dampit Robusta coffee is Apresiocoffee. Therefore, this cafe was chosen as point of interest for investigating the influence of manual brewing techniques (tubruk, drip, French-press, and siphon) and particle size (fine, medium, and coarse) on sensory profiles of Dampit Robusta coffee.

This current study aims to know the effect of different brewing method using particle size according to the standard of Apresiocoffee (Category 1) compared to the different brewing method using the same particle size (Category 2) of Dampit robusta coffee sensory attributes. Benefit of this research output can be used as a reference and practical guidance for barista to brew Dampit robusta.

2. Materials and Methods

2.1. Tools and Materials

The tools were used in this study consist of syphoon, drips, french press, and tubruk glass. Other supporting tools were thermometer, gas stove, glass, digital scale, grinder, teapot, spatula and digital timer.

The materials were used in this study consist of Dampit robusta coffee and mineral water as palate cleanser according to the recommendation from [7] that mineral water can clean the palate for various products. The purpose of using palate cleanser was to avoid carry-over effects and adaption to sensory stimuli.

2.2. Methods

The method used in this study was RATA (Rate-All-That-Apply) method. RATA (Rate-All-That-Apply) method was a method used to determine the intensity of sensory attributes using untrained panelists or consumer panelists with minimum number panelist 70 people [8] 50-100 people [9]. In this study used consumer panelist (104 people) who consumed Dampit robusta coffee at the Apresiocoffee.

The data was analyzed using ANOVA General Linear Model (GLM) in Minitab 16 software with 95% confidence interval with 2 factors, brewing technique (tubruk, french press, drips, syphon) and the type of particle coffee powder (fine, medium, coarse). The meaning of Category 1 and Category 2 and the use of particle size in this study can be seen in the Table 1 and Table 2. The particle size of coffee powder (micron) was tested 3 times replication in Chemical Laboratory of Brawijaya University.

Table 1. Brewing techniques using particle size appropriate with Apresiocoffee standards (Category 1).

Brewing Techniques	Particle size of coffee powder (micron)
Tubruk	Fine (454.94 μm)
French press	Coarse (1,751.56 μm)
Drips	Medium (762.006 μm)
Syphon	Medium (762.006 μm)

Table 2. Brewing techniques using same particle size of coffee powder (Category 2).

Brewing Techniques	Particle size of coffee powder (micron)
Tubruk	Coarse (1,751.56 μm)
Drips	Coarse (1,751.56 μm)
French press	Coarse (1,751.56 μm)
Syphon	Coarse (1,751.56 μm)

3. Results And Discussion

3.1. Panelists response to significantly different attributes to Apresiocoffee standards (Category 1)

The determination of significantly different attributes was made using the General Linear Model (GLM) in Minitab 16 software with 95% confidence interval. The meaning of significantly different was the value of p-value less than 0,05 and denoted by different notation. These significantly different attributes can be interpreted as the panelists can detect the difference in the intensity of the sensory attributes of the difference brewing techniques and the use of particle size. The following attributes were significantly different in Category 1 and Category 2 can be seen in Table 3.

Table 3. The influence brewing techniques and particle size on sensory attributes of Dampit Robusta

Attributes	Techniques	Particle Size	Mean	Grouping
Brown Aroma	F. Press	coarse	3.808± 0.2583	A
	Drips	medium	3.288 ± 0.2583	A B
	Tubruk	fine	3.125 ± 0.2583	A B
	Syphon	medium	2.615 ± 0.2583	B
Bitter Taste	Drips	medium	6.048 ± 0.2287	A
	Tubruk	fine	5.990 ± 0.2287	A B
	F. Press	coarse	5.327 ± 0.2287	A B
	Syphon	medium	5.173 ± 0.2287	B
Acid Aftertaste	Syphon	medium	2.567 ± 0.2373	A
	F. Press	coarse	1.894 ± 0.2373	A B
	Tubruk	fine	1.712 ± 0.2373	A B
	Drips	medium	1.404 ± 0.2373	B
Thick Mouth-feel	Drips	medium	3.154 ± 0.2378	A
	Syphon	medium	2.827 ± 0.2378	A
	Tubruk	fine	2.471 ± 0.2378	A B
	F. Press	coarse	1.952 ± 0.2378	B
Astringent Mouth-feel	Drips	medium	3.106 ± 0.2602	A
	F. Press	coarse	2.500 ± 0.2602	A B
	Tubruk	fine	2.337 ± 0.2602	A B
	Syphon	medium	2.125 ± 0.2602	B

Based on Table 3, it can be seen that there were 5 sensory attributes that significantly different (p-value, 0.05), they were *brown aroma, bitter taste, acid aftertaste, thick mouth-feel,* and *astringent mouth-feel*. Based on mean values, the drips techniques had the highest intensity in *bitter taste, thick mouth-feel*, and *astringent mouth-feel attributes*. Bitter taste in coffee due to in coffee beans contain between 0,8-2,8% caffeine and it contributes to 10 to 30% of the bitter taste of the coffee brews. The caffeine content in Robusta coffee is about two times that of Arabica coffee [10]. Meanwhile, thick mouth-feel attributes associated with density and viscosity in coffee. This attribute is caused by the presence of lipids and dissolved polysaccharides in coffee brew [11].

While the brown aroma attribute, French Press technique had higher intensity than the others. It was allegedly due the coffee brewing left in the french press tool considered as the brewing time (ideally, the coarse particle size accordingly brewing time required for 5,5-8 minutes [3]. It was caused over-extraction of coffee brewing using a french press tool that caused the brown aroma. According to [12], the extraction of some compounds in coffee was dependent upon the time of brewing.

In the *acid aftertaste* attribute, the Syphon technique had the highest intensity. It was allegedly due the syphon process using principal of water vapor pressure. Chlorogenic acids confer *astringency, bitterness, and acidity* to the coffee brew [13]. According to the literature said that the differences in coffee brewing techniques, especially techniques using pressure would have an impact on the concentration of caffeolyquinic acids (CQAs) in coffee extraction. This acid would be responsible for *astringent mouth-feel, bitter flavor, and bitter aftertaste* in coffee [14]. It has been suggested that medium particle size have a large surface, allowing chlorogenic acid more extracted [15].

3.2. Analysis of acceptance test in Category 1

Analysis of Acceptance Test was conducted using 1-Proportion Test on Minitab 16 software. The acceptance results can be seen in Table 4. Table 4 shows the level of acceptance of Dampit robusta coffee sample with drips technique and using medium particle size (78 people) was higher than the level of acceptance of Dampit robusta coffee with tubruk technique and using fine particle size (69 people). It was due to some of the attributes extracted on Dampit robusta coffee using drips technique more extracted and had high intensity (*bitter taste, thick mouth-feel, and astringent mouth-feel*) compared with tubruk technique (*roasted aroma*). This can be reviewed from the GLM results described above.

Table 4. Mean Value of Acceptance Test in Category 1 (p-value 0,000).

Brewing techniques	Particle	Consumers answered Yes (people)	Number of Panelists
Tubruk	Fine	69	104
French Press	Coarse	70	104
Drips	Medium	78	104
Syphon	Medium	75	104

3.3 Analysis of Hedonic Test in Category 1

The hedonic test analysis was done using Friedman Test on Minitab 16 software which consist of 5 scale, that is scale 1 = very dislike, scale 2= dislike, scale 3= rather like, scale 4= like, scale 5= very like. The hedonic test results can be seen in Table 5.

Table 5. Median value of Hedonic test analysis in Category 1 (p-value 0.032).

Brewing techniques	Particle	Panelist (people)	Median
Tubruk	Fine	104	3.000
French press	Coarse	104	3.000
Drips	Medium	104	3.000
Syphon	Medium	104	3.000

Table 5 shows the whole panelist gave the same value, 3. The "3" value can be interpreted that all panelists had a "rather like" preference for all four techniques and the use of particle size according to Apresiocoffee (Category 1) standard presented. It was due to some panelists assume there were some preferred attributes, but there were some attributes that are rather liked by the panelists, so that the panelists gave a "rather like" rating to Dampit robusta coffee.

3.4. Panelists response of significantly different attributes to the same of particle size of coffee powder (Category 2)

The determination of significantly different attributes was made using the General Linear Model (GLM) in Minitab 16 software with 95% confidence interval. These significantly different attributes can be interpreted as the panelists can detect the difference in the intensity of the sensory attributes of the difference brewing techniques and the use of the same particle size of coffee powder. The following attributes that were significantly different in Category 2 can be seen in Table 6.

Based on Table 6, It can be seen that there were 4 significantly different of sensory attributes (p-values<0.05). They were *roasted, nutty, bitter, and rough mouth-feel attributes*. Pyrazines and guaiacol, correlated with higher processing temperature, have been described in literature with sensory notes of *smoky, nutty, hazelnut-like and roasty notes* [16]. It is also suggested that caffeine caused bitter taste in coffee [17]. The bitterness of coffee depends on the concentration of caffeine. While *rough mouth-feel attribute* was caused by astringent component and food particle size [19]. Based on mean values, drips technique also had the highest intensity on the *nutty flavor and bitter taste*.

While the roasted aroma attribute, Syphon technique had higher intensity than the *tubruk* technique. [5]. It said that coffee extraction depends on the particle size of ground coffee. The larger surface area, the higher compounds in ground coffee will be extracted. When the particle size of ground coffee smaller than ideally, it will result over-extraction or under-extraction. Based on current study, the use of the syphon technique with coarse particle size and brewing time requires 4 minutes (ideally, coarse particle size accordingly brewing time required for 5.5-8 minutes [3]. The extraction of some compounds in coffee was dependent upon the time of brewing [12]. The longer brewing time, the higher compounds will be extracted [18]. It also showed that the longer brewing time implies longer contact time between the coffee and water, allowing the higher extraction of some compounds.

Table 6. The influence brewing techniques on sensory attributes of coarse Dampit Robusta

Attributes	Techniques	Particle size	Mean	Grouping
Roasted Aroma	Syphon	coarse	4.231 (±0.2496)	A
	Drips	coarse	4.000 (±0.2496)	A B
	F. Press	coarse	3.673 (±0.2496)	A B
	Tubruk	coarse	3.115 (±0.2496)	B
Nutty Aroma	Drips	coarse	2.067 (±0.2084)	A
	Syphon	coarse	1.423 (±0.2084)	A B
	F. Press	coarse	1.346 (±0.2084)	A B
	Tubruk	coarse	1.202 (±0.2084)	B
Bitter Taste	Drips	coarse	6.327 (±0.1994)	A
	Syphon	coarse	5.933 (±0.1994)	A B
	F. Press	coarse	5.567 (±0.1994)	B
	Tubruk	coarse	5.510 (±0.1994)	B
Rough mouth-feel	Tubruk	coarse	1.8173 (±0.2043)	A
	Drips	coarse	1.7404 (±0.2043)	A
	F. Press	coarse	1.1058 (±0.2043)	A B
	Syphon	coarse	0.9327 (±0.2043)	B

So from the literature can be concluded that the coffee extraction from this syphon technique in the category of under-extraction. However, the panelists can feel the difference the higher intensity of *roasted aroma* attribute using the syphon technique rather than tubruk technique. It was allegedly due to the use of brewing time on the tubruk technique only 1 minute, lower than the use of brewing time on the syphon technique (4 minutes). So, that causes panelists tend to give response of roasted aroma attribute lower than the syphon technique. As it said before, the extraction of some compounds in coffee was dependent upon the time of brewing [12]. The lower of brewing time, the lower compounds will be extracted.

In the rough mouth-feel attribute, the tubruk technique had higher intensity than the others. It was allegedly due to the principle of tubruk technique without using any filter. It means that the coffee powder directly contacted with hot water. Therefore, the possibility of the granule of coffee powder was being brought to the consumer when the study progressed. Thus, the panelists tend to give higher intensity of the rough mouth-feel attribute on the tubruk technique.

3.5 Analysis of Acceptance Test in Category 2

Analysis of Acceptance Test was conducted using 1-Proportion Test on Minitab 16 software. The acceptance test results can be seen in Table 7. Table 7 shows the level of acceptance of Dampit robusta coffee using drips technique and the use of coarse particle size was higher (81 people) than the level of acceptance of Dampit robusta coffee using French press technique and coarse particle size (71 people). It was due to some of the attributes extracted on Dampit robusta coffee using drips technique more extracted and had high intensity (nutty aroma and bitter taste) rather than French press technique.

Table 7. Mean Value of Acceptance Test Analysis in Category 2 (p-value 0.000)

Brewing techniques	Particle	Consumers Answered Yes (people)	Number of Panelists
Tubruk	Coarse	73	104
F. Press	Coarse	71	104
Drips	Coarse	81	104
Syphon	Coarse	79	104

Analysis of Hedonic Test was done using Friedman Test on Minitab 16 software, which consist of 5 scale, scale 1= very dislike, scale 2= dislike, scale 3= rather like, scale 4= like, scale 5= very like. The hedonic test results can be seen in Table 8.

Table 8. Median value of Hedonic test analysis in Category 2 (p-value 0.791).

Brewing techniques	Particle	Panelists (people)	Median
Tubruk	Coarse	104	3.000
French press	Coarse	104	3.000
Drips	Coarse	104	3.000
Syphon	Coarse	104	3.000

Table 8 shows that the whole panelists gave the same value, 3. The "3" can be interpreted that all panelists had a "rather like" preference for all four techniques and the use of same particle size (Category 2). It was due to some panelists assume that there were some preferred attributes.

4. Conclusion

Sensory attributes generated on Dampit robusta coffee Category 1 (difference of brewing technique with the use of particle size according to the Apresiocoffee standard) were *brown aroma, bitter taste, acid aftertaste, thick mouth-feel, and astringent mouth-feel.* While sensory attributes generated in Category 2 (difference of brewing technique with the use of same particle size (coarse)) were *roasted aroma, nutty aroma, bitter taste, and rough mouth-feel.* The acceptance test for Category 1 (especially drips technique with the use of medium particle size) was lower than the level of acceptance Dampit robusta coffee in Category 2 (especially drips technique with the use of same particle size). The level of panelist preferences of Category 1 tended to be the same as all panelists said "rather like" to both categories.

References

[1] Hidayat 2006 Pengalaman pengembangan agribisnis kopi robusta melalui perbaikan mutu dan sistem pemasaran dengan pola kemitraan bermediasi di kawasan "Amstirdam" Kabupaten Malang, Jawa Timur (Robusta coffee agribusiness development experience through quality improvement and marketing system with a partnership pattern in the region of "Amstirdam" Malang Regency, East Java) Penyuluh Pertanian Dinas Pertanian dan Perkebunan Kabupaten Malang [In Indonesian]

[2] Ashari H 2009 Formulasi kopi ginseng (pfaffiapaniculata) cepat saji dengan komposisi kopi robusta pupuan, dampit, dan lampung (Formulation of ginseng coffee (pfaffingkiculata) fast food with the composition of robusta coffee pupuan, dampit, and lampung) Universitas Jember Program Pasca Sarjana [In Indonesian]

[3] Schreiber G B, Maffeo C E, Robins M, Masters M N, Bond A P 1988 Measurement of coffee and caffeine intake implications for epidemiologic research. *Prev. Med* **17** 280-294

[4] Alan K 2011 Development of a coffee lexicon and determination of differences among brewing methods Manhattan Kansas State University

[5] López-Galilea L, De Pena M P, Cid C 2007 Correlation of selected constituent with the total antioxidant capacity of coffee beverages: influence of the brewing procedure *Journal of Agricultural and Food Chemistry* **55** 6110-6117

[6] Petracco M 2001 Technology IV: Beverage Preparation: Brewing Trends for the New Millenium Blackwell Science Oxford

[7] Kemp S E, Hollowood T, Hort J 2009 Sensory Evaluation: A Practice Handbook Blackwell Publishing Singapore

[8] Ares G, Varella P 2014 Novel techniques in sensory characterization and consumer profiling CRC Press New York

[9] Adams J, Williams, Lancaster B, ,Foley 2007 Advantages and uses of check all that apply response compared to traditional scaling of attributes for salty snacks 7th Pangborn Sensory Science Symposium. Minneapolis US, 12-16 August 2007

[10] Heilmann W 2001 Coffee recent development. Technology IV: decaffeination of coffee: brewing trends of the new millenium 108-124.

[11] Mulato S, Suharyanto E 2012 Kopi, Seduhan dan Kesehatan Pusat Penelitian Kopi dan Kako Indonesia Jember [In Indonesian]

[12] Lee T A, Kempthrone R, Hardy J K 1992 Compositional changes in breed coffee as function of brewing time *Journal of Food Science* **57** 1417-1419

[13] Farah A, de Paulis T, Trugo L C, Martin P R 2006 Chlorogenic acids and lactones in regular and water-decafffeinated arabica coffee *J. Agric. Food Chem.* **54** 374-381

[14] Blumberg S, Frank O, Hofmann T 2010 Quantitative studies on the influence of the bean roasting parameters and hot water percolation on the concentrations of bitter compounds in coffee brew *Journal of Agric. Food Chem* **58** 3720–3728

[15] Bell N L, Wetzel C R, Grand A N 1996 Caffeine content in coffee as influenced by grinding and brewing techniques *Food Research International* **29** 785-789

[16] Caporaso N, Genovese A, Canela M D, Civitella A, Sacchi R 2014 Neapolitan coffee brew chemicals analysis in comparison to espresso, moka and American brews *Food Research International* **61** 152-160

[17] Keast R 2008 *Modification of the bitterness of caffeine Food quality and preference* **19** 465-472

[18] Gloess A N, Schonbachler B, Klopprogge B, D'Ambrosio L, Chatelain K, Bongartz A 2013 Comparison of nine common coffee extraction methods: instrumental and sensory analysis *European Food Research and Technology* **236** 607-627

[19] Wijk R A D, Prinz J F 2005 The role of friction in perceived oral texture *Food Quality and Preference* **16** 121-129

IOP Conf. Series: Earth and Environmental Science **131** (2018) 012010 doi:10.1088/1755-1315/131/1/012010

Changes in chlorophyll and polyphenols content in *Camellia sinensis var. sinensis* at different stage of leaf maturity

M I Prawira-Atmaja[1], Shabri[1], H S Khomaini[2], H Maulana[1], S Harianto[1] and D Rohdiana[1]

[1]Department Postharvest and Engineering, Reserach Institute for Tea and Cinchona, Gambung, Bandung, Indonesia
[2]Department Tea and Cinchona Plant Breeding, Reserach Institute for Tea and Cinchona, Gambung, Bandung, Indonesia

E-mail: iqbalprawira06@gmail.com, rohdiana@yahoo.com

Abstract. Chlorophyll and polyphenols are chemical compound related to parameter quality of green tea. We studied the variation of chlorophyll and polyphenol in the development stage of tea leaves (bud, 1st, 2nd, 3rd, and 4th). Five clones of tea *(Camelia sinensis var. sinensis)* from Indonesia and a clone from Japan were used in this study. The results showed that total chlorophyll and total polyphenol content in bud between 1.59-2.15 mg/g (db) and 12.24-14.59%, respectively. The concentration of chlorophyll increased significantly with developments stage of leaf while total polyphenol tended to decrease with leaf maturity. Pearson Correlation analysis showed that chlorophyll content was negatively correlated ($r = -0.83$; $p = 0.05$) with total polyphenol during developmental stage of tea leaves. Results suggests that five clones of tea from Indonesia have similar quality with tea clone from Japan in chlorophyll and polyphenol content. The present study also provides guidelines on application plucking standard to produce high quality of green tea.

1. Introduction

Tea is a beverage that contains many health benefits. According to the nomenclature, tea is divided into two varieties namely *Camelia sinensis var. assamica* from Assam and *Camelia sinensis var. sinensis* originating from China. Both types of varieties have different morphological characteristics. Sinensis varieties of tea plant from Indonesia has been officially released by Minister of Agriculture in 2009 which consisted of 5 clones: GMBS 1, GMBS 2, GMBS 3, GMBS 4, and GMBS 5 [1].

Chlorophyll and polyphenol are chemical compounds related to parameter quality of green tea. Chlorophyll, natural green pigment, has function in photosyntesis metabolism. Chlorophyll a content in fresh tea leaf varies from 1.39 to 5.39 $mg.g^{-1}$ and chlorophyll b 0.77 to 2.06 $mg.g^{-1}$ [2,3]. Chlorophyll content in tea increase gradually as the leaves matured and significantly increase the photosynthesis rate [4]. Polyphenols in green tea were based on the polyphenols content from the fresh leaves. Leaf positions represent a specific stage that affect in primary and secondary metabolites. Generally, young leaves, positioned at apical, considered rich in polyphenols compared with the old leaves. Young leaves contain more tea polyphenols, caffeine, amino acids, and their concentration decrease significantly from the 1st leaf to the 3rd leaf [4]. Therefore, a metabolic balance between chlorophyll and polyphenols contents might present in tea leaves at different stages [5].

The present study aimed to determine the total chlorophyll content and polyphenols at various leaf positions. This research focused on RITC (Research Institute for Tea and Cinchona) clones, GMBS 1 to GMBS 5, compared with Yabukita clone originated from Japan. The leaf position used includes apical bud, first, second, third, and fourth leaves where the leaves indicate leaf maturity level. This study also observed correlation among chlorophyll and polyphenols in development of tea leaf maturity level. The results of this study might serve as a useful guideline on application of plucking standard to produce high quality green tea.

2. Materials and Methods

2.1. Plants material and experimental location

Apical bud and four leaves of *Camellia sinensis var.* sinensis from Research Institute for Tea and Cinchona (RITC) clones (GMBS 1, GMBS 2, GMBS 3, GMBS 4 and GMBS 5) and from Japan (Yabukita) were precisely plucked by hand and immediately separated based on specified criteria (a bud, 1st leaf, 2nd leaf, 3rd leaf, and 4th leaf). All leaves were immediately brought to laboratory to determine the moisture content (ISO1572) and for further analysis. The research was carried out in Gambung tea plantation (Latitude: 7° 08 ' 37 "N, longitude: 107° 30' 54" R, altitude 1200 m) which belongs to Research Institute for Tea and Cinchona (RITC) in West Java. All samples were plucked in January- March 2017 with average rainfall of 218 mm and average minimum and maximum temperature of 16.5 °C and 24.2 °C, respectively.

2.2. Determination of chlorophyll

Chlorophyll was determined according to the method of Ni et.al. [6] with minor modification. Fresh leaves were placed in mortar and grinded finely. Four hundreds (400) mg sample was added with 5 ml of acetone 80% in 15 ml falcon tube, mixed for 5 minutes, and was kept in the dark in a temperature of 4 °C for 15 minutes (chlorophylls degrade under light). Then the mixture was centrifuged for 15 min (at 3,000 rpm), the supernatant was transfered to a new centrifuge tube, and kept in the dark. This treatment was repeated twice and the supernatant was combined so that the total volume was made up to 25 ml using acetone 80%. Shake the mixture thoroughly and measure the absorbance of chlorophyll content using spectrophotometry (Varian carry win UV) with acetone 80% as blank control. Chlorophyll content was calculated using formula:

$$\textbf{Chlorophyll a}_{(mg/g)} = \frac{[12.7 \text{ x } A_{663} - 2.69 \text{ x } A_{645}] \text{ x V}}{1000 \text{ x W}}$$

$$\textbf{Chlorophyll b}_{(mg/g)} = \frac{[22.9 \text{ x } A_{645} - 4.86 \text{ x } A_{663}] \text{ x V}}{1000 \text{ x W}}$$

$$\textbf{Chlorophyll a + b}_{(mg/g)} = \frac{[8.02 \text{ x } A_{663} + 20.20 \text{ x } A_{645}] \text{ x V}}{1000 \text{ x W}}$$

Where V = volume of the extract (ml); W = Weights of fresh leaves (g); A645: absorbance at 645 nm; A663: absorbance at 663 nm.

2.3. Determination of total polyphenols

2.3.1. Preparation and extraction of fresh leaf

Fresh leaves were placed in a mortar and grinded with a pastel finely. Added 2 g of sample with 30 ml boiled methanol 70% heated up for 10 minutes and followed by maceration in oven 60 °C for 2 hours.

The mixture was then sonicated for 30 min using a Sonicator (Bransonic-220 Inst., New York, USA). The extraction mixture was constantly kept cold by adding ice water into the sonicator vessel. The mixture was then passed through a filter with a Whatman No. 1 filter paper to obtain a clear extract. The residues and all glassware were then washed with 70% aqueous methanol and the total volume of the extract was made up to 50 ml in a volumetric flask. One millilitre of the extract was pipetted into a 25 mL volumetric flask and diluted with distilled water. The dilution extract was further used in determining total polyphenol.

2.3.2. Determination of total polyphenol

Folin-Ciocalteu method was used to determine the total polyphenol according to ISO 14502-1:2005 standardization [7]. One millilitre of dilution extract was pipetted into tube flask. The extract was further reacted with 5 mL of Folin-Ciocalteau's reagent (10%) for 5 min to form a blue-coloured solution. Then, 4 mL of sodium carbonate 7.5% (37.5 g was diluted with 500 ml distilled water) solution to stabilize the colour formed. The blue colour was allowed to develop for at least 2 h and its absorbance was then measured at 740 nm using UV-vis Spectro (Varian carry win UV). The amount of total polyphenols was obtained from the standard curve equation of gallic acid solution with a concentration range of 1-100 mg / L (ppm).

2.4. Statistical analysis

The data obtained in the study were analyzed by using one-way ANOVA at a significance level of 95% followed by Duncan's multiple range test (DMRT). Person's correlation analysis was used to assess a correlation between chlorophyll and total phenol. All data analysis was performed using XLSTAT 2014.

3. Results and Discussion

3.1. Profile of chlorophyll in various clones at different stage of leaf maturity

Chlorophyll is natural pigment with green color in leaf. The function of chlorophyll in leaf is for photosyntesis metabolism. Leaf position represents a specific developmental stage that influences both photosynthesis and respiration [8]. Table 1 and 2 show the chlorophyll a and b content of apical bud, the first, second, third, and fourth leaf from various tea clones. Apical bud from various clones has the lowest chlorophyll content ranging from 1.14-1.46 mg/g and 0.44-0.77 mg/g on chlorophyll a and chlorophyll b respectively. Apical bud from GMBS 4 clone has the lowest chlorophyll a and b while those from GMBS 3 and GMBS 5 clones have the highest chlorophyll a and chlorophyll b respectively. Furthermore, fourth leaves from various tea clones show the highest chlorophyll content ranging from 3.40-4.72 mg/g and 1.31-1.68 on chlorophyll a and b respectively. The highest content of chlorophyll a and b are found in Yabukita clones and GMBS 5, respectively.

Table 1. Profile of Chlorophyll a content (mg/g db) in various clones of *Camelia sinensis var.sinensis.*

Leaf position	GMBS1	GMBS 2	GMBS 3	GMBS 4	GMBS 5	Yabukita
Bud	$1.24{\pm}0.13^{a}$	$1.38{\pm}0.45^{a}$	$1.47{\pm}0.02^{a}$	$1.14{\pm}0.31^{a}$	$1.37{\pm}0.21^{a}$	$1.46{\pm}0.06^{a}$
1st leaf	$1.88{\pm}0.42^{a}$	$1.83{\pm}0.65^{a}$	$2.47{\pm}0.27^{b}$	$1.46{\pm}0.36^{a}$	$2.23{\pm}0.37^{ab}$	$2.38{\pm}0.03^{a}$
2nd leaf	$2.77{\pm}0.53^{b}$	$2.77{\pm}0.85^{ab}$	$3.29{\pm}0.29^{bc}$	$2.33{\pm}0.38^{b}$	$2.42{\pm}0.65^{ab}$	$3.58{\pm}0.28^{b}$
3rd leaf	$3.46{\pm}0.50^{bc}$	$3.71{\pm}1.03^{b}$	$3.50{\pm}0.52^{c}$	$3.21{\pm}0.52^{c}$	$3.29{\pm}1.05^{bc}$	$3.59{\pm}1.26^{b}$
4th leaf	$3.58{\pm}0.24^{c}$	$4.10{\pm}1.15^{b}$	$3.98{\pm}0.84^{c}$	$3.40{\pm}0.52^{c}$	$3.87{\pm}0.59^{c}$	$4.72{+}0.24^{c}$

Data are represented as mean ± SD (standard deviation) of three replications. Different letters indicate significant differences between leaf position at $P < 0.05$ according to a Duncan test.

Total chlorophyll of apical bud of GMBS 1 was 1.74 mg/g increased significantly in first, second, third and fourth leaf with 2.57, 3.74, 4.66, and 4.91 mg/g (Figure 1), respectively. Visual observations on the fourth leaves of various clones showing dark green color while apical bud that have low chlorophyll content, has a yellowish green color. Tea leaf positions from various clones significantly ($P < 0.05$) affect chlorophyll a and b and total chlorophyll content. Increased chlorophyll in tea leaves correlates with increasing maturity of tea leaves [1,9]. Increased chlorophyll in the fourth leaf due to an increase in photosynthetic capacity in maturity leaves which contain more starch than in younger leaf (apical bud) [8]. The mean increase in total chlorophyll of all clones in fourth leaf was more than 65% from total chlorophyll of apical bud. The highest increase, 69%, was in the Yabukita clone where the total chlorophyll in the fourth leaf was 6.42 mg/g (Figure 1).

Table 2. Profile of Chlorophyll b content (mg/g db) in various clones of *Camellia sinensis var sinensis.*

Leaf position	GMBS1	GMBS 2	GMBS 3	GMBS 4	GMBS 5	Yabukita
Bud	0.49±0.07^{a}	0.57±0.18^{a}	0.50±0.04^{a}	0.44±0.12^{a}	0.77±0.31a	0.52±0.01^{a}
1st leaf	0.65±0.15^{a}	0.63±0.26^{a}	0.80±0.16^{b}	0.52±0.14^{ab}	1.02±0.61b	0.76±0.04^{a}
2nd leaf	0.92±0.19^{b}	1.02±0.38^{ab}	1.08±0.05^{bc}	0.80±0.12^{bc}	1.18±0.77bc	1.12±0.12^{b}
3rd leaf	1.18±0.12^{c}	1.37±0.37^{ab}	1.17±0.22^{bc}	1.12±0.16^{c}	1.46±0.85c	1.23±0.38^{b}
4th leaf	1.31±0.03^{c}	1.52±0.49^{b}	1.48±0.46^{c}	1.33±0.29^{d}	1.68±0.57	1.63±0.07^{c}

Data are represented as mean ± SD (standard deviation) of three replications. Different letters indicate significant differences between leaf position at P < 0.05 according to a Duncan test.

Chlorophyll a ($C_{55}H_{72}O_5N_4Mg$) and chlorophyl b ($C_{55}H_{70}O_6N_4Mg$) have different functional groups of their chemical structure. In addition, the color of Chlorophyll a is dark green and chlorophyll b is light green. Chlorophyll-a is the main pigment which converts light energy into chemical energy. Chlorophyll-b as accessory pigment acts indirectly in photosynthesis by transferring the light it absorbs to chlorophyll-a [10]. The green color of fresh tea leaves and green tea products is the main role of chlorophyll compounds. Chlorophyll derivatives, chlorophyllides, the water-soluble green pigments derived from chlorophylls through the catalysis of chlorophyllase dissolve during brewing of tea to produce green tea color infusion [11]. Chlorophyll content gives the sensation of grassy taste on green tea product [12,13]. Chlorophyll content in tea leaves is influenced by several factors such as various clones, leaf maturity, and climatic condition [14,15]. Total chlorophyll in various clones of *Camellia sinensis var. sinensis* at different stage of leaf position shown in Figure 1.

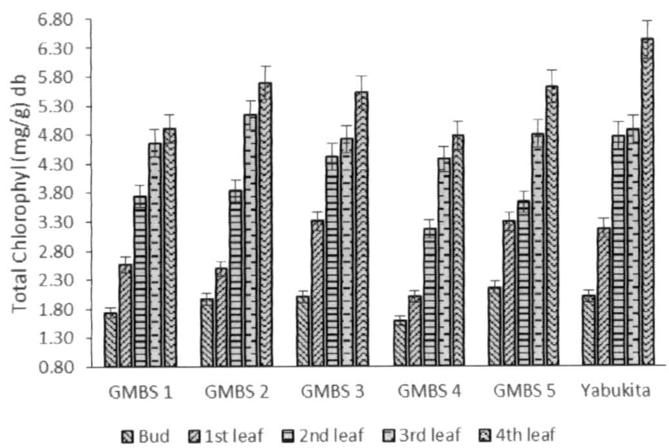

Figure 1. Total chlorophyll in various clones of *Camellia sinensis var. sinensis* at different stage of leaf position.

3.2. Total polyphenol in various clones at different stage maturity of leaf

The results of total polyphenols from bud to fourth leaf in different types of clones are presented in Figure 2. Apical bud from GMBS 2 contains the highest total polyphenols (14.59%) than GMBS 1, GMBS 3, GMBS 4, GMBS 5 and Yabukita with total polyphenols 12.12; 12.98; 13.24; 13.80; and 12.93% respectively. The percentage of total polyphenols in various tea leaf positions tended to decrease significantly of about 34% from first to fourth leaf. Total polyphenols in the fourth leaf ranging from 7.48-9.40% where GMBS 3 has the smallest total polyphenol content. Overall, apical bud contains the highest polyphenols and the concentration decrease significantly with the maturity of leaf. The results found in this study, whereby polyphenols decreased at development stage of leaf is similar with previous study reported by Li et al.[8] and Samanta et al. [5].

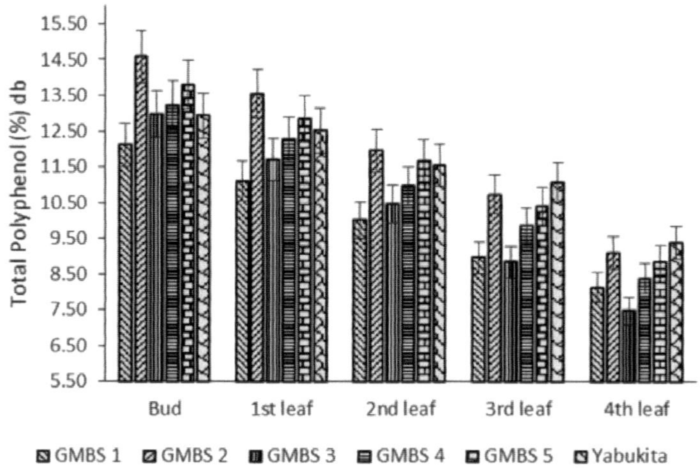

Figure 2. Total polyphenol content in various clones of *Camellia sinensis var. sinensis* at different stages of leaf position.

Fresh tea leaves contain various chemical compounds that play a role in determining the quality parameters of tea products. Total polyphenols in tea leaves predominantly flavan-3-ols (catechins) and flavonol glycosides and phenolic acids and these constitute 20-30% to total component on dry weight basis [16–18]. Phenols have responsibility for characteristics, color and astringent taste of green tea. Polyphenol content of fresh tea leaves is also affected by seasonal variation, leaf handling and harvest methods, leaf maturity, and variety of tea clones [19]. Polyphenols have been reported to have antioxidant activity and may contribute to a reduced risk of cardiovascular disease, cancer, and has many other beneficial effects on health [20,21].

3.3. Correlation analysis between chlorophyll and total polyphenol as influenced by leaf position.

Table 3 shows Pearson correlation coefficient (r) analysis between chlorophyll and Total polyphenol affected by leaf position. Total polyphenol correlated negatively with chlorophyll content ($P < 0.05$). This study showed that the maturity level of leaves contributes to changes in chlorophyll and polyphenol contents in tea leaves. Increasing of chlorophyll content in tea leaves is associated with leaf maturity level. During maturity leaves physiological functions are reduced, even though the photosynthetic capacity increased with the degree of leaf maturity [8]. The old leaves provide nutrients to tea plants and play important roles in the storage of nutrients that are required for the germination and growth of new shoots [22]. Starch is a product generated from photosynthesis in which increased starch correlates with increased chlorophyll in tea leaves [8,23].

The decrease of polyphenolic content in mature tea leaves is probably due to low enzyme activity of Leucocyanidin reductase (LAR) and anthocyanidin synthase (ANS) [15]. Tea catechins, group of flavanol, are considered to be synthesized through phenylpropanoid and flavonoid pathway and stored in the vacuole [24]. The formation of catechins in tea leaves might be affected by the chlorophyll accumulation. MEP (methylerythritol-4-phosphate) pathway that is responsible for the synthesis of precursors of chlorophyll might suppress the catechin formation pathway during tea leaf maturity as both these pathways use common precursor phosphoenol pyruvate [10].

Table 3. Correlation between total polyphenol, chlorophyll a, chlorophyll b, and total chlorophyll parameters as influenced by leaf position.

Variables	Total phenol	Chl a	Chl b	Total Chl
Total Phenol	1			
Chl a	-0.8324*	1		
chl b	-0.7901*	0.9405*	1	
Total Chl	-0.8294*	0.9957*	0.9679*	1

* Significantly different at $P < 0.05$

4. Conclusions

The results of this study showed that chlorophyll content and total polyphenols from various GMBS clones were not significantly different from clone from Japan. Leaf maturity level of all clones influenced the chlorophyll and polyphenol content. Chlorophyll content in tea leaves increase significantly with leaf maturity but negatively correlated with polyphenol contents. Processing of tea using old leaves will affect the decrease of tea quality. The present study also provides guideline on application of plucking standard to produce high quality of green tea.

Acknowledgements
The author would like to thank to Director of Indonesia Research Institute for Tea & Cinchona for the support on this research. The authors also expressed their gratitude to Asri Sopiana Marwah and Siti Patimah for their help and assistance during this research.

References

[1] Sriyadi B 2013 *klon Teh Sinensis Unggul: GMBS 1, GMBS 2, GMBS 3, GMBS 4, GMBS 5* (Bandung)
[2] Mahanta P K and Hazarika M 1985 Chlorophylls and degradation products in orthodox and CTC black teas and their influence on shade of colour and sensory quality in relation to thearubigins *J. Sci. Food Agric.* **36** 1133–9
[3] Yamanishi T 1995 Chemical composition of tea *Food Reviews International* vol 11pp 435–56
[4] Li Z X, Yang W J, Ahammed G J, Shen C, Yan P, Li X and Han W Y 2016 Developmental changes in carbon and nitrogen metabolism affect tea quality in different leaf position *Plant Physiol. Biochem.* **106** 327–35
[5] Samanta T, Kotamreddy J N R, Ghosh B C and Mitra A 2017 Changes in targeted metabolites, enzyme activities and transcripts at different developmental stages of tea leaves: a study for understanding the biochemical basis of tea shoot plucking *Acta Physiol. Plant.* **39**
[6] Ni Z, Kim E-D and Chen Z J 2009 Chlorophyll and starch assays *Protoc. Exch.*
[7] Anon 2005 ISO 14502-1:2005 (E) Determination of substances characteristic of green and black tea- Part 1: Content of total polyphenols in tea-colorimetric method using Folin-Ciocalteu reagent *Int. Stand. ISO*
[8] Li Z, Yang W, Ahammed G J, Shen C, Yan P, Li X and Han W 2016 Developmental changes in carbon and nitrogen metabolism affect tea quality in different leaf position *Plant Physiol. Biochem.* **106** 327–35
[9] Li Y, Chen C, Li Y, Ding Z, Shen J, Wang Y, Zhao L and Xu M 2016 The identification and evaluation of two different color variations of tea *J. Sci. Food Agric.* **96**
[10] Sumanta N, Haque C I, Nishika J and Suprakash R 2014 Spectrophotometric Analysis of Chlorophylls and Carotenoids from Commonly Grown Fern Species by Using Various Extracting Solvents *Res. J. Chem. Sci. Res. J. Chem. Sci* **4** 2231–606
[11] Wang L, Park S and Chung J 2004 The Compounds Contributing to the Greenness of Green Tea *J. Food Sci.* **69** 301–5
[12] Yamanishi T 1995 Processing of tea *Food Rev. Int.* **11** 409–34
[13] Baba R and Kumazawa K 2014 Characterization of the potent odorants contributing to the

characteristic aroma of chinese green tea infusions by aroma extract dilution analysis *J. Agric. Food Chem.* **62** 8308–13

[14] Feng L, Gao M J, Hou R Y, Hu X Y, Zhang L, Wan X C and Wei S 2014 Determination of quality constituents in the young leaves of albino tea cultivars *Food Chem.* **155** 98–104

[15] Wei K, Wang L, Zhou J, He W, Zeng J, Jiang Y and Cheng H 2011 Catechin contents in tea (Camellia sinensis) as affected by cultivar and environment and their relation to chlorophyll contents *Food Chem.* **125** 44–8

[16] Balentine D A, Wiseman S A and Bouwens L C M 1997 The Chemistry of Tea Flavonoids *Crit. Rev. Food Sci. Nutr.* **37** 693–704

[17] Engelhardt U H 2010 Chemistry of Tea *Comprehensive Natural Products II* pp 999–1032

[18] Wang H, Provan G J and Helliwell K 2000 Tea flavonoids: Their functions, utilisation and analysis *Trends Food Sci. Technol.* **11** 152–60

[19] Turkmen N, Sarı F and Velioglu Y S 2009 Factors Affecting Polyphenol Content and Composition of Fresh and Processed Tea Leaves *Akad. Gida* **7** 29–40

[20] Cabrera C, Artacho R and Giménez R 2006 Beneficial effects of green tea--a review. *J. Am. Coll. Nutr.* **25** 79–99

[21] Butt M S, Ahmad R S, Sultan M T, Qayyum M M and Naz A 2015 Green tea and anticancer perspectives: updates from last decade *Crit Rev Food Sci Nutr* **55** 792–805

[22] Li C-F, Zhu Y, Yu Y, Zhao Q-Y, Wang S-J, Wang X-C, Yao M-Z, Luo D, Li X, Chen L and Yang Y-J 2015 Global transcriptome and gene regulation network for secondary metabolite biosynthesis of tea plant (Camellia sinensis). *BMC Genomics* **16** 560

[23] Miyauchi S, Yonetani T, Yuki T, Tomio A, Bamba T and Fukusaki E 2016 Quality evaluation of green tea leaf cultured under artificial light condition using gas chromatography / mass spectrometry *J. Biosci. Bioeng.* 1–6

[24] Zhang L Q, Wei K, Cheng H, Wang L Y and Zhang C C 2016 Accumulation of catechins and expression of catechin synthetic genes in Camellia sinensis at different developmental stages *Bot. Stud.* **57**

IOP Conf. Series: Earth and Environmental Science **131** (2018) 012011 doi:10.1088/1755-1315/131/1/012011

Optimization of instant powdered chicken feet broth's drying temperature and time on pilot plant scale production

N Hidayati and T D Widyaningsih

Department of Agricultural Product Technology, Faculty of Agricultural Technology, Universitas Brawijaya, Malang, Indonesia

E-mail: tridewantiw@ub.ac.id

Abstract. Chicken feet by-product of chicken industries amounted to approximately 65,894 tons/year commonly used as broths. These by-products are potentially produced into an instant form as an anti-inflammatory functional food on industrial scale. Therefore, it is necessary to optimize the critical parameters of the drying process. The aim of this study was to determine the optimum temperature and time of instant powdered chicken feet broth's drying on pilot plant scale, to find out product's comparison of the laboratory and pilot plant scale, and to assess financial feasibility of the business plan. The optimization of pilot plant scale's research prepared and designed with Response Surface Methodology-Central Composite Design. The optimized factors were powdered broth's drying temperature (55°C, 60°C, 65°C) and time (10 minutes, 11 minutes, 12 minutes) with the response observed were water and chondroitin sulphate content. The optimum condition obtained was drying process with temperature of 60.85°C for 10,05 minutes resulting in 1.90 ± 0.02% moisture content, 32.48 ± 0.28% protein content, 12.05 ± 0.80% fat content, 28.92 ± 0.09 % ash content, 24.64 ± 0.52% carbohydrate content, 1.26 ± 0.05% glucosamine content, 0.99 ± 0.23% chondroitin sulphate content, 50.87 ± 1.00% solubility, 8.59 ± 0.19% water vapour absorption, 0.37% levels of free fatty acid, 13.66 ± 4.49% peroxide number, lightness of 60.33 ± 1.24, yellowness of 3.83 ± 0.26 and redness of 21.77 ± 0.42. Financial analysis concluded that this business project was feasible to run.

1. Introduction

The high market demand for Broiler chickens in Indonesia yielded by-products of chicken feet as much as 65,894 tons in 2016 [1]. Chicken feet contain glycosaminoglycan, consisting of glucosamine and chondroitin sulfate which are potentially used as an anti-inflammatory agent as an alternative to anti-inflammation of shark fin bones [2]. In addition, chicken feet is rich in calcium, phosphorus and hydroxyapatite and also contain gelatin, chitin/chitosan and collagen that can potentially cure osteoporosis, reduce aging effects, and act as anti-hypertension [3]. Chicken feet are often used as broths. Research on chicken feet seasoning has proven its function as an anti-inflammatory agent, where edema inflammation's reduction increased from 1st to 5th hours reaching 78.19% with dosage of 200 mg/kg rat body weight [4].

Scaling through the pilot plant is a stage that can provide an approximation of production in industrial scale by identifying critical process conditions, so that the product's quality can be maintained. The critical condition in the production of instant powdered chicken feet broth is the drying process. The drying temperature and time are critical parameters. The drying temperature and time on pilot plant may be different from the laboratory experiment because the amount of materials is multiplied according to the scale-up capacity, different machines and production condition. The aim of this study were to determine the optimum temperature and time of instant powdered chicken feet broth's drying on pilot plant scale, to find out product's comparison of the laboratory and pilot plant scale, and to assess financial feasibility of the business plan.

2. Materials and Methods

Materials used in this research were Broiler chicken feet obtained from Hypermart Malang Plaza and industry. Other ingredients used for instant powder on the laboratory scale were dextrin, sugar, garlic, onion and water. While the used on the pilot plant scale were maltodextrin, sugar, salt, garlic granule, shallot granule, fancy onion powder, pepper, chicken fat and water which were obtained from the factory. The materials used for testing were NaOH (Merck), concentrated H_2SO4 (Merck), HCl 0.1 N, boric acid, PP indicator, methyl red, Kjeldahl tablet, petroleum ether, glucurunolactone standard (Sigma), sodium tetraborate (Sigma), ethanol absolute, carbazole solution (Sigma), glucosamine kit assay (Megazyme), aquades and ice cubes obtained from CV. Kridatama Persada, CV. Panadia and CV. Makmur Sejati.

2.1. Laboratory scale study

Chicken feet powder was made by pressure cooking method for 1 hour ± 2 minutes and then dried for 12 hours ± 5 minutes in cabinet dryer at 60 ± 5°C. Garlic and onions were also dried for 6 hours ± 5 minutes in cabinet dryer at 60 ± 5°C. Then, the ingredients were mixed according to the percentage of the formulation, the mixture was added by water until it didn't stick to the bowl. The mixture was dried for 6 hours ± 5 minutes in cabinet dryer at 60 ± 5°C and sieved with 40 mesh screen. The formulation of powdered chicken feet broth is presented in **Table 1**.

Table 1. Formulation of laboratory scale research.

Material	Percentage (%)
Chicken feet powder	60
Garlic powder	10
Onion powder	10
Salt	18
Dextrin	2

2.2. Pilot plant scale study

Chicken feet powder was made by autoclaving method for 45 minutes and then cooked by using steam for 45 minutes ± 15 seconds under 1 atm pressure. The drying of chicken feet powder was separated into 2 phases, where the first drying was for 10 hours ± 5 minutes and second drying was for 4 hours ± 5 minutes at 60°C in tray oven. Then, the chicken feet powder was mixed with 20% maltodextrin. The production of instant powdered chicken feet broth every batch was 500 g. Seasoning was added afterwards following this composition: 60% chicken feet powder; 8.5% sugar; 1.5% fancy onion powder; 3.2% garlic granule; 3% of shallot granule and 0.5% pepper then the mixture were crushed together. The mixing process was done by inserting salt into the speed kneader machine and adding 60 ± 5°C water as much as 0.6% of the broth powder's amount, then it was mixed for 30 seconds. Next, the mixture was added by 60 ± 5°C chicken fat as much as 0.8% and mixed until 90th seconds. After 1 minute 30 seconds, the mixture was added by ingredients that have been previously crushed and the mixing was done up to 4 minutes.

Instant powdered chicken feet broth's drying process was optimized by using Response Surface Methodology-Central Composite Design. Determination of drying temperature and time based on previous research and through consideration of process conditions at the pilot plant scale. The central composite design for optimization is presented in Table 2. After drying, the powdered broth was sieved with a 10 mesh screen.

Table 2. Pilot plant scale's research design.

Std	Run	Factor 1 A: Temperature (°C)	Factor 2 B: Time (minutes)	Response 1 Water content (%)	Response 2 Chondroitin sulfate content (%)
1	1	55.00	10.00		
2	3	65.00	10.00		
3	9	55.00	12.00		
4	2	65.93	12.00		
5	8	52.93	11.00		
6	4	67.07	11.00		
7	13	60.00	9.59		
8	11	60.00	12.41		
9	7	60.00	11.00		
10	5	60.00	11.00		
11	6	60.00	11.00		
12	10	60.00	11.00		
13	12	60.00	11.00		

2.3. Data analysis
The optimization response was analyzed using Response Surface Methodology-Central Composite Design and processed with Design Expert DX 7.1.5. The result of final product analysis was processed using Microsoft Excel, while to compare the characteristics of chicken feet powder produced in laboratory and pilot plant was done by using paired t-test at 5% significance level (Minitab 16.2.1).

3. Results and Discussion
3.1. Characteristics comparison of chicken feet powder on laboratory and pilot plant scale
The results of analysis from chicken feet powder are presented in Table 3.

Table 3. Comparison of chicken feet powder.

	Analysis	Pilot Plant Scale	Laboratory Scale	Notation
1.	Yield (%)	18.39 ± 1.57	11.37 ± 0.10	*
2.	Proximate:			
	- Water content (%)	5.43 ± 0.40	5.38 ± 0.11	ns
	- Protein content (%)	47.47 ± 1.01	53.49 ± 1.37	*
	- Fat content (%)	15.70 ± 3.05	23.89 ± 2.88	ns
	- Ash content (%)	10.02 ± 0.11	9.32 ± 0.12	**
	- Carbohydrate content(%)	21.38 ± 2.18	7.92 ± 1.58	
3.	Glucosamine content (%)	5.12	11.53	
4.	Chondroitin sulfate content (%)	1.04 ± 0.53	2.10	

Notes:
The amount is an average ± standard deviation of 3 repetitions
ns : no significant difference
* : significant difference ($p < 0.05$)

The yield on the pilot plant scale was higher than the laboratory scale due to the addition of 20% maltodextrin as filler that can increase the amount of solids of the product. The difference in protein levels due to the autoclaving process that causes lower level of nutrients, including protein [5]. A higher level of ash content on the pilot plant scale caused by chicken's nails. Aautoclaving soften the chicken feet most of the broken main bone were included to be dried and added up the mineral content.

The difference was the addition of 20% maltodextrin. The lower chondroitin content on the pilot plant's chicken feet powder was related to the characteristics of chondroitin sulphate which is highly water soluble and unstable to high temperature, while boiling process with high temperature and longer drying time were applied on pilot plant scale.

3.2. Optimization using respon surface methodology

Based on the moisture content response's analysis of variance (ANOVA), it showed that the quadratic model gave a significant effect with p value was 0.0031. While the lack of fit has p value of 0.1549, considered as has no significant effect. The actual equation of the model was as follow:

$$Y = 27.03273 - 0.57437\ X_1 - 1.16712\ X_2 - 0.005\ X_1X_2 + 0.004895X_1^2 + 0.064875\ X_2^2$$

Description: X_1 = Drying temperature (°C)
X_2 = Drying time (minutes)
Y = Response of moisture content

Chondroitin sulphate response's ANOVA showed that linear model gave a significant effect with p value was 0.0040. While the lack of fit has p value of 0.0962, considered no significant effect. The actual equation of the model obtained was as follow:

$$Y = 3.40860 - 0.020602\ X_1 - 0.11990\ X_2$$

Description: Y = Response of chondroitin sulphate content

The relationship between the factors to the response was shown in Figure 1.

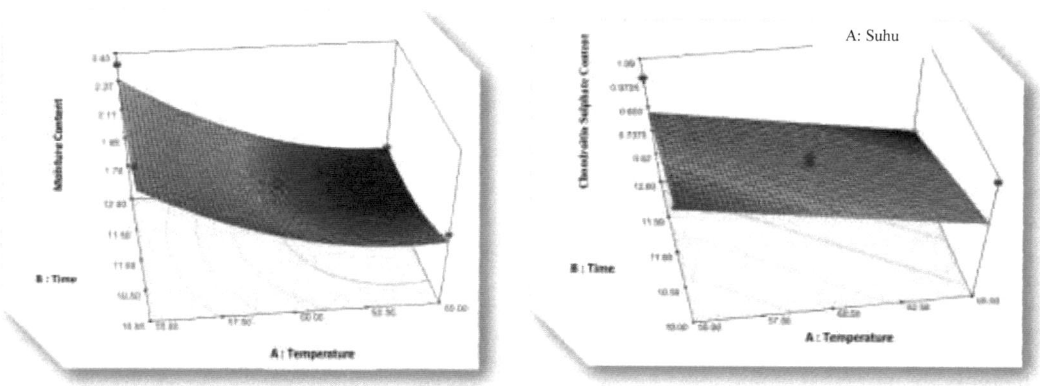

(a) **(b)**

Figure 1. 3D surface of (a) moisture content and (b) chondroitin sulphate content

Figure 1 (a) shows that only the temperature that has a significant effect on the water content response. The curve shows that the higher the temperature, the lower the water content indicated by the gradation of the colour from green-yellow-dark blue. The ability of the material to release water from its surface increase with the increasing drying temperature and the longer drying process, resulting in a lower moisture/water content [6]. Drying time has no significant effect because the time range used in this study is too narrow. Figure 4.8 (b) shows a decreasing linear shape along with the higher temperature and longer drying time. This is because chondroitin sulphate is a non-heat-resistant compound [7].

The solution obtained by using this method is the optimum point selected by Design Expert 7.1.5's calculation results are drying process with the temperature of 60.85°C for 10.05 minutes. Verification was done to assess that the optimum response estimated by the program was confirmed by the actual response. The results of the verification are shown in Table 4.

Table 4. Prediction and verification of instant chicken feet broth on *pilot plant.*

	Factor		Response	
	Temperature (°C)	Time (minutes)	Moisture Content (%)	Chondroitin Sulfate Content (%)
Prediction	60.85	10.05	1.97	0.95
Verification	60.85	10.05	1.90	0.99
PI *low*	60.85	10.05	1.73	0.71
PI *high*	60.85	10.05	2.22	1.19

3.3. Characteristics comparison of instant powdered chicken feet broth on laboratory and pilot plant

The results of analysis from chicken feet powder are presented in Table 5.

Table 5. Comparison of instant powdered chicken feet broths.

	Analysis	Pilot Plant Scale	Laboratory Scale	Notation
1.	Moisture content (%)	1.86 ± 0.04	7.04 ± 0.11	*
2.	Protein content (%)	32.48 ± 0.35	33.89 ± 0.78	ns
3.	Fat content (%)	12.05 ± 0.97	10.93 ± 0.87	*
4.	Ash content (%)	28.92 ± 0.10	23.22 ± 0.09	*
5.	Carbohydrate content(%)	24.64 ± 0.64	24.93 ± 1.58	
6.	Glukosamine content (%)	1.26 ± 0.06	1.06 ± 0.77	ns
7.	Chondroitin sulfate content (%)	1.01 ± 0.22	0.74 ± 0.02	ns
8.	Solubility level (%)	50.87 ± 1.23	47.08	
9.	Water vapor absorbtion	8.59 ± 0.24	9.17 ± 0.30	*
10.	Color			
	- L*	60.33 ± 1.51	59.73 ± 0.75	ns
	- a*	3.83 ± 0.32	5.17 ± 0.25	ns
	- b*	21.77 ± 0.51	25.33 ± 0.23	*
11.	Free Fatty Acid content (%)	0.37		
12.	Peroxide number (mek/kg)	13.66 ± 4.49		

Notes:
The amount is an average ± standard deviation of 3 repetitions
ns : no significant difference
* : significant difference ($p \leq 0.05$)

Differences in moisture content caused by the adding of water until broth mixture didn't stick to the bowl before it was dried on laboratory scale. In addition, the using of different dryers may also be a factor of the difference in moisture content [8].Differences in fat content was caused by an addition of 0.8% chicken fat on pilot plant scale to produce a stronger chicken flavor. While the higher ash content on the pilot plant scale can be due to various ingredients added including granule's form of shallot and garlic. Differences in redness can be due to the addition of filler on the pilot plant's chicken feet powder that resulting in brighter colour. In addition, shorter powdered broth drying on pilot plant may also affect the differences in redness. The free fatty acid content in this product was 0.37% which is quite high and caused by large oil hydrolysis. While the high peroxide number may be due to the analysis of peroxide and FFA numbers performed more than 1 month after the production. It may also be due to long warming process during chicken feet powder's drying and a high content of omega-3 and omega-6, (187 mg and 257.1 mg per 100 g chicken feet respectively). Unsaturated fatty acids bind oxygen to their double bonds to form peroxides [9].

3.4. Financial analysis of instant powdered chicken feet broth

The feasibility of a business project based on financial aspects was indicated by the ability of the business to provide financial benefits. The financial analysis is presented in Table 6.

Table 6. Financial analysis of instant powdered chicken feet broth.

No.	Criteria	Value
1.	Net Present Value	Rp. 5,838,964,104.00
2.	Internal Rate of Return	35.26%
3.	Net Benefit/Cost	2.30
4.	BEP cost	Rp. 3,873,717,400.00
	BEP unit	1,076,033
5.	Pay Back Period	2 years 9 months 21 days

This business project is feasible to run because the value of NPV> 0, IRR value was greater than the prevailing bank interest rate (12%), net B/C ratio was ≥1, PBP was faster than project's duration (10 years), BEP value was below the number of production value [10, 11].

4. Conclusion

The critical parameters set in the pilot plant are suitable, based on the similarity of characteristics with laboratory scale product. The chosen optimum point of instant powdered chicken feet broth's drying process is 60.85°C for 10.05 minutes. Financial analysis stated this business project is feasible to run.

References

[1] Directorate General of Livestock 2016 Broiler chicken production according to province

[2] Widyaningsih T D, Hapsari, D, Wijayanti, N, Dita, S, Dhyantari, O, Milala C T 2015 *Conf. Proc.of Agroindustry and National Workshop FKPT-TPI,* Universitas Brawijaya Malang Indonesia, pp. 5-9

[3] Guimaraes, M, Jose, C, Curvelo, S, Pollana A F 2012 Collagen extraction from chicken feet for jelly production *J.Acta Sci. Maringa* **34** 345-351

[4] Milala C T, 2014 The Influence of Instant Chicken Feet Seasoning as Anti-inflammatory Effect to Male Wistar Rat Induced by Carrageenan Undergraduate Thesis Universitas Brawijaya Malang Indonesia

[5] Sundari, D, Almasyhuri, Lamid, A 2015 The Effect of Cooking Process on Nutrient Composition of Protein Source Food stuff *Libangkes Media* **25** 235-242

[6] Widyaningsih T D, Rukmi W D, Sofia, E, Wijayanti S D, Wijayanti, N, Ersalia, Rochmawati, N, Nangin, D 2016 Extraction of Glycosaminoglycans Containing Glucosamine and Chondroitin Sulfate from Chicken Claw Cartilage *RJLS* **3** 3 181-189.

[7] Marzuki, A, Kasim, S, Arafah, N 2014 The Addition Effect of Sodium Hydroxide to Chondroitin Sulfate Level's Retention Time of Pari's Cartilage (*Taeniura lymna*) *J. Phi.* **9** 239-245

[8] Zikri, A, Erlinawati, Rusnadi, I 2015 Performance Test of Rotary Dryer Based on Drying Thermal Efficiency of Wood Powder for Biopelet Production *J. Chem. Engineering* **21**

[9] Ketaren, S 2005 Food Oil and Fat UI Press Jakarta Indonesia

[10] Burhanuddin, R 2007 Feasibility Study of Slaughter House Establishment in East Kuta Regency Deputy of East Kuta SME Resource Assessment Bali Indonesia

[11] Ibrahim H M Y 2009 Business Feasibility Study Rineka Cipta Jakarta Indonesia

The influence of steaming and a ratio of grated coconut to water on the yield and quality of virgin coconut oil

N L Rahmah, R Istikoma and S Kumalaningsih

Department of Agro-Industrial Technology, Faculty of Agricultural Technology, Universitas Brawijaya, Malang, Indonesia

E-mail : cahyalaila22@gmail.com; nur_laila@ub.ac.id

Abstract. The quality of Virgin Coconut Oil (VCO) is determined by the quality of coconut milk. High quality of coconut milk can be obtained by proper handling of grated coconut as raw material. When coconut was shredded, the lipases are exposed which can hydrolyse the oil resulting free fatty acid (FFA). Steaming is a technique to inactivate lipases. In addition, a ratio of grated coconut to water and steaming duration are important factor to the VCO extraction. Therefore, this study aimed to obtain the best combination of steaming duration and suitable ratio of grated coconut to water in order to produce high quality VCO. The research design was Factorial Randomized Block Design consisted of 2 factors: steaming duration (5; 10; and 15 minutes) and grated coconut to water ratio (1:0; 1:1; 1:2; 1:3; and 1:4 w/v), each treatment was repeated twice. Parameters analyzed were FFA, moisture content, and yield values. The result showed that the best treatment was a treatment with 15 minutes steaming of grated coconut and 1:4 ratio of grated coconut to water. The best treatment VCO had characteristic as follows: FFA 0.054 %, moisture content 0.129 % and yield 17.563 %.

1. Introduction

Coconut has been widely used and all parts of the coconut are highly beneficial toward human life [1]. One of coconut's product is Virgin Coconut Oil (VCO) extracted from old coconut flesh by pressing with or without water addition.VCO contains 93% of long and short chain saturated fatty acid consisted of lauric acid (50,33%); capric acid (10.25%); caproic acid (14.23%); palmitic acid (4.92%); and myristic acid (12.91%) [2].

One of essential factors to be considered in VCO production is good handling of coconut since the quality of the oil produced is determined by the quality of the coconut milk. Grating the coconut disintegrates the cell structure so that lipase can digest oil [3] which leads to premature rancidity. One traditional technique to prevent the deterioration reaction is steaming.

Steaming is one of food processing technique by putting food on vaporizing water in a closed chamber [4]. This process can lengthen the grated coconut shelf life prior to the next process by lipase heat inactivation. However, the longer the steaming causes the damage to the nutritional compounds of the grated coconut

Besides steaming, extracting the coconut milk using certain grated coconut to water ratio also plays essential role in producing high yield and good quality coconut milk. Therefore, this study aimed to find the best combination of steaming duration and grated coconut to water ratio in order to produce high quality VCO.

2. Materials and Methods

2.1. Materials and Instruments

Coconut *Dalam variety* was obtained from Malang district. Materials used in this research were water, plastic bags, 0.05 M NaOH (PA), phenolphthalein indicator (PA), ethanol 95%, aquadest, and filter paper (Whatman 7).

Instruments used on this study were beaker glass (Iwaki), blender (Philips HR2116 - capacity 2 L), electric coconut grater, hotplate (Labinco), steaming apparatus (Jawa - diameter 30 cm), pipette, thermometer, pressing machine, freezer (Samsung), digital scale (Camry), funnel (Herma – 75 mm), and stopwatch.

2.2 Method

This research used Factorial Randomized Block Design consisted of 2 factors: steaming duration (factor A) and a ratio of grated coconut and water (factor B). Factor A was 5, 10, and 15 minutes, whereas Factor B was ratio of grated coconut (kg) to water (L) 1:0; 1:1; 1:2; 1:3; and 1:4. Therefore, there were 15 combinations of treatment, two repetitions for each treatment. Therefore, there were 30 experimental units.

Eleven to twelve months old coconut was obtained from Mergan traditional market supplied from Ngajum, Malang district. First step in producing VCO was separating the coconut flesh from the shell, grating and steaming under 55°C for 5; 10; and 15 minutes.The next step was extracting coconut milk based on the grated coconut to water ratio 1:0; 1:1; 1:2; 1:3; and 1:4 then squeezing to extract milk which then left to stand for 2 hours in order to separate cream from the emulsion. The cream was extracted and frozen for 24 hours. Frozen cream was crushed using blender for 20 minutes until the compact cream was formed. The compact cream was heated until 50°C to separate oil, water, and residue. Finally, the mixture was filtered to obtain VCO. The VCO was analyzed for FFA, moisture, and yield.

2.3 Statistical Analysis

The data was analyzed using Analysis of Variance (ANOVA) test to observe whether there was any major influences due to the agreed factors. If there was any, data was put into 5% of confidence interval Duncan's Multiple Range Test (DMRT). Following the DMRT test, the best treatment was determined and concluded.

3. Results and Discussion

3.1 Free Fatty Acid (FFA)

Steaming duration and a ratio of grated coconut and water affects value of FFA (α=0.05). The highest FFA content was at the combination of 5 minutes steaming and 1:4 grated coconut to water ratio which was 0.068 %. While, the lowest one was at the combination of 15 minutes steaming and 1:0 grated coconut to water ratio which was 0.036 %. This has occurred because oil was hydrolyzed by water and lipase produced FFA, the higher FFA content means the lower the quality of VCO oil [5]. However, the FFA value from those treatments met the SNI 7381:2008 standard of maximum FFA of 0.2 %.

Table 1.The Effect of Steaming Duration on VCO's FFA.

Steaming Duration (minutes)	Average FFA (%)
5	$0.059 \pm 4.240 \times 10^{-6c}$
10	$0.054 \pm 3.210 \times 10^{-6b}$
15	$0.045 \pm 5.270 \times 10^{-6a}$

Note: Numbers followed by a different notation indicates the difference between treatments (DMRT 5%: 2.766×10^{-3})

As shown on Table 1, a 5-minute steaming generates the highest percentage of FFA which is 0.059% whilst the lowest one is in a 15-minute steaming which is 0.045%. The longer steaming duration, VCO's FFA decreases. Steaming for 15 minutes is over than the maximum time required to inactivate naturally occurring of lipase which hydrolyses triglyceride producing FFA and glycerol. Steaming can inactivate the enzyme which is responsible for colour, taste, and nutrients changes [6].

Table 2. The Effect of Grated Coconut to Water Ratio on VCO's FFA.

Grated Coconut to Water Ratio	Average FFA (%)
1:0	$0.044 \pm 5.320 \times 10^{-6a}$
1:1	$0.049 \pm 7.150 \times 10^{-6b}$
1:2	$0.055 \pm 5.100 \times 10^{-6c}$
1:3	$0.056 \pm 3.040 \times 10^{-6d}$
1:4	$0.061 \pm 4.800 \times 10^{-6e}$

Note: Numbers that followed by a different letter indicates the presence of considerable difference (DMRT 5%: 2.766×10^{-3})

Table 2 shows a ratio1:4 generates the highest percentage of FFA which is 0.061% whilst the lowest one was in a ratio1:0 ratio which is 0.044%. The high content of FFA was caused by the presence of excess water that facilitated oil hydrolysis producing glycerol and FFA. Oil hydrolysis is catalyzed by heat, moisture, acid, and catalyst of enzyme or synthetic catalyst. High amount of water used in coconut milk extraction can accelerate oil hydrolysis thus generating more FFA [7].

3.2 Moisture content

Steaming duration and grated coconut to water ratio had considerably affected VCO's moisture(α=0.05). The highest moisture content was on VCO obtained from the treatment of 5 minutes steaming and 1:4 grated coconut to water ratio which was 0.183%. On the other hand, the lowest moisture content was on VCO obtained from the treatment of 15 minutes steaming and 1:0 grated coconut to water ratio which was 0.088%. This has occurred due to water evaporation during steaming and the amount of water used in coconut milk extraction also affects the final products' moisture, moist oil means low quality oil. Moist in the oil tends to shorter the shelf life [5]. However, these two moisture contents met the SNI (Standard National Indonesia) 7381:2008 standard as tolerated maximum of 0.2 %.

Table 3.The effect of steaming duration toward VCO's moisture content.

Steaming Duration (minutes)	Average FFA (%)
5	$0.124 \pm 1.212 \times 10^{-4c}$
10	$0.114 \pm 9.232 \times 10^{-5b}$
15	$0.104 \pm 2.236 \times 10^{-5a}$

Note: Numbers that followed by a different letter indicates the presence of considerable difference (DMRT 5%:8.96 2 x 10^{-3})

Table 3 shows a 5-minute steaming of grated coconut generates the highest percentage of FFA which is 0.124% whilst the lowest one is found in 15 minutes steaming which was 0.104%. The longer steaming duration, the lower the VCO's moisture contents. This observation was supported by the fact that process time and temperature affected the ingredient's moisture [8]. The longer steaming time and the higher the temperature, the more water is evaporated.

Table 4.The effect of grated coconut to water ratio on VCO's moisture content

Ratio of Grated Coconut to Water	Average Moisture Content (%)
1:0	$0.093 \pm 4.710 \times 10^{-6a}$
1:1	$0.102 \pm 3.000 \times 10^{-8b}$
1:2	$0.103 \pm 7.000 \times 10^{-8c}$
1:3	$0.114 \pm 2.660 \times 10^{-5d}$
1:4	$0.160 \pm 7.500 \times 10^{-5e}$

Note: Numbers that followed by a different letter indicates the presence of considerable difference (DMRT 5%:8.962×10^{-3})

Table 4 shows that at a ratio of 1:4, VCO has the highest of moisture content of 0.160% whilst the lowest one is in VCO obtained from treatment at a ratio of 1:0 of 0.093%. The more water used in coconut milk extraction, the higher the VCO's moisture content. This result might be caused by the type of emulsion of coconut milk (O/W) which contains high water. More oil entrapped in the water increases the VCO's yield. Emulsion type of O/W has higher moisture compared to W/O emulsion [9].

3.3 Yield

Interaction between steaming duration and grated coconut to water ratio has considerably affected VCO's yield (α=0.05). The highest yield was at the combination of 15 minutes steaming and a ratio of 1:4 grated coconut to water which was 17.563%. Whilst the lowest one was at the combination of 5 minutes steaming and a ratio of 1:0 grated coconut to water which was 5.562%.

Table 5. Duncan test result for VCO's yield.

Treatment		
Steaming Duration (minutes)	Grated Coconut to Water Ratio (w/v)	Average Yield (%)
5	1:0	5.562^{j}
	1:1	9.471^{g}
	1:2	12.689^{f}
	1:3	12.880^{f}
	1:4	13.723^{e}
10	1:0	7.161^{i}
	1:1	12.138^{f}
	1:2	14.896^{d}
	1:3	16.100^{c}
	1:4	17.204^{a}
15)*	1:0	7.803^{h}
	1:1	12.591^{f}
	1:2	14.980^{d}
	1: 3	16.818^{b}
	1:4)*	*17.563^{a}*

Note: Same letter notation indicates the presence of inconsiderable difference on DMRT test (α=0,05); *) the best treatment

Differences between steaming duration and grated coconut to water ratio as shown on Table 5. These two factors have affected VCO's yield significantly. The result showed that 5 minutes steaming and 1:0 ratio, whilst the highest one in VCO obtained from the combination between 15 minutes of steaming and a ratio of 1:4 which was17.563%. Steaming causes grated coconut's pores to open thus coconut milk can be extracted easily. In addition, high amount of water in the steam can also help in extracting the milk. The high amount of coconut milk results a lot of cream to be produced. Coconut milk extraction efficiency is affected by temperature of extraction and the amount of water [10]. This statement is supported by other studies that the suitable ratio of coconut milk and water in order to produce VCO with 20 % yield was 1:2 [11].

3.4 Determination the best treatment
The results showed that the higher the yield, the lower the oil qualities. The highest yield was obtained from oil extracted with steaming for 15 minutes and a ratio of 1:4 of grated coconut and water. The best quality of the oil was obtained from oil extracted at steaming for 15 minutes and at a ratio of 1:0 of grated coconut and water. The SNI 7381:2008 of VCO standard requires that moisture content and FFA of VCO are a maximum of 0.2%. However, the lowest quality VCO obtained from this research still met the SNI 7381:2008. The best result was determined in accordance with SNI's standards of VCO that had characteristic as follows: 0.054 % of FFA, 0.129 % of moisture content, and 17.563 % of yield.

4. Conclusion
The steaming of grated coconut and a ratio of grated coconut to water affected the VCO's yield, FFA and water content of VCO. The best treatment of the VCO was obtained from the steaming for 15 minutes and a ratio 1:4 of grated coconut to water. It has characteristic as follows: 0.054 % of FFA, 0.129 % of moisture content, and 17.563 % of yield.

References
[1] Setyamidjaja D 1986 Planting Coconut Kanisius Yogyakarta (in Indonesian)
[2] Wirakusumah E S 2008 Fruits and Vegetables for Therapy Penebar Swadaya Jakarta (in Indonesian)
[3] Pontoh J, Surbakti M B, Papilaya M 2008Quality of virgin coconut oil from several methods of manufacture Jurnal Chemical Program**1**160-65 (in Indonesian)
[4] Anonymous2017Steaming Retrieved from https://id.wikipedia.org/wiki/Pengukusan. [Accessed on 17 Januari 2017]
[5] Sudarmadji S, Haryono B, Suhardi1996 Analytical Procedures for Food and Agricultural Materials Liberty Yogyakarta (in Indonesian)
[6] Laily R 2010Processed from Potatoes Kanisius Yogyakarta (in Indonesian)
[7] Susanti N M P, Widjaja IN, Dewi N 2010 Effect of Centrifugation Time on VCO Quality [Minor thesis] Universitas Udayana Bali (in Indonesian)
[8] Prabandari R, Mangalik A, Achmad J, Agustiana 2005 Effect of boiling time of two different types of shrimp on the quality of white shrimp waste (Penaeus indicus) and tiger shrimp (Penaeus monodon) Enviro Scientiae **1**124-28 (in Indonesian)
[9] Anief M 2010 Pharmacetics Gadjah Mada University Press Yogyakarta (in Indonesian)
[10] Cancel LE 1971*Effect of amount and temperature of water in the extractionof coconut milk from coconut pulp* J Agr Univ Puerto Rico 35 167
[11] Ahmad M I 2013 Comparison of Coconut milk and water to the yield, water content and fat free acid virgin coconut oil [Minor thesis] Agricultural Technology Department Universitas Sam Ratulangi-Manado (in Indonesian).

IOP Conf. Series: Earth and Environmental Science **131** (2018) 012013 doi:10.1088/1755-1315/131/1/012013

The bioactive compounds and antioxidant activity of ethanol and ethyl ecetate extracts of *Candi* Banana (*Musa paradisiaca*)

R A Laeliocattleya, T Estiasih, G Griselda and J Muchlisyiyah

Department of Agricultural Product Technology, Faculty of Agricultural Technology, Universitas Brawijaya

Email: deeochalina@gmail.com

Abstract. Banana has various benefits for health. One local variety of banana is *candi* banana (*Musa paradisiaca*). The aim of this research was to study the content of the bioactive compounds of phenolics, flavonoids, tannin, carotenoids and the antioxidant activity of extract ethanol and ethyl acetate of *candi* banana. Powdered *candi* banana was extracted using ethanol and ethyl acetate in an ultrasonic bath. The results showed that the content of phenolics, flavonoids, tannin and carotenoids in ethanol extract were 58.76 ± 3.19 mg/kg, 416.08 ± 18.79 mg/kg, 209.83 ± 15.87 mg/kg and 74.55 ± 4.31 mg/kg, respectively. The content of phenolics, flavonoids, tannin and carotenoids in ethyl acetate extract were 0.83 ± 0.12 mg/kg, 4.31 ± 0.66 mg/kg, 49.97 ± 2.43 mg/kg and 304.40 ± 16.62 mg/kg. While the antioxidant activity (IC_{50}) of ethanol extract and ethyl acetate were 3374.13 ± 123.46 ppm and 40318.19 ± 1014.90 ppm. This research showed that type of solvents of ethanol and ethyl acetate affected the content of bioactive compounds and antioxidant activity of *candi* banana. The antioxidant activity of ethanol extract was higher than that of ethyl acetate extract. It showed that ethanol was a better solvent than ethyl acetate to extract bioactive compounds in *candi* banana.

1. Introduction

Oxidative stress is among the major causative factors in induction of many chronic and degenerative diseases including atherosclerosis, ischemic heart disease, ageing, diabetes mellitus, cancer, immunosuppression, neurodegenerative diseases and others [1]. An antioxidant can be broadly defined as any substance that delays or inhibits oxidative stress [2]. The main characteristic of an antioxidant is its ability to trap free radicals. Antioxidant compounds like polyphenols and flavonoids scavenge free radicals and inhibit the oxidative mechanisms that lead to degenerative diseases [3]. Plants that consist of bioactive compounds considered as good antioxidant since ancient times. Banana has a great source of minerals, vitamins, carbohydrates, and bioactive compounds [4]. Indonesia as one of great producer of banana has many type varieties of banana. One of varieties is called *candi* banana (*Musa paradisiaca*). This variety of banana is rich in bioactive compounds, such as phenol, flavonoid, tannin and carotenoids but not many research explore the content of the bioactive compounds. Furthermore, this type of banana probably have specific antioxidant activity which is beneficial for health as a radical scavenger and it can help to prevent the degenerative diseases.

2. Materials and Methods

2.1. Materials and tools

Tools used in this research were rotary vacuum evaporator, UV-Vis spectrophotometer, automatic cabinet dryer, and column of chromatography. Materials used were ripe *candi* banana, aquadest, ethyl acetate, ethanol (96%), pro-analysis ethanol, methanol, gallic acid standard, tannic acid standard, quercetin, β-carotene, Folin Ciocalteu reagent (10%), Na_2CO_3 (2%), Na_2CO_3 (20%), $NaNO_3$ (5%), $AlCl_3$ (10%), NaOH 1 M, cotton, alumina oxide, Na_2SO_4, petroleum ether : acetone solution (10:1), petroleum ether : acetone solution (1:1), and DPPH 0.2 mM in methanol.

2.2. Methods

2.2.1. Sample preparation

The *candi* banana was peeled, washed and cut into 3 mm of thickness. The cuts were dried in an automatic cabinet dryer for 12 hours at 45°C. The dried sample was ground using dry blender for 2 minutes and sieved using 80 mesh of siever. The *candi* banana powder is ready for the next process.

2.2.2. Sample extraction

The *candi* banana powder of 10 g was added with 200 ml of 96% ethanol or ethyl acetate and put in an ultrasonic bath for 30 minutes at 30°C. After the extraction process, the mixture was filtered using fine filter paper. Filtrate was evaporated using rotary vacuum evaporator at 50 rpm and 40°C.

2.2.3. Total phenolic compounds (TPC) determination

TPC of *candi* banana was determined by Folin-Ciocalteu reagent as described [5] with modification. The absorbance of the mixture was measured using UV-Vis spectrophotometer at 759 nm. TPC is defined as mg of gallic acid equivalent per kg of sample (mg GAE/kg). The formula of TPC calculation is:

$$TPC = \frac{C \times V \times FP}{W} \quad (1)$$

Where: C = phenol concentration (mg/L), V = extract volume (L), FP = dilution factor, W = sample weight (kg)

2.2.4. Total flavonoid compounds (TFC) determination

TFC was determined based on [6] with modification. The absorbance of the mixture was measured using spectrophotometer at 510 nm. TFC is defined as mg quercetin equivalent per kg of sample (mg QE/kg). The formula for this calculation is as follows:

$$TFC = \frac{C \times V \times FP}{W} \quad (2)$$

Where: C = Flavonoid concentration (mg/L), V = extract volume (L), FP = dilution factor, W = sample weight (kg)

2.2.5. Total tannin compounds (TTC) determination

TTC was determined based on [7] with modification. The absorbance of the mixture was measured using UV-vis spectrophotometer at 765 nm. TTC is defined as mg tannic acid equivalent per kg of sample (mg TAE/kg). The formula used to measure this parameter is:

$$TTC = \frac{C \times V \times FP}{W} \quad (3)$$

Where: C= tannin concentration (mg/L), V = extract volume (L), FP = dilution factor, W = sample weight (kg)

2.2.6. Total carotenoids (TC) determination
TC was determined by column chromatography method as described by [8] with modification. Firstly, sample of 1 g was added with 8 mL of petroleum ether : acetone (1:1). The flask was covered with alumina foil and shaken for 10 minutes and then filtered using filter paper. The filtrate obtained was stored at room temperature. The residue was added with 8 mL of petroleum ether:acetone (1:1), covered with alumina foil and shaken for 10 minutes. This treatment was repeated twice. Then, the filtrate was diluted with petroleum ether:acetone (1:1) until total volume of 25 ml. After that, the filtrate was transferred to separator funnel and added with 10 mL of aquadest. The separator funnel was shaken and incubated until the mixture was separated into 2 phases. The lower part (water:acetone phase) was discarded and the upper part (ether phase) was transferred to a tube. Then, 0.5 g of Na_2SO_4 was added for each of 10 mL of ether phase's volume. Then, the mixture was loaded into the column of chromatography. Petroleum ether:acetone (10:1) solvent was added to the mixture in the column until the mixture turned clear. Then the absorbance of the eluate was measured using UV-Vis Spectrophotometer at 450 nm. Total carotenoid is defined as weight of β-carotene (mg) equivalent per kg of sample. The formula for calculation is:

$$TC = \frac{C \times V \times FP}{W} \quad (4)$$

Where: C = β-carotene concentration (mg/L), V = mixture volume (L), FP = dilution factor, W = sample weight (kg)

2.2.7. IC_{50} value of antioxidant activity determination
Antioxidant activity was determined according to [6] with modifications. Extract of *candi* banana was made into 5 different concentration (200, 400, 600, 800 and 1000 ppm). Extract of 3 mL was added with 3 mL of 0.2 mM DPPH. Then, the mixtures were vortexed and incubated for 30 minutes in dark room. After that, the absorbance of the mixture was measured using UV-Vis Spectrophotometer at 514 nm. Radical inhibition was calculated in the following formula:

$$\%\ DPPH\ Radical\ Inhibition = \left(\frac{Blank\ absorbance - Sample\ absorbance}{Blank\ absorbance}\right) \times 100\% \quad (5)$$

IC_{50} value was extrapolated from the curve of Y-axis as inhibition percentage value and X-axis as concentration.

3. Results and Discussion

3.1. TPC
Table 1 shows that TPC of ethanol extract of *candi* banana was higher than that of ethyl acetate. Chemical characteristic and polarity of compounds affect the solubility of the compounds in a particular solvent. An appropriate solvent polarity results a better extraction yield [9]. Furthermore, it can be observed that phenolic compounds tend to be soluble in polar solvent (ethanol).

Table 1. TPC content of *Candi* Banana in ethanol and ethyl acetate extract

Extract	TPC (mg/kg)
Ethanol Extract	58.76 ± 3.19^a
Ethyl Acetate Extract	0.83 ± 0.12^b

3.2. TFC

Table 2 shows that ethanol extract of *candi* banana has higher TPC than that of ethyl acetate extract. Flavonoid is polyphenolic compound that is ubiquitous in nature, comprising a number of hydroxyl groups attached to aromatic ring structures that determine its antioxidative properties [10]. Flavonoid has similar characteristic with other polyphenols which is diluted in more polar solvent such as ethanol.

Table 2. TFC of *Candi* Banana in ethanol and ethyl acetate extract

Extract	TFC (mg/kg)
Ethanol Extract	416.08 ± 18.79^{a}
Ethyl Acetate Extract	4.31 ± 0.66^{b}

3.3. TTC

Tannin are plant polyphenols that can either bind or precipitate proteins. Table 3 shows that TTC of ethanol extract is higher than that of ethyl acetate.

Table 3. TTC of *Candi* Banana in ethanol and ethyl acetate extract

Extract	TTC (mg/kg)
Ethanol Extract	209.83 ± 15.87^{a}
Ethyl Acetate Extract	49.97 ± 2.43^{b}

3.4. TC

Table 4 shows the TC in extract of *candi* banana is higher than that ethyl acetate extract. Carotenoids are lipid-soluble of C40 tetraterpenoids. The majority carotenoids are derived from a 40-carbon polyene chain, which could be considered the backbone of the molecule. These chains are either terminated by cyclic end-groups or complemented with oxygen-containing functional groups [11]. Because carotenoid has diene-conjugated group which is a hydrocarbon, so it was more dilute in nonpolar solvent of ethyl acetate than in polar solvent of ethanol [12].

Table 4. TC of *Candi* Banana in ethanol and ethyl acetate extract

Extract	TC (mg/kg)
Ethanol Extract	74.55 ± 4.31^{a}
Ethyl Acetate Extract	304.40 ± 16.62^{b}

3.5. Antioxidant activity (IC_{50})

Table 5 shows IC_{50} value of ethanol and ethyl acetate extract of *candi* banana were 3374.13 ppm and 40318.19 ppm. Therefore, extract obtained using ethanol had higher antioxidant activity than that of ethyl acetate extract. Extract ethanol contains higher of phenolic, flavonoid and tannin than that of ethyl acetate extract. Phenolic compounds have been found as strong antioxidants in delaying the influence of free radicals and ROS, which are the main causes of chronic human diseases [13]. Phenolic compounds can also inactivate free radicals from lipid deterioration or preventing decomposition of hydroperoxides into free radicals [14].

Table 5. IC_{50} of *Candi* Banana extract obtained using ethanol and ethyl acetate

Extract	IC_{50} (ppm)
Ethanol Extract	3374.13 ± 123.46^a
Ethyl Acetate Extract	40318.19 ± 1014.90^b

4. Conclusion
This research showed that type of solvents of ethanol and ethyl acetate affects the amount of extracted of bioactive compounds and antioxidant activity of *candi* banana. The antioxidant activity of ethanol extract is higher than that of ethyl acetate extract. It showed that ethanol was a better solvent than ethyl acetate to extract bioactive compounds in *candi* banana.

References
[1] Young I S and Woodside J V 2001 Antioxidants in health and disease *J. Clin. Pathol.* **54** 176-186
[2] Yamagishi S and Matsui T 2011 Nitric oxide, A janus-faced therapeutic target for diabetic microangiopathy-friend or foe? *Pharmacol Res.* **64** 187–194
[3] Wu YY, Li W, Xu Y, Jin EH, Tu YY 2011 Evaluation of the antioxidant effects of four main theaflavin derivatives through chemiluminescence and DNA damage analyses *J Zhejiang Univ Sci B.* **12** 744–751
[4] Babu, Melene A, Suriyakala M A, Gothandam K M 2012 Varietal impact on phytochemical contents properties of *Musa acuminata* (Banana). *Journal of Pharmaceutical Sciences and Research* **4** 1950-1955
[5] Singleton V L and Rossi J A 1965 Colorimetry of total phenolics with phosphomolybdic-phosphotungstic acid reagents *Am. J. Enol. Vitic.* **16** 144-158
[6] Sultana B, Anwar F, Asi M R, Chatha, S A S 2008 Antioxidant potential of extracts from different agro wastes: stabilization of corn oil *Grasas y Aceites* **59** 205-217
[7] Tamilselvi N, Krishnamoorthy P, Dhamotharan R, Arumugam P, Sagadevan E 2012 Analysis of total phenols, total tannins and screening of phytocomponents in *Indigofera aspalathoides* (Shivanar Vembu) Vahl EX DC *Journal of Chemical and Pharmaceutical Research* **4** 3259-3262
[8] Cagampang B G and Rodrigues F M 1980 Methods of Analysis for Screening Crops of Approximate Qualities Analytical Services Laboratory Institute of Plant Breeding UPLR
[9] Loganayaki N, Rajendrakumaran D, Manian S 2010 Antioxidant capacity and phenolic content of different solvent extracts from banana (*Musa paradisiaca*) and mustai (*Rivea hypocrateriformis*) *Food Science Biotechnology* **19** 1251-1258
[10] Harborne J B, Baxter H, Moss G P 1999 Phyto-chemical dictionary: Handbook of bioactive compounds from plants (2nd ed.) Taylor & Franci London
[11] Eldahshan O A and Singab A N B 2013 Carotenoids *Journal of Pharmacognosy and Phytochemistry* **2** 225-234
[12] Eugster C H 1995 History: 175 years of carotenoid chemistry. Carotenoids Vol. 1A: Isolation and Analyisis Birkhauser Verlag Basel p1–11
[13] Yu L, Perret J, Harris M, Wilson J, Haley S 2003 Antioxidant properties of bran extracts from "Akron" wheat grown at different locations *Journal of Agricultural and Food Chemistry* **51** 1566-1570
[14] Maisuthisakul P, Suttajit M, Pongsawatmanit R 2007 Assessment of phenolic content and free radical scavenging capacity of some Thai indigenous plants *Food Chemistry* **100** 1409-1418

IOP Conf. Series: Earth and Environmental Science **131** (2018) 012014 doi:10.1088/1755-1315/131/1/012014

Effect of different pretreatments on dried chilli (*Capsicum annum L.*) quality

S B Anoraga[1], I Sabarisman[1] and M Ainuri[2]

[1] Departemen of Bioresources Technology and Veterinary, Vocational College, Universitas Gadjah Mada, Yogyakarta, Indonesia
[2] Departemen of Agro-industrial Technology, Universitas Gadjah Mada, Yogyakarta, Indonesia

E-mail: satriabhirawa@ugm.ac.id

Abstract. Chilli *(Capsicum annum L.)* has significant price fluctuation. When the chilli price is declined, it causes food waste from unsold chilli. Therefore, drying chilli is a solution for this condition. Futhermore, it can be processed for various product like chilli powder, chilli sauce, etc. The aim of this study was to investigate the effect of different pretreatments on dried chilli quality. Chilli was blenched with hot water and steam before drying. The purpose of this pretreatments is to inactivate enzyme that prevents color and vitamin C losses. The quality parameters were moisture content, colour, vitamin C content, and capsaicin. Changes were observed by gravimetri method for moisture content, chromameter in L* a * b * colour model, and iodine titration for vitamin C. After drying for 20 hours at 60°C, chilli with steam blanching pretreatment dried rapidly than other samples. Unpretreated chilli had higher vitamin C content and better color than blanched chilli.

1. Introduction

Chilli (*Capsicum annum L.*) is one of the potential commodities in Indonesia. It has been cultivated over past decades and has high economic value, because it widely used in Indonesian cuisines as raw material [1]. Unfortunately, chilli as commercial vegetables in Indonesia, has significant price fluctuation. When the price is low, it causes food waste from unsold chilli. Drying chilli is a solution for this situation. Futhermore, dried chilli can be processed for various product s like chilli powder, chilli sauce, etc.

Indonesia has standard quality for dried chilli as raw material of food and pharmaceutical industries. Chilies contain capsaicin which has health benefiting and gives strong spicy flavour. Moreover, capsaicin has anti-bacterial, anti-carcinogenic, analgesic and anti-diabetic properties [2]. Chilies are also good sources of vitamin A and C, and rich in β-carotene and minerals that are essential in body requirement [3]. Basically, it is the main reason to dry the chilli in a appropriate method to produce good quality (colour and pungency) dried product [4].

Pretreatment methods can minimize the reduction in the nutritional quality of the dried food products [5]. It normally uses thermal treatments of various time–temperature combinations for the inactivation or decrease in number of microorganisms in order to develop safe products with a long shelf-life. Pretreatments require unit operations of heating (e.g. cooking, boiling, blanching) like final heat treatments such as pasteurization and sterilization [6]. Blanching is the common pretreatment prior to drying of fruits and vegetables. The thermal blanching can improve β-carotene content and inactivate oxidative enzyme which prevent greater losses during drying process [7]. Recomended medium for blanching are hot water and steam. Hot water and steam blanching are adequate to inactivate peroxidase [8]. The right choice of temperature-time combination of blanching can optimize and control the quality of end product. Low temperature blanching in range 55-75°C has made improvement to firming effect for reducing physical breakdown and sloughing during further processing, and maintaining the texture [9].

Therefore, the objectives of this study were to investigate the effect of different pretreatments method of water and steam blanching method on dried chilli quality. Both hot water and steam blanching were assumed the suitable method that can be applied in small medium farmer.

2. Materials and Methods

2.1. Materials

Fresh chilli of good quality were obtained from local market then washed with tap water, packed in plastic bag and stored in a referigerator at a temperature 5°C to reduce the biological and respiratory activities until experiment. The moisture content of fresh chilli was 80-90% (w.b) by using an oven (thermogravimetri) at 105°C.

2.2. Pretreatments

The following treatments used for red chilli before drying are

- Blank (B): Untreated sampel
- Water blanching (WB): chillies were soaked in a hot water bath at 90°C for 3 min and drained to remove excess water
- Steam blanching (SB): chillies were steamed using boilling water for 5 min.

2.3. Drying procedures

A laboratory hot-air dryer (Memmert, UN55, Germany) was used to dry chilli, both treated and untreated samples. The temperature and velocity of the dry air were kept constantly at 60°C and 2.0 m/s respectively. Temperature was monitored using a type-T thermocouple. The drying process was conducted for 20 hours. After drying process finished, the sample was placed in a desiccator for cooling down temperature prior to pack in an aluminum foil, and then kept in a desiccator for further quality evaluations.

2.4. Quality parameters (physical and chemical test)

The effects of pretreatment methods on dried chilli quality were determined. The average moisture ratio (MR) was calculated using Eq. (1):

$$MR=M-M_e/M_0-M_e \qquad (1)$$

Where M is sample moisture content (d.b.), M_0 is initial moisture content (d.b.), and Me is the equilibrium moisture content.

Surface colors of dried chilli were measured using a chromameter. Color was expressed in CIELAB, i.e., L* (whiteness or brightness), a* (redness/greenness) and b* (yellowness/blueness) coordinates. A rectangular CIELAB system (L*, a*, and b*) was measured, and color difference (ΔE) were calculated [10]. Futhermore, vitamin C content was measured according to iodine titration method.

Analysis of variance (ANOVA) was carried out using the SPSS 19.0 software. The results were expressed and plotted as the mean value ± standard deviation (SD). ANOVA tests were performed for all experiments at 95% confidence interval. Mean differences in the treatments were tested for significance using the one way ANOVA and where significant differences were observed, the Duncan Multiple Range test (DMRT) was used to separate the means.

3. Results and Discussion

The physical and chemical qualities of all dried chilli with different pretreatment methods were compared to the fresh chilli. The drying curve (moisture ratio versus time) for the drying of pretreated and untreated chilli at 60°C are shown in Figure 1. The initial average moisture content of fresh chillies were 75-83%. After dried for 20 hours, the final moisture content were significantly different among the samples (P<0.05). Average moisture content of dried chilli were 8.76, 23.03 and 10.203% for SB, WB and B respectively. WB had longest drying time because absorption of moisture and solutions during the blanching [11]. Drying process for steam blanching chilli faster than hot water blanching method. This is in similiar to the result obtained for bell pepper drying [5] that steamed sample has shorter drying time. Moisture content has great impact to physic quality of chilli. Mould should grow easily when moisture content was above 11% and colour loss occure when moisture content was below 4% [12]. This is also recommended for Indonesian dried chilli as regulated according to Indonesian National Standard (SNI 01-3389-1994), that standard quality of dried chili moisture content is below 11%.

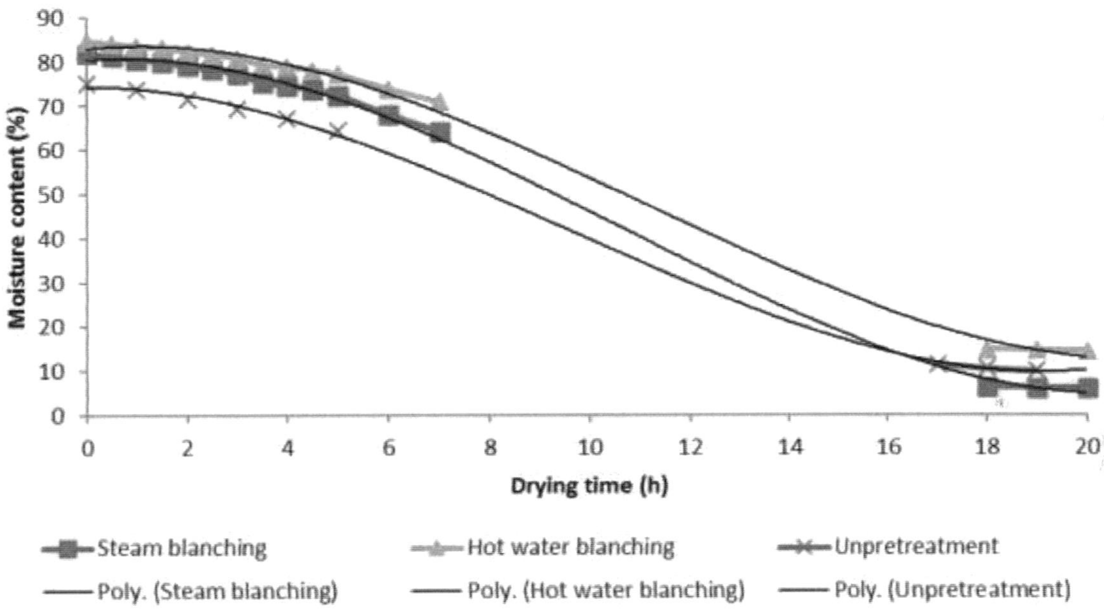

Figure 1. Effects of different pretreatment methods on drying time

The effect of different blanching methods on the colour qualities of dried chilli at 60°C is shown in Figure 2 and Table 1. It was shown that the L* values of all dried chilli ranged from 30.675 to 35.575. The a* values ranged from 23.95 to 32.69 and the b* values ranged from 17.64 to 29.195. The brightness values of dried chilli with water blanching are higher than fresh chilli. The steam blanching pretreatments and untreated sample resulted in lower brightness values than fresh chilli. The values of redness decreased at varying method of pretreatments, except water blanching method. The values of yellowness increased at all method of pretreatment, especially for water blanching method (significantly different, P<0.05). Although, the yellowness value are not desirable for high quality chilli variable [13]. The total color difference (ΔE) was found to be lowest for untreated samples. In the other hand, the dried chilli sample with all pretreatment methods had higher color difference than untreated sample (P<0.05).

Table 1. The color values of fresh and dried chilli

Sample	Lightness (L)	Redness (a*)	Yellowness (b*)	ΔE
Fresh chilli	34.89 ± 0.106[a]	32.69 ± 5.133[b]	17.64 ± 3.75[b]	0
Dried chilli (B)	32.80 ± 0.014[b]	30.35 ± 0.120[ab]	22.55 ± 0.028[b]	5.83 ± 0.019[a]
Dried chilli (WB)	35.58 ± 0.035[c]	32.64 ± 0.177[b]	29.19 ± 0.007[a]	11.58 ± 0.004[b]
Dried chilli (SB)	30.68 ± 0.007[d]	23.95 ± 0.014[a]	20.25 ± 0.057[b]	10.05 ± 0.03[c]

Vitamin C content of dried chilli of all treatments were significantly different ($P<0.05$) and were shown in Figure 2. Untreated chilli has higher vitamin C content than others and water blanching chilli got a big losses of vitamin C content. This losses were caused not only by drying process but also pre-drying treatments. The important parameters that affect the vitamin C degradation were temperature and time [14]. Dried cumari peppers and vitamin C losses occure during convective drying from 32 to 68% in all treatments. It prove that vitamin C degradation increased with increasing temperature in the convective drying [15]. It occured because of destruction or losses of some structure as a effect of heat application [16].

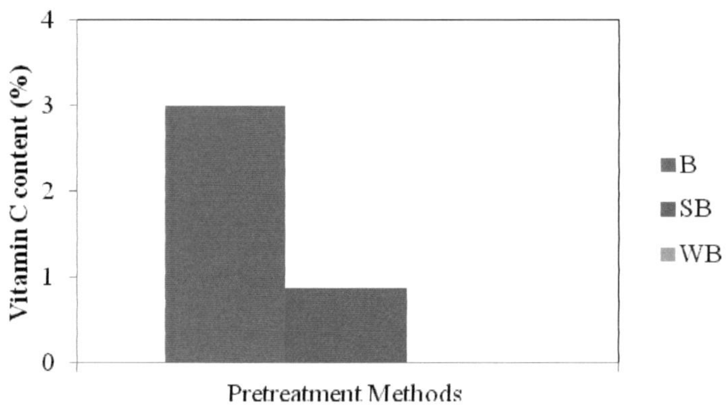

Figure 2. Effects of different pretreatment methods on vitamin C content

4. Conclusion

The drying time for water blanching pretreated chilli was longer than others and significantly different, because water blanching causes moisture absorbtion into the sample. In addition, it had highest color difference and vitamin C losses. Pretreatment methods of SB and WB, affect the lightness, yellowness, color difference, vitamin C content and drying time in all dried chilli. Steam blanching pretreatment was the best predrying method because it can maintain the vitamin C content and color of chilli. Moreover, it had lowest moisture content after drying process for 20 hours.

Acknowledgments

This research was supported by the Research Directorate of Universitas Gadjah Mada under Grant No. 380/DIT.LIT/2017. Authors would like to acknowledge assistance from colleagues, especially by technical staff and support reasearch team.

References

[1] Bhattarai M and Mariyono J 2016 The economic aspects of chilli production in Central Java *Economic Journal of Emerging Markets* **8** 2 85-97

[2] Ajaykumar M T, Sandeep J L, Madhukar B G 2012 Effect of pretreatments on quality attributes of dried green chilli powder *ISCA Journal of Engineering Sciences* **1** 1 71-74

[3] Tavakolipour H and Mokhtarian M 2015 Drying of chili pepper in different conditions. Proceedings of The IRES 4th International Conference, Kuala Lumpur, Malaysia, 16th July 2015, pp 71-74

[4] Bakane P H, Khedkar M B, Wankhade A B, Kolhe R V 2014 Studies on drying of green chilli in dehumidified air dryer *Internat. J. Proc. & Post Harvest Technol.* **5** 2 127-130

[5] Tunde-Akintunde T Y, Oyelade O J, Akintunde B O 2014. Effect of drying temperatures and pre-treatments on drying characteristics, energy consumption, and quality of bell pepper. *Agric Eng Int: CIGR Journal* **16** 2 108 — 118

[6] Deak T 2014 Chapter 17–Thermal Treatment. Food Safety Management pp 423-442. Retrieved from http://www.sciencedirect.com/science/article/pii/B9780123815040000172

[7] Dutta D, Chaudhuri U R, Chakraborty R 2005 Structure health benefits. Antioxidant property and processing and storage of carotenoids *African Journal of Biotechnology* **4** 13 1510-1520

[8] Gupta S, Lakshmi J A, Prakash J 2008 Effect of different blanching treatments on ascorbic acid retention in green leafy vegetables *Natural Product Radiance* **7** 2 111-116

[9] Verlinden B E, Yuksel D, Baheri M, De Baerdemaeker J, Van Dijk C 2000 Low temperature blanching effect on the changes in mechanical properties during subsequent cooking of three potato cultivars *International Journal of Food Science & Technology* **35** 331–340

[10] Kim S, Park J B, Hwang I K 2002 Quality attributes of various varieties of korean red pepper powders (*Capsicum annuum* L.) and color stability during sunlight exposure *Journal of Food Science* **67** 2957–2961

[11] Saengrayap R, Boonlap N, Boonsom U 2016 Effect of pretreatment methods on the color changes during drying of Red Chilli (*Capsicum frutescens* L.) MATEC Web of Conferences **62** 02009

[12] Toontom N, Meenune M, Posri W, Lertsiri S 2012 Effect of drying method on physical and chemical quality, hotness and volatile flavour characteristics of dried chilli *International Food Research Journal* **19** 3 1023-1031

[13] Ergunes, G and Tarhan, S 2006 Color retention of red peppers by chemical pretreatments during greenhouse and open sun drying *Journal of Food Engineering* **76** 446–452

[14] Santos, P H S and Silva M A 2008 Retention of vitamin C in drying processes of fruits and vegetables- a review *Drying Technology* **26** 1421–1437

[15] Reis R C, Castro V C, Devilla I A, Oliveira C A, Barbosa L S, Rodovalho R 2013 Effect of drying temperature on the nutritional and antioxidant qualities of cumari peppers from Pará (*Capsicum chinense* Jacqui) *Brazilian Journal of Chemical Engineering* **30** 2 337-343

[16] Tunde-Akintunde T Y 2010 Effect of pretreatment on drying time and quality of Chilli Pepper. *Journal of Food Processing and Preservation* **34** 595–608

Parametric and non-parametric approach for sensory RATA (Rate-All-That-Apply) method of *ledre* profile attributes

S Hastuti[1,2], Harijono[2], E S Murtini[2] and K Fibrianto[2]

[1]Faculty of Agriculture, Universitas Trunojoyo, Madura, Indonesia
[2]Department of Agricultural Product Technology, Faculty of Agricultural Technology, Universitas Brawijaya, Malang, Indonesia

E-mail: kiki.fibrianto@ub.ac.id

Abstract. This current study is aimed to investigate the use of parametric and non-parametric approach for sensory RATA (Rate-All-That-Apply) method. *Ledre* as Bojonegoro unique local food product was used as point of interest, in which 319 panelists were involved in the study. The result showed that *ledre* is characterized as easy-crushed texture, sticky in mouth, stingy sensation and easy to swallow. It has also strong banana flavour with brown in colour. Compared to eggroll and *semprong*, *ledre* has more variances in terms of taste as well the roll length. As RATA questionnaire is designed to collect categorical data, non-parametric approach is the common statistical procedure. However, similar results were also obtained as parametric approach, regardless the fact of non-normal distributed data. Thus, it suggests that parametric approach can be applicable for consumer study with large number of respondents, even though it may not satisfy the assumption of ANOVA (Analysis of Variances).

1. Introduction

Ledre is traditional food from Bojonegoro, East Java, Indonesia. *Ledre* made from batter mixture of rice flour, tapioca, sugar, coconut milk and banana. The other similar products with *ledre* is *semprong* and eggroll. Compared with *semprong* and eggroll, *ledre* has unique sensory. However, little is known about characteristic sensory of *ledre*. To address this paucity of information, this paper seeks to investigate the attributes sensory of *ledre* using the RATA method.

Profile attributes sensory of *ledre* can analyzed with RATA (rate-all-that-apply). RATA has the potential to improve sample description and discrimination [1]. sensory product descriptions generated by consumers has triggered methodological in relation to sensory product characterization [2]. RATA allows panelist to rate the intensity of selected attributes with 3 intensities (low, medium ang high) [3].

Non-parametric approach has some advantages, which can be used on small data and is free distribution [4]. Parametric approach must satisfy several assumptions such as normality, homogen, linear etc. [5]. Non-parametric can applied when certain assumptions cannot be made about the population. The Friedman test is a useful method of analysis for non-parametric data, especially in human competence research. In Friedman test, the answers of one respondent are ranked. Then all rankings of one competence are summed to gain group results [6]. Parametric (ANOVA), one other hands, incorporated means and variances to determine the test statistic. ANOVA is the most commonly technique for comparing means and it is important research reports.

For small data (n<30), the assumption of normality is particularly worrisome. Non-parametric generally have less power for the same sample than parametric if the data are normal [7].It can be interesting study to use a parametric or non-parametric approach in some cases. As RATA questionnaire is designed to collect categorical data, non-parametric approach is the common statistical procedure but with the large number of panelists, it will be takes a long times and great effort. The objective of this research is comparison non-parametric and parametric approach for consumer study with RATA method.

2. Materials and Methods

2.1. Samples

There were 3 samples *ledre* with different ratio of rice flour and tapioca (Table 1.). *Ledre* was compared with *semprong* and eggroll. A total of 5 samples were produced in a home industry and represented standard recipes.

Table 1. *Ledre* with ratio of rice flour and tapioca.

Ledre	Ratio rice flour : tapioca	
Ledre A	33	1
Ledre B	17	1
Ledre C	2	1

2.2. Panelist

The panelist was composed of 319 untrained panelist with 11 – 45 years of age. The task for the panelist consisted all the attributes, they perceived in the samples and evaluate their intensity on a 3-point scale with all points labeled (1 = ‘‘Low”, 2 = ‘‘Medium”, 3 = ‘‘High”).

2.3. Test procedures

During the sensory evaluations, samples (two pieces) were served in a transparent polypropylen (PP), labeled with a three digit code. The samples were stored at 15°C for 24 h and brought to room temperature (32°C) prior to serving. The samples were presented monadically in a randomized serving order according to a complete block design. Water was used as palate cleansers in between samples. The sensory attributes used in the study is listed in Table 2.

Table 2. Overview of sensory attributes used in RATA.

Category	Attributes	Definitions
Taste	salty	The taste of salt solution
	Sweet taste	The taste of sugar solution
Texture	Easy crack	Easy crack in the first bite
	Easy-crushed	Easy broken from a whole to crumbles
Flavor	Sweet flavor	The flavor of sugar solution
	Banana flavor	The flavor of *Musa textilia*
Physic	Brown color	Intensity of the red color observed in the surface of the *ledre*
Mouth-feel	Stingy sensation	The puncture on the tongue sensation
	Solid sensation	Compact perception, described as ‘‘product remains as a whole”,
	Greasy sensation	Rough and grainy perception in the mouth
	Smoothness	Soft and velvety perception in the mouth
	Dry	Crispy, resistance to flow before saliva modifies the sample
	Sticky in mouth	The layer that remains in the mouth
	Easy to swallow	-

2.4. Data analysis

A mix of non-parametric and parametric statistical analyses were performed in order to address the aims outlined in the introduction. All analyses were performed using SPSS 16.0. Parametric test used analysis of variance (ANOVA) to know the significance and Fisher test to know the difference of 5 samples [8]. Non-parametric test used Friedman test to know the significance and Z test to know the difference of 5 samples. It is calculated as follows [4]:

$$Z\left(1-\frac{\alpha}{k\,(k-1)}\right)\sqrt{\frac{b\,(k+1)}{6}}$$

where Z is probability, α is 0.05, k is the number samples and b is the number of panelists.

3. Results and discussion

The Friedman test showed that assymp.Sig. value was 0.000 (Table 3.). It means that there is significance different between 5 samples.

Table 3. Friedman test.

N	1595
Chi-Square	3.757E3
Df	19
Asymp.Sig	.000

In non-parametric approach, the differences between samples were evaluated by Z value. The comparison was considered significant when the mean of rank difference is larger than critical Z value (112.2). In parametric approach, the different of samples can identified with Fisher test, which the different subcribe (a,b,c) is mean that there was a different between samples (Table 4.)

Table 4. Non-parametric and parametric data comparison.

No.	Attributes	Non-parametric			Parametric		
		Ledre A-B	*Ledre* A-C	*Ledre* B-C	*Ledre* A	*Ledre* B	*Ledre* C
1.	Easy-crused texture	79	83.5	4.5	$2,19^{a}$	$2,22^{a}$	$2,33^{a}$
2.	Banana flavor	11	44	33	2.29^{c}	2.30^{c}	2.37^{c}
3.	Brown color	196.5	339.5	143	1.44^{a}	1.75^{b}	1.97^{c}
4.	Stingy sensation	23	30	53	1.91^{b}	1.94^{b}	2.02^{b}
5.	Sticky in mouth	36	73	109	1.69^{c}	1.80^{cd}	1.86^{d}
6.	Easy to swallow	34	18	16	2.24^{ab}	2.28^{ab}	2.30^{b}
7.	Roll lenght	12	321	309	2.14^{a}	2.15^{a}	2.65^{b}
8.	Taste var	35	20	55	2.13^{c}	2.18^{c}	2.20^{c}
9.	Roll lenght var	7	209	216	2.08	2.09^{b}	2.42^{c}
10.	Easy-crack	89	153.5	64.5	2.34^{c}	2.20^{bc}	2.09^{b}
11.	Sweet flavor	74	56	130	2.17^{c}	2.05^{b}	2.29^{c}
12.	Dry	120	153	33	2.48^{b}	2.28^{a}	2.24^{a}
13.	Solid sensation	59	35	24	1.77^{a}	1.83^{a}	1.85^{a}
14.	Sweet taste	90	47	137	2.29^{bc}	2.16^{b}	2.39^{c}
15.	Smooth texture	76	106	30	2.22^{c}	2.08^{bc}	2.01^{b}
16.	Salty	27	24	51	1.32^{a}	1.36^{a}	1.29^{a}
17.	Greasy sensation	8	33	41	1.32^{a}	1.37^{a}	1.29^{a}
18.	Form var	14	7	21	1.83^{a}	1.83^{a}	1.84^{a}
19.	Packaging var	0	44	44	1.85^{a}	1.83^{a}	1.89^{a}
20.	Form packaging var	31	21	10	1.76^{a}	1.79^{a}	1.78^{a}

Note: a,b,c, is showed significant at $P<0.05$

The rate attribute easy-crushed texture of *ledre* A, B and C below the Z value, it mean that there wasn't different between *ledre* A,B and C. In parametric approach, it also no different between *ledre* A,B, and C showed with the same subscribe between 3 samples. The similar result showed in attributes brown color, the non-parametric and parametric approach showed the similar result. Three samples showed significant different ($P<0.05$) in brown color attributes.

It suggests that parametric approach can be applicable to stand in non-parametric condition, eventhough it may not satisfy the assumption of ANOVA (Asymp.Sig 2-tailed/ normality all attributes was 0.000). Normality test with Kolmogorov-Smornov is that if the significance below 0.05 means the data is not normal [5].

Table 5 shows that *ledre* is characterized as easy-crushed texture, sticky in mouth, stingy sensation and easy to swallow. It has also strong banana flavour with brown in colour. Compared to eggroll and *semprong, ledre* has more variances in terms of taste as well the roll length.

Table 5. Profile attributes sensory of *ledre* with RATA.

No	Attributes	Semprong	*Ledre*	Eggroll
1.	Easy-crused texture	Low	medium	high
2.	Banana flavor	low	high	Low
3.	Brown color	high	medium	Low
4.	Stingy sensation	Low	medium	Low
5.	Sticky in mouth	Low	medium	Low
6.	Easy to swallow	Low	medium	high
7.	Roll lenght	Low	High	Low
8.	Taste var	Low	High	Low
9.	Roll lenght var	low	high	Low

Note: Low : 1 – 1.49; Medium: 1.5 – 2.49; High : 2.5 – 3

4. Conclusion

Based on comparison of parametric and non-parametric data processing, it is concluded that parametric approach can be applicable for consumer study with large number of respondents, even though it may not satisfy the assumption of ANOVA.

References

[1] Meyners M, Jaeger S R, Ares G 2016 On the analysis of Rate-All-That-Apply (RATA) data. *Food Qual Prefer* **49** 1–10

[2] Varela P, Ares G 2012 Sensory profiling, the blurred line between sensory and consumer science. A review of novel methods for product characterization *Food Res International* ***48*** 893–908

[3] Ares G, Bruzzone F, Vidal L, Cadena R S, Giménez A, Pineau B, Hunter D C, Paisley A G, Jaeger S R 2014 Evaluation of a rating-based variant of check-all-that-apply questions: Rate-all-that-apply (RATA) *Food Qual Prefer* **36** 87–95

[4] Sprent P, Smeeton N C 2007 Applied Nonparametric Statistical Methods Chapman and Hall London 30-254

[5] Riduwan 2016 Dasar-dasar Statistika Alfabeta Bandung Indonesia 183-184 [In Indonesian]

[6] Daniel W W 1991 Statistik Nonparametrik Terapan. Gramedia Jakarta Indonesia 187-265 [In Indonesian]

[7] Kaur A, Kumar R 2015 Comparative analysis of Parametric and Non-parametric Test *Journal of Computer and Mathematical Sciences* **6** 6 336-342

[8] Sujarweni W 2015 SPSS untuk penelitian Penerbit Pustaka Baru Press Yogyakarta Indonesia 97-109 [In Indonesian]

Bioactive compounds from palm fatty acid distillate and crude palm oil

T Estiasih[1] and K Ahmadi[2]

[1]Department of Agricultural Product Technology, Faculty of Agricultural Technology, Universitas Brawijaya, Malang, Indonesia
[2]Department of Agro-industrial Technology, Faculty of Agriculture, Tribhuwana Tunggadewi University, Malang, Indonesia

Email: teties@yahoo.co.id, teties@ub.ac.id

Abstract. Crude palm oil (CPO) and palm fatty acid distillate (PFAD) are rich sources of bioactive compounds. PFAD is a by-product of palm oil refinery that produce palm frying oil. Physical refining of palm oil by deodorization produces palm fatty acid distillate. CPO and PFAD contain some bioactive compounds such as vitamin E (tocopherol and tocotrienols), phytosterol, and squalene. Bioactive compounds of CPO and PFAD are vitamin E, phytosterols, and squalene. Vitamin E of CPO and PFAD mainly comprised of tocotrienols and the remaining is tocopherol. Phytosterols of CPO and PFAD contained beta sitosterol, stigmasterol, and campesterol. Tocotrienols and phytosterols of CPO and PFAD, each can be separated to produce tocotrienol rich fraction and phytosterol rich fraction. Tocotrienol rich fraction from PFAD has both antioxidant and cholesterol lowering properties. Bioactive compounds of PFAD silmultaneously have been proven to improve lipid profile, and have hepatoprotector effect, imunomodulator, antioxidant properties, and lactogenic effect in animal test experiment. It is possible to develop separation of bioactive compounds of CPO and PFAD integratively with the other process that utilizes fatty acid.

1. Introduction

Physical refining of crude palm oil (CPO) to produce frying oil passes through several steps including degumming, bleaching, and deodorization. Palm fatty acid distillate (PFAD) is a by-product of deodorization in palm oil refinery. The major components of PFAD are free fatty acid, lipid oxidation products, and other compounds such as tocopherols, tocotrienols, phytosterols, and squalene [1]. PFAD utilization currently is limited to oleochemical industries, while the use of its bioactive compounds is almost neglected. The production of PFAD is 4% based on CPO [2]. It is about 3.66 ton of PFAD is produced from every 100 ton of CPO [3]. PFAD is commonly used as fatty acid source for non-food industries [4] such as soap, feed, and oleochemical industries [5].

According to Liu *et al.* [6], as much as 5-57% of tocotrienols and tocopherols lost from CPO during deodorization. Most of tocotrienols and tocopherols are accumulated in PFAD with the concentration of 0.7-1.0%. Goh and Gee [7] indicated that hydrocarbon in PFAD is squalene as major component and n-alcane ($C_{12}H_{26}$-$C_{36}H_{74}$) as minor component. PFAD contains high amount of squalene that reaches 1.03%, and this quantity is higher than that found in other vegetable oil [8].

PFAD also contains free fatty acids and glycerides 96.1%, and minor bioactive compounds such as tocopherols and tocotrienols (0.48%), phytosterols (0.37%), squalene (0.76%), and other hydrocarbons (0.71%) [1]. Meanwhile the composition in CPO is vitamin E 222 ppm, phytosterols 17,322 ppm, squalene 535 ppm, and β carotene 643 ppm [9]. PFAD has been used integratively as a source of bioactive compounds and free fatty acids for oleochemical industries [10].

Bioactive components of PFAD and CPO are accumulated in unsaponifiable fraction (USF) [9, 11]. USF can be separated from saponifiable fraction by simple saponification. Two fractions are formed during saponification, unsaponifiable fraction that contains bioactive compounds and saponification fraction that rich in free fatty acids or soap [9, 10].

USF from PFAD contains tocopherols 4.05% and tocotrienols 8.04% [12, 13] and phytosterols [14]. Purification of USF produces vitamin E rich fraction with vitamin E concentration as high as 33.88% [15], and phytosterols rich fraction with phytosterols concentration of 17.33% [14]. Our previous study [10] showed that USF from PFAD had vitamin E of 0.80%, phytosterols of 9.16%, squalene of 1.14%, meanwhile polycosanol and co-enzyme Q10 were not detected. USF from CPO contained vitamin E 1.94%, α tocotrienol 0.86%, δ tocotrienol 0.14%, γ tocotrienol 0.94%, total tocotrienol 1.94%, total phytosterols 65.80%, β sitosterol 41.94%, campesterol 4.47%, stigmasterol 19.39%, β carotene 0.054%, and squalen 31.69%.

Vitamin E of PFAD dan CPO is different from other sources that rich in tocotrienols. The composition of vitamin E in PFAD is α tocopherols (20%), α tocotrienols (22%), γ tocotrienols (46%), and δ tocotrienols (12%). Beside rice bran oil, palm oil has high tocotrienols content [16, 17]. Meanwhile the composition in CPO is α tocopherols (0%), α tocotrienols (65%), γ tocotrienols (26%), and δ tocotrienols (9%) [9]. Tocotrienol is an important compound for pharmaceutical product, foods, and food supplements. Tocotrienols have hypocholesterolaemic, antioxidant, antithrombotic, anti atherogenic, anti inflammation, and immunomodulator properties. Tocotrienols are also known to have the ability to reduce blood LDL cholesterol level [18]. Tocotrienols are able to inhibit HMG Co-A reductase, an enzyme that is responsible to convert HMG into mevalonate in cholesterol biosynthesis. This conversion is a rate-limiting step in the cholesterol biosynthesis, and the primary statin target [19].

Phytosterols of PFAD comprises of campesterol (13%), β sitosterol (60%), stigmasterol (24%), and cholesterol (3%) [20]. In CPO, phytosterols consists of campesterol (0.5%), β sitosterol (98.8%), and stigmasterol (0.7%) [9]. Phytosterols lower blood cholesterol level by inhibiting cholesterol absorption. The required dose of phytosterols to have beneficial effect on health is 2-3 g/day [21]. Phytosterols also have the ability to increase breast milk production, as well as anti inflammation, anti cancer, immunomodulatory [22], antithrombotic, and hypocholesterolaemic [23, 24, 25]. So far, the mechanism of cholesterol lowering properties of phytosterol is still unclear, but Bonsdorff-Nikander [26] proposed some theories that phytosterols compete with cholesterol in micelle formation, phytosterols cocrystallize with cholesterols, and phytosterols inhibit ACAT (Acyl CoA:Cholesterol Acyl Transferase) activity in cholesterol synthesis.

Squalene as one of bioactive compounds in PFAD and CPO, has anti cancer and cholesterol lowering properties [20]. Squalene has the ability to increase fecal excretion of bile acid that leads to reduce blood cholesterol level [27, 28, 29]. Other study showed that squalene increase cholesterol level, due to incresed rate of cholesterol synthesis [19]. It is hypothesized that bioactive compounds of USF synergistically affect lipid profile improvement.

2. Bioactive Compounds Of PFAD, CPO, and USF

Bioactive compounds found in PFAD and CPO are vitamin E, phytosterols, and squalene. Loganathan et al. [20] revealed that palm oil contained vitamin E 600-1.000 ppm, phytosterols 300-620 ppm, carotenoid 500-700 ppm, squalene 250-540 ppm, phospholipids 20-100 ppm, co-enzyme Q10 10-80 ppm, and poliphenol 40-70 ppm. Meanwhile, our previous study [30] showed that PFAD contained bioactive compound that comprised of α-tocopherol 8.5-134.62 ppm, α-tocotrienol 11.97-51.63 ppm, δ-tocotrienol 4.56-63.65 ppm, γ-tocotrienol 32.18-117.98 ppm, total vitamin E 64.70-280.63 ppm, β-sitosterol 381.55-3,575.55 ppm, stigmasterol 10.99-1548.32 ppm, campesterol 18.96-1,720.30 ppm, total phytosterols 407.00-6011.72 ppm, squalene 205.73-1,273.64, but coenzyme Q10 and polycosanol were not detected.

Distillate of deodorizer from various vegetable oils had different composition of bioactive compounds. Ceriani and Mirelles [31] reported that soybean oil distillate deodorizer contained tocopherols, and β sitosterol; wheat oil only had tocopherols; meanwhile canola oil contained β sitosterol and tocopherols. Benites et al. [32] also showed that soybean oil distillate of deodorizer contained tocopherols. Furthermore, Khatoon et al. [11] showed that phytosterols were also found in soybean oil distillate deodorizer. Bondioli et al. [33] revealed that olive oil distillate deodorizer had squalene. Distillate deodorizer of dried grain contained phytosterols, steryl ferulic, tocopherols, tocotrienols, and carotenoid.

Generally, phytosterols are the most prominent bioactive compound found in PFAD from several palm oil refineries in Java, Indonesia (Table 1). Phytosterols also the major component bioactive compounds in soybean oil distillate of deodorizer [11] and canola oil, meanwhile tocopherols are dominant in wheat oil [32], and squalene in olive oil [33].

Table 1. Bioactive compounds of PFADs from several palm oil refineries.

Palm Oil Refinery	Vitamin E (ppm)	Phytosterol (ppm)	Squalene (ppm)	Polycosanol (ppm)	Co-enzyme Q10 (ppm)
1	195.60	7476.56	2373.27	nd	nd
2	64.70	407.00	462.87	nd	nd
3	280.63	6011.72	2767.08	nd	nd
4	200.76	2310.52	1380.16	nd	nd
5	172.47	1956.15	2222.41	nd	nd
6	208.82	3915.22	nd	nd	nd

nd = not detectable
Reference: Estiasih et al. (2013) [30]

PFAD, CPO, and their corresponding USF have different concentration and bioactive compound composition. Generally, USF has higher bioactive compounds than PFAD (Table 2). According to Khatoon et al. [11], bioactive compounds of PFAD are accumulated in unsaponifiable fraction. Vitamin E increase 245% in USF as compared to PFAD, phytosterols decrease by 73%, and squalene increase by 13611%. The increase of vitamin E and squalene was related to elimination of free fatty acid and triglycerides, as the major component of PFAD and CPO respectively, during saponification. But, factors that affect the decrease of phytosterols was still unknown. Our previous studies [9, 10] showed that all bioactive compounds increased in USF after PFAD and CPO saponification, including phytosterols with the concentration of 9.18%.

Our previous studies [9, 10] that investigated the optimization of saponification showed that USF from PFAD contained vitamin E of 0.80%, phytosterols of 9.18%, and squalene of 1.14%. USF from optimum conditions for maximum yield comprised of 0.47% vitamin E, 7.77% phytosterols, and 16.26% squalene. The bioactive compounds in USF are expected to synergetically reduce blood cholesterol level in hypercholesterolaemia condition.

The concentration of vitamin E in USF from PFAD was high (800 mg/100 g). Consumption 1 g USF per day will provide 8 mg that meets the daily requirement of vitamin E or 57% of daily reference value [10]. Other sources of vitamin E are rice bran (90.9 mg/100 g), wheat sprout (153.7 mg/100 g), coconut (1.1 mg/100 g), soybean (7.8 mg/100 g), and olive (11.9 mg/100 g) [34].

Phytosterols was the highest concentrated bioactive compounds in USF from PFAD (9,18%). The concentration was high as compared to other sources such as crude corn fiber oil (8.79%), crude corn kernel oil (7.94%), refined corn kernel oil (1.11%), and commercial corn oil (0.74%) [35]. The study of Jiang and Wang [36] showed that a by product of cereal processing (rice bran, wheat bran, wheat sprout, oat bran, and oat hull) contained phytosterols of 9.35 mg/g oil.

Table 2. Bioactive compounds of unsaponifiable fraction of PFAD from the optimum conditions of saponification.

Bioactive Compounds	Concentration	
	ppm	% w/w
Vitamin E	7,968.04	0.80
• α tocopherol	644.11	0.06
• α tocotrienol	1,860.54	0.19
• δ tocotrienol	4,853.78	0.49
• γ tocotrienol	609.61	0.06
• Total tocotrienol	7,323.93	0.73
Phytosterol	91,846.30	9.18
• Beta sitosterol	81,932.59	8.19
• Campesterol	61.87	0.01
• Stigmasterol	9,851.83	0.99
Polycosanol	nd	nd
Squalene	11,436.66	1.14
Co-enzyme Q10	nd	nd

nd = not detected
Reference = Estiasih et al. (2012) [9]

Sabir et al. [37] found the variability of phytosterols in several vegetable oils such as mustard (64 mg/g), corn (23 mg/g), soybean (9 mg/g), rapeseed (5 mg/g), and coconut (0,8 mg/g). Generally, crude vegetable oil contains 1-5 g/kg phytosterols, but rice bran oil contains phytosterols 30 g/kg [38]. In palm oil, phytosterol is a minor component that comprises of campesterol, stigmasterol, and β-sitosterol [17].

Squalene was the bioactive compound in USF with concentration of 1.14% or in 1 g of USF would provide 114 mgsqualene. It is estimated that average daily consumption of squalene in food stuff is 28 mg/day [39]. Olive oil contains 200 to 700 mg squalene/100 g [40].

3. Bioactivity of USF

Hypercholesterolaemia condition was achieved by administration the rats with cholesterol. In USF, the bioactive compounds that were supposed to act as cholesterol lowering agents were phytosterols and squalene, although vitamin E also had a role to reduce cholesterol level. Phytosterols have similar action to cholesterol by micelle formation before being carried by enterocytes. The absorption of phytosterols is inhibited by efflux transporter activity, and this transporter contained a pair of ATP binding protein known as ABCG5 and ABCG8. ABCG5 and ABCG8 transport phytosterols and unesterified cholesterol from enterocytes to digestive lumen [41].

Squalene is the major bioactive compound in USF. Squalene from food is absorbed and converted into cholesterol, and the increase in cholesterol synthesis is not related to the increase of blood cholesterol level but increase of cholesterol excretion in the feces [42]. Squalene was also reported to increase pravastin absorption and a combination of pravastin and squalene was effective to reduce cholesterol [43].

Squalene has a role to increase cholesterol excretion into feses [42]. Several studies showed different effect of squalene on lipid profile. Shin et al. [29] showed the hypolipidaemic effects of amaranth squalen (AS) that were evident in both serum and liver. AS markedly increased faecal excretions of cholesterol and bile acid, and slightly inhibited 3-hydroxy-3-methylglutaryl coenzyme A (HMG-CoA) reductase activity. The preliminary study suggested that the cholesterol-lowering effect of AS may be mediated by the increased faecal elimination of steroids through interference with cholesterol absorption, and that different sources of squalene (plant versus animal) may affect cholesterol metabolism differently. Cholesterol-lowering effect of amaranth is associated, at least in part, with its squalene content. The effect of squalene may be attributed to enhanced excretion of faecal steroids through interference of cholesterol absorption.

Previously, the study of Qureshi et al. [28] showed that all amaranth varieties contain tocotrienols and squalene compounds which are known to affect cholesterol biosynthesis. Serum total cholesterol and LDL-cholesterol were lowered by 10-30% and 7-70%, respectively, in birds fed amaranth-containing diets. HDL-cholesterol was not affected by amaranth supplementation. Activities of liver cholesterol 7 alpha-hydroxylase (the enzyme responsible for cholesterol breakdown into bile acids) were 10-18% higher than those of controls for birds fed most forms of amaranth and its oil, whereas activities of liver 3-hydroxy-3-methylglutaryl coenzyme A reductase (the rate-limiting enzyme for cholesterol biosynthesis) were lowered by about 9% of popped, milled amaranth and its oil. The lack of marked inhibition of this enzyme suggests the presence of some other potent cholesterol inhibitor(s) apart from tocotrienols and squalene in amaranth. Also, the study of de Castro *et al.* [30] showed that amaranth oil, and its squalene, increased the excretion of bile acids but did not have a hypocholesterolemic effect in hamsters fed on a diet containing high amounts of saturated fat and cholesterol.

Another bioactive compound in USF is vitamin E. Tocotrienols down-regulate the liver pathway of HMGCo-A reductase. The advantage of tocotrienols over statin is that tocotrienol does not block coenzyme Q10 pathway as statin does. Therefore, tocotrienols lower cholesterol production without side effects. Gamma and delta tocotrienols subtly affect the genes that code HMG reductase. Delta tocotrienol also uniquely down-regulate SREBP (sterol regulatory element binding protein), a nuclear factor that naturally regulate other genes linked to low density lipoprotein (LDL). This effect may also down-regulate excessive triglyceride synthesis. Other than delta tocotrienols, none of the other seven vitamin Es substantially regulates SREBP [44].

4. Conclusions

CPO and PFAD are the valuable sources of bioactive compounds. Unsaponifiable fraction (USF) of PFAD and CPO contain vitamin E, phytosterols, and squalene, and USF is a potential candiadte for food supplement and fortificant. All of the bioactive compounds in USF are expected to act synergystically. For fortification, USF may be prepared by microemulsification and microencapsulation, as well as by direct fortification into several food products.

Acknowledgements

We are very grateful to Brawijaya University that accomodated this presentation and Ministry of Research, Technology, and Higher Education for INSINAS Research Grant 2017 and Stranas Grant 2017.

References

[1] Gapoor A, Hassan WHW, Sulong M 2002 *Palm Oil Develop.* **36** 13
[2] Rakmi AR, Herawan T 2000 J. Oil Palm Res. **12** 11
[3] Chu BS, Quek SY, Baharin BS 2003 *J. Food Chem.* **8** 295
[4] Cheah KY, Toh TS, Koh PM 2010 *Palm fatty acid distillate biodiesel: next generation palm biodiesel.* Inform AOCS
[5] Nang HLL, Wafti NSA, May CY 2009 *Palm fatty acid distillate* MPOB No 471 MPOB Information Series June
[6] Liu D, Shi J, Posada LR, Kakuda Y, Xue SJ 2008 *Food Rev. Int.* **24** 376
[7] Goh SH, Gee PT 1985 *J. Am. Oil Chem. Soc.* **63** 226
[8] Posada LR, Shi J, Kakuda Y, Xue SJ 2007 *J. Sep. Purif. Technol.* **57** 220
[9] Estiasih T, Mukhlishiyiyah J 2015 Optimizing saponification condition for unsaponifiable matters preparation from crude palm oil *Research Repport of Agricultural Technology Faculty Grant, Brawijaya University Malang*

[10] Estiasih T, Ahmadi K, Widyaningsih TD, Maligan JM 2012 Multi components of bioactive compounds for food supplement and fortificant as an effort to increase economical value of by-product of palm oil refinery *Year 1 MP3EI Report. Research Centre Brawijaya University Malang*
[11] Khatoon S, Rajan RGR, Krishna AGG 2010 *J. Am. Oil Chem. Soc.* **87** 321
[12] Ahmadi K, T Estiasih T 2009 Low temperature solvent crystallization in vitamin E enriched tocotrienols preparation from palm fatty acid distillate and its application for functional foods *Year 1 PHB Report, Research Centre, Tribhuwana Tunggadewi University, Malang*
[13] Ahmadi K, Estiasih T 2010 Low temperature solvent crystallization in vitamin E enriched tocotrienols preparation from palm fatty acid distillate and its application for functional foods *Year 2 PHB Report, Research Centre, Tribhuwana Tunggadewi University, Malang*
[14] Ahmadi K, Estiasih T 2011a Phytosterols separation from palm fatty acid distillate and its utilization for reducing cholesterol *Year 1 PHB Report, Research Centre, Tribhuwana Tunggadewi University, Malang*
[15] Ahmadi K, Estiasih T 2011b *J. Teknol. Industri Pangan* **XXII** 142
[16] Ng MH, Chao YM, Ma AH, Choah CH, Hashim MA 2004 *Lipids* **39** 1031
[17] Puah CW, May CY, Ahngan M, Hock CC 2004 *J. Oil Palm Res.* **16** 57
[18] Lewis J 2001 *Process for the production of tocotrienol* US Patent 6,838,104
[19] Coskun AM, Serteser, Unsal, I 2013 *J. Med Biochem* **32** 16
[20] Loganathan R, Selvaduray KR, Radhakrishnan A, Nesaretnam K 2009 *Palm Oil Develop.* **50** 16
[21] Taşan M, Bilgin B, Geçgel U, Demirci AS 2006 *J. Tekirdag Agric. Fac.* **3** 153
[22] Awad AB, Fink CS 2000 *J. Nutr.* **130** 2127
[23] Carr TP, Ash MM, Brown AW 2010 *Nutr. Dietary Suppl.* **2** 59
[24] Piironen V, Lindsay DG, Mieyyinen TA, Toivo J, Lampi AM 2000 *J. Sci. Food Agric.* **80** 939
[25] Ostlund RE Jr, McGill JB, Zeng CM, Covey DF, Stearns J, Stenson WF, Spilburg CA 2002 *Am. J. Physiol. Endocrinol. Metab.* **282** E911
[26] Bonsdorff-Nikander AV 2005 *Studies on a cholesterol-lowering micro crystalline phytosterol suspension oil* Dissertation University of Helsinki, Helsink
[27] Qureshi AA, Lehmann JW, Peterson DM 1996 *J. Nutr* **126** 1972
[28] Shin DH, Heo HJ, Lee YJ, Kim HK 2004 *British J. Biomed. Sci.* **61** 11
[29] de Castro, Andrade LÍ, Soares RA, Manólio PHN, Saldiva RA, Ferrari, Miguel AM 2013 *Lipids* **48** 609
[30] Estiasih T, Ahmadi K, Widyaningsih TD, Maligan JM, Mubarok AZ, Zubaidah E, Mukhlisiyyah J, Puspitasari R 2013 *Adv. J. Food Sci. Tech.* **5**(9): 1153
[31] Ceriani R, Meirelles AJA 2004 *J. Am. Oil Chem. Soc.* **81** 1059
[32] Benites CI, Chonca VOC, Reis SMPM, De Oliveirra OC 2009. *Chem. Eng.Trans.,* **17** 903
[33] Bondioli P, Mariani C, Lanzani A, Fedeli E, A. Muller 1993 *J. Am. Oil Chem. Soc.* **70** 763
[34] Sheppard AJ, Penington JAT, Weihrauch JL 1993 Analysis and distribution of vitamin E in vegetable oils and foods, In: Vitamin E in Health and Disease. *Marcel Dekker. New York*
[35] Moreau RA, Lampi AM, Kevin BH 2009 *J. Am. Oil Chem. Soc.* **86** 1209
[36] Jiang Y, Wang T 2005 *J. Am Oil Chem. Soc.* **86** 439
[37] Sabir SM, Hayat I, Gardezi DA 2003 Pakistan J. Nutr. **2** 178
[38] Hoed VV, Depaemelaere D, Ayala VJ, Santiwattana P 2006 *J. Am. Oil Chem. Soc.* **83** 315
[39] Ahrens EH Jr, Broucher CA 1978 *J. Am. Diet Assoc.* **73** 613
[40] Liu GCK, Ahrens EH Jr, Schreibman PH, Crouse JR 1976 *J. Lipid Res.* **17** 38
[41] Sudhop T, Lutjohann D, von Bergmann K 2005 *Pharmacol. Ther.* **105** 333
[42] Strandberg TE, Tilvis RS, Miettinen TA 1990 *J. Lipid Res* **31** 1637
[43] Chan P, Tomlinson B, Lee CB, Lee YS 1996 *J. Clin. Pharmacol. 36: 422-427.*
[44] Kidd PM 2009 *Total Health* **30** 18

Comparative study on resistant starch, amilose content and glycaemic index after precooked process in white rice

V N Pratiwi[1,2]

[1]Faculty of Health, University of Nahdlatul Ulama Surabaya, Indonesia
[2]Faculty of Agricultural Technology, Universitas Gadjah Mada, Yogyakarta, Indonesia

E-mail: vieranpratiwi@gmail.com

Abstract. Rice is a staple food and regarded as a useful carbohydrate source. In general rice is high in glycaemic index (GI) and low colonic fermentation. People are aware of the alterations in blood glucose levels or glycaemic index after consuming rice. Resistant starch (RS) and amylose content play an important role in controlling GI. GI and RS content have been established as important indicators of starch digestibility. The aim of this study was to determine the precooked process with hydrothermal (boiling at 80°C, 10 minutes) and cooling process with low temperature (4°C, 1 h) to increase potential content of RS and decrease of glycaemic index of white rice. There were two stages of this research, 1) preparation of white rice with precooked process; 2) analysis of precooked white rice characteristics (resistant starch, amylose content, and estimated glycaemic index). The result of analysis on precooked white rice showed an increased RS content (1.11%) and white rice (0.99%), but the difference was not statistically significant. The amylose content increased significantly after precooked process in white rice (24.70%) compared with white rice (20.89%). Estimated glycaemic index (EGI) decreased after precooked proses (65.63%) but not significant as compared to white rice (66.47%). From the present study it was concluded that precooked process had no significant impact on increasing RS and decreasing EGI of white rice. This may be due to the relatively short cooling time (1hour) in 4°C.

1. Introduction

Rice is a major source of carbohydrates consumed by people in the world and in Asia. In Indonesia, people tend to choose white rice as the main source of carbohydrate with average consumption level according to Indonesian Statistical Bureau is 1,631 kg/capita/week [1]. The main macronutrient content in rice is carbohydrate while dietary fiber is less, leading to a high glycaemic index. Excessive levels of white rice consumption and unhealthy lifestyle will increase risk of Diabetes mellitus (DM). The results of previous studies showed that increasing consumption of white rice was associated with increasing risk of type-2 diabetes [2]. The increased risk was associated with a white rice glycaemic index of 69.96 [3]. A high glycaemic index has been shown to increase the risk of type 2 DM of Japanese men [4]. Therefore, consumption of white rice may increase the risk of type 2 diabetes

IOP Conf. Series: Earth and Environmental Science **131** (2018) 012017 doi:10.1088/1755-1315/131/1/012017

mellitus associated with the white rice glycaemic index [5, 6]. A high portion of white rice consumption significantly effected relative risk of type 2 diabetes mellitus by 1.11 [7]. In other studies it was shown that consumption of white rice more than 1 serving per day was positively correlated with blood pressure, triglycerides, and fasting blood glucose, and negatively correlated with HDL cholesterol [8].

Modification of the components of rice to reduce the glycaemic index can be done by applying of several processing methods. One of them is by precooking process which is associated with heating and cooling. These processes are expected to change the nutrient content in rice, especially in resistant starch (RS) content. This can be explained as a result of temperature changes which are then influences the structure of the rice starch leading to an increased RS content. In the process of heating, gelatinization occurs which can increase solubility and starch digestibility. However, the process of heating and cooling back can lead to the formation of insoluble retrogradation starch [9]. This fact underlies an idea to conduct a research on the comparison of starch resistant content, amylose levels, and glycaemic index of white rice after precooked process. The precooked process was conducted by heating and cooling, and was expected to increase the RS content in white rice.

2. Materials and Methods

The research was conducted at Pusat Studi Pangan dan Gizi (PSPG) laboratory, Laboratory of Pathology, Faculty of Veterinary Medicine, and Laboratory of Food Chemistry and Biochemistry, Faculty of Agricultural Technology, Universitas Gadjah Mada.

2.1. Materials

The material used in this research was white rice of IR 64 variety which is produced by Balai Benih Barongan, Bantul, Yogyakarta. The chemicals for analysis and some enzymes such as α-amylase, pepsin, pancreatin and amyloglucidase as well as glucose kit (GOD-PAP).

2.2. Precooking process

Precooking process was intended to increase the content of resistant starch. The precooking process was carried out by heating the rice at 80°C, then continued by cooling at 4°C. The process of retrogradation (recrystallization) of the starch could lead to the formation of type III resistant starch (RS) [10].

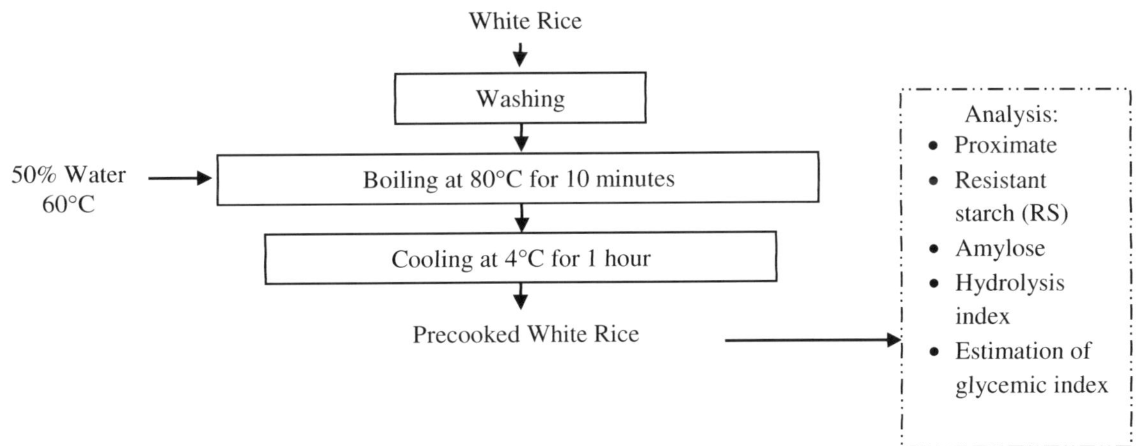

Figure 1. Flow chart of precooked rice processing

2.3. Analysis of precooked white rice

Characterization of the precooked rice include proximate analysis, amylose content, food fiber content, resistant starch content, hydrolysis index, and glycaemic index estimation. Proximate analysis [11] included water and ash content by thermogravimetric method, protein by using micro-Kjeldahl, fat by Soxhlet extraction method, and carbohydrates were calculated by difference [12]. Amylose analysis [10], analysis of dietary fiber [13], starch digestibility analysis, and resistant starch analysis using enzymatic methods (GOD FS) [14], and total starch analysis, hydrolysis index and estimated IG [15].

2.4. Data analysis
The experimental design used in this study was Completely Randomized Design (RAL). The data analysed by using one-way variance analysis (ANOVA). If a significant difference was found then continued with the Duncan Multiple Range Test (DMRT) using SPSS version 17 with 5% confidence interval.

3. Results and Discussion
3.1. Proximate analysis
The chemical composition in precooked white rice is presented in Table 1.

Table 1. Characteristics of precooked white rice.

Component	Precooked white rice (%db)
Ash	0.27 ± 0.02
Fat	0.40 ± 0.04
Protein	9.02 ± 0.55
Carbohydrate by difference	79.90 ± 0.21
Fibers	6.91 ± 0.23

Data is presented in the form of mean ± standard deviation

The chemical composition of precooked white rice was closed to previous studies [3]. Protein content (9.02%) was slightly lower when compared with white rice without precooked processing (10.83%). The fat content (0.40%) was slightly lower as compared to that of white rice with the same variety of 0.56%.

3.2. Amylose and resistant starch levels of white rice precooked
Amylose content (24.70%) increased compared with that in white rice (20.89%). The value of amylose content was in line with that reported by previous studies [16] with the same white rice varieties (24.59%). One-way ANOVA analysis on amylose levels between treatment groups showed significant difference between WR and PWR. RS content increased after the precooked treatment. The RS levels in PWR were 1.11% and thus increased slightly after precooked treatment as compared to WR (0.99%) but no significant difference was found. Amylose and RS content in both raw and precooked rice are presented in Table 2.

Table 2. Amylose and RS content of white rice precooked.

Ingredients	Amylose (%)	RS (%)
White Rice (WR)	20.89 ± 2.40^{a}	0.99 ± 0.11^{a}
Precooked White Rice (PWR)	24.70 ± 0.44^{b}	1.11 ± 0.01^{ab}

Different superscript notations in the same column showed significantly different ($p < 0.05$) using the ANOVA assay and Duncan's follow-up test. Data presented ± standard deviation

Precooking process of rice gives a change effect on the characteristics of a substance such as starch in rice. The results of the present study showed an increase of amylose and RS levels of precooked rice. The heating treatment then continued by cooling turns the starch into an undigested fraction (resistant starch). RS and amylose levels increased after precooked processing, compared with raw rice (results is presented in Table 2). This is in line with the results of previous research reported that process involves heating and cooling modified the starch resistance. Gelatinization process occurs when processing involves heat and can increase solubility and starch digestibility but it can also decrease the content of resistant starch in the material. However, when the heating process followed by re-cooling can lead to the formation of insoluble retrogradation starch [9]. Hydrothermal treatment followed by storage under low temperature (4°C) conditions can increase the formation of resistant starch as a result of recrystallization of amylose and amylopectin structure in rice, more common in amylose [17]. However in the present study, RS level in rice given precooked treatment did not show any significant difference as compared to raw rice. The results of this study are not in line with the other studies aiming to increase RS content such as a 16.8% increase of RS in sukun were reported after boiling process and continued with cooling [18]. The content of the RS that was not significantly different after the precooked process was probably due to a shorter cooling time of the precooking process.

3.3. Calculated Hydrolysis Index (CHI) testing and Estimated Glycaemic Index(EGI) in vitro method
The results of Calculated Hydrolysis Index (CHI) and Estimated Glycaemic Index (EGI) testing between raw and precooked rice is shown in Table 3.

Table 3. Calculated Hydrolisis Index (CHI) and Estimated Glycaemic Index (EGI).

Material	CHI	EGI
White Rice (WR)	48.75 ± 1.54^a	66.47 ± 0.85^a
Precooked White Rice (PWR)	47.03 ± 1.21^a	65.53 ± 0.66^a

Data presented as mean ± standard deviation. Different letters indicates significantly different ($p < 0.05$) between treatments.

Table 3 shows the degree of hydrolysis (calculated hydrolysis index : CHI) and and the estimated glycaemic index: EGI) in white rice raw and after precooking treatment. Test results showed that CHI and EGI decreased after precooked treatment. The EGI in PWR (65.53) was lower as compared to WR (66.47) but not statistically significant. Precooking treatment involved heating and drying causing a decrease in glycaemic index from 69.96 to 51.99 [3]. The precooked process performed on white rice seems to decrease CHI and EGI, but did not differ significantly. The starch retrogradation in rice processing can decrease glycaemic index and starch digestibility [19]. The insignificant changes in CHI and EGI may well associate with RS levels which did not increase significantly either. The retrogradation process may be maximized at 4ºC by increasing storage time up to 24 hours as reported by other studies [19, 20].

4. Conclusion
Precooked treatment of white rice by heating then continued by cooling at 4ºC for 1 hour did not increase RS content and did not decrease estimation of glycaemic index. Future studies may have to increase storing time in low temperature to facilitate intensive retrogradation in white rice to increase RS content and reduce its digestibility.

References
[1] Anonymous 2015 Per Capita Weekly Average Consumption of Several Food Items, 2007-2016 Indonesian Statistical Bureau
[2] Sun Q, Spiegelman D, van Dam R M, Holmes M D, Malik V S, Willett W C, Hu F B 2010 White rice, brown rice, and risk of type 2 diabetes in US men and women *Arch Intern Med* **170** 11 961-969
[3] Widowati S, Santosa B A S, Astawan M, Akhyar 2009 Penurunan indeks glikemik berbagai varietas beras melalui proses pratanak *J. Pascapanen* **6** 1 1-9

[4] Sakurai M, Nakamuraa K, Miurab K, Takamurac T, Yoshitad K, Morikawaa Y, Ishizakie M, Kidof T, Naruseg Y, Suwazonoh Y, Kanekoc S, Sasakii S, Nakagawaa H 2012 Dietary glycaemic index and risk of type 2 diabetes mellitus in middle-aged Japanese men J. *Metabolism: Clinical and Experimental* **61** 1 47-55

[5] Villegas R, Liu S, Gao Y.T, Yang G, Li H, Zheng W 2007 Prospective study of dietary carbohydrates, glycaemic index, glycaemic load, and incidence of type 2 diabetes mellitus in middle-aged Chinese women *Arch Intern Med.* **167** 21 2310-2316

[6] Nanri A, Mizoue T, Noda M, Takahashi Y, Kato M, Inoue M 2010 Rice intake and type 2 diabetes in Japanese men and women: the Japan public health center-based prospective study *Am J Clin Nutr.* **92** 6 1468-1477

[7] Hu E A, Pan A, Malik V, Sun Q 2012 White rice consumption and risk type 2 Diabetes : meta-analysis and systematic review *BMJ* 2012;344:e1454

[8] Mattei J, Hu F B, Campos H 2011 A higher ratio of beans to white rice is associated with lower cardiometabolic risk factors in Costa Rican adults *J. Clinical Nutrition* **94** 869-876

[9] Zavareze E D R, Halal S L M E, de los Santos D G, Helbig E, Pereira J M, Diaz A R G 2012 Resistant starch and thermal, morphological and textural properties of heat moisture treated rice starches with high, medium, and low amylose content *Starch: Biosynthesis Nutririon Biomedical* **64** 1 45-54

[10] Shi M, Chen Y, Yu S, Gao Q 2013 Preparation and properties of RS III from waxy maize starch with pullulanase *J Food Hydrocolloids* **33** 19-25

[11] AOAC 1995 Official of Analysis of The Association of Official Analytical Chemistry AOAC Inc. Airlington

[12] Sudarmadji S, Bambang H, Suhardi 1996 Analisa Bahan Makandan Pertanian Liberty Yogyakarta.

[13] Asp N G, Johansson C G, Hallmer H, Siljestrom M 1983 Rapid enzymatic assay of insoluble and soluble dietary fiber *Journal Agric. Food Chemistry* **31** 3 476-482.

[14] Goni, I., Garcia-Diz, L., Manas, E., Saura-Calixto, F 1996 Analysis of resistant starch: a method for foods and food products *J Food Chemistry* **56** 4 445-449

[15] Frei M, Siddhuraju P, Becker K 2003 Studies on in vitro starch digestibility and the glycaemic index of six different *indigenous* rice cultivars from the Philippines *J Food Chemistry* **83** 395–402

[16] Purwaningsih H, Kristamtini, Widyanti S 2008 Mutu Fisik, Kimia dan Organoleptik Padi Beras Merah Varietas Lokal (Cempo Merah, Mandel dan Segreng) sebagai Plasma Nutfah Padi Provinsi DIY Makalah Seminar Pekan Padi Nasional (PPN) III Sukamandi

[17] Kim W K, Chung M K, Kang N E, Kim M H, Park O J 2003 Effect of resistant starch from corn or rice on glucose control, colonic events, and blood lipid concentrations in streptozotocin-induced diabetic rats *Journal of Nutritional Biochemistry* **14** 3 166–172

[18] Rosida and Ratna Yulistiani 2013 Pengaruh Proses Pengolahan terhadap Kadar Pati Resisten Sukun (*Artocarpus altilis* Park) Research Articel Department Food Technology UPN Veteran Jawa Timur

[19] Ae Wha H, Gwi J H, Woo K K 2012 Effect of Retrogradated Rice on Weigh Control, Gut Function, and Lipid Concentration in Rats *Nutrition Research and Practice* **6** 1 16-20

[20] Tian Y, Jinling Z, Jianwei Z, Zhengjun X, Xueing X, Zhengyu J 2013 Preparation of product rich in slowly digestible starch (SDS) from rice starch by a dual-retrogradation treatment *Food Hydrocolloid* **31** 1-4

IOP Conf. Series: Earth and Environmental Science **131** (2018) 012018 doi:10.1088/1755-1315/131/1/012018

Physical and sensory quality of Java Arabica green coffee beans

W B Sunarharum, S S Yuwono, N B S W Pangestu and H Nadhiroh

Department of Agricultural Product Technology, Faculty of Agricultural Technology, Universitas Brawijaya Malang, Indonesia

E-mail: wbsunarharum@ub.ac.id

Abstract. Demand on high quality coffee for consumption is continually increasing not only in the consuming countries (importers) but also in the producing countries (exporters). Coffee quality could be affected by several factors from farm to cup including the post-harvest processing methods. This research aimed to investigate the influence of different post-harvest processing methods on physical and sensory quality of Java Arabica green coffee beans. The two factors being evaluated were three different post-harvest processing methods to produce green coffee beans (natural/dry, semi-washed and fully-washed processing) under sun drying. Physical quality evaluation was based on The Indonesian National Standard (SNI 01-2907-2008) while sensory quality was evaluated by five expert judges. The result shows that less defects observed in wet processed coffee as compared to the dry processing. The mechanical drying was also proven to yield a higher quality green coffee beans and minimise losses.

1. Introduction

Coffee is one of the most popular beverage worldwide and become the most important traded commodity after petroleum. It's significant impact on the world economy and people welfare has been well-recognized. World Coffee Outlook [1] reported that exporting and re-exporting activities have been growing consistently in the last five decades. Considerable growth has been observed after the year 2000 [1].

Demand on high quality coffee for consumption is continually increasing not only in the consuming countries (importers) but also in the producing countries (exporters). This high quality of coffee could be affected by several factors from farm to cup including the post-harvest processing methods as comprehensively reviewed previously by [2]. Amongst those factors, post-harvest processing method of coffee cherry into green coffee beans was thought to have an important role on the quality.

The impact of this post-harvest processing on coffee quality may vary due to other factors such as environmental involvement that may be specific on certain coffee. This may lead to the choice made by farmers on which processing might be best for the quality of their coffee. This research aimed to investigate the influence of different post-harvest processing methods (natural/dry, semi-washed and fully-washed processing) on physical and sensory quality of Java Arabica green coffee beans cultivated on Arjuno Mountain, UB Forest, Malang, East Java. Expected outcome of this research will be recommendation on best green coffee beans post-harvest processing methods applicable for farmers on Arjuno Mountain.

2. Materials and Methods

2.1. Materials

Coffee cherries (10 kg) used in this research was Java Arabica coffee cultivated on Arjuno Mountain, UB Forest Malang, East Java. Machineries used were locally made such as pulper (75 kg/h), huller (50 kg/h), and electric cabinet dryer (50 kg/h). Other equipment involved includes Sartorius balance, CR 310 photometer (Konica Minolta Imaging, Dietikon, Switzerland), vacuum oven, fermentation buckets with lid.

2.2. Methods

The two factors being evaluated were three different post-harvest processing methods to produce green coffee beans (natural/dry, semi-washed and fully-washed processing) and two drying methods i.e. sun drying and mechanical drying. Sun drying process involves drying on the bamboo racks for 5 days up do 3 weeks while mechanical drying utilized electric cabinet dryer set at ± 40°C for few days up to 2 weeks.

Prior to drying, the harvested red cherries were sorted (floated under water) and follows series of processing based on the methods of choice to produce dried coffee or coffee parchment. For the natural method, whole cherries were dried under sunlight. The other two methods (semi-washed and fully-washed) involved pulping and washing of the red cherries. For semi-washed process, the coffees were dried soon after pulping and washing without fermentation process. For the fully-washed process, the coffees were fermented under water in a closed bucket for ±18 h followed by further washing prior to drying. The coffee husk, parchment, and skin were removed by using dry huller to yield green coffee beans that were analysed further.

Physical quality evaluation on green coffee beans including coffee size and defects was performed based on [3]. Moisture content was analyzed based on [4] method 979.12 gravimetric method II using vacuum oven. Colour (CIE L*a*b*) was measured using CR 310 photometer (Konica Minolta Imaging, Dietikon, Switzerland). Sensory quality was evaluated using descriptive analysis by five expert judges from local coffee industries. Statistical data analysis was performed using Minitab 17 Statistical Software (Minitab Inc., State College, Pennsylvania, USA) and The Unscrambler® X (CAMO, Oslo, Norway, AS).

3. Results and Discussion

Physical evaluation is a fast, easy and applicable technique to assess coffee quality. This evaluation may involves colour, bean size, uniformity, and presence of defects.

As the first important physical criteria to assess green coffee beans quality, colour could be determined objectively using colorimetry based on CIE L*a*b* colour space, where L* measures the lightness, while a* and b* presents coordinates for red/green and yellow/blue, respectively. The chromatic coordinates of the green coffee beans being studied can be seen in Table 1.

Table 1 showed that post-harvest processing factors were found to be significant (α=0.05) on lightness (L*) and yellowness (b*) values of green coffee samples. Besides, there is also a significant difference (α=0.05) between mechanically dried coffee samples and sundried samples on L*, where the mechanical drying produced lighter coffee than its counterpart. However, there is no significant difference on redness (a*) and yellowness (b*) amongst these two drying methods.

Physical assessment results showed general compliance of all samples being studied with quality requirement as stated in [4] such as no living insects, no spoiled aroma, maximum moisture content at

12.5 %, and maximum foreign materials at 0.5 %. The green coffee beans majority were in medium size (± 60 %) while the rest were in small size.

Table 1. Colour of green coffee beans processed under different post-harvest processing and drying methods (n=3)

Post-harvest processing	Drying	L*	a*	b*
Natural	Sun Drying	48.83 ± 3.01	-0.63 ± 1.47	12.80 ± 1.08
Semi-washed	Sun Drying	40.57 ± 4.42	0.37 ± 0.58	7.27 ± 1.60
Fully-washed	Sun Drying	42.97 ± 2.02	-0.83 ± 0.47	10.57 ± 1.06
Natural	Mechanical Drying	49.70 ± 1.20	-0.50 ± 0.80	13.10 ± 0.80
Semi-washed	Mechanical Drying	48.03 ± 1.05	-0.80 ± 0.52	11.87 ± 0.67
Fully-washed	Mechanical Drying	48.27 ± 1.43	-0.60 ± 0.56	11.53 ± 1.03
	n	3	3	3

Data mean ± standard deviation

Further investigation were performed based on this standard, particularly on coffee defects. Figure 1 presented the defective beans from all processing factors. Figure 2 showed the interval plot of total defect value based on processings. The pooled standard deviation was used to calculate this intervals. As can be inferred from this Figure 2, green coffee beans dried under the sun were found to have more defects than the mechanically dried samples where the naturally sundried coffee produced more defective beans.

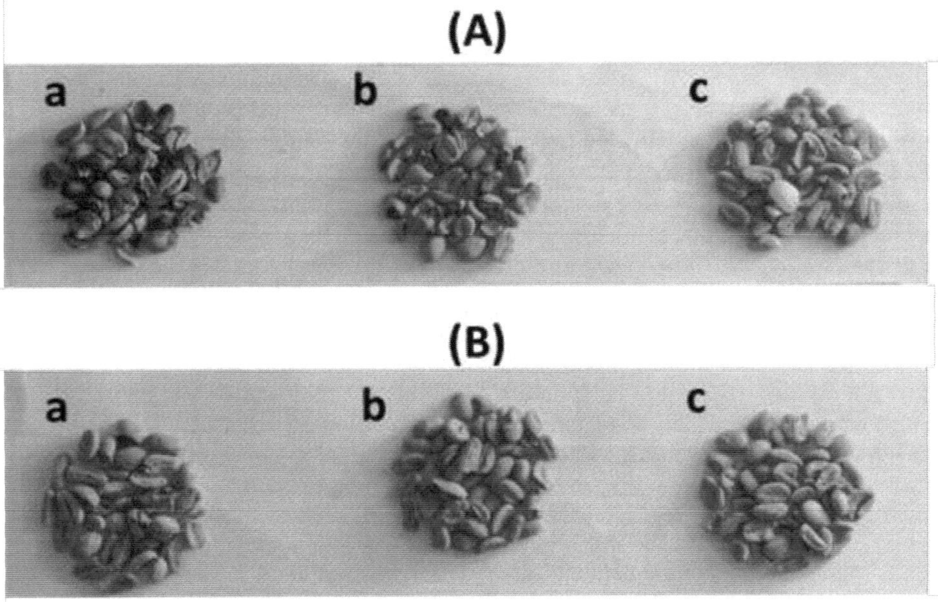

Figure 1. The defective beans from three post-harvest processing method ([a]natural, [b]semi-washed, and [c]fully-washed) dried under sundrying method (A) and machine drying method (B)

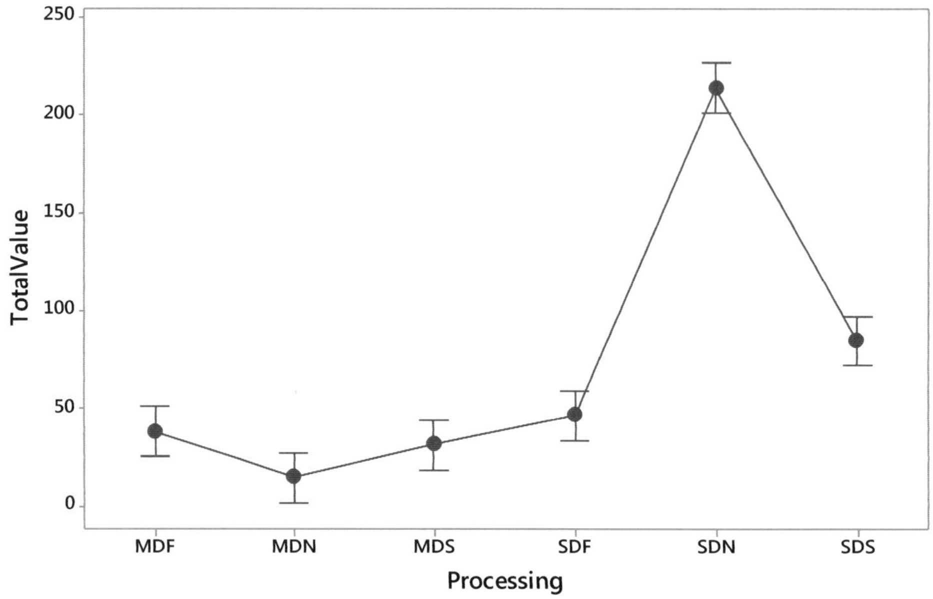

Notes: SDN (Sun drying natural), SDS (Sun drying semi-washed), SDF (Sun drying fully-washed), MDN (Machine drying natural), MDS (Machine drying semi-washed), MDF (Machine drying fully-washed)

Figure 2. Interval plot of coffee total defect value vs 6 processing 95% CI for the mean (n=2)

The data was further explored using PCA to see potential clustering in the dataset as well as relationships between variables (Figure 3). PCA was constructed on data of 20 defect variables measured on coffee processed under three different post-harvest processing and two drying methods (n=2). The PCA bi-plot revealed that defects on the naturally sundried samples were driven by some parameters such as black, partly black, brown, the presence of husk (large, medium, small size), dried cherry, and the presence of holes (one or more holes) on PC 1 (47 %). Semi- and fully-washed samples were clustered mainly due to particular defects on PC 2 (19 %) including spotty beans, broken beans, and beans with coffee parchments (large, medium, small size). Mechanical drying yield less defective beans as compared to the sun drying method. However, it should be noted that this is an initial evaluation on green coffee beans soon after processing and no prior sortation performed. This is beneficial for farmers or producers to check yield and potential losses under different processing method. Green coffee beans for commercial purposes will surely undergo series of sortation to achieve specific quality requirements.

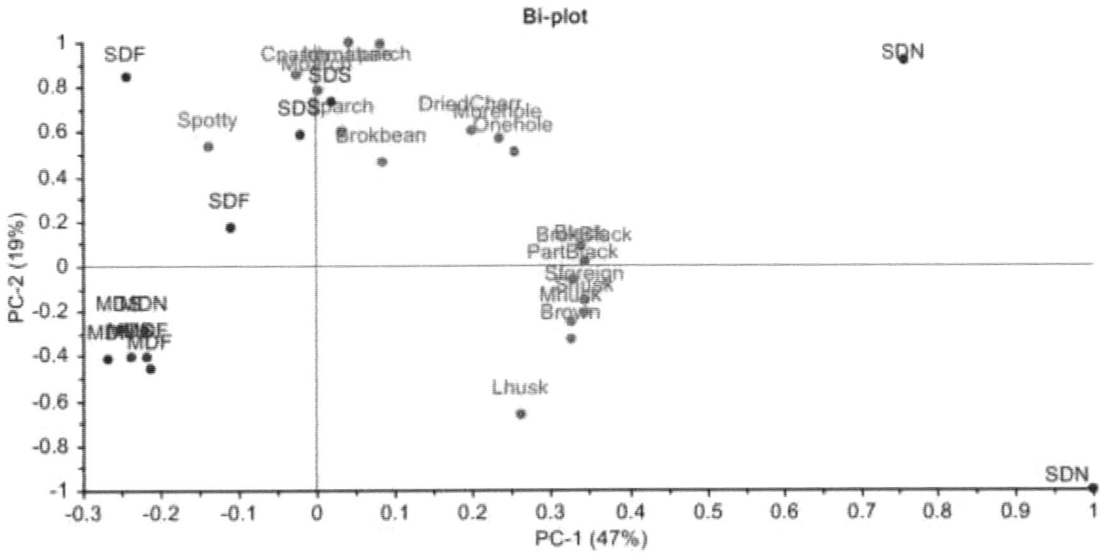

Notes: Black (black bean), PartBlack (partly black bean), BrokBlack (broken black bean), DriedCherry (dried cherry), Brown (brown bean), Lhusk (large size coffee husk), Mhusk (medium size coffee husk), Shusk (small size coffee husk), Cparch (coffee parchment), Lparch (large size coffee parchment), Mparch (medium size coffee parchment), Sparch (small size coffee parchment), Brokbean (broken bean), Immature (immature bean), Onehole (bean with one hole), Morehole (bean with more than one hole), Spotty (spotty bean), Lforeign (large foreign material), Mforeign (medium foreign material), Sforeign (small foreign material). SDN (Sun drying natural), SDS (Sun drying semi-washed), SDF (Sun drying fully-washed), MDN (Machine drying natural), MDS (Machine drying semi-washed), MDF (Machine drying fully-washed)

Figure 3. PCA Bi-plot of 6 coffee samples (n=2) based on 20 variables, PC 1 (47%) vs PC 2 (19%)

Sensory quality of the samples were further explored. Five expert judges were employed to develop and to assess aroma attributes of the green coffee beans. Table 2 presented the definition of the aroma attributes developed during the evaluation process while the results of sensory evaluation can be seen in Figure 4.

Table 2. Description of aroma attributes.

Aroma attributes	Definition
Grassy	Aroma of green grass, wet grass, menthol
Nutty	Aroma of dry peanut and other raw nuts, pea
Earthy	Aroma of earth, wet soil
Fermented	Aroma of overripe fruits, wine
Mouldy	Aroma of mould, wet mushroom, wet wood

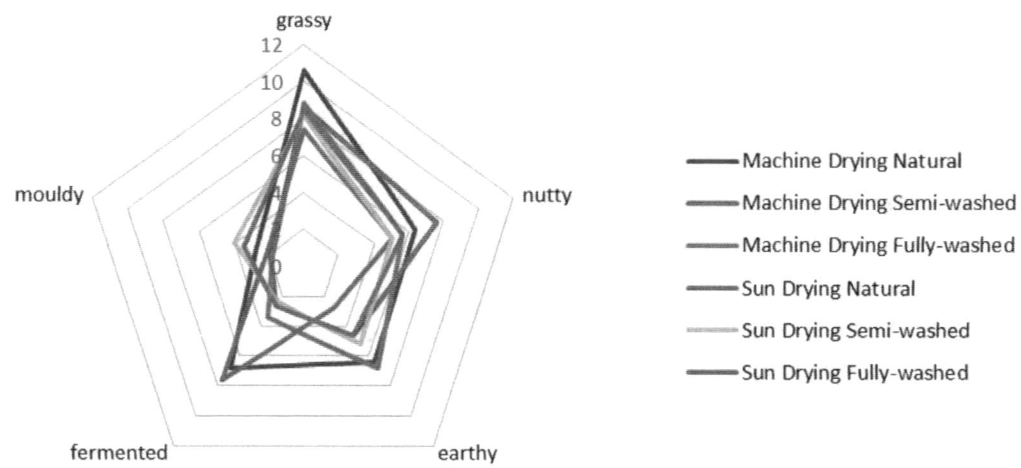

Figure 4. Mean score of aroma of 6 green coffee beans assessed by 5 judges

Figure 4 shows an interesting results where mechanically dried samples were found to be more nutty, earthy and grassy as compared to the sundried samples. Sun drying results in more mouldy character probably due to longer drying time. The naturally processed green coffee beans were found to have the highest fermented and grassy characters than the other two processing. Washing and fermentation produced green coffee with nutty and earthy notes as the most dominant aroma. The aroma of green coffee beans might be one of an important consideration when producing certain products such as green coffee powder. Further common processing of coffee including roasting, grinding, and brewing also contributed to the conversion of coffee aroma. However, it is outside the scope of the current research.

4. Conclusion

In conclusion, post-harvest processing factors were found to influence the physical and sensory quality of green coffee beans. The mechanical drying was proven to yield a higher quality green coffee beans and minimize losses due to coffee defects. The quality of sundried green coffee products is acceptable for commercial purpose but more effort are required such as sortation process. The introduction of water and fermentation process under semi-washed and fully-washed method produced coffee with different physical and sensory characteristics. However, since all of different coffee processing will find its marketplace, the choice of processing should depend on some other factors such as environmental factor, product safety, production capability and capital investment.

Acknowledgments

We acknowledge the Faculty of Agricultural Technology, Universitas Brawijaya, Malang, Indonesia and UB Forest for their ongoing support of this research.

References

[1] ICO 2015 Trade Statistics. Retrieved from http://www.ico.org/trade_statistics.asp. [Accessed 4 November 2015]
[2] Sunarharum W B, Williams D J, Smyth H E 2014 Complexity of coffee flavor: a compositional and sensory perspective *Food Res. Intl.* **62** 315-325
[3] BSN 2008 The Indonesian National Standard on coffee beans (SNI 01-2907-2008)
[4] AOAC 2007 Official methods of analysis 18th ed, 2005 Current through revision 2, 2007 (on line) AOAC International, Gaithersburg, MD

IOP Conf. Series: Earth and Environmental Science **131** (2018) 012019 doi:10.1088/1755-1315/131/1/012019

Shelf life prediction of apple brownies using accelerated method

M H Pulungan, A D Sukmana and I A Dewi

[1]Department of Agro-industrial Technology, Faculty of Agricultural Technology Universitas Brawijaya, Malang, Indonesia

E-mail: hindunmaimunah@yahoo.com

Abstract. The aim of this research was to determine shelf life of apple brownies. Shelf life was determined with Accelerated Shelf Life Testing method and Arrhenius equation. Experiment was conducted at 25, 35, and 45°C for 30 days. Every five days, the sample was analysed for free fatty acid (FFA), water activity (A_w), and organoleptic acceptance (flavour, aroma, and texture). The shelf life of the apple brownies based on FFA were 110, 54, and 28 days at temperature of 25, 35, and 45°C, respectively.

1. Introduction

Brownies are foods from Western countries and now very popular in Indonesia which is actually a kind of pastry. There is one of SMEs (Small and Medium Enterprise) that produce apple brownies in the city of Batu East Java Indonesia. The problem in this SMEs is the industry received many returned brownies products from sales outlets because it has been damaged before expired date. The shelf life is very important as expiring date has to be included on the packaging label. Therefore a re-analysis of the appropriate apple brownies shelf life is required.

Apple Brownies produced by SMEs in the city of Batu contains fat and has a dry texture. According to [1], rancidity is the main problem encountered in foods containing high fat. In addition, degradation mechanism of dry food quality is started by the absorption of water vapour in products [2]. The increasing water content in food lead to deterioration reactions that are mediated by the A_w product [3]. As a result, dried food such as apple brownies becomes moist, loses vitamins and loses aroma causing damage and is no longer acceptable by the consumers [4].

One method used in determining the shelf life of a product is Accelerated Shelf Life Testing (ASLT). The food product damage will be accelerated by placing products in unnormal conditions for establishing the rate of degradation. In this method the Arrhenius approach is used to know the relationship between storage time and temperature, where this linear equation is considered most appropriate for use at high storage temperatures.

2. Method

2.1 Determination of Initial Characteristics of the apple brownies

Preparation of dried apple brownies to be used in the research. apple brownies are needed as many as 18 boxes, each box containing 8 pieces of apple brownies.

IOP Conf. Series: Earth and Environmental Science **131** (2018) 012019 doi:10.1088/1755-1315/131/1/012019

- Preparation and storage of dried apple brownies in 3 storage boxes with temperature 25°C (298K), 35°C (308K), and 45°C (318K).
- Before storage, FFA, A_w and organoleptic test of apple brownies as initial data were performed. Organoleptic tests performed are aroma, taste, and texture using hedonic scale test model.

2.2 Quality Decrease Rate Parameter Selection and Shelf Life Determination of Apple Brownies

- During storage, observations are performed every 5 days for 30 days. Observations were made by taking samples on different packs for each observation and testing the parameters affecting the quality of apple brownies such as; FFA, A_w and organoleptic (flavour, taste and texture).
- The data obtained of FFA and A_w from apple brownies were plotted against time and three equations for different product storage temperatures Y = bx + a will obtained , where x is the storage time (day), Y represents the characteristic value of apple brownies, a is the initial characteristic value of brownies apple, and b is the rate of characteristic change (slope).
- From every linear regression equation is obtained the value of quality decrease rate (*k*).
- Then the ln *k* value is plotted with 1/T (K^{-1}), and the intercept and slope value of the linear regression equation $\ln k = \ln k_0 - (E / R)(1 / T)$ was obtained.
- After the activation energy characteristic of the apple brownies and the constant value k_0 is obtained, the Arrhenius equation is calculated by the formula:

$$k = k_o . e^{-E/RT} \quad \text{...} \quad (1)$$

Where:

k = quality degradation constant
k_o = constant (independent of temperature)
E = activation energy
T = absolute temperature (K)
R = gas constant (1.986 cal / mol K)

With using Arrhenius equation, the rate of reaction (*k*) from the changes characteristic of apple brownies at a predetermined temperature (T) can be calculated. Shelf life determination of apple brownies is done by choosing parameters that have the highest correlation coefficient (R^2). Furthermore, the value of *k* obtained is entered into the order equation of the reaction:

$$At = Ao + kt \quad \text{...} \quad (2)$$

Where:

Ao = A value at the beginning of shelf life
At = A value at end of shelf life
t = shelf life
k = quality degradation constant

Shelf life of apple brownies for each specified temperature will be obtained.

3. Result and Discussion

3.1. Kinetics of Quality Decrease Rate of Apple Brownies Against Free Fatty Acid (FFA)

Free Fatty Acid (FFA) is one of the parameters used to determine the rate of decline in the quality of apple brownies. FFA shows the amount of free fatty acids in a product. Fig. 1 shows that the FFA value of apple brownies increases with the length of storage time at all three storage temperatures. According to [5], an increase in FFA percentage due to oxidation of the product. The oxidation reaction of apple brownies is caused by the contact of oxygen with fat containing in the apple brownies. Fat is derived from margarine, eggs, and chocolate that being composition of apple brownies.

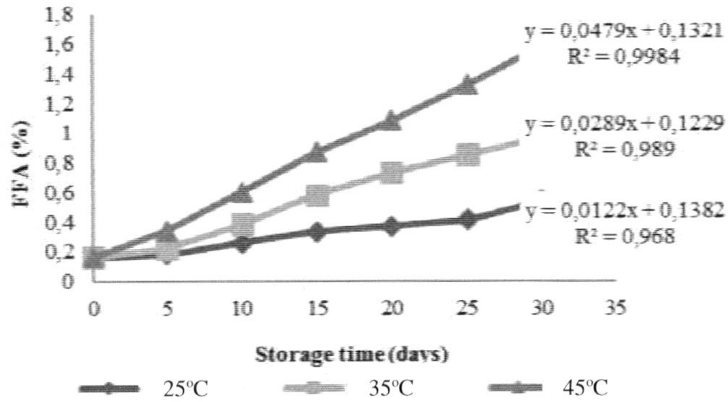

Figure 1. FFA value of apple brownies at three storage temperature conditions

The choice of the reaction order kinetics from decreasing moisture content is done by comparing the value of R^2 in each linear regression equation at the same temperature on order reaction of zero and first order. The chosen reaction order is a reaction order with a larger R^2 value. Table 1 shows that the decreasing rate of apple brownies on FFA based on zero order reaction. This is shown from the value R^2 zero order > R^2 first order at 35°C and 45°C.

Table 1. Linear regression equation for FFA value parameter of apple brownies.

Storage (°C)	Linear Regression Equation		R^2	
	Zero Order	First Order	Zero Order	First Order
25	y = 0.0122x + 0.1382	y = 0.0403x – 1.8202	0.9680	0.9723
35	y = 0.0289x + 0.1229	y = 0.0626x – 1.7055	0.9890	0.9470
45	y = 0.0479x + 0.1321	y = 0.0726x – 1.4825	0.9984	0.9202

3.2. Kinetics of Quality Decreasing of Apple Brownies Against A_w

The A_w of apple brownies is directly proportional to the water content analysis . It because the relationship between water content and A_w is indicated by the tendency that higher water content the higher A_w. Fig. 2 shows that there is a decrease in the value of apple brownies with the length of storage time in three storage temperature conditions. This is appropriate with A_w analysis conducted by [6], against dried apple snack, which also decreased with length of storage time. According to [7], the value of water content is directly proportional to the value of A_w.

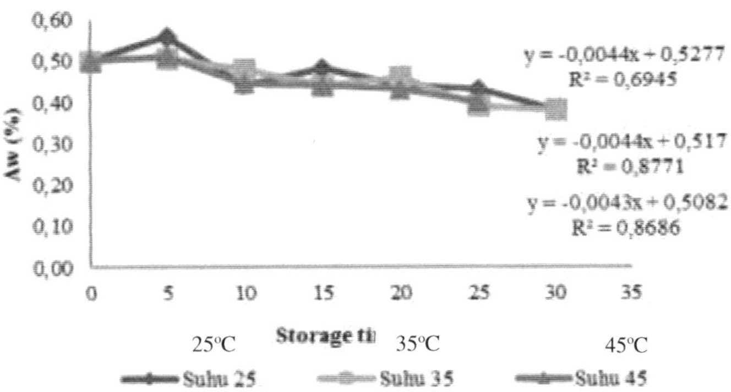

Figure 2. A_w value of the apple brownies at three storage temperature conditions

The determination of the reaction order kinetics from decreasing moisture content is done by comparing the value of R^2 in each linear regression equation at the same temperature on order reaction of zero and first order. The chosen reaction order is a reaction order with a larger R^2 value. Table 2 shows that the A_w decrease rate of apple brownies based first order reaction. It is shown from the value of R^2 first order > R^2 of zero order at temperature 25°C and 45°C.

Table 2. Linear regression equations for the value parameter of the apple brownie A_w.

Storage (°C)	Linear Regression Equation		R^2	
	Zero Order	First Order	Zero Order	First Order
25	y = -0.0044x + 0.5277	y = -0.0096x – 0.6352	0.6945	0.7163
35	y = -0.0044x + 0.517	y = -0.0100x – 0.6524	0.8771	0.8689
45	y = -0.0043x + 0.5082	y = -0.0114x – 0.6578	0.8686	0.9024

3.3. *Shelf Life Determination of Apple Brownies*

Shelf life determination of apple brownies was done based on the parameter of quality which experienced the fastest decline (the highest correlation coefficient (R^2) value). Table 3 shows that the largest parameter of R^2 is the FFA, with the value R^2 = 0.9830 and the activation energy of 12903.836 cal/mol. Therefore, the main parameters in determining the shelf life of apple brownies are measured by FFA value.

Table 3. The Arrhenius equation: correlation coefficient value (R^2) of the apple brownies degradation parameter.

No	Parameter	*Arrhenius* equation	R^2
1	FFA	$\ln k = -6496.4\ (1/T) + 17.444$	0.9830
2	A_w	$\ln k = -809.35\ (1/T) - 1.9455$	0.9052

Table 4. The results of calculating the shelf life of apple brownies store at various temperatures with FFA value parameters.

Storage (°C)	*k* value	Shelf Life	
		Days	Month
25	0.012830	110.68	3.20
35	0.026038	54.54	1.24
45	0.054300	28.09	0.94

Table 4 shows that the shelf life of apple brownies stored at 25°C, 35°C and 45°C are 3 months 20 days, 1 month 24 days, and 28 days, respectively. Therefore, the best way to store apple brownies is at 25°C.

In addition, it is better if the shelf life or expiration date that listed by SMEs on apple brownies packaging sold in the city of Batu is revised to be 3.20 months because the average of temperature in that city is 25°C. The apple brownies on the outlet should be placed away from sunlight to prevent the damage of apple brownies due to light and solar heat. On the other hand, if the producer want to sell apple brownies in areas or cities that have higher temperatures, then the length of shelf life may be modified according to the average temperature in the area. This can help the SMEs expanding market share of apple brownies.

3.4. *Quality Decrease of Apples Brownies Organoleptically*

The overall result of apple brownies organolepticc test showed that the panelist preferences for taste, aroma, and texture of apple brownies had decreased with the length of storage time at all three temperatures. This is related to the increasing FFA and the decreasing A_w value at storage time. According to [8], decreasing the value of A_w on food affect the process of fat oxidation. Therefore, increased free fatty acids in apple brownies be fathomed make the taste will more not good. This is supported by [5] that the presence of free fatty acids can affect the taste and smell of a product. In addition, it is suspected that the decreasing A_w resulting the harder texture of apple brownies. This is also supported by [7] stated that at high temperature, the water in the product tends to evaporate resulting the harder texture of the product.

Therefore, other alternatives are needed to prevent the decreasing quality of apple brownies and make the product has a longer shelf life. One solution is to add oxygen absorber in the form of a sachet in the apple brownies packaging. Oxygen absorber is generally used to absorb oxygen in food so that the oxidation process can be taken slowly and the shelf life of a product can be extended.

4. Conclussion

The shelf life of apple brownies analysed using ASLT at 25, 35 and 45°C were 110 days, 54 days, and 28 days, respectively.

References

[1] Sutanto T 2013 Perbandingan Mutu Minyak Kelapa Yang Di Proses Melalui Pengemasan Dan Pemanasan Sesuai SNI 2902-2011 (Comparison of Coconut Oil Quality In Process Through Packaging And Warming According to SNI 2902-2011) *Jurnal Hasil Penelitian Industri.* **26** 1 1-10

[2] Floros J D dan V Gnanasekharan 1993 Shelf Life Prediction of Packaged Foods: Chemichal, Biological, Physical, and Nutritional Aspects Elsevier Publisher London

[3] Hutasoit N 2009 Penentuan Umur Simpan *Fish Snack* (Produk Ekstruksi) Menggunakan Metode Akselerasi Dengan Pendekatan Kadar Air Kritis dan Metode Konvensional (Determination of Fish Saving Fish Snack (Product of Extrusion) Using Acceleration Method With Critical Approach Water Content and Conventional Method) Skripsi IPB. Bogor

[4] Robertson G L 1999 Shelf Life of Packaged Foods, Its Measurement and Predictioning Developing New Food Products For A Changing Marketplace CRC Press Florida

[5] Jariyah, Rosida, Rambu H L 2015 Efek Sorbitol Terhadap Tekstur dan Daya Simpan Produk Manggul (Sorbitol Effect Against Texture and Shelf Life of Manggul Products) *Jurnal Rekapangan* **9** 2

[6] Saavedra J, Andres C, Lena G, Cesar Q dan Rosa N 2013 Principal Component Analysis of Food Quality: A Case Study of A Dried Apple Cluster Snack *Journal of Food Engineering.* 119: 229-235

[7] Sarastuti M dan Sudarminto S Y 2015 Pengaruh Pengovenan dan Pemanasan Terhadap Sifat-Sifat Bumbu Rujak Cingur Instan Selama Penyimpanan *Jurnal Pangan dan Agroindustri.* **3** 2 464-475

[8] Winarno F G 2000 Kimia Pangan dan Gizi (Food and Nutrition Chemicals) Gramedia Jakarta

IOP Conf. Series: Earth and Environmental Science **131** (2018) 012020 doi:10.1088/1755-1315/131/1/012020

Fish swarm intelligent to optimize real time monitoring of chips drying using machine vision

Y Hendrawan, L C Hawa and R Damayanti

Department of Agricultural Engineering, Faculty of Agricultural Technology, Universitas Brawijaya, Malang, Indonesia

E-mail: yusufhendrawan@gmail.com

Abstract. This study attempted to apply machine vision-based chips drying monitoring system which is able to optimise the drying process of cassava chips. The objective of this study is to propose fish swarm intelligent (FSI) optimization algorithms to find the most significant set of image features suitable for predicting water content of cassava chips during drying process using artificial neural network model (ANN). Feature selection entails choosing the feature subset that maximizes the prediction accuracy of ANN. Multi-Objective Optimization (MOO) was used in this study which consisted of prediction accuracy maximization and feature-subset size minimization. The results showed that the best feature subset i.e. grey mean, $L_{(Lab)}$ Mean, $a_{(Lab)}$ energy, red entropy, hue contrast, and grey homogeneity. The best feature subset has been tested successfully in ANN model to describe the relationship between image features and water content of cassava chips during drying process with R^2 of real and predicted data was equal to 0.9.

1. Introduction

Drying is a vital postharvest process for most agricultural products [1]. The most important processes in flour processing Modified Cassava Flour (MOCAF) is upon drying. Artificial drying is suggested to use due to many advantages such as drying temperature and air flow can be set so that drying process is more quickly and evenly and hygiene materials are also more awake. Aghbashlo et al. [2] uses machine vision method for real time monitoring of food quality during the drying process and found that this method is suitable for observing the changes in the mechanical, sensory and nutritional food. With the existence of this monitoring system, the better characteristic parameters of process at the time of drying and quality evaluation can be achieved. Dutta et al. [3] presented a computer vision based non-destructive method to identify the presence of acrylamide in potato chips with the results have shown accuracy over than 94%. Baigvand et al. [4] observed the used of machine vision system for grading of dried figs and achieved accuracy up to 95.2%. Benalia et al. [5] developed automated systems based on computer vision to improve the quality control and sorting of dried figs. Using models of CIE XYZ, CIELAB, and Hunter Lab colour spaces, in term of the browning index measurement of each fruit by accuracy of 99.5%. Yadollahinia and Jahangiri [6] developed a thin-layer dryer with machine vision system using image processing software for potato slice and shown that the system successfully effective to monitor the shrinkage of potato slice during drying.

In agricultural and biological engineering, researchers and engineers have developed methods such as artificial neural network (ANN) to analyse the operation of food processing [7]. Khazaei et al. [8] presented a new method for predictive modelling of grape drying process for on-line monitoring and controlling of this process. In the area of food quality control, ANN has been successfully applied to predict the quality of agricultural raw material [9-11]. Azadbakht et al. [12] investigated the application of ANN in intelligent drying process to energy analysis of fluidized bed dryer for potato cubes. A large share of energy utilization in the food industry.
Feature selection entails choosing the feature subset that maximizes the prediction accuracy of ANN. Features which do not appear relevant singly may become highly relevant when taken with others [13]. Many studies have proved the benefit of using nature inspired algorithms for feature selection techniques [14-18]. Hendrawan and Al Riza [19] has successfully implemented fish swarm intelligent (FSI) to select relevant features as the input of ANN to predict water content of a plant. The objectives of this study were modelling image features to predict the water content of cassava chips during drying by using ANN and optimizing the feature selection technique to select relevant image features subset by using FSI.

2. Materials and Methods

2.1. Materials and equipment

Five hundred samples of cassava chips in various water contents were used. As a mean of manipulating their physiological status, the samples were subjected to different water states. Water states was determined by drying the sample at 70°C in tray dryer. Water status was defined as the average amount of water available for each sample of data acquisition in percentage of its initial dry weight [20]. The drying experiments were performed in a tray dryer as shown in Figure 1, which can be controlled for drying in the temperature range of 50-70°C. Prior to starting of each experiment cassavas were washed, peeled, and sliced into chips of 1 mm thickness with a mechanical slicing machine.

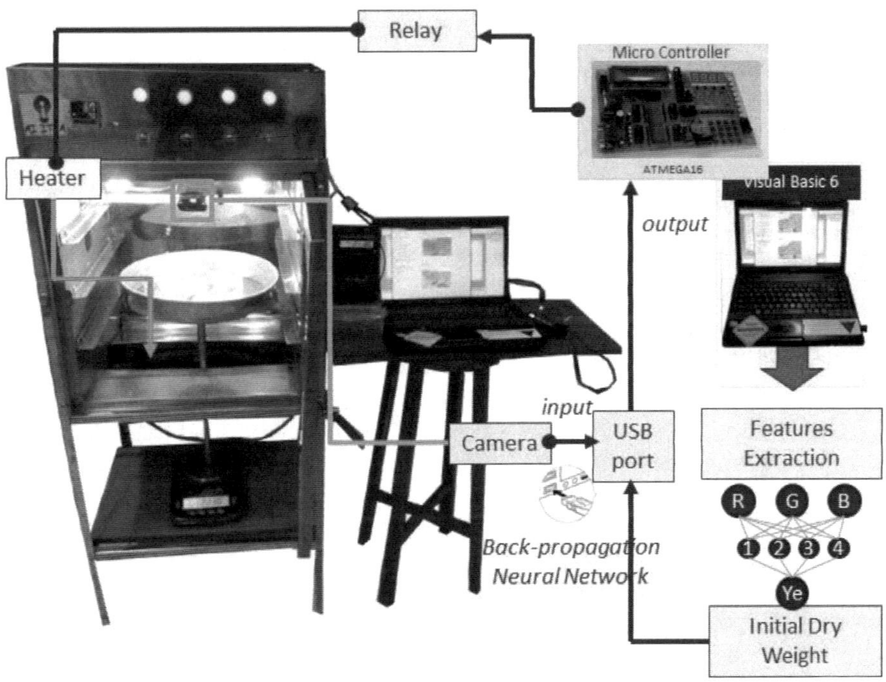

Figure 1. Modified tray dryer using computer vision

2.2. Model of study

First process was image acquisition, in which the cassava chip images were captured using digital camera (Logitec HD Webcam C270, Japan) placed at 300 mm perpendicular to the sample surface and connected to the USB port of a computer with Intel core i7 processor. The digital camera was used for image acquisition which provides images in BMP format. Images were captured with its maximum resolution (1280 x 720). Imaging was done under controlled and well distributed light conditions. Light was provided by two 22W lamps (EFD25N/22, National Corporation, Japan). Light intensity over the cassava chip surface was uniform at 300 lux in the centre of the region during image acquisition. During drying image of the cassava chips were captured continuously by computer vision system. After drying had finished, total of 500 image data which varies at different water contents were acquired. Image analysis was performed according to a software specially developed for this purpose using Visual Basic 6.0. Image features which consist of colour features (CFs) and textural features (TFs) were extracted from each image data. Modelling was done using back propagation neural network (BPNN) to describe the relationship between image features (CFs and TFs) and water content of cassava chip. Selection process for selecting relevant image features was done using nature-inspired approaches *i.e.* FSI. Multi objectives optimization (MOO) concerns optimization problems with multiple objectives [21]. The fitness was calculated using equation 1, 2, and 3 where $MSE_{(x)}$ is the mean square error of validation-set data of BPNN using only the expression values of the selected image features in a subset x, where $IF_{(x)}$ is the number of selected image features in x. f_t is the total number of image features, $weight_1$ and $weight_2$ are two priority weights corresponding to the importance of the accuracy and the number of selected image features, respectively, where $weight_1$ = 0.6 and $weight_2$ = 0.4. In this study, the accuracy was more important than the number of selected image features in a feature-subset.

$$function_1 = weight_1 \times MSE_{(x)} \quad (1); \quad function_2 = weight_2 \times \frac{IF_{(x)}}{f_t} \quad (2); \quad fitness(x) = function_1 + function_2 \quad (3)$$

2.3. Colour features (CFs)

CFs include colour mean value which can be described according to equation 4 [22] where: colour value can be defined as the range of each colour space in the pixel *i.e.* red, green, blue, grey, hue, $saturation_{(HSL)}$, $saturation_{(HSV)}$, $lightness_{(HSL)}$, $value_{(HSV)}$, $X_{(XYZ)}$, $Y_{(XYZ)}$, $Z_{(XYZ)}$, L, $a_{(Lab)}$, $b_{(Lab)}$, $C_{(LCH)}$, $H_{(LCH)}$, $u_{(Luv)}$ and $v_{(Luv)}$. M is the total number of pixels in the image. The total number of CFs are 19 features.

$$colour\ mean\ value = \frac{1}{M}\sum_{i=1}^{M} colour\ value \quad (4)$$

2.4. Textural features (TFs)

The textural analysis can be considered as one of applicable techniques for extracting image features [23]. The Colour Co-occurrence Matrix (CCM) procedure consists of three primary mathematical processes: (1) the image is transformed from RGB colour representation to other colour representation such as grey [24], HSL and HSV [25], Lab and XYZ [26], LCH [27] and Luv [28]; (2) generation of Spatial Gray-Level Dependence Matrices (SGDMs) [29], resulting in one CCM for each colour space, the CCM was calculated based on normalization value; and (3) determination of ten Haralick Textural Features [17]. Based on the results of preliminary observation in various combination of angle ($\theta = 0$, $\theta = 45$, $\theta = 90$, $\theta = 135$) and distance ($d = 1$, $d = 2$, $d = 3$), it was showed that combination of angle ($\theta = 0$) and distance ($d = 1$) performed better than the other combination of θ and d to identify cassava chip water content. Therefore, in this study, TFs were extracted at those values of θ and d. A total of 190 TFs were extracted *i.e.* 10 TFs each for R, G, B, grey, hue, $saturation_{(HSL)}$, $saturation_{(HSV)}$, $lightness_{(HSL)}$, $value_{(HSV)}$, $X_{(XYZ)}$, $Y_{(XYZ)}$, $Z_{(XYZ)}$, $L_{(Lab)}$, $a_{(Lab)}$, $b_{(Lab)}$, $C_{(LCH)}$, $H_{(LCH)}$, $u_{(Luv)}$, $v_{(Luv)}$. Therefore, the total image features (f_t) which were extracted from both CFs and TFs are 209 features.

2.5. Fish swarm intelligent (FSI)

The steps of the proposed FSI [30] were as follows: (1) initialisation of FSI parameters. The maximum iteration was 500. The number of fish population (N_{fish}) was 70, the crowded parameter (C_p) was 0.3 and the leaping value (*leap*) was 10 based on the results of preliminary runs; (2) generated the location of each fish randomly [0, 1] (e.g. $fish_i$: 0,1,1,0,0,0,0,1,0,0,1,0,..... f_t); (3) evaluated the fitness of $fish_i$ using MSE of BPNN; (4) updated the individual solution $F(fish_i)$; (5) found the best solution ($fish_{best}$); (6) calculated visual scope (vis_{scope}) using equation 5 and 6; (7) calculated central point of the population ($cent_{point}$) using equation 7 and 8, e.g. $cent_{point}$: 0,0,0,1,1,0,0,1,0,0,.......f_t; (8) repeated steps 8.1 to 8.4 for those $fish_i$ with partial solutions; (8.1) if the visual scope of $fish_i$ was empty ($vis_{scope} = 0$) then $fish_{i+1}$ would generate its location randomly. Otherwise it went to *condition1*; (8.2) *Condition1*: If the visual scope of $fish_i$ was crowded ($vis_{scope} > C_p$) then *searching* which means $fish_i$ will generate random location $fish_{rand}$. If $F(fish_{rand})$ was better than $F(fish_i)$ then there would be a two points crossover process between $fish_{rand}$ and $fish_i$. Otherwise it went to *condition2*; (8.3) *Condition2*: 1th *process*: If the $F(cent_{point})$ was better than $F(fish_i)$ then *swarming* which means there would be a two points crossover process between $cent_{point}$ and $fish_i$, but if the $F(cent_{point})$ was not better than $F(fish_i)$ then *searching*. The output of the 1st *process* was $fish_{y1}$. 2nd *process*: If the $F(fish_{best})$ was better than $F(fish_i)$ then *chasing* which means there would be a two points crossover process between $fish_{best}$ and $fish_i$. But if the $F(fish_{best})$ was not better than $F(fish_i)$ then *searching*. The output of the 2nd *process* was $fish_{y2}$; (8.4) updated vis_{scope}, $cent_{point}$ and $fish_{best}$ using equation 9; (9) If the $F(fish_i)$ equal to $F(fish_{i+1})$ then updated the *leap* factor. If the *leap* factor reached the threshold point ($leap = 10$), then $fish_{i+1}$ would move randomly; (10) updated the best feature-subset; (11) the search would terminate if the maximum iteration had been reached.

$$M = [M_{jk}] = \frac{\|F(fish_j) - F(fish_k)\|}{2} \quad (5); \quad vis_{scope} = median\{M_{jk} \in M : 1 \le j < k \le N_{fish}\} \quad (6)$$

$$features_{ft} = \frac{\sum_{i=1}^{N_{fish}} features_{(i)}}{N_{fish}} \quad (7); \quad features_{ft} \begin{cases} 1 & if\ features_{ft} \ge 0.5 \\ 0 & otherwise \end{cases} \quad (8)$$

$$fish_{i+1} \begin{cases} fish_{y1} & if\ q(F(fish_{y1})) > q(F(fish_{y2})) \\ fish_{y2} & otherwise \end{cases} \quad (9)$$

3. Results and Discussion

The main assumption underlying this study was that changes in the external appearances and surface structure of a food product which is caused by water stress can be detected by visible light imaging techniques. The type of water content measurement performed on cassava chips is the measurement of wet base water content. Measurement of moisture content was done three times for each category of dryness level of B cassava chips for wet (range 1), SK for semi-dry (range 2) and KO for optimal drying (range 3). Based on the measurement of moisture content of the wet base, each grade category of drought level of cassava chips had different water content. The water content in each category was averaged. The average value of the water content of each category of drought-level cassava chips was shown on the graph in Figure 2 and Figure 3.

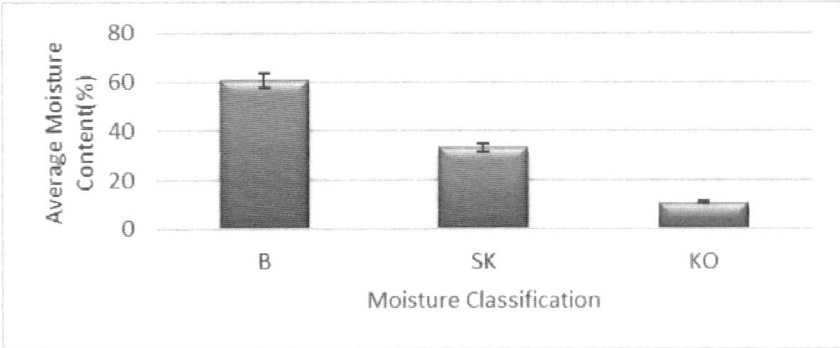

Figure 2. Moisture classification of cassava chips during drying

In Figure 3 typical sample images of cassava chips used in this study were shown where Figure 3a, 3b and 3c were described wet chips, semi-dry chips and dry chips, respectively. Figure 3a to 3c were not distinguishable. Therefore, the application of intelligent approaches to solve this problem. BPNN model performance was tested successfully to describe the relationship between cassava chips water status and image features. It indicated that colour and texture could be good indicators to predict water content in cassava chips during drying. Figure 4 shows that Lab-L has lowest error value compared with other colour parameter that is with MSE value of 0.000247 and absolute relative error (ARE) value of 68.99%. While for the highest error was obtained by Lab-a with MSE value of 0.004197 and ARE value of 91.55%. BPNN results of training and validation on the lowest MSE value of Lab-a was achieved by using hidden layer 20, learning rate 0.4 and momentum 0.1. Figure 5 showed that the MSE value of the best validation result was the energy value with the MSE value of 0.00006 and ARE value of 6.46%, and for the highest value obtained by maximum probability value with MSE of 0.004197 with ARE value of 31.87%. BPNN results of training and validation on the lowest MSE value was energy which was achieved by using hidden layer 20, learning rate 0.4 and momentum 0.1. Figure 6 showed the plot of best normalized fitness values of MOO using FSI. It showed effective in iteration process since the iteration is getting lower through all iterations. Figure 7 showed the training performance of BPNN using FSI and it showed effective in iteration process since the iteration is getting lower through all iterations. By using FSI optimization, the MSE was smaller compare to the method without feature selection optimization. The result of training-set data MSE had the lowest value of 0.0062 with the validation-set data MSE of 0.0135. The R^2 (0.9011) showed acceptable to use the model of ANN to predict the moisture content of cassava chips during drying.

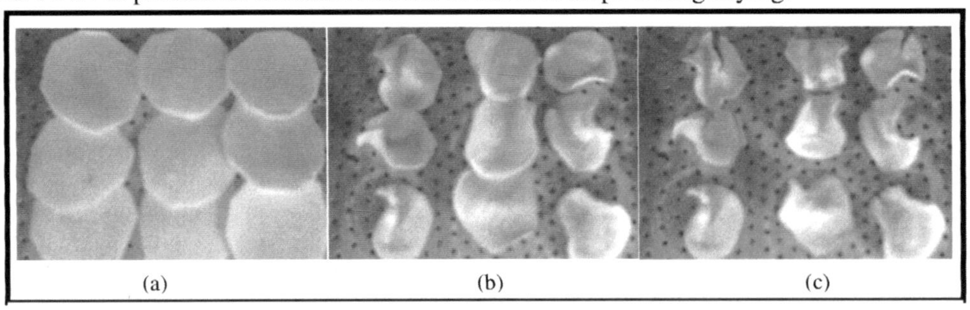

Figure 3. The appearance of cassava chips during drying in different moisture content: (a) wet / range 1; (b) semi-dry / range 2; (c) optimum drying / range 3

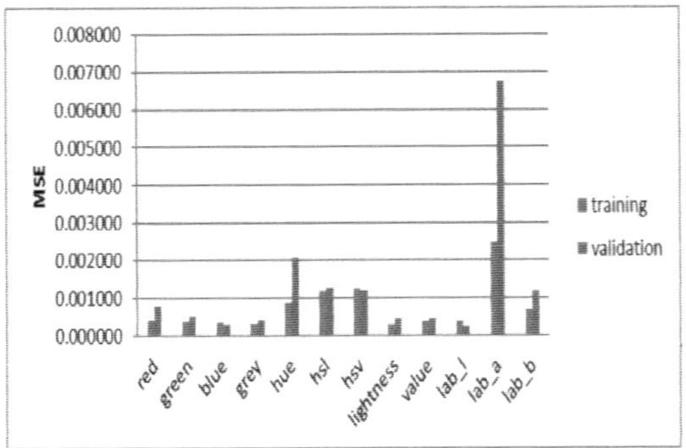

Figure 4. BPNN training and validation results using colour features (*hidden layer* 20, *learning rate* 0.4 and momentum 0.1)

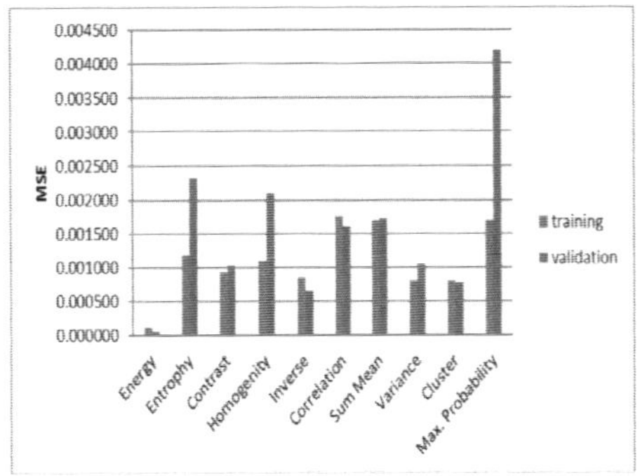

Figure 5. BPNN training and validation results using textural features (*hidden layer* 20, *learning rate* 0.4 and momentum 0.1)

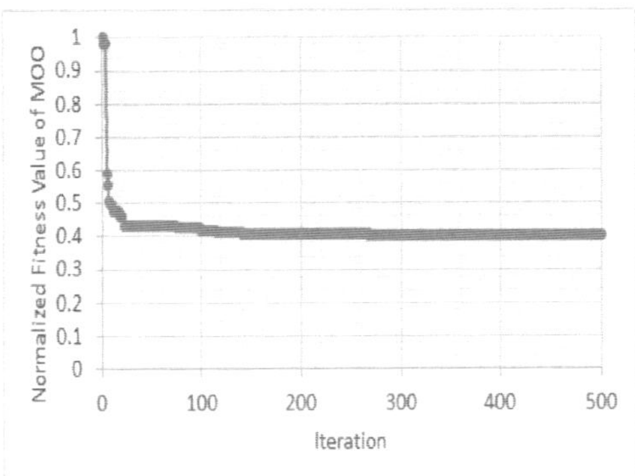

Figure 6. Plot of best normalized fitness values of MOO using FSI

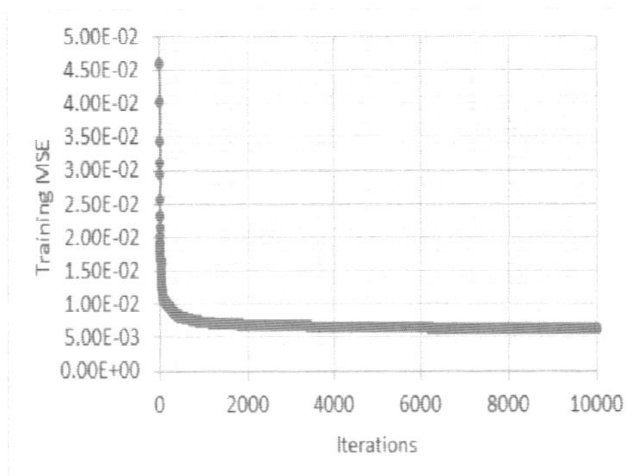

Figure 7. Training performance of BPNN using FSI

Table 1 shows the results of the optimum weights from the input layer to the hidden layer which were obtained from the BPNN model. Table 2 shows the results of the optimum weights from the hidden layer to the output layer which were obtained from the BPNN model. Therefore the structure of ANN model was shown in Figure 8.

Table 1. The weights from the input layer to the hidden

V(x,y)		x					
		1	2	3	4	5	6
y	1	-1.16497	0.564407	-1.46367	-1.46152	1.920421	-2.56927
	2	-4.91792	-2.62822	-1.0272	2.87E-02	2.5128	-8.30325
	3	-1.44197	0.626267	-2.71323	-0.75411	0.521782	-2.32564
	4	-0.23283	-0.98239	-0.35654	-1.40021	5.38E-02	-0.54782
	5	2.148937	8.70E-03	2.159154	0.729724	-0.60778	3.108686
	6	-3.97563	7.56E-02	-4.26567	1.088974	-0.39299	-4.30092
	7	0.65281	-1.82747	-5.96E-02	-0.94782	-1.73359	1.039046
	8	9.90E-02	-1.85041	0.745667	-1.49335	-0.60912	0.304743
	9	2.407663	0.409768	1.047853	1.551232	-1.94237	4.3001
	10	0.147574	-0.99473	-0.87917	-1.09507	-0.56111	-0.17694

Table 2. The weights from the hidden to output layer.

W1	-6.84702
W2	-8.07537
W3	-5.61476
W4	-4.62656
W5	2.345198
W6	-6.67257
W7	-5.82548
W8	-4.54159
W9	3.812706
W10	-5.93728

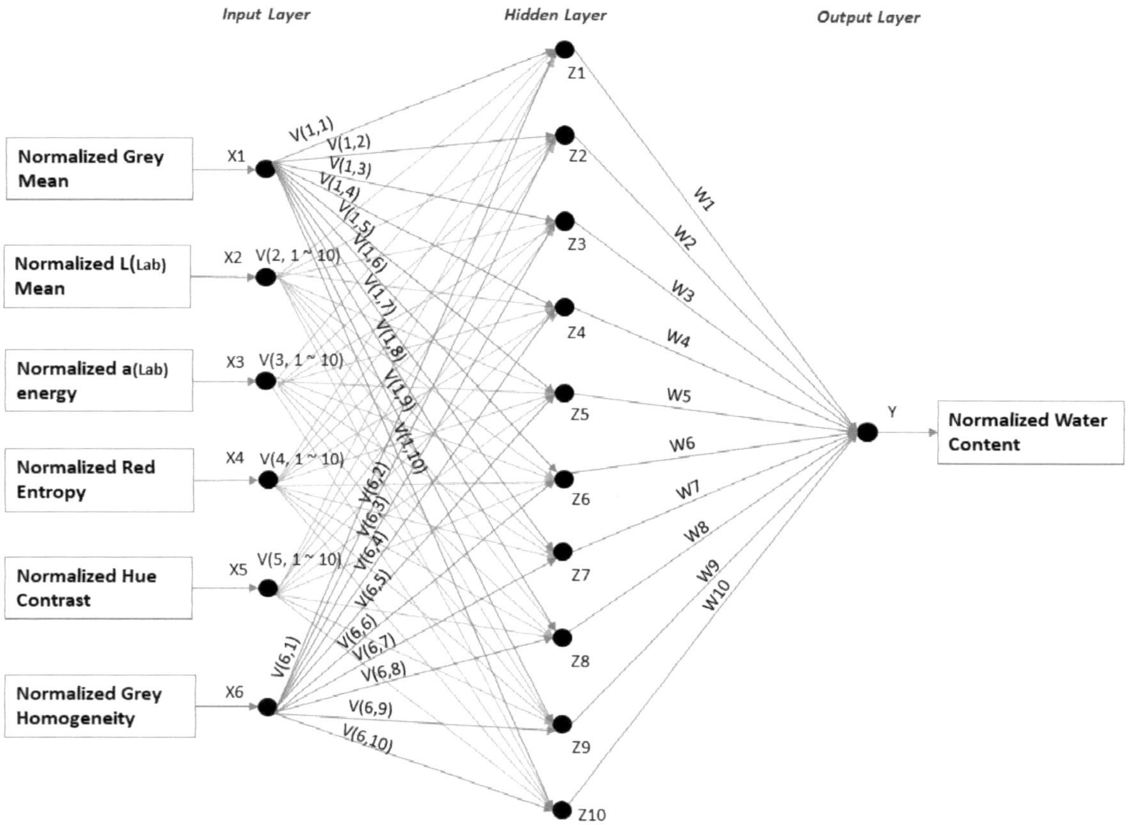

Figure 8. The ANN model

4. Conclusion

Back-propagation Neural Network (BPNN) has been tested successfully to describe relationship between image features and moisture content in cassava chips during drying. Feature Selection methods using fish swarm intelligent improved the BPNN performance for prediction. Overall, there is a significant difference between methods using feature selection and methods without feature selection.

Acknowledgments
Authors wishing to acknowledge assistance or encouragement from colleagues, special work by technical staff or financial support from KEMRISTEKDIKTI for STRANAS Project to accomplish this research.

References
[1] Barzegar M, Zare D, Stroshine R L 2015 An integrated energy and quality to optimization of green peas drying in a hot air infrared-assisted vibratory bed dryer *J. Food. Eng.* **166** 302-315
[2] Aghbashlo M, Hosseinpour S, Ghasemi-Varnamkhasti M 2014 Computer vision technology for real time food quality assurance during drying process *Trends Food Sci. Technol.* **39** 76-84
[3] Dutta M K, Singh A, Ghosal S 2015 A computer vision based technique for identification of acrylamide in potato chips *Comput. Electron. Agr.* **119** 40-50
[4] Baigvand M, Banakar A, Minaei S, Khodaei J, Khazaei N B 2015 Machine vision system for grading of dried figs *Comput. Electron. Agr.* **119** 158-165
[5] Benalia S, Cubero S, Montalban J M P, Bernardi B, Zimbalatti G, Blasco J 2015 Computer vision for automatic quality inspection of dried figs (Ficus carica L.) in real time *Comput. Electron. Agr.* **120** 17-25
[6] Yadollahinia A, Jahangiri M 2009 Shrinkage of potato slice during drying *J. Food. Eng.* **94** 52-58
[7] Huang Y, Lan Y, Thomson S J, Fang A, Hoffmann W C, Lacey R E 2010 Development of soft computing and applications in agricultural and biological engineering *Comput. Electron. Agr.* **71** 107-127
[8] Khazaei N B, Tavakoli T, Ghassemian H, Khoshtaghaza M H, Banakar A 2013 Applied machine vision and artificial neural network for modeling and controlling of the grape drying process *Comput. Electron. Agr.* **98** 205-213
[9] Ni H, Gunasekaran S 1998 Food quality prediction with neural networks *Food Technology* **52** 60-65
[10] Xie G, Xiong R 1999 Use of hyperbolic and neural network models in modelling quality changes of dry peas in long time cooking *J. Food. Eng.* **41** 151-162
[11] Singh R R B, Ruhil A P, Jain D K, Patel A A, Patil G R 2009 Prediction of sensory quality of UHT milk – a comparison of kinetic and neural network approaches *J. Food. Eng.* **92** 146-151
[12] Azadbakht M, Aghili H, Ziaratban A, Torshizi M V 2017 Application of artificial neural network method to exergy and energy analysis of fluidized bed dryer for potato cubes *Energy.* **120** 947-958
[13] Gheyas I A, Smith L S 2010 Feature subset selection in large dimensionality domains *Pattern Recogn.* **43** 5-13
[14] Hendrawan Y, Murase H 2010 Neural-Genetic Algorithm as feature selection technique for determining sunagoke moss water content *Engineering in Agriculture, Environment and Food (EAEF); Asia Agricultural and Biological Engineering Association (AABEA) J.* **3** 1 25-31
[15] Hendrawan Y, Murase H 2011a Non-destructive sensing for determining Sunagoke moss water content: Bio-inspired approaches *Agricultural Engineering International: CIGR J.* **13** 1 1-18
[16] Hendrawan Y, Murase H 2011b Bio-inspired feature selection to select informative image features for determining water content of cultured Sunagoke moss. *Expert Syst. Appl.* **38** 11 14321-14335
[17] Hendrawan Y, Murase H 2011c Neural-intelligent water drops algorithm to select relevant textural features for developing precision irrigation system using machine vision *Comput. Electron. Agr.* **77** 2 214-228
[18] Hendrawan Y, Murase H 2011d Neural-discrete hungry roach infestation optimization to select informative textural features for determining water content of cultured Sunagoke moss *Environ. Control Biol.* **49** 1 1-21

[19] Hendrawan Y, Al Riza D F 2015 Machine vision optimization using Nature –inspired algorithms to model Sunagoke moss water status. *Int. J. .Adv. Sci. Eng. Inform. Technol.* **6** 2088-5334

[20] Ondimu S S, Murase H, 2008 Comparison of plant water stress detection ability of colour and gray-level texture in Sunagoke moss *T. ASABE.* **51** 3 1111-1120

[21] Handl J, Douglas B K, Knowles J 2007 Multiobjective optimization in bioinformatics and computational biology *IEEE/ACM Trans. Comput. Biol. Bioinform.* **4** 2 279-292

[22] Hendrawan Y, Murase H 2009 Precision irrigation for Sunagoke moss production using intelligent image analysis *Environ. Control Biol.* **47** 1 21-36

[23] Haralick R M, Shanmugam K, Dinstein I 1973 Textural features for image classification *IEEE T. Syst. Man Cyb.* **3** 6 610-621

[24] Rotterman Y, Porat M 2006 Colour image coding using regional correlation of primary colours *Image Vision Comput.* **25** 637-665

[25] Angulo J, Serra J 2007 Modelling and segmentation of colour images in polar representations *Image Vision Comput.* **25** 475-495

[26] Leon K, Mery D, Pedreschi F, Leon F 2006 Colour measurement in L*a*b* units from RGB digital images *Food Res. Int.* **39** 1084-109.

[27] Kim M C 2008 Comparative colour gamut analysis of xvYCC standard *Displays.* **29** 376-385

[28] Palm C 2004 Colour texture classification by integrative Co-occurrence matrices *Pattern Recogn.* 37 965-976

[29] Pydipati R, Burks T F, Lee W S 2006 Identification of citrus disease using colour texture features and discriminant analysis *Comput. Electron. Agr.* **52** 49-59

[30] Fernandes, E M G P, Martins, T F M C, Rocha, A M A C 2009 Fish swarm intelligent algorithm for bound constrained global optimization. In the proceedings of the international Conference on Computational and Mathematical Methods in Science and Engineering, CMMSE, 30 June-3 July 2009.

IOP Conf. Series: Earth and Environmental Science **131** (2018) 012021 doi:10.1088/1755-1315/131/1/012021

Implementation of hazard analysis and critical control point (HACCP) in dried anchovy production process

A D P Citraresmi and E E Wahyuni

Department of Agro-industrial Technology, Faculty of Agricultural Technology, Universitas Brawijaya, Malang, Indonesia

E-mail: ardanezz@gmail.com

Abstract. The aim of this study was to inspect the implementation of Hazard Analysis and Critical Control Point (HACCP) for identification and prevention of potential hazards in the production process of dried anchovy at PT. Kelola Mina Laut (KML), Lobuk unit, Sumenep. Cold storage process is needed in each anchovy processing step in order to maintain its physical and chemical condition. In addition, the implementation of quality assurance system should be undertaken to maintain product quality. The research was conducted using a survey method, by following the whole process of making anchovy from the receiving raw materials to the packaging of final product. The method of data analysis used was descriptive analysis method. Implementation of HACCP at PT. KML, Lobuk unit, Sumenep was conducted by applying Pre Requisite Programs (PRP) and preparation stage consisting of 5 initial stages and 7 principles of HACCP. The results showed that CCP was found in boiling process flow with significant hazard of *Listeria monocytogenesis* bacteria and final sorting process with significant hazard of foreign material contamination in the product. Actions taken were controlling boiling temperature of 100 – 105°C for 3 - 5 minutes and training for sorting process employees.

1. Introduction

Indonesia is the largest archipelago in the world, therefore, fish become one of enormous marine commodities. Fish production in Indonesia is steadily increasing up to 2.87% annually [1]. Anchovy is one of the promising fish commodities. Anchovy production was increased up to 11.89% in 2000-2010. Moreover, exported anchovy was also elevated up to 50% in 2015 [2]. Anchovy contribute significantly to the economy [3]. It was due to its simple catching operation, processing, as well as a high selling price [4]. Anchovy is delicious and nutritious with a relatively high protein and calcium content [5]. Anchovy can be considered as perishable food. Its small size renders it to be more susceptible to contamination and quality degradation [6]. Furthermore, in order to assure the dried anchovy product's quality, good handling and proper processing methods are required [7]. Those are including raw material handling and production process. A proper production technique and knowledge are also required to prevent either harmful or deteriorating bacteria contamination in dried anchovy product [8]. Therefore, quality assurance system is important to be implemented. One of the widely used quality assurance system is Hazard Analysis and Critical Control Point (HACCP)

HACCP is a quality assurance system to identify, assess and control potential hazard as well as prevention-focused control system [9]. HACCP purpose is to prevent known potential hazard (biological, chemical and physical hazard) as well as hazard occurrence risk reduction by controlling each critical point of production process [10]. HACCP is applied in every step of the production (process) of dried anchovy [11].

2. Materials and Methods

This research was conducted in PT. Kelola Mina Laut Lobuk, Sumenep in February 2017. Materials used were fresh anchovy, salt, ice block and plastic packaging (PP plastic). Equipments used were conveyor, steam boiler, boiler machine, vacuum machine, blower, dryer, vibrator, sizing machine, scales, trolley, basket, sorting desk, brush, pallet, tray, and cold storage. Methods used were consisted of: interview and observation, documentation for supporting data as well as microbiological and chemical assay. This study used HACCP method divided into initial and application stages. The initial stage consisted of 1. Creating a HACCP team. 2. Preparing a product description. 3. Identifying of consumer. 4. Arranging process flowchart. 5. Verifying process flowchart. The HACCP system was based on seven principles as follow: 1. Conduct a hazard analysis. 2. Determine the critical control points (CCPs). 3. Establish target levels and critical limit(s). 4. Establish a system to monitor the CCPs. 5. Establish the corrective action to be taken when monitoring indicates that a particular CCP is not under control. 6. Establish procedures to verify that the HACCP system is working effectively. 7. Establish documentation concerning all procedures and keep records in accordance to these principles and their application [12]

3. Results and Discussion

3.1. The initial stages of HACCP method in dried anchovy production

3.1.1. Assembling HACCP team

HACCP team members consisted of: food safety team leader, business manager, head of certificate and documentation, factory manager, production supervisor, head of quality control and sanitation, head of procurement division, head of Human Resource and Development and head of technical division. HACCP team is responsible for arranging HACCP plan, supervising HACCP implementation as well as verifying and implementing HACCP system.

3.1.2. Product description and identification of consumer

Product description of dried anchovy can be seen in Table 1.

Table 1. Description product of dried anchovy.

Product and Species Name	dried anchovy (*stolephorus sp)*
Composition	anchovy and salt
Processing Stages	reception, washing, boiling, drying, sorting, sizing, final sorting, product weighing, packaging and labelling, storing under low temperature (cold storage), loading and export
Packaging Type	packed in pp plastic bag (polypropylene) then put into master carton (mc), with packing 5 kg per mc
Storage and shelf life	the product is stored in cold storage (minimum temperature is -10^{0}c). the shelf life up to 24 months (at -10^{0}c)
Label Specification	company name, product type, product size, net weight, area code, and production date
Purpose of Product Use	general consumption (infant excluded)
Instructions for Use	ready to be cooked or fried before serving

Source: PT. KML unit Lobuk, Sumenep (2017)

3.1.3. Arranging process flowchart

Flowchart of dried anchovy production and the CCP determination is shown in Figure 1.

IOP Conf. Series: Earth and Environmental Science **131** (2018) 012021 doi:10.1088/1755-1315/131/1/012021

3.1.4. Verifying process flowchart

Verification was conducted by reviewing the production process, coordinating with production operator to adjust the flowchart with actual condition in the field as well as doing a sample assessment in order to confirm the arranged flowchart diagram precision as compared to actual condition in the field. If there is flowchart application unconformities found, then flowchart modification or refinement must be conducted. However, if the flowchart has successfully been applied to the process and verified, then the documentation process regarding the flowchart application should be made. Quality standard of semi-dried anchovy based on SNI 01-3461-1994 is shown in Table 2.

Table 2. Quality standard of semi-dried anchovy.

Type of Test	Unit	Quality Req
Organoleptic, min		7
- Mold		-
Microbiology		
- ALT, Max	colony/gram	$2x10^5$
- *Escherichia coli*	APM/gram	<3
- *Salmonella* and Vibrio cholera**	Per 25 gram	Negative
- *Staphylococcus aureus* positive coagulation, max	colony/gram	100
Chemical Content		
- Water	% weight / weight	30-60
- Ash content and Arsenic, max	% weight / weight	1
- Salt, max	% weight / weight	15
- Tin / Sn, max	mg/kg	40
- Lead / Pb and Mercury,max	mg/kg	0,5
- Zinc, max	mg/kg	100,0
- Copper, max	mg/kg	20,0
Physical : Net weight		In accordance with the label

Source: National Standardization Agency (1994) [13]

3.2 The application of HACCP principles in dried anchovy production

3.2.1 Hazard Analysis

HACCP team needs to consider the probability of each identified hazard and its severity level which can be determined by observing the effect to the consumer health or company reputation. The result of the identification will be used to determine the control action should be performed for each corresponding hazard possibilities. Since dried anchovy product is attached to an E hazard category, its hazard level can be categorized as low (Category 1). An E Hazard category is defined as every hazard occurrence due to the distribution or product handling errors by consumer. Potential hazard occurred in distribution process was caused by inappropriate storage temperature (-10 °C). This error may affect product freshness and shelf life. Inappropriate storage temperature may result in reduction in product shelf life and elevated deterioration level when the products reach the consumer due to product quality degradation.

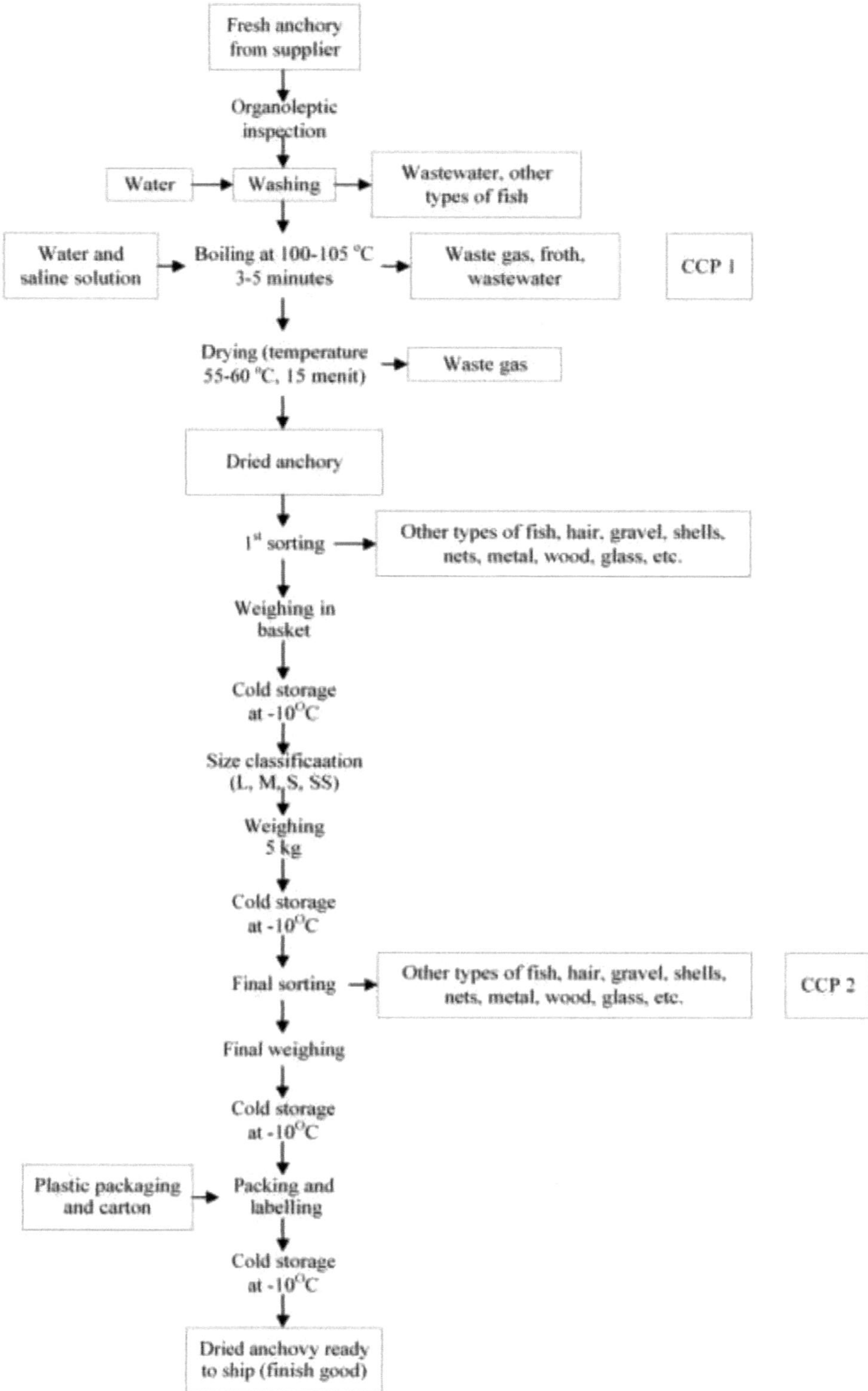

Figure 1. Flowchart diagram of dried anchovy production CCP determination

Table 3. Hazard analysis of dried anchovy handling and processing.

CCP	Significant Hazard	Critical Limit	Monitoring					Correction action	Verification	Documentation
			What?	When?	Where?	Who?	How?			
Boiling	*Listeria monocytogenes*	boiling temperature not less than the standard (100 - 105°C)	Boiling temperature and time	Every 1 hour	At boiling panel	QC Staff	Visual temperature and time monitoring in boiling panel	If the temperature and time used does not match to the standard, every batch production must be repeated	Review the monitoring process and final product pathogenic bacteria assessment as well as regular thermometer calibration	Control form of boiling and drying
Final sorting	Pebble, glass, plastic, metal, woods etc.	No foreign contaminant detected	Foreign contaminant	Continually as the process undergoes	Final assessment chamber	Examiner	Visually examining part by part of the product	Taking the foreign contaminant	Reviewing the data	Control form of foreign contaminant
	Residues from other types of fish	No mixed species	Mixed with other fish species	Every single *batch*	*Packing room*	Examiner	Visually examining	Taking the other types of fish	Reviewing the control data	Control form of foreign contaminant

Source: PT. KML Lobuk, Sumenep (2017)

3.2.2. Determining the Critical Control Point (CCP)

Based on the flowchart of dried anchovy production and its CCP determination (Figure 1), CCP of dried anchovy production process lies to its boiling and final sorting process. The potential hazard occurred at boiling process was microbiological contamination. Hazard may occur due to improper temperature of boiling process (standard: 100-105 °C for 3-5 minutes). If the temperature used less than 100 °C, it will lead to pathogenic bacteria growth, especially *Listeria monocytogenesis* [14]. This pathogen may cause listeria foodborne illness (listeriosis) [15]. If the temperature increases beyond 105 °C, the product may become mushy and fragile. In other words, it will be physically deteriorated. Proper controlling actions for this process are temperature and duration in hour basis. This controlling action will be conducted for every batch. This can be done by a proficient operator using temperature control instrument and thermometer at boiling machine.

In final sorting process, there is significant hazard in the form of foreign contamination to the product. The foreign contaminant can be found as metal flakes etc. If there is inconsistent process in the final sorting process, then it must be repeated by sorting operator. Sorting supervisor is responsible for checking the sorted product by analysing the sample. The dried anchovy was sorted on a white – coloured board to easily find and remove contamination.

3.2.3. Establishing critical limit, monitoring procedure determination, and establishing correction actions

CCP limit of the dried anchovy product is described in detail for each CCPs. Temperature and contaminants are the major factors which significantly affect the product quality and safety. Monitoring is done by using 4W+1H rule (*What, Where, Who, When (Frequency),* and *How*). Boiling temperature and time are supervised using temperature control and temperature measurement using thermometer in the boiling machine. Contamination of foreign materials is checked by contaminant documentation. Every contaminant must be documented as the evidence for the anchovy supplier. It must be conducted in order to evaluate the process as well as to prevent the similar incidence in the future. Correction actions are obligatory to be done when there is lost control found in monitoring process. Critical limit of boiling process is 100-105^{0}C and critical limit of final sorting process is that the products are free from contaminant, etc. If there are any unconformities in the process, for example boiling temperature deviation, the operator must be informed to do immediate adjustment. If the unconformities occur in the final sorting process, the sorting supervisor must order other sorting process until the products are free of contaminant. Limit determination of dried anchovy product, monitoring procedure determination, and correction actions are shown in Table 3.

3.2.4. Establishing procedure of HACCP verification process

Internal verification is done by food safety team leader every 6 months. Verification is conducted by examining the conformity between manual book and the actual condition in the field. External verification is conducted by the authority, or HACCP system certification agency in a year basis.

3.2.5. Establish proper documentation and record keeping

Documentation of dried anchovy product processing include every remark about CCP, critical limit, monitoring record, correction action conducted in response to occurred unconformities and remarks about verification process. The documentation is conducted in monitoring process. The documentation data are shown to the food safety inspector while the company undergoes external audit. These data may also be used by the operator to supervise the HACCP implementation process.

4. Conclusion

In order to maintain the dried anchovy product's quality and safety, PT. KML Lobuk Unit, Sumenep ensure the implementation of HACCP analysis. Based on the research, dried anchovy production process CCPs lies in the boiling and final sorting process. Prevention action are conducted by the

IOP Conf. Series: Earth and Environmental Science **131** (2018) 012021 doi:10.1088/1755-1315/131/1/012021

company using thermometer on the boiling machine and controlling temperature at 100-105°C for 3-5 minutes, as well as ensuring the proper re-sorting process which is done by sorting operators.

References

[1] Ministry of Marine and Fishery 2010 Kelautan dan Perikanan dalam Angka 2010 Pusat Data Statistik dan Informasi Jakarta Indonesia
[2] Ministry of Marine and Fishery 2016 Kelautan dan Perikanan dalam Angka 2016 Pusat Data Statistik dan Informasi Jakarta Indonesia
[3] Erwin L T 2011 Gemar Makan Ikan: 25 Resep Masakan Olahan Ikan Teri PT. Gramedia Pustaka Utama Jakarta Indonesia
[4] Djazuli, N 2014 Badan Karantina Ikan, Pengendalian Mutu, dan Keamanan Hasil Perikanan Kementrian Kelautan dan Perikanan Republik Indonesia
[5] Rahayu S M 2012 Pengaruh Kosentrasi Garam dalam Proses Perebusan Ikan Teri Nasi *(Stolephorus sp.)* Setengah Kering dan Pendugaan Umur Simpannya dengan Metode Akselerasi Departemen Teknologi Hasil Perairan Fakultas Perikanan dan Ilmu Kelautan Institut Pertanian Bogor Indonesia
[6] Nadal, A, Coll, A, Cook, N and Pla, M 2007 *J. Microbiol. Methods* **68** pp 623 – 632
[7] Priyono, A 2011 Pedoman Teknis Penerapan Sistem Jaminan Mutu dan Keamanan Hasil Perikanan BKIPM Jakarta Indonesia
[8] Muchtadi T R, Sugiyono 1989 Teknologi Proses Pengolahan Pangan Departemen Pendidikan dan Kebudayaan, Direktorat Jenderal Pendidikan Tinggi, Pusat Antar Universitas Pangan dan Gizi IPB Bogor Indonesia
[9] Tjahja and Darwin, K 2006 Sistem Jaminan Mutu Industri Pangan IPB Press Bogor Indonesia
[10] Surono 2016 Pengantar Keamanan Pangan untuk Industri Pangan Deepublish Yogyakarta Indonesia
[11] Sutrisno, A, Basith, A, and Wijaya N A 2013 *Jurnal Manajemen dan Organisasi* **4** 2 pp 73-90.
[12] Safefood 360^0 Whitepaper 2014 Developing a HACCP Plan Safefood 360^0 Inc New York USA
[13] SNI 01-3466-1994 1994 Persyaratan Mutu Ikan Teri Nasi Setengah Kering Badan Standarisasi Nasional Jakarta Indonesia
[14] Rivoal, K, Quéguiner, S, Boscher, E, Bougeard, S, Ermel, G, Salvat, G, Federighi, M, Jugiau, F and Protais, J 2010 *Int. J. Food Microbiol.* **138** 56 – 62
[15] Churchill R L T, Lee, H, and Hall J C 2006 *J. Microbiol. Methods* **64** 141 – 170

IOP Conf. Series: Earth and Environmental Science **131** (2018) 012022 doi:10.1088/1755-1315/131/1/012022

Application of multi attribute failure mode analysis of milk production using analytical hierarchy process method

A L Rucitra

Department of Agro-industrial Technology, Faculty of Agricultural Technology, Universitas Brawijaya, Malang, Indonesia

E-mail: andanrucitra@gmail.com

Abstract. Pusat Koperasi Induk Susu (PKIS) Sekar Tanjung, East Java is one of the modern dairy industries producing Ultra High Temperature (UHT) milk. A problem that often occurs in the production process in PKIS Sekar Tanjung is a mismatch between the production process and the predetermined standard. The purpose of applying Analytical Hierarchy Process (AHP) was to identify the most potential cause of failure in the milk production process. Multi Attribute Failure Mode Analysis (MAFMA) method was used to eliminate or reduce the possibility of failure when viewed from the failure causes. This method integrates the severity, occurrence, detection, and expected cost criteria obtained from depth interview with the head of the production department as an expert. The AHP approach was used to formulate the priority ranking of the cause of failure in the milk production process. At level 1, the severity has the highest weight of 0.41 or 41% compared to other criteria. While at level 2, identifying failure in the UHT milk production process, the most potential cause was the average mixing temperature of more than 70 °C which was higher than the standard temperature (≤70 ° C). This failure cause has a contributes weight of 0.47 or 47% of all criteria Therefore, this study suggested the company to control the mixing temperature to minimise or eliminate the failure in this process.

1. Introduction

In the 21st century, the needs for quality products is critical. Therefore, it is necessary to maintain and control the product quality as a guarantee to the consumer that the marketed product has good quality [1]. This is crucial for companies to increase customer's interest and satisfaction level.

Currently, the development of dairy products is rapidly increasing, as indicated by a lot of milk brands in the market. Consumer preference and loyalty are the keys to the company's success to win the competition.

Pusat Koperasi Induk Susu (PKIS) Sekar Tanjung East Java is one of the modern dairy industries that established by the cooperation of six cooperatives. Ultra High Temperature (UHT) milk is produced by PKIS Sekar Tanjung. The main problem faced by PKIS Sekar Tanjung is the failure of the production process that is not accordance with the standard. Milk is a very sensitive product and easily damaged, therefore a research to identify the most potential failure in the production process is required. This is important to find an action or strategy to be carried out for minimising the failure immediately. Thus, PKIS can produce high quality and competitive products. Therefore, this study aimed to apply Multi

Attribute Failure Mode Analysis (MAFMA) on the milk production process by using Analytical Hierarchy Process (AHP)" method for measuring the weight of criteria.

2. Research Methods

This research was conducted at PKIS Sekar Tanjung, located at Jalan Raya Puntir Martopuro Village Purwosari Subdistrict. Data processing and analyses were carried out in Computational Laboratory and System Analysis, Department of Agroindustrial Technology, Faculty of Agricultural Technology, Universitas Brawijaya, Malang.

2.1. Research procedure

The first step, in this study, was to conduct the preliminary survey to find out the real condition of PKIS Sekar Tanjung. This was then followed by a survey using structured questionnaires and interview to obtain baseline data for a problem formulation. The next step was an in-depth interview with an expert respondent, the head of the production department, who know better about the problem in the company. The identification of the problems was aimed to enable researchers to determine the steps needed in this study. MAFMA is one of the methods used to eliminate or reduce the possibility of failure when viewed from the root of the failure cause [2]. Therefore, this method can be used to minimise or prevent the failure. The formulation of the priority ranking of causes of failure is carried out by AHP approach [3].

2.2. MAFMA method.

MAFMA is a method that integrates Failure Mode and Effect Analysis (FMEA) with the economic aspect [2]. This method aims to identify the causes of overall failure to determine the most potential causes of failure. This method can be applied by identifying process functions, identifying potentially failed products and processes, analyzing the effects, identifying the causes of failure and the variables in the process, and ranking each failure in determining priorities for corrective action, and documenting the results. Determination of the highest weight for each criterion was carried out using the AHP approach based on severity, occurrence, detection, and expected cost.

2.2.1. Severity

Severity is a rank that shows a serious effect that comes from the failure mode (Table 1).

Table 1. Severity rating criteria.

rank	1	2	3	4	5	6	7	8	9	10
criteria	neglible severity	mild severity		moderate severity			high severity		potential safety problems	

2.2.2. Occurrence

Occurrence shows the frequency of a problem that cause of failure, as shown in Table 2.

Table 2. Occurrence rating criteria.

Degree	Frequency of occurrence	Rating
Remote	0.01 per 1000 item	1
Low	0.1 per 1000 item	2
	0.5 per 1000 item	3
Moderate	1 per 1000 item	4
	2 per 1000 item	5
	5 per 1000 item	6
High	10 per 1000 item	7
	20 per 1000 item	8
Very High	50 per 1000 item	9
	100 per 1000 item	10

2.2.3. Detection

Detection is a control tool applied to detect the cause of failure, as can be seen in Table 3.

Table 3. Detection rating criteria.

Rating	Criteria	Frequency of detection
1	Preventive methods are very effective	0.01 per 1000 item
2	Possible causes are very low	0.1 per 1000 item
3		0.5 per 1000 item
4	Possible causes are moderate	1 per 1000 item
5		2 per 1000 item
6		5 per 1000 item
7	The probable cause is still high.	10 per 1000 item
8		20 per 1000 item
9	The probable cause is very high	50 per 1000 item
10		100 per 1000 item

2.2.4. Expected cost

Expected cost is an assessment of cost aspects. To identify the cause of failure, it is necessary to integrate conventional aspects of popular FMEA procedures with economic considerations [3].

2.3. AHP

The AHP method provides a framework for addressing situations involving multiple intuitive, qualitative, and quantitative criteria on the MAFMA. Therefore, after obtaining the highest rating of the criteria, both corrective and evaluation actions can be taken.

2.4. Data analysis

2.4.1. Weighting of scoring criteria (Level 1)

Score of criteria used were based on the questionnaires filled by the head of the production department. The collected data were analysed using AHP with the following steps:

- Create a hierarchy. Hierarchy is used to explain the problem in a smaller section. Hierarchy consists of several levels, the top level is the main goal, the second level is the criteria, and the last level is the risk to be assessed based on the impact occurred. Factors and alternatives to the cause of failure are arranged in a hierarchical structure
- Develop an opinion matrix from respondents. The score of criteria, sub-criteria, and alternatives as a critical function is essential for the corresponding element of the higher level. Both qualitative and quantitative criteria can be compared using informal ratings to gain score and priority. For qualitative criteria, AHP pairwise comparisons are used to determine simple weight and rank. So the analysis can concentrate on two factors at a time. The verbal assessment is translated into scores through the discrete use of a nine-point scale.
- Calculate the priority vector or vector eigen (VP). The normalization matrix is obtained by dividing the value of each matrix cell in pairs of criteria by the total of each column. The score of VP) is obtained by dividing the total value of normalization of each criterion by the total value of normalization of all criteria.

$$VPi = \frac{\sqrt[n]{\prod_{k-1}^{m} a\ ij\ (k)}}{\sum_{i=1}^{n} \sqrt[n]{\prod_{k-1}^{m} aij\ (k)}}$$

Where:

$a\ ij$: the value of the comparison scale between criterion i and j

VPi : score ratio of each variable with total weight of variable

Determining the value of λ max used the formula below:

$$\lambda \text{ max} = \Sigma \text{ (number of columns to-}j\ x\ VPi \text{ for } i = j)$$

Where: λ is the number of times the weighted matrix per criterion with *VP* per criterion

Determining the consistency index (CI) is carried out by the formula below:

$$\text{CI} = \frac{\lambda maks-n}{n-1}$$

Where: n is the number of criteria

- Calculate the value of consistency ratio (CR). The consistency ratio is the benchmark consistency of the paired comparisons. The consistency ratio is the results of the consistency index (CI) and random index (RI), by the following formula:

$$\text{CR} = \frac{CI}{RI}$$

  Testing is done by finding the value of consistency ratio for the respondent. If the consistency ratio is less than or equal to 0.1 means the assessment is valid.
- Determine the alternative weights of each pairwise comparison

2.4.2. Assessment the alternative (Level 2)
Assessment the alternative at level 2 are used to determine the most potential impact on the production process from the hierarchy level alternatives based on the questionnaire filled by the head of the production department to determine the level of the industry's capability in addressing the cause of failure with a pairwise comparison method.

3. Results and Discussion

PKIS Sekar Tanjung is a primary cooperative that holds milk products from six cooperatives in East Java. This company produces various types of milk products such as UHT milk, pasteurized milk and fresh milk.

3.1. Potential failure identification strategy in production process
The proposed identification process fits the failure identification criteria. The criteria are used based on MAFMA to determine the most potential failure in the production process. MAFMA is based on the AHP technique, which considered the four different factors of severity, occurrence, detection, and expected cost as decision attributes, possible causes of failure as decision alternatives, and the selection of failure cause as the goal decision. The goal, attributes and alternatives formed a three-level hierarchy, in which the pairwise comparison matrix was used to estimate the attribute weights and the local priorities of the failure cause with respect to the expected cost attribute [4].

Hierarchical structure consists of 3 levels, the main goal is the most potential failure occurs in the production process. Then the criterion level (Level 1) is based on the criteria on MAFMA (i.e. severity, occurrence, detection, and expected cost). Alternative level (Level 2) as an initial preventive measure of the various types of failures that occur.

The hierarchy of failure of the milk production process in PKIS Sekar Tanjung can be seen in Figure 1. The figure shows that at level 1, the severity indicated as the serious effects of failure mode has the highest impact with a weight of 0.41. This was followed by the detection indicated as a control tool used to detect the cause of failure with a value of 0.29. Occurrence indicated as the value of the frequency a problem occurs, was in the third rank with a value of 0.22. While, the expected cost indicated as the evaluation of the cost aspect ranks fourth with a value of 0.07. This result confirmed that the severity contributed as the most potential failure cause in the milk production process in PKIS Sekar Tanjung, with the value of 41%. Several measures can be done PKIS Sekar Tanjung to overcome their problem of the non-standardised milk production process. The alternative stratgey is therefore needed for the company to achieve and follow the standard process for producing UHT milk. Also, based on the consideration and the assessment of potential causes of failure from the head of the production department, the cost factor was selected as the main problem faced by the company compared to the severity. This is in accordance with Hetharia [5] who stated that severity factor is not a big problem in the company.

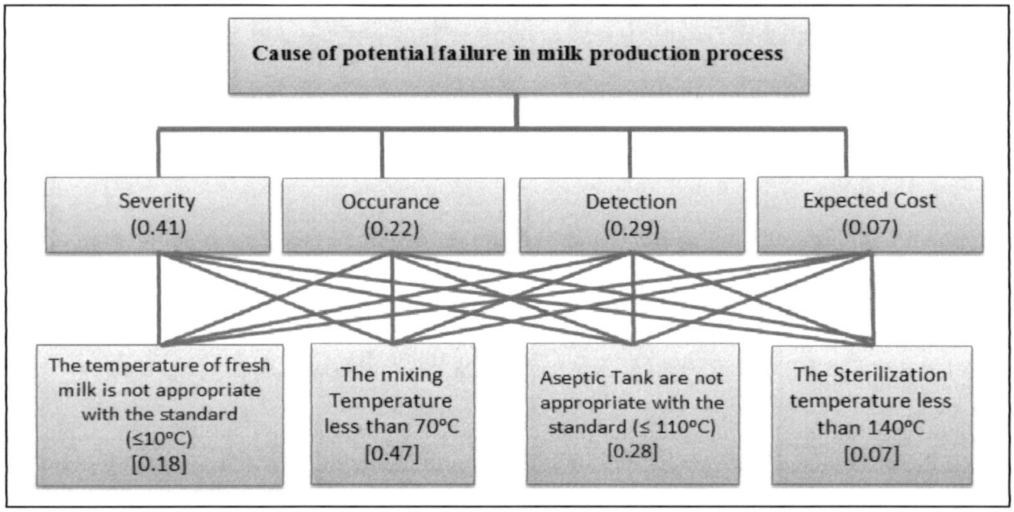

Figure 1. The hierarchy structure of potential failure causes in the production process

The results and criteria from level 1 were used to determine the strategy to minimise or remove the failure. The study found that there were four potential causes of failure in the UHT milk production process, include the mixing temperature is not standard (i.e. less than 70 °C), the fresh milk temperature is not standard, the steam barrier on the aseptic tank is not standard, and the sterilisation temperature is less than the standard of 140 °C.

As shown in Figure 1, the highest weighting criterion (Level 2) was the mixing process temperature of less than 70 °C with a value of 0.47. In the second rank was the potential failure cause of the aseptic tank temperature less than 110 ° C, with a weight value of 0.28. The temperature of fresh milk of less than 10 °C contributed as the third failure cause, with the weight value of 0.18. In the last place was the sterilization temperature of less than 140 °C with a weight of 0.07. These results indicated that the most potential failure cause in the milk production process in PKIS Sekar Tanjung was the inconsistency of the mixing process temperature. In this company, the mixing temperature was higher than 70 °C, which is not meeting the standard for UHT milk processing of ≤ 70 °C. Mixing is a stage where the formulated additive ingredients were mixed together with the fresh milk to form a milk emulsion with the appropriate and standardised composition. The mixing process is a critical in the production process [6].

The process of mixing in PKIS Sekar Tanjung begins by heating water at a temperature of 75°C. This temperature determines the solubility rate of the dairy product. Before the water reaches the optimum temperature the required additive ingredients such as sugar, salt, dyes and stabilizers cannot be mixed. If the additive ingredients were added below the optimal temperature, this will affect the milk solubility level and will eventually affect the next process. In this company, the solubility level was tested regularly in the QC physics-chemical laboratory. Then, after the desired (i.e. standard) temperature is reached, a predertemined volume of pasteurised milk is added.

3.2. CR results

The CR is the value used to determine the level of consistency of respondents in assessing each potential identification criteria of failure in the milk production process at PKIS Sekar Tanjung. The hierarchy CR value should be $\leq 10\%$, otherwise the quality of information should be improved by improving pairwise comparison questions [6]. The results showed that the CR at each level was 0.01 (Level 1) and 0.07 (Level 2). It can be seen that the CR values at each level were smaller than 0.1, indicating that the calculation was consistent and did not require repetition.

4. Conclusion
This study demonstrated that MAFMA combined with AHP approach can be used to determine the potential cause of failure in the UHT milk production process. Despite the severity criteria has the highest weight value of 0.41, based on the expert evaluation, the expected cost criteria is considered as potential failure cause in this company. Further investigation revealed that the high temperature (i.e. 75 ^{o}C) in the mixing process was selected as the most potential cause in the milk prodcution process failure. Therefore, the company need to apply several measures on keeping the mixing temperature within the standard value of ≤ 70 ^{o}C, to minimise and remove the mixing process failure.

References
[1] Dorothea W A 2004 Pengendalian Kualitas Statistik: Pendekatan Kuantitatif Dalam Manajemen Kualitas Statistik (Quality Control Statistics Quantitative Approach in Quality Management) Andi Publisher Yogyakarta Indonesia 370 [In Indonesian]
[2] Susila W R, Munadi E 2007 Penggunaan analytical hierarchy process untuk penyusunan prioritas (Use of analytical hierarchy process for preparation of priorities) *J. Informatika Pertanian* **16** 2 983-998 [In Indonesian]
[3] Braglia M 2000 Multi attribute failure mode analysis *Int. J. Qual. Reliab. Manage.* **17** 9 1017-1033
[4] Shaghaghi M, Rezaie K 2012 Failure mode and effects analysis using generalized mixture operators *J. Opt. Ind. Eng.* **5** 11 1-10
[5] Hetharia D 2009 Penerapan fuzzy analytic hierarchy process dalam metode multi atribute failure mode analysis untuk mengidentifikasikan penyebab kegagalan potensial pada proses produksi (Application of fuzzy analytic hierarchy process in multi attribute failure mode analysis method to identify potential failure causes in production process) *J@TI Undip* **4** 2 106-113 [In Indonesian]
[6] Saaty T L 1993 Decission Making for Leaders: The Analitycal Hierarchy Process for Decission in Complex World RWS Publications Washington USA 292

Evaluation of halal assurance system (HAS) implementation on bakery products processing in small and medium enterprises (case study in X Bakery Batu, East Java)

C G Perdani, N U Chasanah and Sucipto

Department of Agro-industrial Technology, Faculty of Agricultal Technology, Universitas Brawijaya, Malang, Indonesia

E-mail: cgadizza@gmail.com

Abstract. Small and Medium Enterprises (SMEs) have an important role to catch opportunities in the global halal market. To achieve halal standards, bakery SMEs need to implement Halal Assurance System (HAS). The objective of this research was to evaluate HAS applied by SMEs in bakery product processing. Evaluation was conducted by audit method in accordance to Indonesian Halal Assurance System (a guideline regarding the Fulfillment of Criteria of Halal Assurance System in Processing Industry – namely HAS 23101). The result of the evaluation shows that 44% of HAS criteria have been met. Halal team is the key of HAS management, therefore the team needs to be institutionally established. Halal policy need to be documented and announced to the public. Regular training will improve understanding and awareness of the importance of HAS responsibilities among stakeholders. With regard to bread ingredients, 85% products hold halal certificate from MUI (Indonesian *Ulema* Council). Halal Control Point (HCP) of production process can be derived from contamination of *haram* (not permitted by Islamic Law) and *najis* (dirty according to Islamic Law) in term of material, equipment, and packaging process. It requires top management commitment and cooperation of a good halal team to implement a sustainable halal system.

1. Introduction

The consumption of bakery products is increasing in the last 10 years in Indonesia, therefore the assurance of halal is needed as Indonesia is the largest moslem country in the world. Evidences were shown that during 2011 - 2015 the level of white rice consumption in Indonesia decreased by 3.07% per serving [1]. According to the average rate of consumption of processed wheat products such as bread, it was reported to be slowly increasing from 2009 with the number of 0.121 kg per capita until 2013 with the number of 0.144 kg per capita [2]. Announcement of halal certified bakery products will facilitate the Muslim community to choose which products are halal guaranteed [3]. Bakery or patisserie products are classified into 4 groups, namely bread, cake, pastry, and cookies. The bakery product studied classified as bread group [4]. General differences between bread with other patisserie class lies on the fermentation process. Bread requires certain amount of yeast as a *leavener* ingredient.

Halal control point (HCP) in the processing of bread does not apply for the ingredients alone, but also for the processing. Koswara [5] reported, the ingredients often used in the making of bread are wheat flour, yeast, sugar, salt, fat in the form of shortening/margarine/butter, milk, and eggs. Bread is often added by several additives to improve the quality such as emulsifier, flavor, coloring, bread improver, and preservatives. With the emergence of various and complex bread ingredients it is important to record the control point and halal status. Baking consists of 4 general processes; dough mixing, fermentation, forming, and roasting [4]. In general, halal control points on the processing can be derived from contamination of unlawful or *najis* materials

The problem raised from this research was the extension of Indonesian Ulema Council (MUI)'s halal certificate which is still constrained and has not been acted upon by the company. From this point, it is necessary to study the general obstacles faced by the small and medium enterprises to apply the halal assurance system. The purpose of this research was to evaluate the consistency of halal assurance system in the companies by case study.

2. Materials and Methods

This study used an audit method according to Indonesian Halal Assurance System Guidelines regarding the Fulfillment of Criteria of Halal Assurance System in Processing Industry (6). Questionnaire used in the form of check list of HAS audit which refers to Indonesian Halal Assurance System Guidelines (6). The checklist form contains closed questions to conduct monitoring or evaluation of Halal assurance system (HAS) implementation by the company. Questionnaire filling was done by interview to selected respondents. Interview method was chosen to collect more valid data and facilitate the respondent in understanding each question in the questionnaire. Respondents were obtained from internal company employees. At least 2 persons are taken from each part or division within the company to become respondents. Respondents must have at least 2 years of experience or duration of working at the company.

3. Results and Discussion

3.1 Fulfillment of Halal Assurance System Criteria

1. Halal Policy

Halal policy in the form of a written statement contains a commitment by the top management to consistently produce halal products (6). Based on the results of filling in the audit check list form, 28.5% of respondents answered "Yes" or "No", and the rest reported that they did not know the absence or the availability of halal policy. Respondents who answered "Yes" said that the halal policy set by the company leaders was the use of materials that have been halal labeled. The respondents who gave the answer "No" explained that the company did not have a written/documented policy of HAS. There was only policy that was spoken by the business owner that all materials used must have a halal label of MUI. Furthermore, all respondents said that there had never been any of halal policies implemented and communicated to all stakeholders.

2. Halal Management Team

Malaysia Halal Certification Body stipulates that the applicant company must appoint an internal halal team of 2 Muslims, locals, and committed to halal management and work full-time in the company (7). Indonesian Halal Certification body added that the appointment of halal management team must come from the company's permanent employees and accompanied by written evidence (6). Based on the internal audit check-list results, 43% of respondents reported that there was no halal management team that was formed systematically. The remaining 57% said not knowing about the presence or absence of halal management team. Once further explored, all the needs of halal assurance systems are mostly handled by the owners of the company. In case of the need for assistance, the owner will delegate one or more persons who are entrusted to handle the implementation of halal assurance system.

3. Training and Education

Guidelines for Halal Assurance Management System of Malaysia Halal Certification requires the management of the company to provide adequate and sustained training on halal assurance system in which all workers must be trained to apply halal system with full sense of responsibility (8). The result of filling the audit check list was 43% of respondents stated that there were employees who attended the HAS external training, 14% reported "No" and 43% stated that they were not aware of any external training. Based on the interviews with company owners, internal training of employees related to halal assurance system had never been held. The owner felt that it was sufficient to convey

IOP Conf. Series: Earth and Environmental Science **131** (2018) 012023 doi:10.1088/1755-1315/131/1/012023

orally the halal rules / policies applied by the company, namely the guarantee that the materials used must have MUI halal logo or certification.

4. Materials

In determining halal status, all baking materials are grouped based on 4 (four) categories, namely vegetable ingredients, animal ingredients, microbial ingredients, and other ingredients. The halal control points of the bread ingredients are identified using a halal critical point decision tree and the results of the identification of each category of ingredients. There are 12 halal control points of vegetable ingredients: wheat flour, granulated sugar, refined sugar, palm oil, jam, brown grains, cocoa powder, sesame oil and fat derivatives (shortening, margarine, baker's fat, white fat, and butter oil substitute). Wheat flour is made from raw wheat, but it is often added / fortified with additives such as vitamin A to enrich nutritional value [10]. Vitamin A is usually coated to dissolve easily and not easily damaged during storage [11]. Coating material used may be derived from halal materials such as gum, but may also be derived from a material that is doubtful of its halal such as gelatin. Apart from the potential for such a prohibition, wheat flour used by the company has been traced to be halal as indicated by MUI's halal logo on the packaging material. Based on the tree chart of the sesame decision is indeed a vegetable material, but to make sesame as a food mixture requires simple processing. Sesame seeds in very simple foods, ie washed thoroughly and dried and then mixed into the food [12]. In the drying process, sesame seeds are only roasted without oil, so in general there is no addition of food additive.

Dye for bread production is a mixture of organic and inorganic materials, so it needs to be studied further for the critical point. As viewed from the compositions information on the packaging label consisting of sorbitol and inorganic dyes. Sorbitol or sugar alcohol can be found in fruits [13]. Sorbitol includes a natural sweetener of extraction or enzymatic isolation in which the most commonly used enzymes are the alpha-amylase and gluco-amylase enzymes [14]. One of the sources of this alpha-amylase enzyme comes from animals, especially pigs [3]. However, many of these alpha-amylase enzymes are derived from microorganism products and if their manufacture does not involve haram elements it may be allowed to be used. The inorganic dye composition that exists in dark green and egg yolk is Tartrazin, Briliant Blue, and Ponceau 4R. Three materials are inorganic dyes so that there is no problem for halal status. Since the dye material used by the company has been halal certified by MUI [15].

5. Products

Indonesian Halal Certification Body declare that the brand or product name should not use a word that leads to something forbidden or worshiped that is not in accordance with Islamic Shari'a Law [6]. In addition, this also applied to a word associated with the sensory profile of the product. It should not have a tendency of smell or taste that leads to haram products. Based on the results of the inspection, all respondents agree that the brand or trade name used by bakery products in this company does not use a name that is contrary to Islamic Shari'a Law. In addition to trademarks, on the packaging of bread products the name of the variant are mentioned. The name of the variant is adjusted to the shape and physical character of the product. For example, cream ball is round with stuffing cream, for bread braids are shaped like a braid hair. All respondents also stated that the characteristics or sensory profiles of each available product do not have a tendency to smell or taste that lead to haram products based on the fatwa of the Indonesian Ulema Council (MUI). Various fresh bread and sweet breads have the flavor and aroma of bread in general and nothing else are suspicious.

6. Production Facilities

Based on the information obtained, all production facilities used in the company have never changed. There is only one additional equipment for water distillation. The water distiller used in this bread production process uses a reverse osmosis system that can process and filter water to be safer from contaminants such as chlorine. The halal production process starts from the use of sterile machines and equipment, the use of clean and holy water, and also pays attention to the sanitation of workers to produce a high quality and safe product [9]. Another important aspect in the production process is the sanitation and health of employees. A healthy employee will prevent disease contamination into the product during the production process. Therefore, it is necessary to check the health of employees on a regular basis. It is important to provide protection to employees while protecting the company's assets [9].

3.2 Evaluation of Halal Assurance System Implementation

The evaluation was conducted to find out the conformity and consistency of the implementation of HAS in the company with the rules set by Indonesian Halal Certification Body [6]. The results showed that the implementation of HAS in the company should be improved to get halal certification. Evaluation of the implementation of HAS criteria at bakery company is shown in **Table 1.** The result of the evaluation on halal assurance system in this company found as many as 44% criteria in HAS 23101 assessed have been fulfilled by the company. The percentage of evaluation value is obtained from the number of scored criteria divided by the total score of the criteria.

Table 1. Fulfillment of halal assurance system criteria.

No	Criteria	Score	Fulfillment of HAS Criteria
1	Halal Policy	1	-
2	Halal Management Team	1	-
3	Training and Education	1	-
4	Materials	1	-
5	Products	1	√
6	Production Facilities	1	√
7	Written Procedures of Critical Activity		
	a. Selection of New Materials	1	-
	b. Purchase of Materials	1	√
	c. New Product / Product Development Formulation	1	√
	d. Inspection of Coming Materials	1	√
	e. Production	1	-
	f. Laundering Production Facilities and Additional Equipment	1	√
	g. Storage and Handling of Materials and Products	1	√
	h. Transportation	1	√
8	Tracing ability (*Traceability*)	1	-
9	Handling of Unsuitable Product Critera	1	-
10	Internal Audit	1	-
11	Management Review	1	-
	Total Number of Criteria Scores	18	

Description: √ : has been fulfilled
- : unfulfilled

4. Conclusion

The halal management team is the key to the implementation of HAS management, therefore the team needs to be institutionally established and the halal policy needs to be written to guarantee that constraints such as the Indonesian Ulema Council (MUI) extension can be resolved properly. All materials included in the halal control point mostly have halal certification except for fine granulated sugar, jam, flavor, and chocolate chip. The halal control point of the production process can be derived from contamination of haram and unclean materials, workers, environmental sanitation, and packaging process. The result of HAS evaluation found around 44% criteria in HAS 23101 have been fulfilled. It takes top management commitment and multidisciplinary cooperation of a good halal team to implement a sustainable halal assurance system.

References

[1] Pusat Data dan Sistem Informasi Pertanian. 2015. Food consumption statistic in 2015 (*Statistik konsumsi pangan 2015*).Retrived from www.epublikasi.setjen.pertanian.go.id. [Accessed on April 1st 2017]

[2] Badan Pusat Statistik. 2015. Statistics on the number of production and consumption of food in Indonesia Period 2009 – 2013 (*Statistik jumlah produksi dan konsumsi pangan di Indonesia periode 2009 – 2013*).Retrived from www.bps.go.id [Accessed on March 8th 2017]

[3] Apriyantono, A. 2015. Critical point halal bakery products and cakes (*Titik kritis kehalalan produk bakery dan kue*).www.antonapriyantono.net/titik-kritis-kehalalan-produk-bakery-dan-kue

[4] Yulianto and Putri N. 2016. The role of pastry section in reminiscent of income at RosIn Hotel Yogyakarta (*Peranan pastry section dalam meningatkan pendapatan di RosIn Hotel Yogyakarta*). *Jurnal Khasanah Ilmu* **7** (1): 3 – 4.

[5] Koswara, S. 2009. Bread processing technology (*Teknologi pengolahan Roti*).Retrived from www.tekpan.unimus.ac.id/wpcontent/uploads/2013/07/Teknologi-Roti-Teori-dan Praktek.pdf (Accessed April 11th 2017)

[6] Lembaga Pengkajian Pangan Obat-Obatan dan Kosmetika Majelis Ulama Indonesia (LPPOM MUI). 2013. HAS 23101 Guidelines for fulfilling the criteria of halal assurance systems in the processing industry (*HAS 23101 Pedoman pemenuhan kriteria sistem jaminan halal di industri pengolahan*). Penerbit LPPOM MUI. Jakarta.

[7] Jabatan Kemajuan Islam Malaysia (JAKIM). 2015. Manual procedure for Malaysia halal certification (3rd Revision) 2014. Firdaus Press. Selangor, Malaysia.

[8] Halal Hub Division. 2012. Guidelines for halal assurance management system of Malaysia halal certification.Retrived from www.halal.gov.my/v4/images/pdf/halalassurance system.pdf (Accessed July 27th 2017)

[9] Salhah, A., Agus S., and Widodo H. 2011. Employee obedience analysis in periodic medical examination at PKU Muhammadiyah Hospital Yogyakarta City (*Analisis ketaatan karyawan dalam pemeriksaan kesehatan berkala di Rumah Sakit PKU Muhammadiyah Kota Yogyakarta*). *Jurnal Kesehatan masyarakat* **5** (1): 11.

[10] Roswiem, A. P. 2015. Pocket Books Halal Products: Food and Drink (*Buku saku produk halal:makanan dan minuman*).Penerbit Republika. Jakarta

[11] Lembaga Pengkajian Pangan Obat-Obatan dan Kosmetika Majelis Ulama Indonesia. 2014. Halal Critical control point in bread product (*Titik kritis kehalalan bakery*). Retrived from www.halalmui.org/mui14/index.php/main/detil_page/48/2141 (accessed 21th March 2017).

[12] Juanda, D., dan Bambang C. 2009. Sesame, cultivation technique and farmer analysis (*Wijen, teknik budidaya dan analisis usaha tani*). Penerbit Kanisius. Yogyakarta.

[13] Jan, K. N., Abhishek D. T., Shubhendra S., Diksha S., and Surendra P. S. 2017. Enhanced Sorbitol Production Under Submerged Fermentation Using Lactobacillus Plantarum. *Journal Applied Food Biotechnology* **4** (2): 85.

[14] Dewi, H, K., Debra A. P., dan Arief W. 2013. Pre design of sorbitol plant from tapioca flour with catalytic hydrogenation (*Pra desain pabrik sorbitol dari tepung tapioka dengan hidrogenasi katalitik*). *Jurnal Teknik Pomits* **2** (1): 85.

[15] Yaqin, A. 2014. Halal in modern era: Discussion of halal haram food products, drugs and cosmetics around us (*Halal di era modern: kupas tuntas halal haram produk pangan, obat dan kosmetik di Sekitar Kita*). Majelis Ulama Indonesia Provinsi Jawa Timur. Surabaya.

IOP Conf. Series: Earth and Environmental Science **131** (2018) 012024 doi:10.1088/1755-1315/131/1/012024

Inventory control of raw material using silver meal heuristic method in PR. Trubus Alami Malang

D M Ikasari, E R Lestari and E Prastya

Department of Agro-industrial Technology, Faculty of Agricultural Technology, Universitas Brawijaya, Malang, Indonesia

E-mail: dhitamorita@gmail.com; thamauree@ub.ac.id

Abstract. The purpose of this study was to compare the total inventory cost calculated using the method applied by PR. Trubus Alami and Silver Meal Heuristic (SMH) method. The study was started by forecasting the cigarette demand from July 2016 to June 2017 (48 weeks) using additive decomposition forecasting method. The additive decomposition was used because it has the lowest value of Mean Abosolute Deviation (MAD) and Mean Squared Deviation (MSD) compared to other methods such as multiplicative decomposition, moving average, single exponential smoothing, and double exponential smoothing. The forcasting results was then converted as a raw material needs and further calculated using SMH method to obtain inventory cost. As expected, the result shows that the order frequency of using SMH methods was smaller than that of using the method applied by Trubus Alami. This affected the total inventory cost. The result suggests that using SMH method gave a 29.41% lower inventory cost, giving the cost different of IDR 21,290,622. The findings, is therefore, indicated that the PR. Trubus Alami should apply the SMH method if the company wants to reduce the total inventory cost.

1. Introduction

The cigarette industry is one of the industries that contribute greatly to Indonesian income. According to the Directorate General of Customs and Excise [1], the country's income from cigarette is IDR. 76.3 trillion or 72.89% of the total Indonesian income of IDR 104.7 trillion in 2013. However, the local cigarette industries in Indonesia have decreased in number every year. This declined is possibly due to tight competition, tight regulations related to cigarettes, and an increase of annual excise tariffs. Thus, the cigarette industries need to resolve those threats, for example, by reducing the production costs and improving the operational efficiency [2].

PR. Trubus Alami Malang is one of the companies producing cigarettes with brand SKT (Sigaret Kretek Tangan) Trubus Alami cigarettes. Currently, the demand for SKT Trubus Alami cigarettes is fluctuated every month. Therefore, the material requirement planning and the raw material inventory control in PR. Trubus Alami are currently estimated based on the previous requirements. By doing this practice, thus the company often experienced an excess in raw material inventory which may lead to the waste problems. The optimal and efficient methods to control the inventory is therefore critical.

Lot-sizing heuristic methods can be used to tackle the those inventory problems in PR. Trubus Alami. One of the lot-sizing heuristic methods is Silver Meal Heuristic (SMH), which is a method used to determine the lot size of the orders by carefully examined the inventory costs. Each productive system manager knows that finding the optimal trade- off between reducing inventory and decreasing the frequency of production/ replenishment orders allows a great cut- back in operations costs [3]. This method is more appropriately applied under fluctuating demand conditions [4]. This research was aimed at comparing the total inventory cost calculated using the method applied by PR. Trubus Alami and using the SMH method.

2. Research Methods

In this researh, the demand forecasting was carried out using five time series methods include multiplicative decomposition, additive decomposition, moving average, single exponential smoothing, and double exponential smoothing. The selection of the forecasting methods was carried based on their error values, calculated using the smallest error comparison analysis of the Mean Abosolute Deviation (MAD), Mean Squared Deviation (MSD), and Mean Absolute Percentage Error (MAPE) [5].

2.1 Product demand forecasting

The sales data used in this study was from the period of Juli 2014 to June 2016 (96 weeks), which was then used as a initial data to forecast the product demand. Product demand forecasting was estimated using the demand data of SKT Trubus Alami from the period of July 2016 to June 2017. The product demand forecast was then interpreted as the raw material requirement, and listed as a bill of materials (BOM).

2.2 Calculation of raw material requirement

A BOM was used to describe the requirement of the raw material in the company needed to produce SKT Trubus Alami.

2.3 SMH

The SMH finds the lowest average cost per period [6] . The main equation used is as follows :

$$\frac{TRC(T)}{T} = \frac{C + Ph\sum_{k=1}^{T}(k-1)Rk}{T}$$

Where:

C	= ordering cost per period	TRC (T)	= total relevant cost in period T
T	= addition time in period	Ph	= holding cost per period
P	= purchase cost per unit	Rk	= demand level of period k

2.4 Safety stock calculation and reorder point

2.4.1 Safety stock.

Safety stock is defined as inventory that is carried to prevent stock out and back order situations. [7], was carried out using the following formula [8] :

$$S = k\sqrt{LT(\sigma d^2)}$$

Where:

SS	= safety stock	LT	= lead time
k	= safety factor	σd	= standard deviation

2.4.2 Reorder point.

The calculation of reorder point, aimed at identifying the rate of re-ordering of the raw materials, was performed using the following formula [9]:

$$ROP = d \times L + SS$$

Where:

ROP	= reorder point	L	= lead time
d	= demand per unit time	SS	= safety stock

2.5 Comparison method

The obtained data were analysed by comparing the total inventory cost, which was counted using the SMH method and the company's method.

3. Results and Discussion

3.1 Demand Forecasting Results

Figure 1 shows the product's sales of SKT Trubus Alami Cigarettes from July 2014 to June 2016. The figure indicates that the cigarette sales were fluctuated, following the seasonal pattern. Despite a sharp decrease in week 51 with the value of less than 100,000 packages in general the average sales every week was in the range of 200,000 - 600,000 packages.

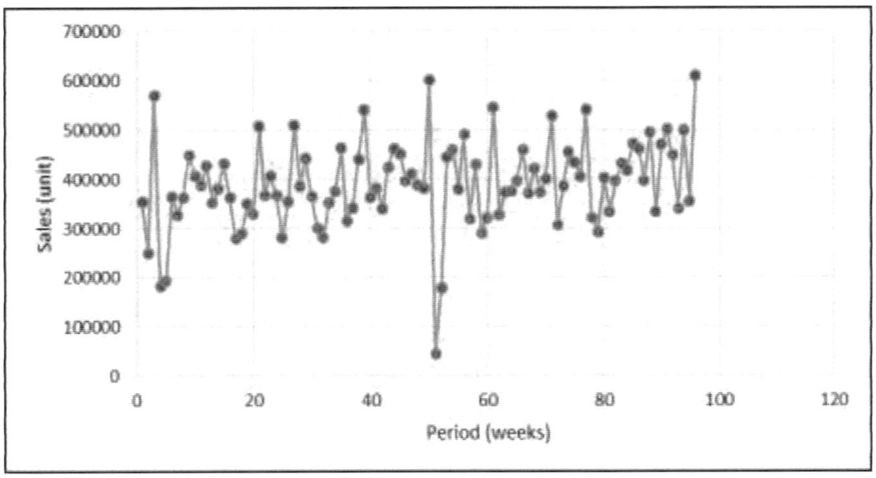

Figure 1. Sales of SKT Trubus Alami Cigarettes from July 2014 to June 2016

The error values for each forecasting method can be seen in Table 1. The table shows that the additive decomposition method had the smallest MAD, MSD and MAPE values compared to other methods. This was because the additive decomposition is an appropriate method for the time series data that is influenced by seasonal patterns and trends [10]. Therefore, the additive decomposition method was selected to be used for forecasting the product demand based on the sales values, and the result is shown in Table 2.

Table 1. Comparison of forecasting error values.

Forecasting Method	Error Value		
	MAPE (%)	MAD	MSD
Multiplicative Decomposition	19	57,018	5,956,036,618
Additive Decomposition	19	56,849	5,836,706,578
Moving Average	25	62,734	7,558,919,593
Single Exponential Smoothing	25	64,656	7,953,587,513
Double Exponential Smoothing	33.59	85,469	14,131,100,000

Table 2. Additive decomposition forecasting results.

Period	Sales (packages)	Period	Sales (packages)	Period	Sales (packages)	Period	Sales (packages)
1	429,105	13	551,408	25	379,830	37	443,336
2	470,927	14	416,007	26	473,052	38	507,369
3	470,341	15	407,780	27	481,001	39	620,200
4	421,311	16	449,387	28	435,401	40	421,273
5	411,647	17	411,535	29	575,295	41	469,683
6	475,680	18	525,286	30	427,891	42	482,965
7	588,511	19	277,995	31	420,454	43	515,221
8	389,584	20	336,191	32	414,331	44	520,090
9	437,995	21	555,259	33	460,794	45	583,097
10	451,276	22	492,337	34	502,615	46	447,695
11	483,532	23	471,488	35	502,029	47	439,468
12	488,402	24	506,344	36	452,999	48	481,076

3.2 BOM of SKT Trubus Alami Cigarettes

BOM of SKT Trubus Alami cigarettes can be seen in Table 3. The table shows that there were 13 ingredients required to produce the cigarette products. From these materials, nine materials were purchased, while four materials were resulted from the production process. The data from BOM and the demand forecast for SKT Natural Trubus Cigarettes were then used to determine the requirement of each raw material from July 2016 to June 2017.

Table 3. BOM of SKT Trubus Alami Cigarettes.

Item	Number of Unit	Unit	Description	Item	Number of Unit	Unit	Description
Cigarettes pack	1	package	Product	Clove	0.005073	kg	Purchase
OPP Plastic Packaging	1	piece	Purchase	TSS	0.016909	kg	Process result
Etiquette paper	1	sheet	Purchase	Sauce	0.000343	kg	Purchase
Cigarettes	12	pieces	Process result	TSL	0.017135	kg	Process result
Glue	0.000326	kg	Purchase	Fine Tobacco	0.010559	kg	Purchase
Ambri paper	12	sheets	Purchase	Rough Tobacco	0.005280	kg	Purchase

TSG	0.023765	kg	Process result	Virginia Tobacco	0.001760	kg	Purchase

3.3 Inventory Cost

In this study, the inventory costs calculated include ordering and holding costs. The ordering costs made by the company consisted of transportation costs, loading and unloading costs, and administrative costs (Table 4). The holding costs in the company consisted of lighting costs, PBB (or Land and Building Tax) fees, and cleaning costs (Table 5). Based on Table 4, it can be seen that both tobacco and clove materials were ordered by the company in a single order because they were bought from the same suppliers. While, other raw materials (i.e. sauce, glue, ambient paper, etiquette paper, and OPP plastic packages) were ordered twice a week.

3.4. Results of the Inventory Calculation in The Company

The results found that the quantity for the ordering and the safety stock of each material were based on the average end product demand of 80,000 packages / day or 480,000 packages / week. The inventory cost incurred by the company from July 2016 to June 2017 was IDR 72,400,538. In this case, the inventory control of raw material applied by the company was relied on a constant-time inventory. Also, the company applied 1 week of safety stock for tobacco and clove materials, as well as a minimum of 0.5 weeks of safety stock for sauce, ambient paper, etiquette paper, OPP plastic packaging, and glue.

Table 4. Ordering cost.

Ingredient	Ordering Cost (IDR per order)		Ingredient	Ordering Cost (IDR per order)	
	1x order	Per ingredient		1x order	Per ingredient
Fine Tobacco	577,789	269,107	Ambri Paper	312,500	312,500
Rough Tobacco		134,554	Etiquette Paper	312,500	312,500
Virginia Tobacco		44,851	OPP Plastic Packaging	345,846	345,846
Clove		129,277	Glue	312,500	312,500
Sauce	312,500	312,500			

Table 5. Holding cost.

Ingredient	Storage Cost (IDR/week)	Unit	Ingredient	Storage Cost (IDR/week)	Unit
Fine Tobacco	7.49088	kg	Ambri Paper	0.00233	sheet
Rough Tobacco	7.49088	kg	Etiquette Paper	0.03349	sheet
Virginia Tobacco	7.49088	kg	OPP Plastic Packaging	0.03349	pieces
Clove	5.22044	kg	Glue	51.78490	kg
Sauce	16.28491	kg			

3.5. The Results of SMH Method Calculation

Using the SMH method, the requirement of tobacco and clove materials was separately calculated, resulted in an ordering cost for three types of tobacco and clove materials at a value of IDR 577,789. The limit of the order capacity of 14,400 kg per order was also considered when calculating the cost for tobacco & clove materials, as shown in Table 6.

Table 6. SMH calculation result.

Combination Period	Lot Size Cumulative (kg)	Total Cost (IDR)	TRC/T (IDR)
1	4,531	577,789	577,789
1, 2*	9,504	615,039	307,520
1, 2, 3	14,470	689,447	229,816
3	4,967	577,789	577,789
⋮	⋮	⋮	⋮
47. 48*	9,720	615,842	307,921

Note: * = optimal

Table 6 indicates that the ordering of combinations of period 1.,2 was more optimal than that of periods 1,2,3, although the TRC / T value of the period 1,2,3 was smaller than that of the period 1.,2. This is because the lot size booking combination of period 1,2,3 has exceeded the order limit of 14,400 kg. SMH calculation can be performed even though the lowest cost value has not been reached. The total inventory cost resulted from SMH method was IDR 49,186,661, lower than that of calculated using the company's method**.** This indicated that applying SMH method can generate an effective ordering period and an optimal order amount, with a minimum total cost per period.

3.6. Calculation Result of Safety Stock

Lead time of tobacco and cloves was 1.67 weeks and for other materials was 2.33 weeks. The value of service level used in the company was 95%. The results of the safety stock calculation for each raw material were 1,421 kg of fine tobacco, 710 kg of rough tobacco, 237 kg of Virginia tobacco, 683 kg of cloves, 55 kg of sauce, 1,910,379 sheets of ambri paper, 159,198 sheets of etiquette paper, 159,198 pieces of OPP plastic packaging, and 52 kg of glue.

3.7. Calculation Result of Reorder Point

Reorder point of each raw material was 9,625 kg of fine tobacco, 4,813 kg of rough tobacco, 1,604 kg of Virginia tobacco, 4,624 kg of cloves, 427 kg of sauce, 14,963,332 sheets of ambri paper, 1,246,944 sheets of etiquette paper, 1,246,944 pieces of OPP plastic packaging, and 407 kg of glue.

3.8. Comparison Result of Inventory Methods

By using the SMH method, the ordering cost was IDR 35,736,711, the holding cost was IDR 15,373,165 and inventory cost was IDR 51,109,875. While, the calculation results using the company's method demonstrated that the ordering cost was IDR 66,034,200, the holding cost was IDR 6,366,337 and inventory cost was IDR 72,400,538. The total savings that can be made with SMH method was 21,290,663 or 29.407%, possibly due to a lower cost of ordering using SMH method was much lower than the cost of ordering using the company's method. Furthermore, the savings on SMH orders cover the larger holding costs of the company. This is related to the frequency of orders made on the SMH method. This study revealed that the ordering frequency obtained by SMH is lower than the company. The SMH is the optimum method for the company than others methods. This is because SMH method combines several booking periods in one booking period. The high of ordering frequency can have an impact on the lower cost of ordering, but it may result in higher holding costs due to the high amount of material stored [11].

4. Conclusion

IOP Conf. Series: Earth and Environmental Science **131** (2018) 012024 doi:10.1088/1755-1315/131/1/012024

The PR. Trubus Alami undertakes inventory control by ordering every 1 week of fine tobacco, rough tobacco, Virginia tobacco, and cloves with a safety stock of 1-week requirement. The inventory control of raw material for production of SKT Trubus Alami Cigarettes calculated using SMH method has a different ordering frequency for each raw material. The findings confirmed that the ordering frequency by SMH method is smaller than that of by the company's method, indicating that SMH method can be further used to minime the inventory cost paid by the company.

References

[1] Saraswati N, Topowijoyo, Yaningwati F 2015 Analisis *DU point system* sebagai salah satu alat mengukur kinerja keuangan perusahaan (Studi pada perusahaan rokok yang listing di bursa efek Indonesia tahun 2011-2013) (DU point system analysis as tools to measures company financial performance (case study of cigarettes companies listed in Indonesian stock exchange from 2011-2013)) *J. Adm. Bisnis* **23** 1 1-9 [In Indonesian]

[2] Pahlevi R, Handayani S 2013 Pengukuran kinerja keuangan perusahaan menggunakan analisis rasio keuangan dan metode *Economic Value Added* (EVA) (Studi pada perusahaan rokok yang terdaftar di bursa efek Indonesia periode tahun 2009-2011) (Measurement of company financial performance using financial ratio analysis and economic value added method (case study of cigarettes companies listed in Indonesian stock exchange from 2009-2011)) *J. Adm. Bisnis* **2** 1 118-127 [In Indonesian]

[3] Baciarello L, Avino M, Onori R and Schiraldi M 2013 Lot sizing heuristics performance *Intl. J. of Eng. Business Mgt* **5** 6

[4] Mbota W, Tantrika M, Eunike A 2015 Perencanaan persediaan bahan baku dan bahan bakar dengan *dynamic lot sizing* (Studi Kasus: PT Holcim Indonesia Tbk, Tuban Plant) (Inventory planning for raw material and fuel using dynamic lot sizing (case study of PT Holcim Indonesia Tbk, Tuban Plant) *J. Rekayasa dan Manajemen Sistem Industri* **3** 1 178-188 [In Indonesian]

[5] Sukerti N 2015 Peramalan menggunakan deret waktu S-curve dan quadratic trend model (Forecasting using time series S-curve and quadratic trend model) *Konf. Nas. Sistem & Informatika* **2** 1 592-597 [In Indonesian]

[6] Lembke R 2002 Average cost lot sizing heuristics : revised stopping conditions with increased intuition *Production and Inventory Mgt J.* **1** 6-8

[7] Radasanu A 2016 Inventory management, service level and safety stock *J. of Public Administration, Finance and Law* **9** 145-153

[8] Taryana N 2008 Analisis pengendalian persediaan bahan baku pada produk sepatu dengan pendekatan *lot sizing* dalam mendukung sistem MRP (Analysis of raw material inventory control on shoes product using lot sizing approach to support MRP system) Undergraduate Thesis Institut Pertanian Bogor Bogor Indonesia [In Indonesian]

[9] Das A 2016 An Introduction to Operations Management: The Joy of Operations Routledge New York USA

[10] Madinah N, Sumantri Y, Azlia W 2015 Penentuan metode *lot sizing* pada pengadaan bahan baku kikir dan mata bor (Studi Kasus: PT X, Sidoarjo) (Identification of lot sizing method on raw material inventory planning for metal files and drill bit) *J. Rekayasa dan Manajemen Sistem Industri* **3** 3 505-515 [In Indonesian]

[11] Jauhari A 2006 Model persediaan terintegrasi pada sistem *supply chain* yang melibatkan pemasok pemanufaktur dan pembeli (Integrated inventory model in supply chain system involving manufacturing suppliers and consumer) *J. Ilmiah Teknik Industri* **5** 2 82-88 [In Indonesian]

Implementation of lean manufacturing for frozen fish process at PT. XYZ

D T Setiyawan, H R Pertiwijaya and U Effendi

Department of Agro-industrial Technology, Faculty of Agricultural Technology, Universitas Brawijaya, Malang, Indonesia

E-mail: danangtriagus@ub.ac.id

Abstract. PT. XYZ is a company specialized in the processing of fishery products particularly in frozen fish fillet. The purpose of this research was to identify the type of waste and determine the recommendations of minimizing waste Lean manufacturing approach was used in the identification of waste by describing the Value Stream Mapping (VSM) and selecting tools in the Value Stream Analysis Tools (VALSAT). The results of this research showed that the highest waste that generated was the defect of leak packaging on fillet products with an average of 1.21%. In addition to defect, other insufficiencies were found such as: unnecessary motion, unnecessary overhead, and waiting time. Recommendations for improvements that given include reduction of time at several stages of the process, making production schedules, and conducting regular machine maintenance. VSM analysis shows reduced lead time of 582.04 minutes to 572.01 minutes.

1. Introduction

PT. XYZ is a company specialized in the processing of fishery products, particularly in freezing of fish. The products of this company are: red snapper, anggoli, grouper, nike anchovy, *kapasan* which later will be processed into fillet, whole round, whole gutted, whole gilled and gutted, and whole gilled gutted and scaled fish products. The products are exported to foreign countries such as China, Vietnam, Malaysia, America, European Union and Australia. PT. XYZ aim to improve productivity. One of the efforts is minimizing the waste that generated during the production process. Among several types of waste that generated is a defect in the form of leak packaging of fillet product. The company target is a zero defects, but the defect still exceeds the expected target.

The approach that can be used to identify and minimize waste is lean manufacturing. Lean manufacturing is a quality management philosophy that focuses on reducing waste and eliminating unnecessary steps in the operations [1]. The methods that can be used in lean manufacturing are Value Stream Mapping (VSM) and Value Stream Analysis Tools (VALSAT). VSM is a visual presentation of the current process flow, which is then analyzed for its improvement to produce future state map [7]. While VALSAT serves as a method to

IOP Conf. Series: Earth and Environmental Science **131** (2018) 012025 doi:10.1088/1755-1315/131/1/012025

analyze the most dominant waste generated in the production process and to provide recommendations for improvement [3].

2. Research method

The scope of the study was limited to red snapper fillet products. There were 12 types of waste observed. However, this study did not analyze the production cost. The study was conducted until giving a recommendation of improvement. The data collection was divided into two, namely primary and secondary data collection. There were several stages in the present study as follow:

1. Determination of the standard time

At this stage, observations were conducted by controlling the processing time using a stopwatch. The formula for calculating cycle time, normal time, and standard time has been described elsewhere[6]:

Cycle time formula:

$$Ws = \frac{\sum Xi}{N}$$

Note:

Xi = observation time

N = number of observations

Normal time formula:

Wn = Ws x Rf

Note:

Ws = cycle time

Rf = rating factor

Standard time formula

$$Wb = Wn \text{ x } \frac{100\%}{100\% - \%\, allowance}$$

Note:

Wb = standard time (default time)

Wn = normal time

2. Creation of Current State Map
3. Identification of Waste
4. Selection of VALSAT Tools
5. Creation of Future State Map

3. Results and Discussion

One of the fish products manufactured by PT. XYZ is a frozen fish fillet. The type of fillet that observed was red snapper fillet with natural cut and skin-on weighing 6-10 oz (170.1-226.8 gram).

3.1 Determination of the Standard Time

The data required in the standard timing was observation of 14 stages of the process measured by using a stopwatch. The first step was to select a trained worker to be observe at each stage of the process. One calculated cycle is the production process in a master carton pack containing ± 20 fillets or equivalent to ± 10 fish weighing ± 4.54 kg. Data cycle time (Ws), normal time (Wn), and standard time (Wb) are shown in Table 1.

Table 1. Data of Ws, Wn, Wb.

No	Process	Cycle Time (minutes)	Normal Time (minutes)	Standard Time (minutes)
1	Materials Sorting	1.45	1.65	2.02
2	Washing 1	2.38	2.72	3.27
3	Scaling	2.96	3.47	4.20
4	Filleting	9.41	11.39	13.72
5	Separation of thorns	3.36	3.90	5.54
6	Trimming	8.94	10.55	12.64
7	Sizing	0.91	1.01	1.24
8	Washing 2	1.57	1.79	2.17
9	Wrapping	3.29	3.75	4.29
10	Vacumming	1.34	1.65	2.03
11	Arrangement of fish	0.62	0.68	0.78
12	Freezing	480.00	480.00	480.00
13	Metal Detecting	0.33	0.37	0.40
14	Secondary Packaging	1.34	1.53	1.75
	Total	517.9	524.46	534.05

Source: Processed Primary Data (2016)

3.2 Creation of Current State Map

Current state map was made based on the current condition of the red snapper fillet production process. Movement of material through various processes/facilities during manufacturing is shown in this map [4]. Current state map production process of red snapper fillet at PT. XYZ is shown in Figure 1. The analysis performed when making the current state map was by grouping activity time during production process into value added (VA), non-value added (NVA), and necessary but non-value added (NNVA). The main key in grouping these activities is to take the consumer perspective that can also be used to differentiate waste [9]. Current state map has a total time of VA of 530.39 minutes and NVA (including NNVA) of 51.65 minutes.

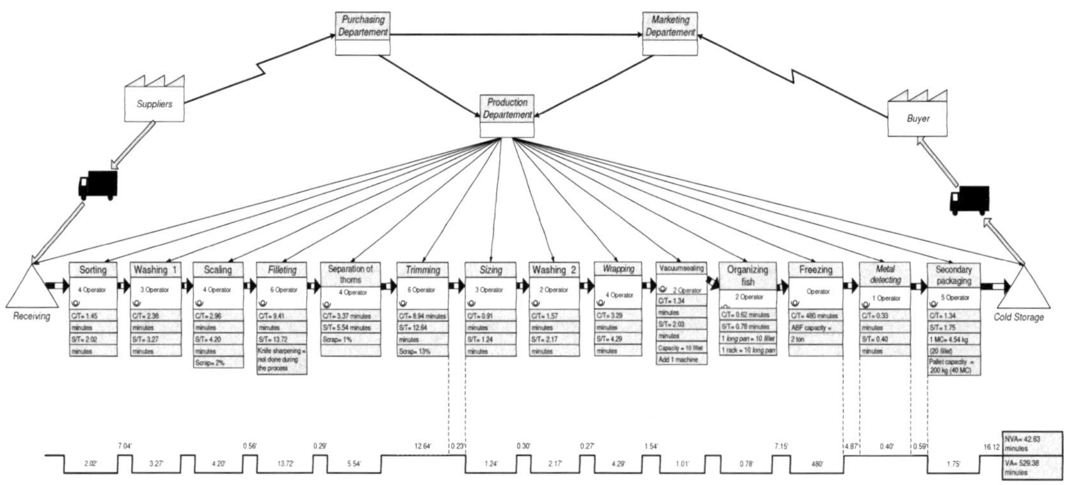

Figure 1. Current State Map

3.3 Identification of Waste

The waste identification was performed by recasting the results of the 12 waste questionnaires that had been distributed to the plan manager, the quality assurance manager, the head of production department, and the supervisors of the production lines. Wastes that have the highest value was the most dominant waste occurs ie defect with an average value of 2.67. While the waste with the lowest value was overproduction worthed an average of 0. The result of questionnaires distributed to respondents is presented in Table 2.

Table 2. Result of questionnaires

No.	Waste/inefficiencies	Mean	Rangking
1	Overproduction	0.00	8
2	Waiting	0.78	3
3	Transportation	0.67	4
4	Inappropriate processing	0.56	5
5	Unnecessary inventory	0.56	5
6	Unnecessary motion	1.22	2
7	Defect	**2.67**	**1**
8	Power and energy	0.78	3
9	Human Potential	0.44	6
10	Environmental pollution	0.56	5
11	Unnecessary overhead	1.22	2
12	Inappropriate design	0.22	7

Source : Processed Primary Data (2016)

3.4 Selection of VALSAT Tools

The selection of tools at this stage was based on the correlation value between each tool with the dominant waste. The tools chosen in this research were the overall supply chain effectiveness mapping. Overall supply chain effectiveness mapping provides a measure of the total effectiveness of each process area or part of the supply chain. Overall supply chain effectiveness mapping can be calculated using the following formula [8]:

Supply Chain Effectiveness= *Components Available for use x On Time Delivery Performance x Quality of Incoming Goods*

The value of component available for use and on time delivery performance of each process was obtained from Gantt charts based on the data of measured time during observation. While quality of incoming goods is obtained from the observation, expert opinion and historical data company. The overall supply chain effectiveness of the red snapper fillet production process is presented in Figure 2. Some stages of the process have a low percentage of effectiveness. This was due to the inaccuracy of time by means of cart displacing between processes and the waiting times. In addition, the presence of vacuum plastic packaging leaked in the vacuum process can cause the percentage of subsequent processes to be low.

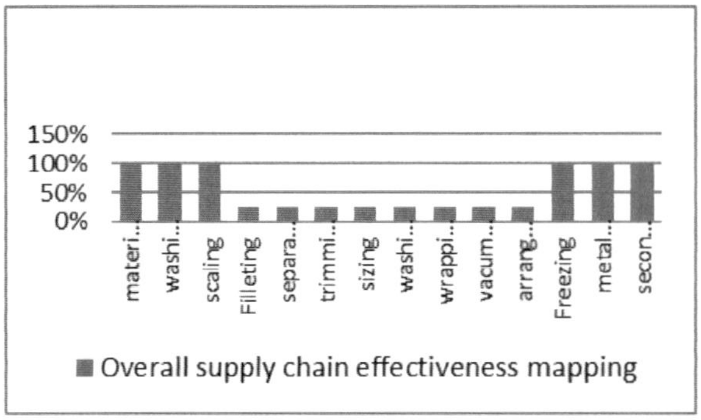

Figure 2. *Overall supply chain effectiveness* of red snapper fish fillet

3.5 Causes of Waste Analysis

Analysis of the cause of waste can be done after viewing the current state map or through VALSAT tools selected. The dominant waste, which was the defect due to a leaky vacuum plastic. The average percentage of the leaky vacuum plastic on the quality filter mapping was 1.21%. As a result of this waste, there was a re-processing that will certainly increase the production lead time. Fishbone diagram of the defect is shown in Figure 3.

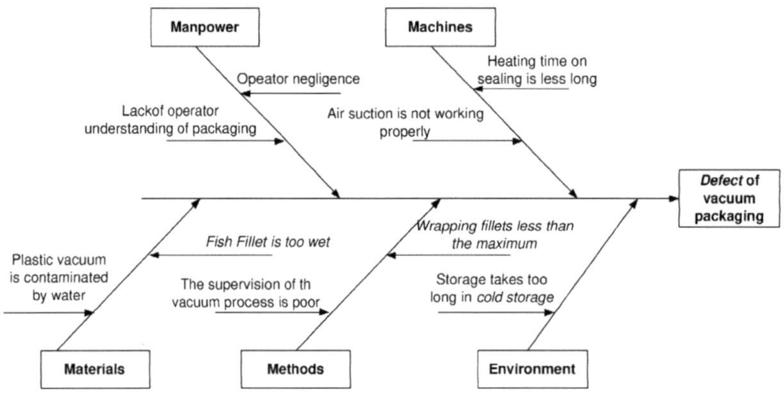

Figure 3. Fishbone diagram of defect

3.6 Creation of the future state map

The futures state map shows the output of the proposed improvements through an overview of the current state map [10]. The main focus of the creation of future state maps was the minimization of time on non-value added activities. At the current state map it was recorded that the total time of non valued added was 51.65 minutes. After the recommendation of improvement with certain assumptions, the total time of non value added was reduced to 42.63 minutes. Details of time reduction can be seen in Table 3. Future state map of red snapper fillet process is shown in Figure 4.

IOP Conf. Series: Earth and Environmental Science **131** (2018) 012025 doi:10.1088/1755-1315/131/1/012025

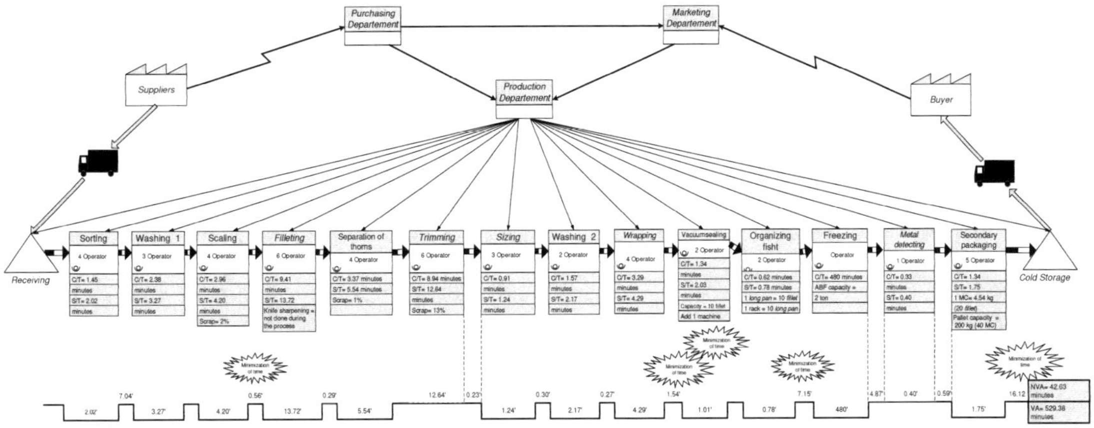

Figure 4. Future State Map

The following is an explanation of the proposed reduction time:

a. Waiting time due to knife blade activity is done when the process resulted in the accumulation of raw materials of fish to be in filleting.

b. Stacking on vacuum processing can be overcome by adding 1 vacuum sealer machine. The addition of vacuum sealer machine is based on brainstorming and discussion with the company to avoid bottle neck. The waiting time for the previous vacuum process will be reduced to 1.19 minutes (2.39 minutes: 2), while the vacuum processing time will be reduced from 2.03 minutes to 1.01 minutes (2.03 minutes: 2). If the capacity of red snapper fish fillet in 1 day is 80 MC, then the amount of time deducted for such activity is 95.53 minutes (1.19 minutes x 80 MC) and the amount of vacuum time is reduced to 81.09 minutes (1.01 minutes x 80 MC).

c. In the secondary packaging process with 4 operators, there is full pallet waiting activity before being taken to cold storage room. The full pallet waiting time is 17.50 minutes. Recommendations are given to optimize the workforce that is not currently performed in production activities.

Table 3. Details of reduction time.

No	Activity	Current state time (minutes)	Future state time (minutes)	Reduction time (minutes)	Reduction time (minutes)
1	Sharpen the fillet knife	1.26	0.00	1.26	10.08
2	Waiting for the vacuum sealer is ready	2.39	1.19	1.19	95.53
3	Vacuum	2.03	1.01	1.01	81.09
4	Waiting for the push rack is full	6.14	3.07	3.07	49.12
5	Waiting for the pallet is full	17.50	14.00	3.50	7.00
	Total	29.32	19.27	10.03	

4. Conclusion

From the research, there are 12 identified waste. Wastes with the highest ratings were defect (2.67), unnecessary motion (1.22), unnecessary overhead (1.22), waiting (0.78), and power and energy (0.78). The percentage defect of leaked primary packaging was 1.21%. Identified waste occurred due to the uncertain arrival of raw materials, the timing of delay, leaky vacuum plastic packaging, and the exhaustion and negligence of workers during the production process. Recommendations for improvements that given include reduction of time at several stages of the process, making production schedules, and perform regular machine maintenance. VSM analysis shows reduced lead time of 582.04 minutes to 572.01 minutes.

IOP Conf. Series: Earth and Environmental Science **131** (2018) 012025 doi:10.1088/1755-1315/131/1/012025

References

[1] Black, K 2010 Business Statistics for Contemporary Decision Making John Wiley & Sons, Inc Canada

[2] Hines, P, Nick R, John B , David B, and David T 1998 Value Stream Management *International Journal of Logistics Management*, 9(1): 25-42

[3] Intifada, G S and Witantyo 2012 Minimasi *Waste* (Pemborosan) Menggunakan *Value Stream Analysis Tool* untuk Meningkatkan Efisiensi Waktu Produksi (Studi Kasus PT Barata Indonesia, Gresik) Jurnal Teknik POMITS, 1(1): 1-6 [in Indonesian]

[4] Kumar, M , J Antony, R K Singh, M K Tiwari, and D Perry 2006 Implementing the Lean Sigma Framework in an Indian SME: A Case Study *Production Planning & Control*, 17(4): 407-423

[5] Melton, T 2005 *The Benefits of Lean Manufacturing: What Lean Thinking has to Offer the Process Industries Chemical Engineering Research and Design*, 83(A6): 662–673

[6] Setiyawan, D T , Sudjito S , and Rudy S 2013 Minimasi *Waste* untuk Perbaikan Proses Produksi Kantong Kemasan dengan Pendekatan *Lean Manufacturing* JEMIS, 1(1): 8-13 [in Indonesian]

[7] Sukmoro, W 2010 *Turning Loss into Profit* PT Gramedia Pustaka Utama Jakarta [inIndonesia]

[8] Taylor, D and David B 2001 *Manufacturing Operations and Supply Chain Management* Thomson London

[9] Teichgräber, U K and Maximilian de B 2012 *Applying Value Stream Mapping Techniques to Eliminate Non-Value-Added Waste for The Procurement of Endovascular Stents European Journal of Radiology* **81** 47-52

[10] Tyagi, S , Alok C , Xianming C , and Kai Y 2015 *Value Stream Mapping to Reduce the Lead-Time of a Product Development Process Int. J. Production Economics*. 160: 202–212

The impact of Indonesia's food law reform on the concept of food sovereignty in Indonesia

E O H Soetoto

Graduate Studies of Law, Universitas Bhayangkara Jakarta Raya, Indonesia

E-mail: erwin.hermansyah@gmail.com

Abstract. Food sovereignty, since first articulated in 1996, evaluations of the practice and potential of this emerging concept and its sub-elements got fast popularity, which first emerged from civil society publications. Indonesia experienced food law reform from Law number 7 of 1996 to Law number 18 of 2012.The purposes of the work were: (1) to understand the reform process from the old to the new Indonesia's food law, (2) to investigate the impact of Indonesia's food law reform on the concept of food sovereignty in Indonesia. The approach was doctrinal research. The paper used secondary data with primary, secondary, and tertiary legal materials. The scientific innovation of this paper is the use of legal systems to strengthen the food sovereignty in Indonesia, suggesting that the law is needed to ensure the food sovereignty in Indonesia. The results show that there is a paradigm shift in the Indonesia's food law from food security and food self-sufficiency towards food sovereignty and in the content materials of new Indonesia's food law which include the concept of food sovereignty. The present reasearch concluded that there is a reform in Indonesia's food law and there is an impact - in term of legal substance, legal structure, and legal culture- of Indonesia's food law reform on the concept of food sovereignty in Indonesia.

1. Introduction

The food sovereignty approach can be distinguished as an "epistemic shift" in which value relations, approaches to rights, and a shift from an economic to an ecological calculus concurrently challenge the rules and relations of a corporate or neoliberal food regime. [1] One of the most salient shifts has been in the value relations, justification regimes, and frames around the concept of food security as it has been challenged by a new conceptualizations of food sovereignty. [2-6]

Food security—frames as a universal ideal to prevent world hunger—emerged as a post-World War II development principle enshrined in the 1948 Universal Declaration of Human Rights and the 1966 International Covenant of Economic, Social, and Cultural Rights. As defined by Food and Agriculture Organization [7], food security exists when "all people, at all times, have physical, social, and economic access to sufficient, safe, nutritious food that meets their dietary needs and food preferences for an active and healthy life."

As critiqued by several authors writing from a food sovereignty perspective, this definition of food security treats food as a problem of insufficient trade rather than hunger by privileging access to food rather than control over systems of production and consumption. In this concept, food is a tradable commodity rather than a right, and hunger simply a problem of distribution [8,9]. Food sovereignty, since first articulated in 1996, evaluations of the practice and potential of this emerging concept and its sub-elements were quick in forthcoming, first emerging from civil society publications [10-12].

Before going into those definitions and contradictions, though, it is worth constrasting food sovereignty with the concept against which it has traditionally been ranged – food security. The Food and Agriculture Organization of the United Nations (FAO) has done a fine job of tracking the evolution of 'food security' [13], but it is useful to be reminded that the official definition in 1974 of 'food security' was the availability at all times of adequate world food supplies of basic foodstuffs to sustain a steady expansion of food consumption and to off-set fluctuations in production and prices. [14]

Via Campesina's position was t a discussion of internal political arrangements and was a necessary part of the substance of food security. Indeed, food sovereignty is declared as a logical precondition for the existence of the food security: Long-term food security depends on those who produce food and care for the natural environment. As the stewards of food producing resources we hold the following principles as the necessary foundation for achieving food security. Food is basic human right. The right can only be realized in a system where food sovereignty is guaranteed. Food sovereignty is the right of each nation to maintain and develop its own capacity to produce its basic food respecting cultural and productive diversity. We have the right to produce our own food in our own territory. Food sovereignty is precondition to genuine food security. [15]

The term has changed over time, just like 'food security', but while it is possible to write an account of the evolution of 'food security' with reference to changing international politics, it is much harder to make coherent the changes with 'food sovereignty'. From the core of the 1996 definition, consider this one, written six years later: Food sovereignty is the right of people to define their own food and agriculture; to protect and regulate domestic agricultural production and trade in order to achieve sustainable development objectives; to determine the extent to which they want to be self reliant; to restrict the dumping of product in their markets; and to provide local fishes-based communities the priority in managing the use of and the rights to aquatic resources.

Food sovereignty does not negate trade, but rather, it promotes the formulation of trade policies and practices that serve the right of the people to safe, healthy, and ecological sustainable production. [16] Law No. 18 of 2012 on Food is one of the products of legislation as a consequence of Indonesia's membership in WTO. These conditions, poses certain problems for Indonesia as a sovereign state. On the one hand do not want ostracized internationally that participate in international association with a member of the WTO, on the other hand have to defend the sovereignty of the nation and the welfare of the whole society, including traditional farmers.

Indonesia experienced food law reform from Law number 7 of 1996 to Law number 18 of 2012. [17,18] The purposes of the present work were : (1) to understand the reform process from the old to the new Indonesia's food law, (2) to investigate the impact of Indonesia's food law reform on the concept of food sovereignty in Indonesia.

2. Materials and Methods

The approach which is used in this paper is doctrinal research. The model of this research is normative juridical or literature legal research. In a normative legal research, the collection of research materials is done in libraries, the official websites which can be a reference to the data.

The paper uses secondary data with primary, secondary, and tertiary legal materials. The secondary data which is used in this paper is the literature from the library. The primary legal materials are Regulation Legislation such as Indonesia's Law number 7 of 1996 and Indonesia's Law number 18 of 2012, both are on Food, while the secondary legal materials sourced from theory, research and scientific opinion, literature books related to the research topic and journals, and the third legal materials are dictionaries related to the topic of study such as law dictionary and the encyclopedia.

The data obtained were analyzed by using qualitative data analysis method where all the data that have been obtained were systematically compiled and analyzed from the perspective of researcher and then pulled towards a conclusion. Therefor, the nature of this research was an analytical perspective.

There are three aspects of legal system: substance, structure, and legal culture. [19] The legal system is used as a scientific innovation to increase the food sovereignty in Indonesia.

3. Results and Discussions

The results show that there are paradigm shifts in the Indonesia's food law from food security and food self-sufficiency into food sovereignty and in the content materials of new Indonesia's food law have concluded the concept of food sovereignty. The doctrinal research, in this case the normative juridical research method is a research method concerning the legal system and the principles of law.

3.1. Legal Substance, Legal Structure, and Legal Culture

The first discussion is the analysis and evaluation regarding the compatibility with legal system, such as legal substance, legal structure, and legal culture to increase food sovereignty in Indonesia. The first analysis and evaluation is about legal substance regarding Article 53 Law No. 18 of 2012 on Food. This law is about prohibition of hoarding food, delegate government regulations to further regulate, and government regulations further stipulated in trade minister regulation, but that trade minister's regulation has not yet been issued, so that article 53 has not been implemented optimally. The mandate of article 151 Law No. 18 of 2012 on food to establish a food agency has been implemented at the latest November 2015, but at this time the institution is not yet established.

The second analysis and evaluation is about legal structure regarding the government agency that handles food has not been established, which is directly under and accountable to the president, as mandated by Article 126 of the Food Law. This institution is expected to reduce the inconsistent handling of food sovereignty issues among government sectors in agriculture, trade, and State-Owned Enterprises (SOEs). The third analysis and evaluation is about legal culture regarding increase in population, with a population of 252 million but the proportion of the poor is still large at around 11.4 per cent, the level of dependence on rice is still high, even increases in each region (where according to data 2014 per capita consumption of 114.8 kg). As I mentioned earlier, that in Indonesia the regulation on food has been stipulated in Law No. 18 of 2012 which was previously regulated in Law No. 7 of 1996.

We need to understand that food security and food sovereignty are not the same things. The emergence of the concept of food sovereignty is the answer to the failure of food security that has been applied so far. Referring to the previous law on food, Law No. 7 of 1996, the concept of food security implemented by the government is limited to "the condition of the fulfillment of food for the household which is reflected in the availability of adequate food, both quantity and quality, safe, even , and affordable. " In the new law, the Law No. 18 of 2012 commands that food provision is carried out to meet basic human needs that provide benefits equitably, evenly and sustainably based on national food sovereignty, food self-sufficiency, and food security. In the law number 18 of 2012 on food explains that the organization of food aims to improve the ability of producing food independently, providing food diverse, and meet the requirements of security, quality, and nutrition for public consumption, increasing the level of food sufficiency, especially food with reasonable prices and affordable according to the needs of the community. Therefore there is a paradigm shift in the Indonesia's food law from food security and food self-sufficiency towards food sovereignty.

3.2. Legislation Principles

The second discussion is the analysis and evaluation regarding the compatibility with legislation principles. Evaluation is performed on Law No. 18/ 2012 concerning regulations on food. Food law consists of 154 articles and it applies entirely. Every article has complied with all principles which are: A. Clarity of formulation principle, that every legislation should fulfil legislation preparation technical requirement, the systematics, diction or terminology, and clear legal language, and easy to understand, so it does not generate any kind of interpretation, B. Content material principle, that every legislation should reflect content material principles as arranged in article 6 Law No. 12 of 2011 about formulation of legislation [20], those principles are: a. Protection: Legislation content material should function as the protector for public tranquility. b. Humanity: Every legislation content material should reflect protection and respect for human right and prestige and dignity of every citizen and population of Indonesia dan propotionately. c. Nationality: Every legislation content material should reflect the nature and character while maintaining the principle of the unitary state of the Republic of Indonesia or plural nation of Indonesia. d. Kinship: Every legislation content material should reflect deliberations for consensus in taking any decision. e. Archipelagic: Every legislation content material

IOP Conf. Series: Earth and Environmental Science **131** (2018) 012026 doi:10.1088/1755-1315/131/1/012026

always pays attention to the interests of the entire territory of Indonesia and material content of legislation made in the region is part of the national legal system based on *Pancasila* (Indonesia's basic principles and way of life) and the Constitution of the State of Indonesia Year 1945. f. *Bhineka Tunggal Ika* (Unity in Diversity principle of Indonesia): Every legislation content material should pay attention to the diversity of population, religion, ethnicity, and class, special condition of area, and culture in society, nation, and state life. g. Justice: Every legislation content material should reflect proportional justice for every citizen. h. Equal status in law and government: Every legislation content material should not contain things that are distinguishable based on background, including religion, ethnicity, race, class, gender, or social status. i. Order and legal certainty: and / or every legislation content material must be able to realize order in society through guarantee of legal certainty. j. Balance, Elegance, and Harmony: Every legislation content material should reflect balance, elegance, and harmony must reflect balance, elegance, harmony between the interests of individuals, society, the interests of the nation, and the state. Beside that, the concept of food sovereignty includes the regulation of food institutional, local food rules, and the rules of food insecurity. These concepts have been embodied in the new food law. Thus, the content materials of new Indonesia's food law have concluded the concept of food sovereignty.

4. Conclusions

Based on the purposes of the work and the foregoing discussion two conclusions can be settled as follows: First, that there is a reform in Indonesia's food law because there is a paradigm shift in the Indonesia's food law from food security, and food self-sufficiency into food sovereignty. Second, there is an impact – in term of legal substance, legal structure, and legal culture- of Indonesia's food law reform on the concept of food sovereignty in Indonesia because the content materials of the new Indonesia's food law have concluded the concept of food sovereignty. Some suggestions are given related to the discussion in this paper: first, because the paradigm has shift from food security into food sovereignty, so there must be a national food sovereignty council. Second, this council should have the authority to coordinate, organize and direct cross-ministries / sectors in various food-related policies, and this council should be directly under and accountable to the president.

References

[1] McMichael, Philip 2009 A Food Regime Genealogy *Journal of Peasant Study* **36** 1 139-169

[2] Boyer, Jeff 2010 Food Security, Food Sovereignty, Local Challenges for Transnational Agrarian Movement: Honduras Case *Journal of Peasant Study* **37** 2 319-351

[3] Fairbairn, Madeleine 2010 Framing Resistance: International Food Regimes and the Root of Food Sovereignty Pp. 15-31 in *Food Sovereignty: Reconneting Food, Nature and Community,* ed. Hannah Wittman, Annette Desmarais, and Nettie Wiebe. Halifax: Fernwood.

[4] Gonzales, Humberto 2010 Debate on Food Security and Agrofood World Governance *International Journal od Food Science and Technology* **45** 7 1345-1352

[5] Mooney, P.H., and S.A. Hunt 2009 Food Security: The Elaboration and Contested Claims to a Consensus Frame *Rural Sociology* **74** 4 469-497

[6] Windfuhr, Michael, and Jennie Jonsén 2005 *Food Sovereignty: Towards Democracy in Localized Food Systems.* Bourton-on-Dunsmore, UK: FIAN-International; ITDG Publishing.

[7] Food and Agriculture Organization (FAO) 2010 The State of Food Security in the World 2010: Addressing Food Insecurity in Protracted Crises. Rome: Food and Agriculture Organization of the United Nations.

[8] McMichael, Philip 2004 Development and Social Change: A Global Perpective. Thousand Oaks, CA: Pine Forge Press.

[9] Patel, Raj 2010 What Does Food Sovereignty Look Like? Pp. 186-196 in *Food Sovereignty: Reconneting Food, Nature and Community,* ed. Hannah Wittman, Annette Desmarais, and Nettie Wiebe. Halifax: Fernwood.

[10] Bagambanya, Jean Bakole 1999 The Food Security Initiative: An Interview with Jean Bakole Bagambanya *Multinational Monitor* **20** (1/2) 27-29

[11] Menezes, Francisco 2001 Food Sovereignty: A Vital Requirement for Food Security in the Context of Globalization *Development* **44** 4 29-33

[12] Roberts, Wayne 2003 The Whole Enchilada *Alternative Journal* **29** 4 3.

[13] Food and Agriculture Organization (FAO) 2003 Trade reforms and food security: conceptualising the linkages. Rome: Commodity Policy and Projections Service, Commodities and Trade Division

[14] United Nations 1975 Report of the World Food Conference, Rome 5-16 November 1974. New York: Unites Nations

[15] Via Campesina 1996 The right to produce and access to land. Voice of the Turtle. Retrieved from http://www.voiceoftheturtle.org/library/1996%20Declaration%20of%20Food%20-Sovereignty.pdf. [Accessed on 2 October, 2017]

[16] Peoples Food Sovereignty Network 2002 Statement on peoples' food sovereignty. Retrieved from http://www.peoplesfoodsovereignty.org/statements/new%20statement/statement_01htm. [Accessed on 2 October, 2017]

[17] Republic of Indonesia 1996 Law Number 7 of 1996 on Food

[18] Republic of Indonesia 2012 Law Number 18 of 2012 on Food

[19] Friedman, Lawrence M., Grant M. Hayden 2016 *American Law: An Introduction*, third edition. London: Oxford University Press

[20] Republic of Indonesia 2011 Law Number 12 of 2011 on Formulation of Legislation

IOP Conf. Series: Earth and Environmental Science **131** (2018) 012027 doi:10.1088/1755-1315/131/1/012027

Firm performance model in small and medium enterprises (SMEs) based on learning orientation and innovation

E R Lestari, F L Ardianti and L Rachmawati

Department of Agro-industrial Technology, Faculty of Agricultural Technology, Universitas Brawijaya, Malang, Indonesia

E-mail: endahlestari24@yahoo.com

Abstract. This study investigated the relationship between learning orientation, innovation, and firm performance. A conceptual model and hypothesis were empirically examined using structural equation modelling. The study involved a questionnaire-based survey of owners of small and medium enterprises (SMEs) operating in Batu City, Indonesia. The results showed that both variables of learning orientation and innovation effect positively on firm performance. Additionally, learning orientation has positive effect innovation. This study has implication for SMEs aiming at increasing their firm performance based on learning orientation and innovation capability.

1. Introduction

In late 1990's, innovation became an essential agenda for managers to discuss. Innovation commonly was perceived as an imperative requirement of successfulness and sustainability of an organization. Furthermore, it turned into an interesting topic due to role in explaining sustained competitive advantage. Innovation as the development and use of new ideas or behaviors in organizations can be instigated in a new product, service, method of production (technical innovation) or a new market, and organizational structure or administrative system [1]. Innovation is critical since it makes an organization able to adapt to market change and meet new demand. Innovation also helps an organization to respond environmental changes [2]. It reflects the importance of innovation for organization in a recent business environment to help an organization facing both internal and external environment's turbulence [3-5].

Many researchers have found evidence for a relationship between innovation and performance [6-8]. Furthermore, Farrell [9] and Giniuniene and Jurksiene [10] argued that learning orientation is a source of firm's competitive advantage. In addition, learning orientation is important for a firm's competitive advantage and literature has generally focused on the effects of learning orientation on financial performance [11]. Finally, Keskin [12] indicates that the majority of empirical studies concerning innovation and its drivers focus on large scale companies in developed countries, while ignoring, to some extent, small firms in developing countries. Simon et al. [13] assert that the existence of innovation products is important for small industries. They argue that to satisfy customers, small industries must offer quality products. Many obstacles for small industries to innovate include limited financial resources, human resource quality, and risk that will arise if it fails [14, 15]

This study examined the relationships between learning orientation, innovation, and firm performance in the SMEs apple product-based business in Batu City, Indonesia. The hypothesis research model is developed and tested using structural equation modeling. This paper does not just testify the effect of learning orientation on firm performance, but also tests how this mechanism works through analyzing the mediating effect of innovation.

2. Literature review and hypothesis

2.1. Innovation and firm performance

Innovation has been conceptualized as an important way to achieve superior performance of a firm. Innovation tends to be strategic activities of a firm to face the change of internal and external environment. According to some researchers, a firm cannot avoid innovation if the firm wants market develop and retains competitive advantage as well as enters new markets [16]. Innovation is often related to an organization's ability to survive and surpass the competitors; thus, innovation becomes the critical determiner of firm performance. If a firm has a capacity to innovate, then the capacity will enable firms to develop competitive advantage and achieve greater results [7, 17].

Most of the empirical studies on the relation between innovation and performance provide evidence that this relation is positive [5, 12, 18-24]. In addition, some studies arrive at conflicting conclusions [5]. Meanwhile, as Simpson et al. [25] point out, innovation is an expensive and risky activity, with positive outcomes on firm performances, and also with negative outcomes, such as increased exposure to market risk, increased costs, employee dissatisfaction or unwarranted changes. For instance, Wright et al. [26], using a sample of small businesses, find that product innovation does not affect performance in benign environments, but has a positive effect on performance in hostile environments. Although some empirical studies provide different evidence, most theories and most of the research findings suggest that there is a positive relationship between innovation and firm performance.

H1. Innovation will positively affect firm performance

2.2. Learning orientation and firm performance

Organizational learning is the process by which the firm develops new knowledge and insights from the common experiences of people in the organization, and has the potential to influence behaviors and improve the firm's capabilities [8, 27]. Organizational learning is a basis for gaining a sustainable competitive advantage and a key variable in the enhancement of organizational performance [28]. Furthermore, Wang [11] conceptualizes learning orientation as those firm values that influence a firm's approach to acquiring information. They emphasize the importance of planned processes in allowing firm learning to lead to the achievement of common organizational goals. In accordance with Calantone et al. [18], the term learning orientation is defined as a firm commitment to learning, shared vision, open-mindedness and intra-organizational knowledge sharing. A learning orientation helps a firm to acquire, disseminate and share information 11. Some studies provide evidence of a positive relationship between organizational learning and firm performance. For instance, Nybakk [29] find that learning orientation has a direct effect on organizational performance. Other studies have found similar results [2, 5, 12, 30-33].

In sum, empirical findings are consistent with theory and provide evidence that supports the positive relationship between organizational learning and performance. In this study, a learning organization defined as an organization with a learning orientation.

H2. Learning orientation will positively affect firm performance.

2.3. Learning orientation and innovation

Learning orientation has been described as the adoption of a basic learning process. As a firm becomes larger, commitment to learning plays a crucial role in updating its assets and capabilities concerning its key activities [11]. Learning orientation is thought to be an important antecedent on firm innovation

[18, 34, 35]. Clearly, most empirical researches on the relationship between learning orientation and innovation is conducted on large firms [12], which have more of the resources needed for innovation and can take on a larger degree of risk. Nevertheless, smaller firms have other advantages, such as a less extensive bureaucracy [36], and can access resources by collaborating with other.

Some qualitative and quantitative studies show that organizational learning enhances innovation [4, 5, 7, 37-39]. Similarly, previous studies focus on one phase of the organizational learning process or on one type of innovation, mainly product or process innovation. For instance, Yli-Renko et al. [40] find a positive relationship between knowledge acquisition and product innovation.

Although the above-mentioned studies focus on different aspects of the relationship between organizational learning and innovation, most find a positive relationship between them.

H3. Learning orientation will positively affect innovation.

2.4. The mediating effect of innovation

From the earlier discussions, it can be postulated that learning orientation positively brings about a capacity to innovate. Some studies provide evidence of a positive relationship between organizational learning and firm performance. Some articles suggest that organizational learning allows the company to develop capabilities that enhance innovation and that innovation is what positively affects performance [7, 41-45].

Based on the literature examining the effect of learning orientation on innovation and the effect of innovation on financial performance, one can logically assume that learning orientation will have an indirect positive impact on financial performance via innovation. Garcia-Morales et al. [2] studied farming, manufacturing, construction and service firms in Spain and found that learning orientation both directly and indirectly influence the financial performances of large firms and SMEs through innovation. The positive effect of organizational learning on both innovation and performance is greater for smaller firm [5].

H4. Learning orientation will positively affect firm performance via innovation

3. Method

This study was based on a survey of owner SMEs apple product-based business in Batu City, Indonesia. The data were obtained using a structured questionnaire was developed and pilot tested before the formal data collection. The final data tested by Partial Least Square (PLS). The five-point Likert-type scales ranging from 1 (totally disagree) to 5 (totally agree) were used throughout the questionnaire. A total of 47 usable questionnaires was received from SMEs. Orientation learning was measured using 16-items adapted from several studies [5, 7, 12, 18] and innovation was measured using six items adapted from three studies [5, 17, 18]. The questions for measuring financial performance of the firm measured using four items used by Jimenez-Jimenez and Sanz-Valle [5] and Ar and Baki [18].

4. Results and Discussion

First, to evaluate the validity of the measurement model, convergent validity and discriminate validity were assessed. The result of measurement model is shown that indicates factor loadings in the measurement model are all greater than 0.50. The result shows that our model meets the convergent validity criteria. Furthermore, all constructs in the measurement model were judged as fulfill discriminate validity criteria (loading factor > cross loading). Composite reliability coefficients for the constructs in the measurement model are all greater than 0.06, suggesting that a high internal reliability existed in the constructs. Moreover, the measurement model shows acceptable fit values for AFVIF, GoF, SPR, RSCR, SSR and NLBCDR revealing that the model does a good job in explaining the relationships between latent variables and observed variables.

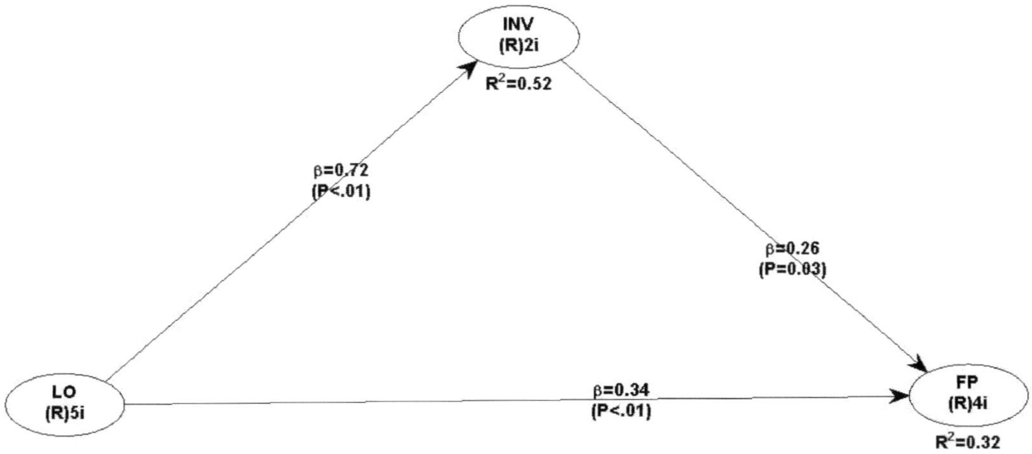

Figure 1. Research model and testing result

Table 1. Hypothesis testing.

Variable correlations	Path Coefficient	*P* value	Description
Learning orientation => Innovation	0.72	<0.01	Supported
Learning orientation => Firm performance	0.34	<0.01	Supported
Innovation => Firm performance	0.26	0.03	Supported
Learning orientation =>Innovation => Firm performance	0.19	0.02	Supported

For H1, we examined the influence of innovation on firm performance. As shown in Table 1 and Figure1, the effects of innovation on firm performance have a value path coefficient of 0.72 ($p < 0.01$), hypothesis 1 supported. For H2, we examined the effect of learning orientation on firm performance. The results also show that the effects of learning orientation on firm performance have a path coefficient value of 0.34 ($p < 0.01$), providing support H2. For H3, we examined the effect of learning orientation on innovation. The effects of learning orientation on firm performance, have a path coefficient value of 0.26 ($p < 0.05$), providing support H3. As to the indirect effect, the finding support for H4, showing that the learning orientation has a positive effect on both performance and innovation. These results seem to reflect that innovation partially mediates the relationship between learning orientation and firm performance.

The finding of this study provides additional evidence to the previous study that innovation has positive effects on firm performance [5, 12, 18-24]. Also, the findings show a positive relationship between learning orientation and firm performance [2, 5, 12, 30- 32] and between learning orientation on innovation [4, 5, 7, 37-39]. It was found that a set of knowledge-questioning values via open-mindedness, shared vision, commitment, and knowledge sharing, facilitate firms to try out new ideas, seeks out new ways to do things, develop and launch new product/service and to be creative in its methods of operations [12].

The partial mediation of innovation was detected in the relationship between learning orientation and firm performance. The result of this study confirmed that learning orientation is important to innovation and firm performance. Implementing a learning orientation will help a company to increase its innovation and improve firm performance. This study was also in line with other studies [2, 5]. This

IOP Conf. Series: Earth and Environmental Science **131** (2018) 012027 doi:10.1088/1755-1315/131/1/012027

result may imply that learning orientation influence firm performance mainly by facilitating innovation. The findings generally suggest that when members of an organization acquire knowledge via learning process, that organization acquires the ability to be innovative. The finding is consistent with the work of [43], which indicates that the absorptive capacity of the firm is linked to the absorptive capacity of the people in the firm [4].

5. Conclusion

The study suggests that learning process in organization facilitates innovation. Therefore, an organizational hoping to enhance firm performance through innovation should improve its learning process. The study has practical implications. The relationship between learning orientation, innovation, and firm performance may provide a guide as to how companies should achieve better performance by increasing commitment to learning, promote acquisition of knowledge, then share and interpretation knowledge within the firm, and should try to keep knowledge and information in the best and most efficient way.

References

[1] Damanpour F, Wischnevsky J D 2006 Research on innovation in organization: Distinguishing in generating from innovation adopting organizations *J. Eng. Technol. Manage.* **23** 4 269-291

[2] Garcia-Morales V J, Liorens-Montes F J, Verdu-Jover A J 2007 Influence of personal mastery on organizational performance through organizational learning and innovation in large firms and SMEs *Technovation* **27** 9 547-568

[3] Damanpour F 1991 Organizational innovation: A meta-analysis of effects of determinants and moderators *Acad. Manage. J.* **34** 3 550-590

[4] Hult T G M, Hurley R F, Knight G A 2004 Innovativeness: Its antecedens and impact on business performance *Ind. Market. Manage.* **33** 5 429-438

[5] Jimenez-Jimenez D, Sanz-Valle R 2011 Innovation, organizational learning, and performance *J. Bus. Res.* **64** 4 408-417

[6] Damanpour F, Szabat K A, Evan W M 1989 The relationship between types of innovation and organizational performance *J. Manage. Stud.***26** 6 587-601

[7] Hurley R E, Hult G T M 1998 Innovation, market orientation and organizational learning: an integration and empirical examination *J. Market.* **62** 42-54

[8] Slater S F, Narver J C 1995 Market orientation and the learning organization *J. Market.* **59** 3 63-74

[9] Farrell M A 2000 Developing a market-oriented learning organization *Aust. J. Manage.* **25** 2 201-222

[10] Giniuniene J, Jurksiene L 2015 Dynamic capabilities, innovation and organizational learning, interrelations and impact on firm performance *J. Soc. Behav. Sci.* **213** 985-991

[11]Wang C L 2008 Entrepreneurial orientation, learning orientation, and firm performance *Entrepreneurship Theor. Pract.* **32** 4 635-637

[12] Keskin H 2006 Market orientation, learning orientation, and innovation capabilities in SMEs *Eur. J. Innov. Manage.* **9** 4 396-417

[13] Simon M, Houghton S M, Aquino K 2000 Cognitive biases, risk perception, and venture formation: how individuals decide to start companies *J. Bus. Venturing* **15** 2 113-134

[14] Madrid-Gujiarro A, Garcia D, Auken H V 2009 Barriers to innovation among Spanish manufacturing SMEs *J. Small Bus. Manage.* **47** 4 465-488

[15] Nassar M L, Faloye D O 2015 Barrier to innovation in developing countries' firms: Evidence from Nigerian small and medium scale enterprises *Eur. Sci. J.* **11** 19 196 -213

[16] Stock G N, Greis N P, Fischer W A 2002 Firm size and dynamic technological innovation *Technovation* **22** 9 537-549

[17] Damanpour F, Walter R M, Avellaneda C N 2009 Combinative effects of innovation types and organizational performance: A longitudinal study of service organization *J. Manage. Stud.* **46** 4 650-675
[18] Ar I M, Baki B 2011 Antecedents and performance impacts of product versus process innovation. Empirical evidence from SMEs located in Turkish science and technology parks *Eur. J. Innov. Manage.* **14** 2 172-206
[19] Calantone R J, Cavusgil S T, Zhao Y 2002 Learning orientation, firm innovation capability and firm performance *Ind. Market. Manage.* **31** 6 515-524
[20] Lopez-Nicolas C, Merono-Cerdan A L 2011 Strategic knowledge management, innovation and performance *Int. J. Inform. Manage.* **31** 6 502-509
[21]Rhee J, Park T, Lee D H 2010 Drivers of innovativeness and performance for innovative SMEs in South Korea: Mediation of learning orientation *Technovation* **30** 1 65-75
[22] Rosenbusch N, Brinckmann J, Bausch A 2011 Is innovation always beneficial? A meta-analysis of the relationship between innovation and performance in SMEs *J. Bus. Venturing* **26** 4 441-457
[23] Sok P, O'Cass A, Sok K M 2013 Achieving superior SME performance: Overarching role of marketing, innovation, and learning capabilities *Aust. Market. J.* **21** 3 161-167
[24] Thornhill S 2006 Knowledge, innovation and firm performance in high-and low-technology regimes *J. Bus. Venturing* **21** 5 687-703
[25] Simpson P M, Siguaw J A, Enz C A 2006 Innovation orientation outcomes: The good and the bad *J. Bus. Res.* **59** 10/11 1133-1141
[26] Wright R E, Palmer J C, Perking D 2005 Types of product innovations an small business performance in hostile and benign environments *J. Small Bus. Strat.* **15** 2 33-44
[27] Senge P M 1990 The Fifth Discipline: The Art and Practice of the Learning Organization Doubleday New York USA 424
[28] Brockmand B, Morgan F 2003 The role of existing knowledge in new product innovativeness and performance *Decis. Sci.* **32** 2 385-419.
[29] Nybakk E 2012 Learning orientation, innovativeness and financial performance in traditional manufacturing firms: a higher-order structural equation model *Int. J. Innov. Manage.* **16** 5 1-34
[30] Che-Ha N, Mavondo F T, Mohd-Said S, 2014 Performance or learning goal orientation: implications for business performance *J. Bus. Res.* **67** 1 2811-2820
[31] Kitapci H, Aydin B, Celik V 2012 The Effects of Organizational Learning capacity and innovativeness on financial performance: An Empirical study *Afr. J. Bus. Manage.,* **6** 6 2332-2341
[32] Ramirez A M, Garcia-Morales V J, Rojas R M 2011 Knowledge creation, organizational learning and their effects on organizational performance *Inzinerine Ekonomika-Eng. Econ.* **22** 3 309-318
[33] Wang Z, Wang N 2012 Knowledge sharing, innovation, and firm performance *J. Expert Sys. Appl.* **39** 8899-8908
[34] Alegre J, Chiva R 2008 Assessing the impact of organizational learning capability on product innovation performance: an empirical test *Technovation* **28** 6 315-326
[35] Liao S H, Fei W C, Liu C T 2008 Relationship between knowledge inertia, organizational learning and organizational innovation *Technovation* **28** 4 183-195
[36] Thompson V A 1965 Bureucracy and innovation *Adm Sci Q.* **1** 1-20
[37] Forrester R H 2000 Capturing learning and applying knowledge: An investigation of the use of innovation teams in Japanese and American automotive firms *J. Bus. Res.* **47** 1 35-45
[38] Salavou H, Lioukas S 2003 Radikal product innovation in SMEs: the dominance of entrepreneurial orientation *Creativity Innov. Manage.* **12** 2 94-108
[39] Yeung A C L, Lai K, Yee R W Y 2007 Organizational learning, innovativeness and organizational performance: a qualitative investigation *Int. J. Prod Res.* **45** 11 2459-2477

[40] Yli-Renko H, Autio E, Sapienza H J 2001 Social capital, knowledge acquisition, and knowledge exploitation in young technology-based firm *Strat. Manage. J.* **45** 1 2459-2477
[41] Baker W E, Sinkula J M 1999 The sinergetic effect of market orientation and learning orientation on organizational performance *J. Acad. Market. Sci.* **24** 4 411-427
[42] Bolivar-Ramos M T, Garcia-Morales V J, Garcia-Sanchez E 2012 Technological distinctive competencies and organizational learning: effects on organizational innovation to improve firm performance *J. Eng. Technol. Manage.* **29** 331-357
[43] Cohen W M, Levinthal D A 1990 Absorptive capacity: A new perspective on learning and innovation *Admin. Sci. Q.* **35** 1 128-152
[44] Gomes G, Wojahn R M 2017 Organizational learning capability, innovation, and performance: study in small and medium-sized enterprises (SMEs) *J. Revista de Administracao* **52** 163-175
[45] Serna M D C M, Martines J E V, Martines J V 2016 The impact of learning orientation on innovation and performance in SME's in Mexico *Journal IRMBR.* **5** 1 48-64

IOP Conf. Series: Earth and Environmental Science **131** (2018) 012028 doi:10.1088/1755-1315/131/1/012028

Assessment of good manufacturing practice for small scale food industry in Malang region, East Java, Indonesia

I Purwantiningrum , W Widyhastuty, J Christian and N Sari

Food Science and Technology Study Program, Department of Agricultural Product Technology, Faculty of Agricultural Technology, Universitas Brawijaya, Malang, Indonesia

E-mail: airdni@ub.ac.id; airdni@yahoo.com

Abstract. Enhancing food safety in developing countries, such as Indonesia, poses more challenges, especially those of the small- and medium-scale. Various food safety systems are available and readily implemented in the food industry. However, to ensure the effectiveness of such systems, pre-requisite programs should be applied prior to the implementation of food safety system. One of the most acknowledged pre-requisite program is Good Manufacturing Practices (GMP). The aim of this study is to assess the GMP compliance of some small-scale food companies in East Java. Three types of traditional food product were selected, include tempe chips, palm sugar, and instant herbal drink. A survey involving three companies for each type of traditional food was conducted. Data was obtained through observation and assessment based on tabulated criteria in GMP criteria. In essential, the result revealed the compliment level of the food companies being surveyed. There was different level of compliment between each type of the food industry, where the palm sugar industry had the lowest level of compliment compared to the other two. This difference is due to the food safety awareness, social and cultural influences, and also knowledge on food safety and hygiene practice.

1. Introduction

Food safety has become more and more essential both in developed and developing countries due to the expanding food trade, affecting health, economic and sustainability aspects [1, 2]. Enhancing food safety in developing countries, such as Indonesia, poses more challenges due to the situation of the food companies, especially those of the small- and medium-scale [3, 4]. These condition emerges due to improper processing of food, lack of awareness and responsibility toward handling and food additive uses, and lack of knowledge upon food preparation and production environment requirements [5].

Good Manufacturing Practices (GMP) is one of the well-known food safety program, developed by Food and Drug Administration-USA, as a guideline for the food, cosmetics and drug producers, handlers, retailers, etc.[6, 7]. GMP includes policy, procedures and methods as guidance for complying quality standards and hygiene. It provides guidelines for producing food that complies specific requirements to aid food producers in the manufacture and distribution of unadulterated and no misbranded foods. These guidelines are located in the Code of Federal Regulations 21CFR part 110 as Current Good Manufacturing Practice in Manufacturing, Packing or Holding Human Food. It serves as the standard by which the government determines if the food is adulterated [7, 8]. Thus, it is critical that every food manufacturer and distributor understand the regulations and develop systems and programs that prove adherence to the letter of this law [9–11].

Various criteria are included in the standard, including personal hygiene, food production facilities sanitation and design, process control and pest control. These criteria are qualitatively described in the standard and used as the basis for GMP compliment assessment. This study focused on assessing the real condition of small-scale food companies in East Java based on GMP criteria issued by Food and Drug Control Agency (BPOM). This can be very important for setting up the basis understanding for the real condition of small-scale food companies and hence the action plan of the sequential steps to achieve food safety for both the government agencies and the food companies.

2. Materials and Methods

Investigation was carried out on many aspects of plant sanitation based on the standard issued by BPOM and Ministry of Health. Three types of traditional food were selected, include palm sugar (PS), tempe chips (TC), and instant herbal drink (HD). Codes were applied rather than actual company names of each type of food industry to protect the reputation of the company. Alphabetical scores were given to sanitary observation on 3 classes of complimentary condition: G-good, F- fair, I-insufficient. These classifications were developed accordingly to the issued standard criteria. The classification is quantified into numerical scale for each class as 3= good, 2=fair, and 1=insufficient. Data was obtained through observation and evaluation of each company performance were tabulated.

3. Results and Discussion

Assessment of GMP according to the standard criteria [12] can be seen on Table 1. The figures show the quantified class of condition where bigger number shows the higher compliment to the standard.

3.1. *Production environment*

Production environment varied between industry types, but generally those were not located in swampy areas or near centers of pollution. All the industries visited were located close to or inside residential areas. PS industries were located in rural areas, while TC and HD industries were located in sub-urban or urban areas. Most of the industries surroundings were free from bushes and weeds except for PS1. Waste and disposal container were mostly present at adequate amount; however, some were not fully covered. Normal drainage facilities were found in all industries, except for PS1 where water was logged during visit. Thus, a special precaution should be made due to the possibility of contamination round the location [13].

3.2. *Building and facility*

The industries visited were all single-story building. There were many false ceiling in all PS, as well as in TC1 and HD3. This open ceiling and cracks enable rodents and reptiles to harbor and also allows mold growth. The walls in all TC and HD industries had smooth surfaces however the opposite was found in PS industries which was constructed from plaited bamboo mats. The floor was made of impervious material except for the PS industries which used no floor but direct hardened soil. The surfaces of all interior should be designed to maximize clean ability and protection toward contamination [13, 14]. Doors and windows in all industries were designed to open outward the production area.

3.3. *Production equipment and utensils*
Production equipment and utensils should be made of stable, non-corrosive and easy to clean material, such as stainless-steel knife, melamine surfaces, and stainless-steel cooking pans. Production lay out were in sequence accordingly to process flow. Process flow is essential to help prevention of cross-contamination in food premises [8–10, 15]. Production equipment were all functioned well, however, maintenance schedule was not established. Cleaning and sanitizing were also inadequately scheduled. Although some utensils in PS industry were very traditional and had shown wear over time, they showed no evidence of leakage, cleavage and other damages that allow contamination to the produced sugar, as it has high osmotic pressure against microbial growth [10].

3.4. Water supply
Water supply for the food industries investigated were all obtained from boreholes or municipal water sources. These water resources were supplied sufficient amount and quality for food processing and other purposes in the premises. Thus, awareness of the need to use potable water in their processing were shown in all industries. Clean and contamination-free water is essential in food processing, as water is one of the primary areas of concern in food contamination sources [15].

Table 1. Good Manufacturing Practice Assessment of Some Small-Scale Food Industries in East Java, Indonesia

Food Industry	PE	BF	PEq	WS	HSF	PsC	PHS	PC	L	S	MR	RD	PTr	TS
PS1	1	1	2	2	1	1	2	1	1	2	2	1	1	18
PS2	3	1	3	2	1	2	2	1	1	2	3	1	1	23
PS3	2	1	3	1	1	2	2	1	1	2	1	1	1	19
PS Average	2	1	3	2	1	2	2	1	1	2	2	1	1	21
TC1	2	1	3	2	1	2	2	3	1	3	3	2	3	28
TC2	2	2	3	3	2	3	3	3	1	3	2	2	1	30
TC3	2	3	2	2	1	3	3	2	1	3	3	1	2	28
TC Average	2	2	3	2	1	3	3	3	1	3	3	2	2	30
HD1	3	2	3	2	2	2	3	3	2	3	3	2	3	33
HD2	2	3	3	3	2	3	3	2	1	3	2	1	2	30
HD3	2	1	2	2	3	3	3	3	3	3	3	2	1	31
HD Average	2	2	3	3	2	3	3	3	2	3	3	2	2	33

Note:
PE=Production Environment
BF=Building&Facility
PEq=Production Equipment
WS= Water Supply
HSF=Hygiene & Sanitation Facility
PsC= Pest Control
PHS=Personal Hygiene&Sanitation
PC=Process Control
L= Labelling
S= Storage
MR= Monitoring & Responsibility
RD= Record & Documentation
PTr=Personal Training
TS= Total Score

3.5. Hygiene and sanitation facility
All the industries provided toilet facilities for their staff. Some toilets had their floors and wall tile. No soap was provided in the industries. Hand washing facilities were provided in TC and HD industries but not in PS industries. Hands may become soiled with a wide variety of contaminants during normal food

processing operations [16], which can be a transmitting agents for contamination of food and food contact surfaces [17]. Hygiene and sanitation facilities are essential to ensure that building and equipment always in clean condition to prevent cross contamination, especially those of staff and personnel [7, 11].

3.6. Pest control
All the industries visited showed no animals or pets surrounding the plants. Pests (rodents, insects, and other animals) are carrier of biological contaminants that decreases food quality and safety [3 , 4, 15, 18]. TC and HD industries showed prevention measures for pest infestation. However, probabilities of pest infestation were found high enough due to the surroundings condition especially in palm sugar industries. False ceilings in the premises contributed probability of pest infestation, as well as the opened door of the production area [19].

3.7. Personal hygiene and sanitation
All personnel in the industries visited were in good health and showed no symptoms of sickness. However, periodical health check was not conducted in all industries. Working apparels were in clean condition except for PS industries and TC1. No faulty habits (chewing, smoking, wearing jewelry, etc.) were found in all premises. Personal health and hygiene plays an important role in assuring food safety and quality, as people and environment can easily contaminate food [9, 10, 15].

3.8. Process control
TC and HD industries had determined their raw material specification, but the PS industries had their raw material as given from nature. Process formulation and flow was established and consistently applied to daily operation. Packaging specification were established in TC and HD industries, but inadequate product information was specified in TC industries. Moreover, no packaging was used in PS industries. According to Lelieveld et al. [14], the operations and processes used in manufacture should, with the premises, equipment, materials, personnel and services provided, be capable of consistently yielding finished products that conform to their specifications and are suitably protected against contamination or deterioration.

3.9. Labelling
No product labels were found in PS industries. Production date and code, expired or best before date as well as net weight were found only in HD labels. Food label should be clear and informative to enable consumer to choose, store, process and consume food [14].

3.10. Storage
Raw material and finished product were stored separately to prevent cross contamination and to increase handling efficiency [18]. All industries visited showed this separation, proofing an encouraging awareness of storage essential [20]. Labels and packaging material were stored at separate premise in TC and HD drink industries. This practice is also evidence of GMP compliance [21].

3.11. Monitoring and responsibility
In most industries visited, the owner takes responsibility to monitor production process and directly involved in daily operation except for PS1, PS3, TC2 and HD2 where monitoring was not periodically conducted. In a small scale food industry, the manager who takes responsibility for all production process is the owner [4, 22].

3.12. Record and documentation

Most of the industry investigated had no record and documentation procedure for the entire production process except for TC1, TC2, HD1 and HD3. The four industries recorded material receiving and finish product yield, however, the records were very simple and did not cover other information such as production code and date [23]. Inconsistency of record were shown in TC2 and HD1, as well. Record were not kept or well documented that may cause difficulties in recall [4, 22].

3.13. Personal training

In all industries investigated there were no formal training provided for personnel except for TC1 and HD1. This was reasonable as there were little specific skill needed to run the process. Skills and knowledge were acquired through daily operation practice and dissemination from local Health Agency. Personnel holds significant role in labor-intensive food industry [4, 22]. Sufficient personnel should be available depending on the size and type of business, at all levels with the ability, training, experience and other qualifications [14, 18, 24].

4. Conclusion

There was difference level of compliance between each type of the food industry investigated. All HD industries and TC2 showed high level of compliance, while PS industries had the lowest level of compliance of the standard compared to the other two type of food products.

The findings demonstrated that more focus for improvement should be made especially for: process control, hygiene and sanitation facility, labelling, record and documentation, as well as personnel training. Improving hygiene and sanitation standard will significantly contribute to food safety in the small-scale food industries. Personnel skill and knowledge improvement will also enhance the small-scale industry competitiveness to global market.

References

[1] King T, Cole M, Farber J M, Eisenbrand G, Zabaras D, Fox E M, Hill J P 2017 Food safety for food security: Relationship between global megatrends and developments in food safety *Trends Food Sci. Technol.* **68** 160–175
[2] Henson S, Caswell J 1999 Food safety regulation: an overview of contemporary issues *Food Pol.* **24** 6 589–603
[3] Heijden K A 1999 International Food Safety Handbook Marcel Dekker New York USA 811
[4] Ehresman T 1995 Small business success through TQM ASQC Quality Press Milwaukee USA 255
[5] Purwantiningrum I 2011 Production process control in small-scale tempe chips industry phase I: field observation *J. Agric. Food Technol.* **1** 9149–152
[6] Dias M A C, Sant'Ana A S, Cruz A G, José de Assis F F, de Oliveira C A F, Bona E 2012 D On the implementation of good manufacturing practices in a small processing unity of mozzarella cheese in Brazil *Food Control.* **24** 1–2 199–205
[7] FDA 2015 Code of Federal Regulations Title 21- Part 110 Current Good Manufacturing Practice, Hazard Analysis, and Risk-Based Preventive Controls for Human Food. United States of America. Retrieved from https://www.federalregister.gov/documents/2015/09/17/2015-21920/current-good-manufacturing-practice-hazard-analysis-and-risk-based-preventive-controls-for-human [Accessed on 28 August 2017]
[8] Newslow D 2013 Food Safety Management Programs CRC Press Boca Raton USA 364
[9] David A S, Norah F S 1991 Principles and Practices for the Safe Processing of Foods Butterworth-Heinemann Ltd Oxford UK 456

[10] Heinz H J 2013 Principles and Practices for the Safe Processing of Foods Butterworth-Heinemann Ltd Oxford UK 572

11] Forsythe S J, Hayes P R 1998 Cleaning and Disinfection: Methods In Food Hygiene, Microbiology and HACCP Springer New York USA pp. 327–363.

[12] Kementerian Perindustrian RI 2010 Peraturan Menteri Perindustrian Republik Indonesia Nomor: 75/M-IND/PER/7/2010 tentang Pedoman Cara Produksi Pangan Olahan yang Baik (Good Manufacturing Practices) (In Indonesian)

[13] Schmidt R H, Erickson D J 2008 Sanitary Design and Construction of Food Equipment UF/IFAS Extention **1** 1–8

[14] Lelieveld H L M, Holah J, Gabric D 2016 Handbook of Hygiene Control in the Food Industry Woodhead Publishing Duxford UK 756

[15] Marriott N, Gravani R B 2006 Principles of Food Sanitation Springer Boston USA 420

[16] Pellegrino R, Crandall P G, O'Bryan C A, Seo H S 2015 A review of motivational models for improving hand hygiene among an increasingly diverse food service workforce *Food Control.* **50** 446–456

[17] Taylor J, Purdon C 2016 Responsibility and hand washing behaviour *J. Behav. Ther. Exp. Psychiatry.* **51** 43–50

[18] Cramer M M 2013 Food Plant Sanitation: Design, Maintenance, and Good Manufacturing Practices CRC Press Florida USA 306

[19] Trafialek J, Drosinos E H, Kolanowski W 2017 Evaluation of street food vendors' hygienic practices using fast observation questionnaire *Food Control.* **80** 350–359

[20] Egwari L O, Ayepola O O, Adekeye B, Adegbayi T, Olurunshola S, Kuye A 2016 Evaluating the effect of hand washing and sanitization on the microbial burden of the hand *Int. J. Infect. Dis.* **45** 277

[21] Grob K, Stocker J, Colwell R 2009 Assurance of compliance within the production chain of food contact materials by good manufacturing practice and documentation - Part 2: Implementation by the compliance box: call for guidelines *Food Control.* **20** 5 483–490

[22] Foster S T 2001 Managing Quality: an Integrative Approach Prentice Hall New Jersey USA 476

[23] Zanin L M, da Cunha D T, de Rosso V V, Capriles V D, Stedefeldt E 2017 Knowledge, attitudes and practices of food handlers in food safety: An integrative review *Food Res. Int.* **100** 53–62

[24] Sani N A, Siow O N 2014 Knowledge, attitudes and practices of food handlers on food safety in food service operations at the Universiti Kebangsaan Malaysia *Food Control.* **37** 1 210–217

IOP Conf. Series: Earth and Environmental Science **131** (2018) 012029 doi:10.1088/1755-1315/131/1/012029

Transforming ex-small scale mining land as farming areas for sustainable development and poverty alleviation

I W Nampa[1,4], J E R Markus[2,4], I W Mudita[2,4], R L Natonis[3], W Bunga[2] and N R Kaho[2]

[1] Department of Agribusinesses, Faculty of Agriculture, Nusa Cendana University, Kupang, Indonesia
[2]Department of Agro-Technology, Faculty of Agriculture, Nusa Cendana University, Kupang, Indonesia
[3]Agriculture and Estate Crop Service, East Nusa Tenggara Province, Indonesia
[4]Center of Excellent (CoE) Archipelagic Dry-Land Agriculture (ALRIC) Nusa Cendana University, Kupang, Indonesia

E-mail: wayannampa@staf.undana.ac.id

Abstract. When the price of manganese ores in 2012, mining activities declined or even terminated. Ex-miners lose an important source of income, but they did not have any other alternative except going back to slash and burn cultivation, producing enough only for their own food. Their hope for a better live was gone and at the same time they faced stigmatisation as causing environmental degradation from the rest of the community. We carried out this case study to followex-miners in the Tubuhue village who organised themselves to do post-mining rehabilitation by turning the former mining site into an area of productive farming. In-depth interview, field observation and focus group discussion were conducted from 2015 to 2017. We found that during the period of mining boom, slash and burn cultivation decrease significantly but began to increase after no mining activities. Various social transformations took place along with this land use change, but the most important was the miners' decision to do mining as an organised activity. A strong leader of this organization played a pivotal role in turning the former mining site into an area of productive sedentary farming. This was carried out by organizing the ex-miners into farmers groups and together, constructing drip and sprinkler irrigation networks to water their crops using rain water collected in the mining holes that they had turned into small check-dams. The leader expected that this farming could provide an alternative for ex-miners to obtain cash income to limit them going back doing swidden farming.

1. Introduction

The small-scale manganese mining provided the drought-prone swidden farmers of West Timor an opportunity for obtaining instant cash income. By simply digging in their own land, a small-scale miners could obtain 50 to 150 kg of manganese ores a day with a price of IDR 800 to IDR1,500 per kg. For an average ores of 100 kg a day and an average price of IDR 1,150 per kg, one would receive Rp 115,000 cash income a day, a sum of money far larger from what a person could earn from any other job in rural areas . Swidden farmers used the cash income to improve their living condition while they continued to relay on swidden farming to produce their food. According to Dara , the significance of the economic benefits derived from artisanal small-scale mining activities can be identified by the changes in their capacity to fund their basic needs such us food, water and sanitation, adequate housing, and basic education. This way, Dara [1] continued, small-scale manganese mining created alternative livelihoods.

When in 2012 the price of manganese ores crashed, those alternative livelihoods were gone. There was no other alternative available except going back to their abandoned swidden farm to produce food for their own household and to raise cattle, pig, or chicken to obtain much smaller cash income [1–3]. Those who could not afford to do so, relied on rice-for-the-poor ration provisioned by the government [4,5]. Their hope for a better future suddenly gone with the wind that crashed the price of manganese ores. What even worse, those abandoned swidden farming for mining had faced stigmatization as causing environmental degradation from the rest of the community.

This article presents a case study about the impact of artisanal small-scale manganese mining on the environment and the struggle of ex-miners to continue their life after the collapse of the mining in 2012. This research was part of a larger multi-years research project on environmental and social impacts of small-scale artisanal mining in Eastern Indonesia. The research in West Timor was focused on artisanal small-scale mining involving local communities. In the final year, research in West Timor was more focused on small-scale mining sites in Noelmina catchment.

2. Materials and Methods

We carried out this case study in the village of Tubuhue, sub-district of west Amanuban, South Central Timor District, as part of a larger research project encompassing the whole area of Noelmina catchment. The village was once the most vibrant artisanal small-scale manganese mining location in the catchment during the mining boom in 2008-2012 and where ex-miners has successfully organised themselves to turn former mining sites into a centre of horticultural crop production.

Tubuhue, one of villages within the catchment where artisanal small-scale manganese mining, was most vibrant during the peak of manganese mining in the region. We particularly interested to this village because those involved in mining activities in this village had been from the beginning aware of the impacts of their activities and mobilized efforts to make the mining beneficial for their community.

For the purpose of this case study, we collected social and physical data during the period of 2015-2017. For the collection of social data consisting of working processes and human behaviors, we followed three farmers groups which members consisted of those formerly involved in in artisanal small-scale manganese mining on site. During our several visits, we discussed reasons they engaged in group farming in ex-mining sites, observed their day-to-day farming practices, and invited members of the farmer groups to attend focus group discussion we carried in the village and in Kupang. For land use analysis, we downloaded path 110 and row 67 of Landsat 8-OLI (Operational Land Imager) image year 2013 and 2017 and Landsat 5 TM (Thematic Mapper) year 2009 from USGS website (https://speclab.cr.usgs.gov). We choose the path and row of the image for reasons of reasonable cloud-free cover (less than 10% cloud cover) and adequate spectral information from the intensity over 80% irradiation as shown in Figure 1.

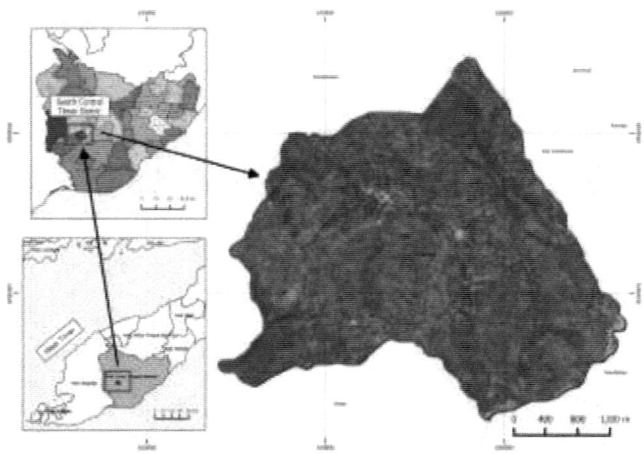

Figure 1. Research site and its land cover characteristics

We employed a thematic analysis technique to analyse data resulting from in-depth interview, field observations, and focus group discussion. We did the analysis in a naturalistic approach by treating the data in their natural setting along a continuum of data reduction, data display, and drawing conclusion [6]. We followed the recommendation of UNCCD (United Nations Conventions to Combat Desertification) in analysing land cover change by using Normalized Difference Vegetation Index (NDVI) with three cluster NDVI score for vegetation classification (Table 1). NDVI had been proven useful for mapping of vegetation, vegetation typology and land use analysis [7,8]. We carried out this spatial analysis using SAGA (System for Automated Geo-scientific Analyses) 5 and Quantum GIS 2.18 open-access GIS software.

Table 1. NDVI scores and their land cover characteristics.

Score	Land cover characteristics
≤ 0.2	Barren soil, open ground, and rock outcrops
0.2 – 0.3	Grassland, shrubs, agricultural areas, and dry forest
0.3 – 1	Closed vegetation (forest)

3. Results and Discussion

We present and discuss this research finding in three sections. We start with land use transformation in the first section, social transformation in the second section, and finally, mining activities in the last section. The explanation of those sections is as follows.

3.1. Land use transformation

In analysing land use change, we used the 2009 land cover as the baseline data, the 2013 land cover as measures of mining impacts, and the 2017 land cover as measures of impacts of the shift from swidden farming to more sedentary farming activities. As shown in Figure 2, the 2009 land cover was dominated by open land and shrubs, indicating the dominance of swidden farming prior to the start of artisanal small-scale manganese mining. The 2013 land cover showed a shift from open land and shrub to shrubs and forest, indicating significant reduction of swidden activities because during the period from 2009 to 2013, people engaged more in mining than in swidden farming. During this period, the area of open land had been also significantly reduced from 206 ha to only 0.19 ha. This land cover change indicated that the artisanal small-scale manganese mining had become a more attractive alternative to swidden farming. Being small-scale, these mining activities did not necessary open a large track of land that contributed significantly to land cover change. In fact, the mining was carried out by simply collecting ores already exposed above ground or by digging small pits under the forest trees.

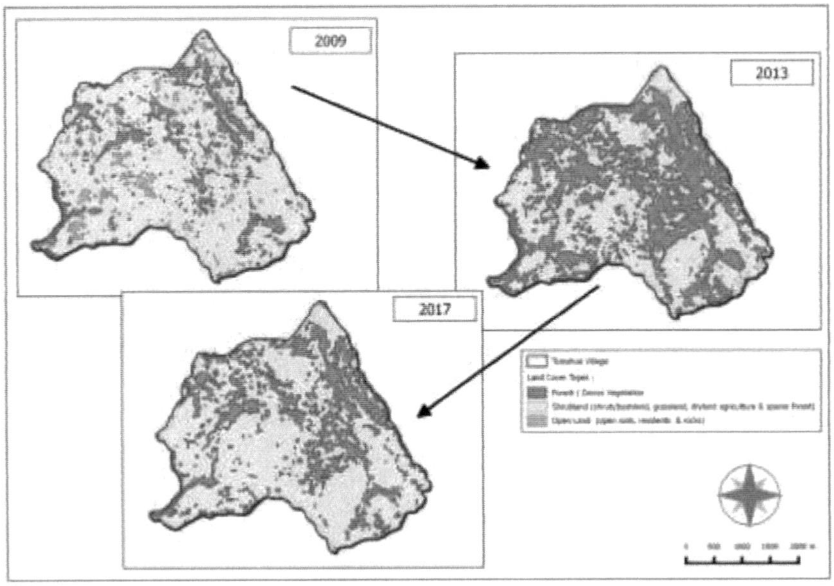

Figure 2. Land-cover condition of the year 2009, 2013, and 2017 indicating land-use change in Tubuhue village from prior to the beginning to the end of mining peak activities, and to present. (Source: Primary data, 2017)

The increasing area open land and shrub in the 2017 land cover indicated that swidden farming again regained its dominance. After the end of the mining boom, members of local communities returned to practicing swidden farming. Despite the fact that some ex-miners choose to organize themselves to farm at the former mining sites, the size of the site was not large enough to accommodate all ex-miners. In addition, the access to the former mining sites was not free for all of those formerly involved in mining activities in different sites. It involved negotiation with the traditional owner of the land where farming in the former mining site was located. Hence the only available alternative for the majority of the ex-miners was to go back to their former locations of swidden farming.

What is apparent from this analysis is that blaming artisanal small-scale manganese mining as causing environmental degradation is not fully justified. Because base on observation, in-depth interview and spatial analysis (Figure 2), swidden farmers had obtained a better income from mining, they abandoned swidden farming, allowing shrub to grow in their abandoned swiddens. Meanwhile, no other shrubs and forest areas needed to be opened for swidden. In combination, the abandonment of swidden land and no further clearing of shrubs and forest to open new swidden farming had contributed to the reduction of the areas of open land and the increase of shrub and forest land during 2009-2013. When the mining collapsed in 2012, most ex-miners returned to practice swidden farming by reclaiming their abandoned swidden or opening new one by clearing shrubs and forest, thereby increasing the area of open land and shrubs in 2017. Digging holes for mining could trigger soil erosion as pointed out by Riwu Kaho et al. [9], but the process could have been much slowed down by the growth of grasses and other vegetation within and around the mining site.

3.2. Social transformation

Land-use transformation taking place during and after the manganese mining boom in Tubuhue and in other villages in Noelmina Catchment linked to the social dynamics of local communities. Livelihood changes cause social change. Such social dynamics were evolved around efforts to survive a harsh physical environment characterised by infertile soil and dry climate. Soil in the region is mostly

IOP Conf. Series: Earth and Environmental Science **131** (2018) 012029 doi:10.1088/1755-1315/131/1/012029

limestone-derived and hence its fertility is generally low. The climate is characterized by a short wet period, usually by the West monsoon during December to March, and followed by a long dry months (Figure 3). In such a harsh environment, the focus of people in using the available land is to produce food enough for their own consumption. Large track of land is available, but there are issues of complicated land tenure, low soil fertility, and lack of rainfall that is intricately interconnected in causing low household food production leading to famines.

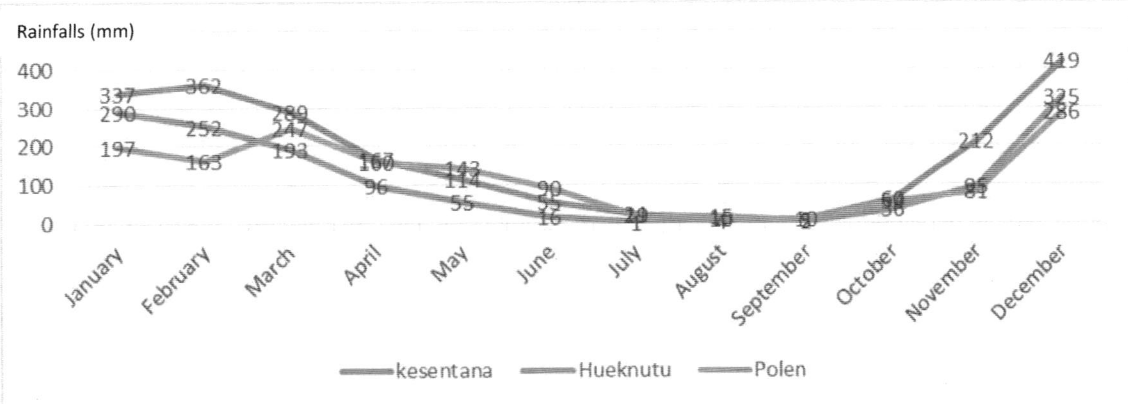

Figure 3. Monthly rainfalls based on the average of ten year data (January 2007 to December 2016) from Climatology Station of Lasiana-Kupang (Indonesia Agency for Meteorology, Climatological and Geophisics, BMKG)

Swidden agriculture had been practiced for generations without much change, despite government efforts to modernize agriculture [10]. Communal land tenure prevented long-term investment even for such an effort as simple as soil conservation because land that someone occupied for swidden in a particular year would not be necessarily available for the same person thereafter. Those involved in swidden farming by clearing and burning a particular track of land gained their access based on kinship relation and through negotiation with the clan that traditionally entitled to that track of land. Soil fertility was generally low, not simply because of the geological history of West Timor as an uplifted limestone sediment, but also because the pressure of continuous swidden practices that, because of population increase, allowed only a short fallow period barely enough for the soil to regain its fertility. In anticipation to low and unpredictable rainfalls, farmers relied on low-yielding crop varieties that survived the unpredictable rainfalls better than high-yielding varieties.

Swidden farmers in West Timor planted maize once a year during the rainy season by intercropping with other crops, usually with pulses and squash. From being cultivated for generations, maize, pulses, and squash had evolved to produce a large number of landraces adapted to dry condition. According to local farmers, the productivity of such landraces was much lower than that of the introduced varieties, but their capability to withstand drought was much better. Local farmers told us that they practiced intercropping both to maximize land productivity and to guarantee they still had other crop to harvestin case one crop failed. In addition, they also developed locally specific coping mechanisms such as collecting bee honey and tamarind pods from nearby shrubs or forest and also raised cattle, pig, and chicken that they could sell any time in case they need cash. They treated manganese mining initially as another form of this coping mechanism. However, since the income they obtained from the mining was much higher, then they spent most of their time for mining. Later they neglected and even abandoned their swidden cropsand doing other coping mechanism activities.

When manganese mining reached its peak, income from mining reached IDR 500,000 to IDR 1,000,000 per week. This was a large sum of money according to the local standard, although according to Dara [1], this amount was just part of 2.8-3% shares of the world price of manganese ores at that time. To local swidden farmers, however, this changed their life considerably. From the income they obtained from mining, they were able to pay tuition fees for their children, to rebuild their houses, and to pay other day-to-day household expenditure. Markus (2017) mention miners on the Neolmina Catchment area (included Tubuhue village), used their income for household needs (32%), children school fee (19%), donation to church/village (17%), house reparing (11%), furniture, jewelry, land, motorcycles (8%), farming tools (5%), bought Cattle (4%), and bridge well (4%). In Tubuhue, where manganese mining was organized by a local leader, the income from the mining was used to send children from the village to go to school in the district capital and to pay for tuition fees for those graduated from high schools to attend nursing college in the provincial capital. Similar benefits were also obtained from manganese mining on other villages as pointed out by Dara [1] as follows:

> *Most of the respondents highlighted that the mining activities have created an alternative livelihood for the local people, although they realize that the benefits are too small compared to time and efforts that have been spent on mining activities. The significance of the economic benefits derived from ASM activities can be identified by the changes in their capacity to fund their basic needs such us food, water and sanitation, adequate housing and basic education.*

What then made the mining in Tubuhue different from those in other villages was leadership. In Tubuhue, the mining was organized by a charismatic local leader who was able to convince members of the local community to carefully plan mining sites, manage the marketing of the ores, and use part of the income from the mining for improving local public infrastructure. A church now standing magnificently in the village was built from the income of manganese mining contributed by miners.

That artisanal small-scale mining has a potential for driving local development has also been pointed out by Hilson [11]. However, he also highlighted several drawbacks that could arise, the most common one was land tenure conflict. Such land tenure conflict had killed one miner in a village in Kupang District. Another important issue associated with artisanal small-scale mining found during the conduct of this study, as also common in other locations and countries [2,12–14], was its legal status. All artisanal small-scale manganese mining sites in the district of South Central Timor was by law illegal because for reasons not clearly explained, the district government adopted a moratorium policy. Nevertheless, despite being illegal, the mining continued to take place [1,15]. In fact, it was this illegal status that allowed negative impacts to take place. On site gambling and prostitution, child abuse, unhealthy mining practices, smuggling, etc., had been reported of taking place in illegal mining sites [16,17]. Fortunately, those negative impacts did not arise in Tubuhue, thank to the charismatic local leader that had successfully organizing the miners in the village.

One important issue we worried about was whether those who have been involved in the mining for quite some time and earned easy money would be willing to return to farming. As we have discussed in the section about land-use transformation, it was true that shrubs and forest reclaimed the landscape of the village and this reclaiming could be associated with the decrease in swidden farming activities. However, as indicated by results of our interviews, those involved in the mining did not completely abandoned their swidden. As put forward by an ex-miner (MBS; anonymous), "doing swidden or mining is all about digging holes in the ground. The only difference between them is whether you dig to plant seeds or to remove rocks from the soil". Another ex-miner told us that they engaged in manganese mining simply to take an opportunity available for them to earn additional income to pay tuition fees, house improvement, and "to buy fish or chicken so we can eat better food". Base on good organisation practices for mining resulting in social transformation, from swidden farmers, mining, and the and there are ultimately intensive horticultural cultivation utilizes mining land.

3.3. Post-mining activities

When the price of manganese ores crashed in 2012, mining was no longer feasible. Those who had for a while enjoyed better income had to return to their previous real life. One ex miner told us that the period of mining boom was "like a sweet dream" and the crash of the boom "like a sweet dream turn into a nightmare". Not only losing additional income, the ex-miners also faced stigmatization as causing environmental degradation from other community members who did not join mining activities (base on FGD and in-depth interview data). In fact, many people blamed artisanal small-scale mining as causing erosion and landslides, ignoring the fact that due to its geological history, the characteristic of soil in West Timor is intrinsically prone to erosion and that swidden has worsened this situation. Fortunate enough, ex-miners in Tubuhue had a local leader that they could turn to for advice.

The local leader was the one that had from the beginning organized miners in the village. When the mining collapsed, he asked ex-miners to voluntarily helped him to turn a number of big ex-mining holes into a check dam to collect rainwater and later to channel water from a spring nearby using propylene pipes (Figure 4A) and then leveling the ground around the check dam to plant vegetables (Figure 4B). "At the beginning only few ex-miners participated", the leader recalled, "but as our vegetable flourished and they saw us sold our produce in the market, more and more of them come to me asking to join". He then told us that he formed a farmer group consisting of those ex-miners who joined him in the business. As more and more ex-miners joining, he asked them to form two more farmer groups and to expand the farming area beyond the mining site. At the same time, because they needed to distribute the water from the check dam into a larger area, they developed a network of drip and sprinkler irrigation systems out of additional pipeline they laid out from a spring located above their farming site.

There they (farmers/ex-miners) planted different kinds of annual vegetable and fruit crops as well as and spices and herbs. "At the beginning we plant only 'sawiputih' (a cultivar of *Brassica rapa*)", an ex-miner member of the earliest farmer group told us, "but then we realized that we could get more profit from planting chilies". He continued telling us that because they planted at the right timing, "we harvested our chilies when the price was very high". From selling the harvest members of the groups were again able to have enough money to pay tuition fees for their children and to buy fish or chicken for their kid to eat better food. However, the price of chili pepper was very fluctuated, causing them to obtain a large profit from one harvest and from to lose money from another. It was from such an experience that they no longer relied on one crop. Instead, they planted different kind of vegetables, fruit, and spice crops either each in a separate block or intercropped two or more of them in one block (Figure 4C).

A B C

Figure 4. Land Rehabilitation and Agribusinesses activity at Tubuhue, A: The check dam constructed from mining holes, B: Leveling and preparing the ground around the check dam, and C: Annual vegetable and fruit crop and spices and herbs planted in various cropping patterns (Source: Researcher documentation, 2016)

The role of a local leader with a far-reaching vision is indispensable in the whole process of transformation taking place in Tubuhue. This local leader is the owner of the land where the mining took place during the booming period. Instead of asking those who mined in his land to pay him, he asked them to contribute a small portion of their income to pay the cost of the village church construction and the tuition fees of children he sent to high school and to nursing college. Similarly, he did not as those who did the farming in the former mining site to pay him some money. "I am more than happy to see if someday more and more people in this village abandoned swidden and do sedentary farming like what we do here", he explained. He continued by telling us that he had a dream that someday he could see more forest in his village. When we show and explain to him the map in Figure 2, he only said, "I wish the manganese boom were a bit longer".

4. Conclusion

Referring to our objectives in doing this case study and based on the results and discussion we have presented, we concluded that artisanal small-scale mining in Tubuhue has contributed to the dynamic ol local land use and social-economic transformation. The organized artisanal small-scale mining and post-mining land rehabilitation practised in the village can be used as a model for solving rural poverty as well as for organising environmentally friendlier artisanal small-scale mining operations. The way the ex-miners turned their nightmare into a new opportunity by developing a flourishing horticultural farm in their former mining site could serve as way to convince swidden farmers to abandon their practices of clearing and burning a track of land every year.

Acknowledgment

Funding for this case study was partially obtained from DFAT, Australia, as part of the funding for a larger research project, and from Nusa Cendana University, through its project phase-out facilitation programme. We are grateful to local leaders and members of community in the village of Tubuhue for their hospitality and willingness to share their life experience with us.

References

[1] Dara M J R 2014 Understanding risk and benefits of Informal (Artisanal small-scale) mining of manganese by peasants in Indonesia: A study on resource management and livelihood options in Kupang and TTS, West Timor, NTT IRGSC IRGSC Working Paper No 12

[2] Hobgen, S, Fishe,r R, Riwu Kaho, N, Beavis, S, Markus, J, Lio, A 2016 Community Mining in Indonesia - Minimising harm, maximising benefits Conference program and abstract book.

[3] Markus, J E R, Natonis, R L, Nampa, I W, Mudita, I W, McWilliam, A, Stacey, N 2017 Dampak Sosial Pertambangan Mangan di DAS Noelmina (Pengalaman Perempuan Penggali Mangan di Timor Barat) (Social Impacts of Manganese Mining in DAS Noelmina (Experiences of Manganese Women Miners in West Timor). Paper presented at Report Back "A Government Partnerships for Development project (GPFD) Artisanal Mining for Development in Eastern Indonesia 20 September 2017 Kupang Indonesia [In Indonesian]

[4] Apu R T, Pellokila M R, Nainiti S P 2017 Strategi Pemenuhan kebutuhan pangan pokok rumah tangga petani di Desa Noelbaki Kecamatan Kupang Tengah Kabupaten Kupang (Strategy to provide food requirement of farmer households in Noelbaki Village Kupang Tengah District Kupang Regency) on household *Excell Media Komun Agribisnis* **6** 1 85–91 [In Indonesian]

[5] Likadja Y M, Seran, S, Bernandina, L 2016 Ketersediaan dan konsumsi pangan Pokok Rumah Tangga Petani di Kecamatan Nekamiese Kabupaten Kupang (Availability and Food Consumption of Farmer Households in Nekamiese District Kupang Regency). In: Strategi Pengembangan Pertanian lahan kering Berkelanjutan dalam Mengatasi Kerawanan Pangan Kupang (In Sustainable Agricultural Development Strategy on Drylands to Tackle Food Insecurity in Kupang) Fakultas Pertanian Universitas Nusa Cendana pp. 297–307 [In Indonesian]

[6] Sugiyono 2009 Metodologi Penelitian Kuantitatif Kualitatif dan R&D (Quantitative Qualitative Research Method and Research & Development) Alfabeta Bandung Indonesia [In Indonesian]

[7] Balulescu B, Buresin M, Tecar L, Herbi M 2013 Processing and interpretation of satelite images-Landsat *Res. J Agric. Sci.* **45** 4 1-7

[8] Nuarsa I W, Nishio F, Hongo C 2011 Rice yield estimation using Landsat ETM+ data and field observation *Res. J Agric. Sci.* **4** 3 45-56

[9] Riwu Kaho, N, Hobgen, S, Natonis R L, Fisher, R, Nuban S R 2017 Soil erosion rates in the manganese mining areas of West Timor. Report back of the Government Partnerships for Development project. Artisanal mining for development in eastern Indonesia Kupang
[10] Benu F L, Mudita I W 2013 Revisitasi Lahan Kering - Diskusi Ringan Seputar Lahan Kering dan Pertanian Lahan Kering [Dryland Revitalisation - Small Discussion of Dryland and Dryland Farming) JP II Publishing House Jakarta Indonesia 189 [In Indonesian]
[11] Hilson G M 2006 The Socio-Economic Impacts of Artisanal and Small-Scale Mining in Developing Countries Taylor & Francis UK 738
[12] Hentschel, T 2003 Artisanal and Small-scale Mining: Challenges and Opportunities. IIED 94
[13] Leach R M 1971 Role of manganese in mucopolysaccharide metabolism *Fed. Proc. Fed. Am. Soc. Exp. Biol.* **30** 3 991–994
[14] Siegel S, Veiga M M 2009 Artisanal and small-scale mining as an extralegal economy: De Soto and the redefinition of "formalization" *Resour. Poli.* **34** 1 51–56
[15] Erb M 2016 Mining and the conflict over values in Nusa Tenggara Timur Province, Eastern Indonesia *Extr. Ind. Soc.* **3** 2 370–382
[16] Banchirigah S M 2006 How have reforms fuelled the expansion of artisanal mining? Evidence from sub-Saharan Africa *Resour. Policy.* **31** 3 165–171
[17] Kitula A G N 2006 The environmental and socio-economic impacts of mining on local livelihoods in Tanzania: A case study of Geita District *J Clean Prod.* **14** 3 405–414

IOP Conf. Series: Earth and Environmental Science **131** (2018) 012030 doi:10.1088/1755-1315/131/1/012030

Re-design of apple pia packaging using quality function deployment method

M H Pulungan, N Nadira and I A Dewi

Department of Agroindustrial Technology, Faculty of Agricultural Technology Universitas Brawijaya, Malang, Indonesia

E-mail: hindunmaimunah@yahoo.com

Abstract. This study was aimed to identify the attributes for premium apple pia packaging, to determine the technical response to be carried out by Permata Agro Mandiri Small and Medium Enterprise (SME) and to design a new apple pie packaging acceptable by the SME. The Quality Function Deployment (QFD) method was employed to improve the apple pia packaging design, which consisted of seven stages in data analysis. The results indicated that *whats* attribute required by the costumers include graphic design, dimensions, capacity, shape, strength, and resistance of packaging. While, the technical responses to be conducted by the SMEs were as follows: attractive visual packaging designs, attractive colors, clear images and information, packaging size dimensions, a larger capacity packaging (more product content), ergonomic premium packaging, not easily torn, and impact resistant packaging materials. The findings further confirmed that the design of premium apple pia packaging accepted by the SMES was the one with the capacity of ten apple pia or 200 g weight, and with rectangular or beam shape form. The packaging material used was a duplex carton with 400 grammage (g/m^2), the outer part of the packaging was coated with plastic and the inside was added with duplex carton. The acceptable packaging dimension was 30 cm x 5 cm x 3 cm (L x W x H) with a mix of black and yellow color in the graphical design.

1. Introduction

Packaging is one of the factors that can influence a selling price of a product on the market. In Malang, one of the food products known as Malang's special souvenirs is apple pia. The apple pia is packed with plastic for primary packaging and duplex carton for secondary packaging. Currently, the secondary packaging of apple pia has an unattractive shape, appearance and color, as well as has a small capacity (i.e. each packaging contains 6 apple pia). Therefore, this packaging was classified as a small packaging and this was made due to its market segmentation of middle to lower class customers. In recent years, the apple pia SMEs, particularly Permata Agro Mandiri SME, want to expand their market segmentation to middle and upper class customers. In addition, there are market demands for the SME to make a more attractive and larger packaging. Therefore, to expand market segmentation, to fulfill the costumer desire and to increase the apple pia sales, the apple pia packaging design is necessary to be improved and re-designed.

Thus, changing the current packaging into a premium packaging is necessary for attracting middle and upper class customers. For that puposes, the premium apple pia packaging design was made in a larger size to increase the capacity. Also, a stronger material and a more interesting graphical design than the previous packaging will be used, aiming at increasing the consumer buying interest in apple pia.

In the research, Quality Function Deployment (QFD) method was used to obtain attribute level of consumer interest, which then will be adjusted with the technical responses from the manufacturer. QFD method can reduce the length of product development cycle, as well as can improve the quality and the marketing of products in the market [1]. In addition, with QFD method consumers can be involved prior the re-designing process of apple pia packaging to ensure that the new packaging meets the consumers' needs. The purposes of this study are to identify the costumer interest attributes for premium apple pia packaging, to determine technical response to be performed by the SME, and to design a new packaging of apple pia acceptable by the SME.

2. Method

The attribute level of customer interest (*whats* or needs) questionnaire was given to consumers who have purchased/consumed apple pia with the age of 17 years old or over. The questionnaires was distributed in two Malang's souvenirs store (i.e. Lancar Jaya and Burung Swari), with a total of 40 questionnaires were filled. The questionnaire results were tested using the validity and reliability test.

2.1 Identification of Variable

Identification of variables aimed at determining customer attributes (*whats*) and technical capabilities possessed by Permata Agro Mandiri SME to meet their customer needs. The customer attribute of Permata Agro Mandiri SME was obtained through an interview during the preliminary survey with the owner. This activity, conducted after *whats* attributes were identified, was carried out to identify the consumers' satisfaction level of the existing apple pia products, particularly on the customer needs and interests of the SME. The *whats* attributes can be seen in Table 1.

Table 1. *Whats* attributes

No.	Attributes	Technical Response (*Hows*)
1.	Graphical Design	Innovation for a new premium apple pia packaging
2.	Dimension	The compatibility of the packaging size to the product
3.	Capacity	Premium apple pia packaging capacity
4.	Shape	Ergonomic premium apple pia packaging
5.	Strength	Premium apple pia packaging is not easily torn
6.	Resistance	Premium apple pia packaging is impact resistant

2.2 Data Analysis

Data analysis for re-design of apple pia packaging was carried out using QFD method. QFD is a way to improve the quality of goods or services by understanding the needs of consumers and then connecting with the technical responses in its production process [2]. The use of QFD method enable to increase customer satisfaction, reduce implementation time, improve teamwork, and assist the company in re-developing a product or service [3].

According to Suryaningrat et al. [4], the stages of QFD data analysis are as follows:

1. Voice of customer collecting phase
2. Phase of creating house of quality
3. Making a technical responses
4. Determining the relationship of technical responses and customer needs
5. Determination of technical relations
6. Calculation of technical weight and priority
7. Calculation of benchmarking and target values

3. Result and Discussion

3.1 Creating House of Quality

The result of creating house of quality is shown in Figure 1.

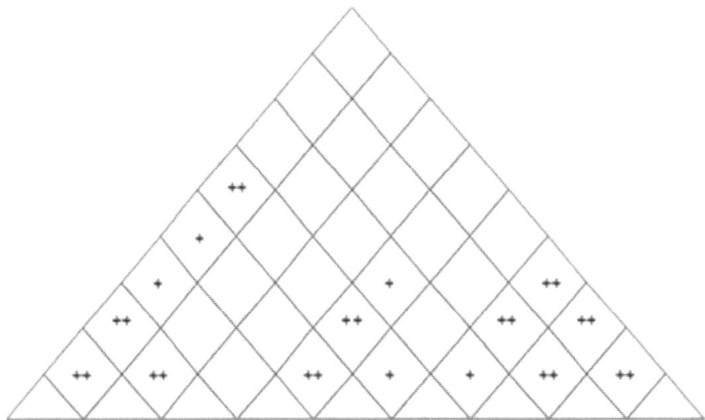

Hows / Whats	1 Atractive visual packaging design	2 Atractive colours	3 Clear image and information	4 Size dimension suitable for capacity	5 Larger capacity (more product content)	6 Ergonomic premium pacakaging	7 Thcicker pacakaging materials	8 Not easy torn pacakaging materials	9 Impcat resistance pacakaging materials	Importance to customer	Customer satisfaction	Improvement ratio	Goal	Raw Weight	Normalised raw Weight
Graphic design	●	○	●			△				4.43	2.33	1.97	4.58	12.00	0.246
Dimension	○			●	○	○	△			4.03	3.25	1.23	3.98	5.95	0.112
Capacity				●		○				3.95	3.23	1.28	4.30	6.07	0.114
Shape	●		△	○		●				3.73	3.43	1.18	4.05	5.31	0.10
Strength						△	●	△	●	4.45	2.65	1.72	4.55	11.43	0.216
Resistance						△	●	●	●	4.50	2.63	1.67	4.40	11.27	0.212
Technical Responce weight	183.45	39.27	123.12	124.11	77.79	119.69	210.70	112.92	204.57						
Priority	3	9	5	4	8	6	1	7	2						
Recent packaging	2.93	2.33	2.44	3.27	3.25	3.22	2.67	2.63	2.64						
New packaging	4.27	4.58	4.53	4.13	4.23	4.16	4.45	4.42	4.48						
Target	5	5	5	5	3	5	4	5	5						

Figure 1. House of Quality

Table 2. Sales Point

No.	Attributes	Importance to Customer	Sales Point
1.	Graphical Design	4.43	1.5
2.	Dimension	4.03	1.2
3.	Capacity	3.95	1.2
4.	Shape	3.73	1.2
5.	Strength	4.45	1.5
6.	Resistance	4.50	1.5

Table 2 shows that there were three attributes have a high sales point value of 1.5, namely graphical design, strength and resistance of packaging. This result indicated that those attributes have been highlighted as the marketing focus for re-designing the packaging of apple pia. Bangun [5] stated that attributes with a high sales point value (i.e. 1.5) need to be selected as the focus of marketing and profit improvement for the company.

For the priority of re-designing packaging, the top three weight of technical responses were thicker packaging materials, impact resistant packaging materials and attractive visual packaging graphical designs (Figure 1). These technical responses have a large contribution to meet consumer needs. This is in agreement with Putri [6] who said that a priority shall be given based on the relationship of technical response, the service attributes and the attributes weight. Therefore, a technical response with a higher weight compared to other technical responses needs to be prioritized first due to its major contribution to the product quality.

For benchmarking values, the technical responses of larger capacity (more product contents) and thicker packaging materials were higher than the target values. These results confirmed that changes in the new packaging is acceptable and further improvement is not recommended. However, for other technical responses, the benchmarking values were lower than the target values, indicating that the new packaging improvement is still needed to fit with the SME targets and requirement. According to Rahmatika [7] the value of the target becomes a reference to make changes.

3.2 Designing of New Packaging

Designing of the new packaging was conducted based on the data analysis results of QFD method and House of Quality. In re-designing a new packaging, the relationship between *whats* and *hows* attributes and relationships between *hows* attributes should be considered. Attributes of *whats* and *hows* with strong relationships should be prioritised, followed by those with moderate and weak relationships. Similarly, for *hows* attributes with strong positive and positive relationships should also be prioritised.

The graphical design of premium apple pia packaging was made of two color choices: a mix of blue and pink, and a mix of yellow and black. These color were selected because many consumers of apple pia are young adults who prefer a modern and exclusive look. Furthermore, some new information (i.e. halal logo and year of the SME establishment) were added into the new packaging design. Also, a transparent section with triangular shape was made on the top of the new packaging.

The packaging dimension was designed with the size of 30 cm x 5 cm x 3 cm (L x W x H). In terms of the packaging capacity, it was increased to 200 g per package, containing 10 apple pia. The shape of the premium apple pia packaging was remained as a rectangular or beam, but the packaging material was replaced by a duplex carton with 400 g/m^2. The outer part of the premium packaging will be wrapped by plastic and the inner part will be added with a duplex carton functioned to protect against shock or collision with other objects.

3.3 User Verification

User verification was carried out to find out the response from the Permata Agro Mandiri SME to the new apple pia packaging designed by the researcher. For graphical design, two recommendation of the color choices were given and should be chosen by the SME. The results showed that a mix of black and yellow color was selected by the SME for the premium apple pia packaging While, for other packaging attributes, only one recommendation was provided, and those have been agreed by the SMEs. The comparison of the current and the premium packaging design are shown in Figure 2 and 3. The differences between the current and the new packaging design are the color combination, the image, the information on packaging, and the dimensions of packaging.

Figure 2. Current packaging design of the apple pia

Figure 3. New premium packaging design of the apple pia

4. Conclussion

The *whats* attributes (i.e. graphical design, dimensions, capacity, shape, strength, and resistance) provided by the consumers contributed to the re-designing process of the premium apple pia packaging. In combination with the identified technical responses, a new design for premium packaging of apple pia was created and accepted by the Permata Agro Mandiri SME. This new packaging design composed of a mix of black and yellow color as graphical design, a bigger size dimension enable to contain 200 g capacity (or equvalent with 10 apple pia in each package), a rectangular or beam shape, a 400 g/m^2 duplex carton material, as well as additional packaging materials such as plastic for wrapping the packaging and a duplex carton as an inner part for protecting the product.

References

[1] Krstic J 2014 Using Possibility of QFD Method for Development of the "Ready-To-Go" Package *Acta Graphica* **25** 1-2 37-46
[2] Arif M 2016 *Bahan Ajar Rancangan Teknik Industri* ((Yogyakarta: Deepublish)
[3] Besterfield D C, Besterfield-Michina, G H Besterfield, and M Besterfield-Sacre 2003 *Total Quality Management : Third Edition* (New Jersey: Pearson Education, Inc.)
[4] Suryaningrat, I B Djumarti, E Ruriani and I Kurniawati 2010 Aplikasi Metode *Quality Function Deployment* (QFD) Untuk Peningkatan Kualitas Produk Mie Jagung (Application of *Function Deployment* (QFD) to quality improvement of corn noodle) *Agrotek* **4** 1 8 – 17 (In Indonesian)
[5] Bangun A, Ginting R, and Tarigan U 2013 Analisis Kualitas Pelayanan Jasa Pendidikan dengan Menggunakan *Quality Function Deployment* (QFD) (Quality analysis of educational services using *Quality Function Deployment* (QFD)) *E-Journal Teknik Industri* **3** 1 47 – 51 17 (In Indonesian)
[6] Putri A R 2012 Analisis Perencanaan Strategi Peningkatan Kualitas Pelayanan Konsumen dengan Metode *Quality Function Deployment* (QFD) (Analysis of strategic planning to improve the quality of educational services using *Quality Function Deployment* (QFD)) *Jurnal Industria* **4** 1 41 – 52 (In Indonesian)
[7] Rahmatika I N 2008 Penerapan *Quality Function Deployment* (QFD) Untuk Mengetahui Tingkat Kepuasan Konsumen Produk Biskuit Di PT. Arnott's Indonesia (Application of *Function Deployment* (QFD) to identify customer satisfaction level in PT. Arnott's Indonesia) (Bogor: Institut Pertanian Bogor Thesis) (In Indonesian)

IOP Conf. Series: Earth and Environmental Science **131** (2018) 012031 doi:10.1088/1755-1315/131/1/012031

Institutional analysis of milkfish supply chain using interpretive structural modelling (ISM) (case study of UD. Bunda Foods, Sidoarjo District)

R L R Silalahi, S A Mustaniroh, D M Ikasari and R P Sriulina

Department of Agro-industrial Technology, Faculty of Agricultural Technology, Universitas Brawijaya, Malang, Indonesia

E-mail: rizkylrs@ub.ac.id

Abstract. UD. Bunda Foods is an SME located in the district of Sidoarjo. UD. Bunda Foods has problems of maintaining its milkfish's quality assurance and developing marketing strategies. Improving those problems enables UD. Bunda Foods to compete with other similar SMEs and to market its product for further expansion of their business. The objectives of this study were to determine the model of the institutional structure of the milkfish supply chain, to determine the elements, the sub-elements, and the relationship among each element. The method used in this research was Interpretive Structural Modeling (ISM), involving 5 experts as respondents consisting of 1 practitioner, 1 academician, and 3 government organisation employees. The results showed that there were two key elements include requirement and goals elements. Based on the Drive Power-Dependence (DP-D) matrix, the key sub-elements of requirement element, consisted of raw material continuity, appropriate marketing strategy, and production capital, were positioned in the *Linkage* sector quadrant. The DP-D matrix for the key sub-elements of the goal element also showed a similar position. The findings suggested several managerial implications to be carried out by UD. Bunda Foods include establishing good relationships with all involved institutions, obtaining capital assistance, and attending the marketing training provided by the government.

1. Introduction

There are many milkfish-processing SMEs in Sidoarjo City. One of their processed milkfish products is smoked milkfish, currently known as Sidoarjo City specialty product. In Sidoarjo, there are 13 SMEs of smoked milkfish, one of them is UD. Bunda Foods. Not only smoked milkfish, UD. Bunda Foods also produces milkfish presto, milkfish *otak-otak*, and milkfish nugget. This SME has five (5) employees and the production capacity is about 80 kg/month.

The goals of effective supply chain activity are to have precise amount of specific products, to be available in the right place, and to cost with the right range of price [1]. The supply chain is related to the product or service flow and transformation from raw material providing phase, until the final product reaches the consumer [2]. UD. Bunda Foods obtained their raw material (i.e. milkfish) from the breeders and the sellers in the market. However, the problems of maintaining the milkfish's quality assurance and developing marketing strategies are faced by UD. Bunda Foods. Those problems need to be tackled if UD. Bunda Foods want to market its product, enabling the SME to compete with other similar SMEs and to expand the business.

The supply chain organisation is defined as a well-structured series of related institutions or enterprises governed by any means of strict rules and structure to control, organize and maintain the flow of the money, materials, and information from its upstream to the final user [3]. Therefore, this research is aimed to determine the structure model of smoked milkfish supply chain organisation, to identify the elements, the sub-elements, and the interaction among those elements. To achieve this, Interpretive Structural Modelling (ISM) method was employed.

2. Research Method

This research was carried out in UD Bunda Foods in Penatarsewu Village, Tanggulangin, Sidoarjo. The collected data were analysed in Agroindustry Management Laboratory, Dept. of Agroindustrial Technology (AIT), Faculty of Agriculture Technology (FAT), Universitas Brawijaya (UB), Malang. There were five expert respondents involved, include 1 SME owner, 1 lecturer form Dept. of AIT – FAT-UB, 1 Division Head of Marine and Fisheries Agency, 1 staff of Industry and Trading Agency of Sidoarjo City, and 1 Division Head of Micro Business Supervisor Agency from Cooperatives and Micro Business of Sidoarjo. Data collected from these respondents are ISM scoring with VAXO questionnaire and verbal interview to obtain company profile and milkfish business condition.

2.1. Organization analysis

In this phase, the problem limitation includes the boundary analysis of only two elements out of nine organisation program elements. Those nine elements are community involvement, program requirement, main obstacles, possible alteration, program goals, measurement parameters, planning organisation activity, measurement activity to evaluate the result of each activity, and institution involvement [4]. Therefore, the selected elements to be analysed were requirement and goal element. These elements were determined through interview with the expert respondents. Then, each element was divided by several sub-elements as seen in Table 1.

Table 1. Elements and sub-elements of supply chain organization.

Element	Sub-element
Requirement	1. Raw material continuity
	2. Proper marketing strategy
	3. Production capital
	4. Skilled employee
Goal	1. Raw material continuity assurance
	2. Product quality assurance
	3. Expanding marketing area
	4. Increasing business income

2.2. Data analysis

The Interpretive Structural Modelling (ISM) method was used to analyse the data of smoked milkfish supply chain organisation in UD. Bunda Foods. This method was used to analyse the system elements as a graphical representation, showing a direct relationship among elements and the hierarchical arrangement. Generally, ISM technique was divided into two parts: element classification and hierarchy arrangement [5]. ISM implementation process steps are as follows: 1) variable identification, 2) contextual relationship determination, 3) structural self-interaction (SSI) matrix development using pairwise comparison, 4) matrix transformation into Reachability Matrix, 5) transitivity assessment, 6) final reachability matrix arrangement, 7) directed graph withdrawal, and 8) transitive link deletion [6].

IOP Conf. Series: Earth and Environmental Science **131** (2018) 012031 doi:10.1088/1755-1315/131/1/012031

3. Results and Analysis

3.1. Profile of UD. Bunda Foods

UD. Bunda Foods is an SME of smoked milkfish, which its success is supported by its effective supply chain activity, from the raw material to the final product. Sole proprietorship (UD) is a self-owned business that requires no business partner [7]. If there is a person or a party supporting the business activity, they will not be in the same position as the owner. UD. Bunda Foods, founded by Lita Eka Yufidaf in 2014, has been producing smoked milkfish with the brand name of 'Bitha". This SMEs has five employees with the production capacity of 80 kg per month. Usually, SMEs have 5-19 employees [8].

The supply chain organisation of smoked milkfish in UD. Bunda Foods can be seen in Figure 1.The figure shows that the milkfish breeders received information flow about fresh and stout milkfish request and money flow of milkfish purchase from UD. Bunda Foods. Since milkfish is herbivore, its cultivation has comparative advantages [9], such as it can easily be cultivated either using intensive or semi-intensive system with relatively lower cost [10].

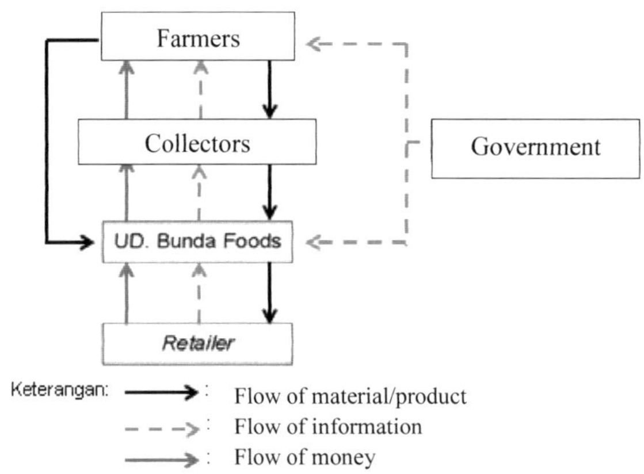

Figure.1 Structure of smoked milkfish supply chain organisation

Milkfish collector receives information flow about ≥ 60 ounce-sized fresh milkfish request and money flow to fulfill that request. The seller or collector is an independent marketing institution which involved in the distribution channel with specific motive [11]. Their main motive is to earn some profits from the difference in the buying and the selling price.

In the supply chain organisation, UD. Bunda Foods distributes smoked fish product to the retailer. Retailer, as the final activity of business, is required to distribute products to consumers [12]. UD. Bunda Foods receives information flow from retailers about their selling capacity, the consumer demands, and money flow from smoked fish sales.

In addition, the government also provides information for each party in the supply chain organisation. The government provides the breeder with information about good quality for milkfish to be harvested. Government participation is essential to develop small-scale business. The government might participate by providing capital in the form of money or production instrument [13]. The collector is provided with information about determining the corresponding price for each milkfish size and quality. UD. Bunda Foods also receives information about exhibition events to sell the smoked fish product. The retailer receives information about strategic retail location to increase smoked fish sales.

3.2. Requirement element analysis

Figure 2 shows that the sub-elements of the requirement element are consisted of raw material continuity, proper marketing strategy, and production capital. This confirmed that business activity in UD. Bunda Foods was successfully operated because of its raw material continuity assurance from the supplier, proper marketing strategy of direct selling, and adequate production capital. Another influencing factor to the SME business success was skilled employees. Skilled employee might enhance production processes by helping the owner to process the raw material into products. Production factors involved in the business development, such as capital, employee, raw material, transportation, energy resource and marketing should be combined all together [14].

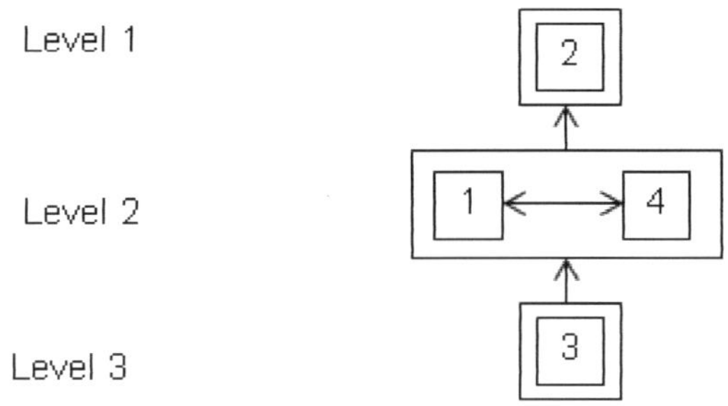

Figure 2. Diagram of requirement element structural model

Figure 3 shows that the sub-element of raw material continuity, proper marketing strategy and production capital are fitted to the *Linkage* sector quadrant. The results suggested that those sub-elements have strong propelling force and dependency to other sub-elements. This means that if the production capital cannot be provided, then the raw material cannot be acquired which may hindrance the marketing process. Furthermore, this occurrence may have negative impacts on the product sales. The figure also indicates that the sub-element 4 (i.e. skilled employee) is fitted in the *Dependent* sector quadrant. This means that this sub-element is influenced by other sub-elements within the *Linkage* sector. Major factors influencing the industry development include capital, employee, raw material, transportation, energy resource and marketing [15].

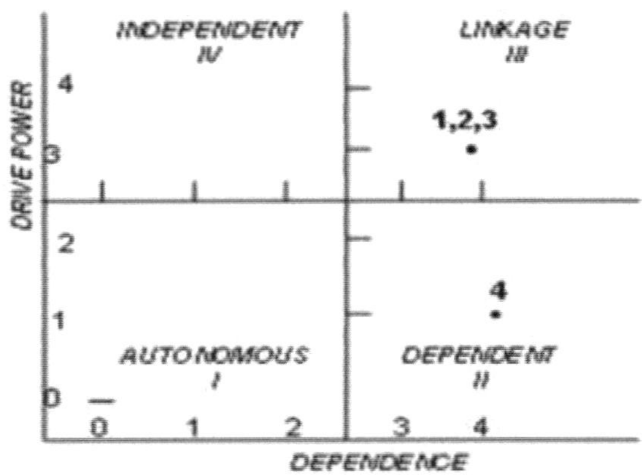

Figure 3. DP-D matrix of requirement element

3.3. Goal element analysis

As can be seen in Figure 4, all the sub-elements of the goal element were the key sub-elements. Those sub-elements are the goals wanted to be achieved by UD. Bunda Foods, which will enable them to maintain their business. Business success means the businesses completed or achieved their goals [16]. A small-scale business success is influenced by its company performance, which is considered as one of the businessman goals.

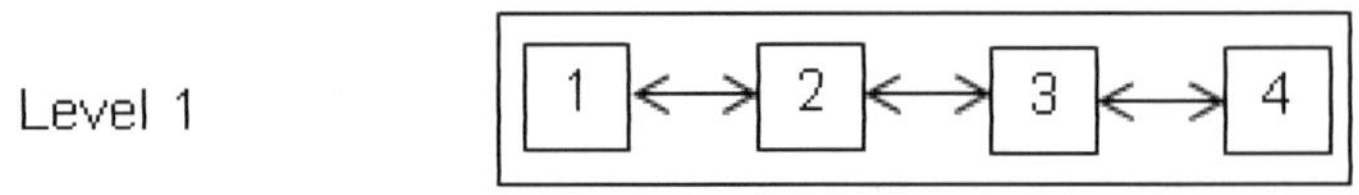

Figure 4. Diagram of goal element structural model

Figure 5 shows that all the sub-elements of the goal element were in the *Linkage* sector quadrant. This means that those sub-elements are strong, either in propelling force or dependency to other sub-elements. Therefore, the product quality can be assured if the industry is able to assure their raw material continuity. Furthermore, if the product quality can be assured, the marketing area and income can also be increased. Product quality is an obligatory factor to ensure business success [17].

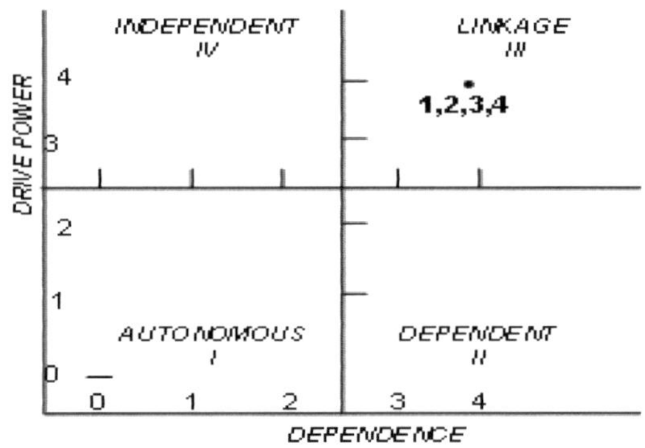

Figure 5. Matrix of goal element DP-D

3.4. Managerial implication

The findings from observation and ISM analysis suggested the following managerial implication, include:

1. UD. Bunda Food may be able to maintain its business if they have good relationship with seller, collector, retailer, and related government agencies.
2. UD. Bunda foods requires proper marketing strategy, such as direct selling with support and supervision from government to maintain their business persistence.

4. Conclusion

Parties involved in the structure model of UD Bunda Foods smoked milkfish organization include breeder, collector, UD. Bunda Foods, retailer and government. Several sub-elements of the requirement element were raw material continuity, proper marketing strategy and production capital which belong to *Linkage* sector. The findings confirmed that these sub-elements influences the sub-element (i.e. skilled employee) in *Dependent* sector. The sub-elements of the goal element, such as raw material continuity assurance, product quality assurance, marketing area expansion and business income improvement, also occupy the same position in *Linkage* sector.

References

[1] Bateman T S, Snell S A 2008 Manajemen (Management) 7th Edition Salemba Empat Jakarta Indonesia 440 [In Indonesian]
[2] Furqon C 2014 Analisis Manajemen Dan Kinerja Rantai Pasokan Agribisnis Buah Stroberi Di Kabupaten Bandung (Supply chain management and performance analysis on strawberry agribusiness in Bandung Regency) *Image* **3** 2 109-126 [In Indonesian]
[3] Siagian Y M 2005 Aplikasi Supply Chain Management dalam Dunia Bisnis (Supply Chain Management Application in Business) Grasindo Jakarta Indonesia 233 [In Indonesian]
[4] Marimin 2004 Teknik dan Aplikasi Pengambilan Keputusan Kriteria Majemuk (Decision Making Technique and Application) Grasindo Jakarta Indonesia [In Indonesian]
[5] Kholil E, Sutjahyo S H, Soekarto S H 2008 Pengembangan model kelembagaaan pengelola sampah kota dengan metode ISM (Interpretive Structural Modeling) (studi kasus di Jakarta Selatan) (Institutional model development on garbage management using Interpretive Structural Modeling method (case study in South Jakarta) *J. Transdisiplin Sosiologi, Komunikasi, dan Ekologi Manusia* **2** 1 31-48 [In Indonesian]
[6] Hussain M, Awasthi A, Tiwari M K 2015 Interpretive structural modeling- analytic network process integrated framework for evaluating sustainable supply chain management alternatives *Appl. Math. Mod.* **40** 3671-3687
[7] Purnamasari I D 2010 Panduan Lengkap Hukum Praktis Populer: Kiat-Kiat Cerdas, Mudah dan Bijak Mendirikan Badan Usaha (Complete Guide to Popular Practical Law: Smart, Easy and Wise Tips for Establishing a Business Entity) Mizan Pustaka Bandung Indonesia [In Indonesian]
[8] Nurseto T 2004 Strategi Menumbuhkan Wirausaha Kecil Menengah Yang Tangguh (Strategy on growing tough SMEs) *J. Ekonomi & Pendidikan* **1** 1 96-105 [In Indonesian]
[9] Murtidjo B A 2002 Budi Daya dan Pembenihan Bandeng (Cultivation and Breeding Milkfish) Kanisius Yogyakarta Indonesia 113 [In Indonesian]
[10] Pasaribu A M 2004 Kajian Sistem Modular Pada Usahatani Ikan Bandeng (*Chanos-Chanos forskal*) di Sulawesi Selatan (Study of modular system on milkfish (*Chanos-Chanos forskal*) farming business in South Sulawesi. *J. Pengkajian dan Pengembangan Teknologi Pertanian* **7** 2 187-192 [In Indonesian]
[11] Lentera T 2003 Cermat & Tepat Memasarkan Gurami (Meticulous & Precise Marketing of Gurami) PT. AgroMedia Pustaka Jakarta Indonesia 97 [In Indonesian]
[12] Vhia O, Benarto I 2008 Pengaruh Kualitas Pelayanan Ritel Terhadap Kualitas Hubungan (Studi Kasus Zara Mall Taman Anggrek) (Influence on the quality of retail service to the relationship quality (case study of Zara Mall Taman Anggrek) *DeReMa Jurnal Manajemen* **3** 2 173-187 [In Indonesian]
[13] Afifah R Z, Setiawan A H 2012 Analisis bantuan modal dan kredit bagi kelompok pelaku usaha mikro oleh Dinas Koperasi dan UMKM Kota Semarang (Studi Kasus: KPUM di Kelurahan Pekunden, Kecamatan Semarang Tengah (Anaylisis of capital and credit support for SMEs from cooperatives and SMEs agency in Semarang City (case study of KPUM in Pekunden Sub-district, Semarang tengah District) *Diponegoro Journal of Economics* **1** 1 1-15 [In Indonesian]
[14] Suroyah I A 2016 Analisis faktor-faktor yang mempengaruhi nilai produksi industri kecil tenun ikat di Kabupaten Jepara (Studi Kasus di Desa Troso, Kecamatan Pecangaan, Kabupaten Jepara) (Analysis of factors influencing production capacity of *tenun ikat* in Jepara Regency (case study in Troso Village, Pecangan District, Jepara Regency) *J. Pendidikan dan Ekonomi* **5** 11-5 [In Indonesian]

[15] Rosmiati 2012 Analisis Program Bantuan Modal Kredit Usaha Penguatan Ekonomi Masyarakat (KUPEM) oleh Pemerintah Kota Jambi terhadap Pengembangan Usaha Mikro Kecil dan Menengah di kota Jambi (Anaylis oof mcirocredit programme on strenghtening the small and medium scale enterprises in Jambi City) *Mankeu* **1** 3 239-244 [In Indonesian]
[16] Suarmawan K A 2015 Analisis faktor-faktor yang mempengaruhi keberhasilan usaha mikro dan kecil (Studi pada usaha kerajinan Ingka di Desa Bulian, Kec. Kubutambahan (Analysis of factors influencing SMEs' success (case study in Ingka handycraft business in Bulian Village, Kabutambahan District) *J.Jurusan Pendidikan Ekonomi (JJPE)* **5** 1 1-10 [In Indonesian]
[17] Sutrisno, Rahab I, Jaryono 2010 Peningkatan kinerja UKM melalui improvement process Di Kabupaten Purbalingga (SMEs' performance improvement with improvement process in Purbalingga Regency) *Ekobis* **11** 1 407-414 [In Indonesian]

IOP Conf. Series: Earth and Environmental Science **131** (2018) 012032 doi:10.1088/1755-1315/131/1/012032

Green productivity analysis at tofu production (case study of UD Gudange Tahu Takwa Kediri)

R Septifani, P Deoranto and I Jannah

Department of Agro-industrial Technology, Faculty of Agricultural Technology, Universitas Brawijaya, Malang, Indonesia

Email: riskaseptifani@ub.ac.id

Abstract. UD. Gudang Tahu Takwa (GTT) is one of tofu Small and Medium Enterprises (SMEs). In Kediri, there are many tofu SMEs, each of them competes to survive in the market by improving and increasing productivity. The purposes of this study were to determine the most influenced factors towards productivity level and to determine the best wastewater treatment alternatives to be applied. An approach to improve productivity while lowering environmental impacts is known as Green Productivity (GP). The GP approach can be applied by using several tools such as mass balance, causal diagram, Environmental Performance Indicator (EPI), and Net Present Value (NPV). The results show the lowest productivity was washing process with the value of 17%. The EPI index total was -3.94 indicating that the environmental performance was still under the quality standard. The findings further suggested two alternatives to be implemented to improve productivity level, include replacement of suppliers and procurement of filtration equipment. The most influenced factor on productivity level was the volume of wastewater generated. Therefore, the second alternative was selected, due to its implementation could increase the productivity level of washing process by 79%, increase the EPI index to -3,00 and reduce the NPV value to IDR 1,582,204. The suggestion from this research is that UD GTT should buy the filtering equipment, and then applied it to treating the wastewater.

1. Introduction

Tofu is a typical Kediri food made from soybeans and it is very much sought after by people as a typical Kediri hand gifts. Therefore, the surrounding community takes an advantage of it by opening business opportunities. The data from the Industry and Trade Department (Disperindag) Kediri shows that there are 149 units of small and medium enterprises (SMEs) of *Takwa* tofu [1]. The large numbers of these SMEs have motivated the owners to continue their survival in the market competition by increasing the productivity level. Productivity is the ratio between the results achieved (output) and the overall resources used (input) [2]. Increased productivity is crucial in every production process, improving both the output quality and the efficiency of resources. Increasing the production of tofu can cause environmental problems to its surrounding environment because of the disposal of materials and energy that burden the environment.

UD. Gudange Tahu Takwa (GTT), established in 1990, is one of tofu SMEs. UD. GTT is located at Toyoresmi Village Ngasem District of Kediri Regency. This SME sells a variety of processed tofu such as *takwa* tofu, round tofu, tofu stick, tofu chips, tofu crackers, crispy tofu, *emping* tofu, and

Sumedang fried tofu. Its main product is *takwa* tofu, with a production capacity of 250 kg per day. During the production process in UD. GTT, both solid waste and wastewater were generated. The solid waste produced is further processed as cattle feeds. However, wastewater was directly disposed without any treatments, thus contributing to damaging the surrounding environment and the neighbourhood community. The wastewater is generated from washing, soaking, and tofu pressing process. The problems of waste in UD. GTT required further measures, for instance by applying waste management in the production process, which is expected to positively affect the productivity of the company.

One of the approaches to improve productivity while lowering environmental impacts is the Green Productivity (GP) method. The GP is a strategy to increase corporate productivity and environmental performance simultaneously in socio-economic development as a whole. The application of GP to SMEs is considered relevant because it begins with a strategy to improve productivity and environmental performance by analyzing inputs, processes, and outputs. The GP shows the way to reduce the practical impact on the environment that can lead to cost savings and risk reduction. The GP also produces significant benefits for increased productivity [3].

Productivity measurement is carried out by using the result of mass balance calculation in each production process. The GP approach can be applied by using several tools such as cause and effect diagram and Net Present Value (NPV). Factors that affect the level of productivity can be identified using a cause and effect diagram. Cause-and-effect diagrams are simple graphical methods to hypothesise the cause and effect chain [4]. The determination of the best improvement alternative in handling waste and increasing productivity use NPV. The NPV is a tool that compares the costs and benefits of some alternatives [5]. The purpose of this research is to determine the factors that most influence the level of productivity of tofu production process and the best alternative for waste treatment to improve the environmental performance at UD. GTT.

2. Materials and Methods

2.1. Materials and instruments

This research was conducted at UD. GTT from March to April 2017. The data processing is carried out at the Computational and System Analysis Laboratory, Department of Industrial Technology of Agriculture, Faculty of Agricultural Technology, Universitas Brawijaya, Malang. Instruments used in this research were mass balance, Environmental Performance Indicator (EPI), Fishbone diagram and NPV.

2.2 Method

The followings are the steps of data processing [6]:

1. Calculating mass balance and measurement of productivity
2. Determining the EPI

   The measurement of EPI Index is useful to minimize waste generated in the production process and increase its productivity [6]. If the positive EPI value indicates the company's environmental performance is good. Conversely, if the value of EPI is negative, then the company's environmental performance is not good [5]. The measurement of the EPI index can be formulated as follows:

   $$\text{Index EPI} = \sum_{i=1}^{k} Wi.Pi$$

   (Where, k = number of waste criteria proposed, Wi = weight of each criterion, Pi = deviation from each criterion between normal waste standard and laboratory analysis result)

   The EPI shows the environmental performance shown by analyzing the results of the search waste test from UD. GTT and wastewater quality standards. The parameters used are biochemical oxygen demand (BOD), chemical oxygen demand (COD), total suspended solids (TSS), and pH. Those four parameters were used to determine the level of wastewater pollution in UD. GTT.

The weight value is obtained from the results of the questionnaire filled by two experts and processed using pairwise comparison

3. Identifying the cause of the problem by using a Fishbone diagram
4. Selecting the best alternative solutions by using NPV

3. Results and Discussion

3.1. Results of mass balance and productivity calculation

The calculation of mass balance requires input and output in each production process. Inputs used are soybeans, water and vinegar, while the resulting output is tofu and waste. Waste generated is in the form of wastewater and solid waste (tofu dregs). The process that distinguishes *takwa* tofu and the ordinary tofu is in the process of coloring by adding turmeric. Coloring is carried out by boiling the tofu that has previously been moulded on a wood stove. Productivity calculation is obtained from the comparison between output and input in each production process. The steps of tofu production are washing, soaking, draining, milling, boiling, filtering, forming, and coloring. The highest level of productivity was found on the soaking process with the value of 100%. This is because soy soaking does not produce waste. Residual water soaking discharged in the draining process. The lowest productivity level was in the washing process counted for 17%, due to a high water usage. A large amount of water is used because the soybean contains too much dirt. Compared to all production processes, the washing process is the only process having the productivity level below 50%. Productivity level that falls into the high category should be above 50% [7].

3.2. The measurement of EPI

Table 1 shows that the EPI index on COD and TSS parameters show a negative value, then the EPI index value in COD is also negative (-3,85). A negative EPI index value means that the productivity performance on the environment is not good. The laboratory results showed that the COD concentration exceeded the standard effluent quality, possibly due to the addition of vinegar in the filtering process. Similarly, the TSS concentration also exceeded the standard values. The result shows that the deviation value obtained was -2,60, thus the value of the EPI index on TSS is also negative (-0,49). The high value of TSS is possibly due to the use of soybeans that contain lots of epidermis, and twigs or gravel. The total value of EPI index was also negative with value of -3,94, meaning the environmental performance of UD. GTT is still below the quality standard and the environmental pollution is still high. The quality of wastewater generated from UD. GTT did not meet the environmental quality standards. The EPI index is necessary because it reflects the impacts of an environmental activity and its reducing efforts [8]. The high value of COD causes the absence of aquatic biota and disturb the balance of aquatic ecosystems, while the high level of TSS will reduce the depth of sunlight penetration into the water and it also reduces the value of water and affects the organisms living in it [9].

Table 1. Result of calculation of EPI index.

Parameter	Weight (Wi) (a)	Standard Quality of Raw Liquid Waste[7] (b)	Result of Lab Analysis (c)	Retention (Pi) (%) (d) = ((b-c)/b)*100%	EPI Index (e) = a*d
BOD	0.34	50.00	1.36	0.97	0.33
COD	0.22	100.00	1,851.60	-17.52	-3.85
TSS	0.19	200.00	720.00	-2.60	-0.49
pH	0.25	6.00	4.04	0.33	0.08
	Total of EPI Index				-3.94

3.3. Identify problems and causes

In the previous discussion, it was found that the washing productivity level is still low and the COD and TSS values still exceeded the standard. Therefore, the Fishbone diagrams were used to identify each problem. It was found that there is no standard for the soybean, whether it is clean or not. UD. GTT has not prepared any quality standards of soybean so that it depends on the expertise of the employees. Therefore, to overcome this problem, the standard quality of soybean should be made. It was also found that the used soybeans still contain dirt, causing a large amount of water needed in the washing process. Another cause is due to the use of conventional machine which is manually operated using a regular tub. This manual method requires a lot of water in the washing process. The high amount of water consumption is equal to the resulted wastewater [10]. Another cause is the addition of a high concentration of vinegar, which can cause undesirable odor. Such condition may lead to the excessive amounts of chemicals in the wastewater. High COD concentration will be followed by an increased need for oxygen to break down organic matter. The tofu wastewater that continuously discharged without treatment into the river will affect its water quality [11].

3.4. Compilation of alternatives and repair solutions

a. Alternative 1

Alternative 1 was carried out by choosing the suppliers with good soybean seed. The characteristics of good soybean seeds include the soybeans should be clean, having large seeds and free of epidermis, gravel, twigs. Soybean with such characteristics can be obtained from Jalan Raya Mojo Mlati Village, Mojo Subdistrict, Kediri regency. The price of the soybean was IDR 7,500.

b. Alternative 2

Alternative 2 was carried out by procurement of filtration equipment and filter basin. This filtration equipment is made in a tube with a diameter of 100 cm and a height of 150 cm. The main ingredients in this filtration tool are water hyacinth and banana peel [11]. Water Hyacinth can absorb nutrients, organic compounds and chemical elements, while banana peels can bind heavy metals and liquid waste. Other additives are gravel, sand, charcoal and woven cotton. Each material is sequently arranged inside the tube, while the filter basin is built with a volume of 600 L. The filter materials of this filtration equipment must be replaced each day to avoid any clogging or blocking. Tofu wastewater is left in a reservoir pool for 12 hours (in a cold condition), functioned as a sedimentation tank to reduce the concentration of dissolved solids. Then it is flown into the pipe, and into the filtration tube. The absorption capacity of this filtration device is 1 liter per 2 minutes [11].

The water hyacinth can be utilised as an absorbent material due to its ability to absorb toxic substances in the environment. The advantages of water hyacinth are reducing pollutants, purifying wastewater by biofiltration, absorbing heavy metals such as Pb, Cd, Hg, Zn, Fe, Mn, Ni, Au, Co and Sr, absorbing organic compounds and other contents [12]. Also, the utilisation of banana peels as an absorbent material can increase their economic value. Banana peels contain nitrogen, sulfur and organic compounds such as carboxylic acids [9]. Banana peel that has been dried and cut is useful to bind to the compound of the mixture contained in the wastewater. Charcoal is wood that has been carbonized with partial combustion with little air. Charcoal is useful for cleansing water because it releases chlorine and other harmful substances. Charcoal is wrapped in cloth so that there is no powder scattered everywhere. Charcoal should be sterilized in boiling water first, then dried under the sun before used. Sand is useful for filtering dissolved solids (size of 0,2 to 0,8 mm), while gravel works for final filters before wastewater is disposed into the river.

3.5. Financial analysis of any alternative using the NPV

The selected alternatives are expected to overcome the environmental problems faced by UD. GTT. Financial feasibility analysis combined with NPV was used to select the best alternative.

Table 2. NPV alternative 1.

Interest rate (12%) (a)	Total Saving (IDR) (b)	Invest (IDR) (c)	Operational and Maintenance (IDR) (d)	PV Benefit (IDR) (e)=(b*a)	PV Cost (IDR) (f)=c+(d*a)	NPV (IDR) (g) = (e-f)
0.89	51,261,224.88	43,800,000	0	45,771,148	43,800,000	1,971,148

Table 3. NPV alternative 2.

Interest Rate (12%) (a)	Total Saving (IDR) (b)	Invest (IDR) (c)	Operational and Maintenance (IDR) (d)	PV Benefit (IDR) (e)=(b*a)	PV Cost (IDR) (f)=c+(d*a)	NPV (IDR) (g) = (e-f)
0.89	13,277,506.36	4,785,000	5,872,590	11,810,840	10,028,636	1,582,204

3.6. Alternative selection

After determining the NPV value, the amount of waste, the productivity, and the EPI index in each alternative, next step is choosing the best alternative. This study shows that the chosen alternative was alternative 2, which may increase the productivity level to 79% and the EPI index of -3,00. In terms of the NPV value, however, the alternative 1 is more beneficial (IDR 1,971,148) compared to that of the alternative 2 (IDR 1,582,204). The alternative 2 was chosen because it has a better productivity level and the EPI index than alternative 1. The GP has several objectives in order to improve the environmental quality and production of waste reduction, material management, pollution prevention and product enhancement [13].

Table 4. Comparison of alternative result 1 and alternative 2.

Alternatives	NPV (IDR)	Amount of Waste			Productivity (%)		EPI Index	
		Before (L)	After (L)	Percentage(%)	Before	After	Before	After
Alternative 1	1,971,148	58.94	30.14	48.46	17	29	-3.94	-3.25
Alternative 2	1,582,204	58.94	17.59	75.00	17	79	-3.94	-3.00

4. Conclusion

The findings confirmed that the productivity level in UD. GTT was mainly caused by the amount of wastewater generated from the production process. Specifically, the washing process has poor environmental performance as indicated by its lowest productivity level, measured at 17%. This was due to a high amount of water used to wash the soybeans. To tackle this problem, it was suggested that UD. GTT to apply alternative 2 (i.e. the procurement of filtration equipment). Despite a high NPV value, by selecting this strategy, UD. GTT might experience an increase in the productivity level to 79% and in the EPI index to -3,00. This study further demonstrated that the GP approach is an applicable tool to help SMEs in the decision-making process aiming at improving their environmental performance.

References

[1] Mustaniroh A S, Susinggih W, dan Nasrul N 2006 Analisis aspek pemasaran dalam pengembangan usaha tahu takwa usaha tahu takwa di Kediri (Market analysis to improve the performance of business on tofu ‘takwa’ industri in Kediri) *J. Teknologi Pertanian* **7** 2 135-140 [In Indonesian]

[2] Umar H 2002 Riset Sumber Daya Manusia dalam Organisasi (Research on Human Development in Organisation) PT Gramedia Pustaka Utama Jakarta Indonesia [In Indonesian]

[3] Asian Productivity Organization 2006 Handbook on Green Productivity APO Tokyo Japan 341

[4] Evans J R dan William M L 2007 Pengantar Six Sigma (Introduction to Six Sigma) Salemba Empat Jakarta Indonesia 410 [In Indonesian]

[5] Margaretha F 2007 Manajemen Keuangan Bagi Industri Jasa (Service Industry Financial Management) Grasindo Jakarta Indonesia [In Indonesian]

[6] Singgih M L dan Ketut R D 2015 Evaluasi dan Perbaikan Kinerja Lingkungan dan Peningkatan Produktivitas menggunakan Metode *Green Productivity* di Pabrik Gula (Evaluation and Improvement of Environmental Performance and Productivity Improvement using Green Productivity Method in Sugar Factory) Undergraduate Thesis Institut Teknologi Sepuluh November Surabaya Indonesia [In Indonesian]

[7] Peraturan Menteri Lingkungan Hidup Republik Indonesia Nomor 5 Tahun 2014 Baku mutu air limbah (Liquid water quality standards) Retrieved from http://ditjenpp.kemenkumham.go.id/arsip/bn/2014/bn1815-2014.pdf. [Accessed on 17 Oktober 2017] [In Indonesian]

[8] Gaspersz V 2000 Manajemen Produktivitas Total: Strategi Peningkatan Produktivitas Bisnis Global (Total Productivity Management: Global Business Productivity Improvement Strategy) PT Gramedia Pustaka Utama Jakarta 305 [In Indonesian]

[9] Hidayat N 2016 Bioproses Limbah Cair (Wastewater Bioprocess) Andi Offset Yogyakarta Indonesia [In Indonesian]

[10] Kaswinarni F 2007 Kajian Teknis Pengolahan Limbah Padat dan Cair Industri Tahu (Technical Review of Solid Waste and Wastewater Treatment at *Tofu* Industry) Thesis Ilmu Lingkungan Universitas Diponegoro Semarang [In Indonesian]

[11] Hudha M, Jimmy dan Muyassaroh 2014 Studi Penurunan COD dan TSS Limbah Cair Industri Tahu menggunakan Proses Elektrokimia (A study of COD and TSS Removal on Industrial Tofu Wastewater using Electrochemical Process) *J. Teknik Kimia* **1** 3 185-191 [In Indonesian]

[12] Utami A R 2014 Pemanfaatan Enceng Gondok sebagai Katalis dalam Pengolahan Limbah Cair Industri Tahu secara Anaerobik (Utilization of Water Hyacinth as Catalyst in Anaerobic Treatment of Industrial Wastewater) Undergraduate Thesis Institut Teknologi Sepuluh November Surabaya [In Indonesian]

[13] Singgih M dan Mera K 2008 Peningkatan Produktivitas dan Kinerja Lingkungan dengan Pendekatan *Green Productivity* pada Rumah Pemotongan Ayam XX (Improvement of productivity and environmental performance using green productivity approach in XX chicken slaughtering centre) *J. Teknik Industri* **2** 3 11-20 Undergraduate Thesis

Developing cluster strategy of apples dodol SMEs by integration K-means clustering and analytical hierarchy process method

S A Mustaniroh, U Effendi, R L R Silalahi, T Sari and M Ala

Department of Agro-industrial Technology, Faculty of Agricultural Technology, Universitas Brawijaya, Malang, Indonesia

E-mail: asmaul_m@yahoo.com

Abstract. The purposes of this research were to determine the grouping of apples dodol small and medium enterprises (SMEs) in Batu City and to determine an appropriate development strategy for each cluster. The methods used for clustering SMEs was k-means. The Analytical Hierarchy Process (AHP) approach was then applied to determine the development strategy priority for each cluster. The variables used in grouping include production capacity per month, length of operation, investment value, average sales revenue per month, amount of SMEs assets, and the number of workers. Several factors were considered in AHP include industry cluster, government, as well as related and supporting industries. Data was collected using the methods of questionaire and interviews. SMEs respondents were selected among SMEs appels dodol in Batu City using purposive sampling. The result showed that two clusters were formed from five apples dodol SMEs. The 1stcluster of apples dodol SMEs, classified as small enterprises, included SME A, SME C, and SME D. The 2ndcluster of SMEs apples dodol, classified as medium enterprises, consisted of SME B and SME E. The AHP results indicated that the priority development strategy for the 1stcluster of apples dodol SMEs was improving quality and the product standardisation, while for the 2nd cluster was increasing the marketing access.

1. Introduction

In Batu City, a high apple production, counted at 17.05 tons/ha, supports apple-processing industry to develop [1]. The apple (*Malus sylvestris*) is then processed into different products, such as apple chips made from the sub-grade apples [1]. Currently, the implementation of industrial clusters is critical due to the demands of global industry competitiveness, leading to the transformation of individual competencies into institutional clusters [2]. Apple-processing SMEs, especially apple dodol SMEs, have some problems such as poor product standardisation, limited market access, limited capital access, poor knowledge on marketing and business strategy, limited partnership with other organisations, and non-conducive business atmosphere. Recently, SMEs have been mainly characterised based on their activity, area, and business size. Such clustering method is considered as less appropriate because each group often has a different number of group members, thus special coaching technique is required [3].

To facilitate the participation of related parties/stakeholders in tackling the SMEs problems, clustering and classification processes become essential [2]. Cluster defined as a business group which consists of some businesses in an area with similar business sector [4]. Furthermore, the SMEs clustering may bring various benefits such as higher job employment, easier access to capital sources and supplier, specific service input, as well as information and knowledge transfer [3]. The development of industrial cluster is widely explained using Diamond Porter Model, which consists of four main factors and two additional factors. The main factors are (1) condition, (2) related and supporting industry, (3) strategy, structure and competition and (4) demand condition, while the additional factors are (1) opportunity and (2) government [5].

2. Materials and Methods

2.1. Problem limitation

The problem limitations of this research were as follows:

1. This research object only consisted of the registered and actively operated apple dodol SMEs, based on the data acquired from the Department of Cooperatives, SMEs, Industry and Trade (DISKOPERINDAG).
2. Apple dodol SMEs were clustered based on several characteristics include monthly production capacity, operating period, investment value, average sale per month, total asset of business unit, and total number of workers.
3. Factors used to develop the SMEs consisted of industrial cluster, government, as well as related and supporting industry.

2.2. The sampling methods

In this research, the sample population was targeted from all apple dodol SMEs in East Java. However, the population of apple dodol SMEs in Batu City was selected due to the easy access for data, information and a site visit. The purposive sampling method was employed to choose the apple dodol SMEs sample as respondents. The main selection criteria were that the SMEs still exists and operates over the last 3 years. The selected samples were consisted of five apple dodol SMEs, namely Agro Citra Abadi Industry (SME A), CV. Bagus Agriseta Mandiri (SME B), Petik Apel Farm (SME C), Barokah Jaya (SME D), and UD. Harum Manis (SME E).

Similarly, the expert respondents for analysing the strategy formulation were also selected using the purposive sampling method. Three experts were chosen include the practitioners from the SMEs Association in Batu City (i.e. Guyub Rukun Agawe Santosa/GRAS and Association of Batu City Entrepreneur/APKB) and the government (i.e. DISKOPERINDAG staff). Those experts were selected due to their competencies to guidance the apple dodol SMEs in Batu City. The data were collected using questionnaires with Likert scale (1-5) to assess the SMEs responses from Disagree to Agree.

2.3. K-means clustering methods

The classification processes was carried out using *k-means clustering* method with the following steps [6]:

1. Determining Cluster Number
2. Calculating the centroids of the K clusters
3. Calculating minimum distance among data and centroid using *Euclidean Distance*
4. Classifying data based on the minimum distance
5. Re-calculating the centroid, so that no object is moved

2.4. AHP approach

The AHP analysis was conducted using a complementary comparison scale of 1-9. The steps are as follows [7]:

1. Problem decomposition by arranging hierarchy.
2. Arranging composite opinion matrix.

3. Arranging decision element priority.
4. Calculating λ max value and Consistency Index (CI).
5. Calculating Consistency Ratio (CR)

3. Results and Discussion

The results show that, in analysing the K-means clustering data, two iterations were needed to reach the optimum values. Two clusters SMEs were identified from the process, as seen in Table 1. The table illustrates that three SMEs (i.e. SME A, SME C, and SME D) were part of the 1st cluster, while other SMEs (i.e. SME B and SME E) were in the 2nd cluster.

Table 1. The SMEs characteristics.

Characteristics	1st cluster (SMEs A, C, and D)	2nd cluster (SMEs B and E)
	Cluster Center	
Monthly production capacity (kg)	1,016.67	963.5
Operating period (year)	4	12.5
Investment value (IDR)	12,940,666	373,500,000
Average sale per month (IDR)	31,333,333	25,850,000
Total asset possession (IDR)	47,666,666	360,000,000
Total workers (people)	10	37

Table 1 also shows that each cluster has different characteristics. For instance, in terms of monthly production capacity, the 1st cluster was higher than the 2nd cluster. This was due to the strategy applied by the 1st cluster was solely focused on producing apple dodol, while the 2nd cluster was able to diversify various types of apple products (i.e. apple wingko, apple crisp, concentrated apple juice beverages). Furthermore, compared to the 1st cluster, the 2nd cluster SMEs was supported with their own outlet to independently market their products, either for retail or wholesaler purpose. All SMEs clusters tend to have a flexible production capacity depended on the market demand. However, a lack of capital support for all SMEs clusters has limited the production capacity. Therefore, they did not have a regular and constant production capacity. This condition made their selling price to be less competitive compared to other SMEs producing larger quantities of similar products.

A business can be maintained because of its market flexibility and competitiveness to deal with many business competitions. More experience means better skill and knowledge to analyze the probability which may occur as a consequence of the chosen decision [8]. A common problem to be dealt by SMEs is limited initial investment or owned capital. This study revealed that apple dodol SMEs have limited capital for expanding their business scale, particularly for the 1st cluster which classified as small scale. Capital is an obligatory factor to be provided by any SMEs due to its direct impact to the business, especially in supplying materials for production purpose. Initial investment or capital is an active profit on approved and measured debt based on a standard accountancy [9]. Investment is the initial capital that must be issued by the business owner for the purposes of purchasing the needs associated with the production process. Initial investment has a critical influence on the production process wherein the initial investment is the input for financing the production process [10]. For any industry, the initial capital investment usually comes from two sources, include the owner investment and the profit acquired during the industrial production operation period. A less production capacity of the 1st cluster was due to a small number of manpowers and the limited capital owned by the SMEs, causing no additional funding can be added either to increase the raw materials capacity or to increase the production capacity. Production capacity is related to the income level of a business [11]. According to the Ministry of Coperatives and SMEs in 1995, as stated in Indonesian regulation (UU No.9/1995) and Indonesian Bank (SK Dir BI No.30/45/Dir/UK on 5 January 1997), the 1st cluster of apple dodol SMEs is characterised as small-scale industry, while the 2nd cluster is as medium-scale industry. This characterisation is based on several indicators, such as: annual sales, total asset, and total productive workers. Small-scale industry has an asset value indicator of less than IDR 200 million (excluding land and building) and the number of workforce between 6-19 people. While, medium-scale industry has an asset value indicator of less than IDR 600 million (excluding land and building) and labor force of more than 20 people. The total number of laborers owned by the SMEs has also a positive impact on their business income, in which having a higher amount of the productive laborers will create a more established industry [12]. The government or private parties may involve in developing the SMEs by facilitating knowledge and information technology transfer as well as establishing better information technology [13].

In this study, the hierarchy structure from industry development strategy determination consisted of four levels, include industrial cluster development, factor, purpose and alternative strategy. The industrial cluster development arrangement for each cluster employed the same hierarchy and expert respondents. The hierarchy of apple dodol SMEs cluster development for the 1st cluster can be seen in Figure 1. The figure shows that in the 1st cluster, the most contributing factor in level 1 is the government with a score of 0.56. While, in level 2, the purpose to increase product quality has the highest contribution score with the value of 0.40. In the case of alternative strategy (level 3), the quality improvement and product standardization strategy has the most contribution with a score of 0.44.

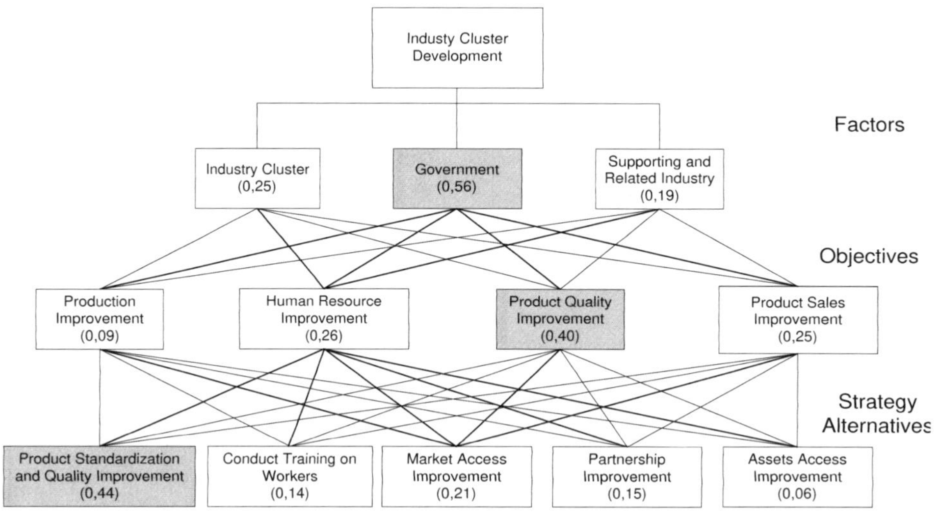

Figure 1. Hierarchy of apple dodol industrial cluster improvement for 1st cluster

This study revealed that the government has the biggest priority score of 56% in developing the apple dodol SMEs cluster. Therefore, in recent years, the Batu City government has established an Integrated Business Service Center (PLUT) facility, functioning as the central data office and the business help center. This facility exists to help either operating or starting up the business as well as to assist the SMEs in marketing and selling their products to several offices. This is relevant to the vision of SMEs Cooperatives Service and Trade Industry in Batu City to the establishment of strong, independent, growing and competitive cooperatives, SMEs, industry and trading. The government wants them to be the main contributor of Batu City economic activities, with the use of an equitable market mechanism [14].

To improve the SMEs in Batu City, the government established an Association of Batu City Entrepreneur (or APKB). APKB establishment is a group-based approach to solve problems faced by the entrepreneurs in Batu city. In the future, this approach is expected to make the problem solving process become easier. In addition, the government also established SMEs network, known as "*Guyub Rukun Agawe Santosa*" (or GRAS), under supervision of Batu City

DISKOPERINDAG and Department of Youth, Sport, and Labor (DISPORANAKER). The government facilitates the GRAS-participating SMEs based on related government policy and priority as well as business management supervision. The facilities provided can be an outlet to market their products or even machinery and production instruments. In comparison to public parties, the government should provide SMEs with public goods rather than private goods, to ensure a fair responsibility from the government. Furthermore, the government efforts to develop industrial cluster should be related to the facility compliance for small business development.

The top priority variable for development strategy was the product quality improvement. Taking this strategy in practice allows the apple dodol SMEs to improve their products. A better product quality produce by the SMEs will increase their consumer satisfaction, sales volume, competitiveness level, market, and price. A product can be considered as high in quality if the product has no defect [15]. The results show that the product quality improvement strategy and the product standardisation have the highest weight/priority. Product quality improvement strategy is essential to facilitate in getting business partners and loyal consumers. Consumer's loyalty can be acquired if the consumer satisfied with the product or the services, thus they have a strong commitment to buy and use the products/services continuously [16]. Therefore, this study suggested the apple dodol SMEs to maintain and improve their product quality.

The hierarchy of apple dodol SMEs cluster development for the 2nd cluster is shown in Figure 2. The figure shows that, at level 1, the related and supporting industry factor has the most contribution to the cluster development with a score of 0.61. While, in level 2, it can be seen that the product quality development has the most contribution to the purpose priority determination with a score of 0.37. In level 3, the market access enhancement strategy has the highest contribution to the alternative strategy of industrial cluster development, with a score of 0.28. This study indicated that the industrial cluster has 61% contribution, calculated based on the inputs required by SMEs to compete with other relevant industries. The achievement of business targets/objectives can be well acquired if the company is able to optimise their resource utilisation include man, money, methods, materials, machines and markets [17].

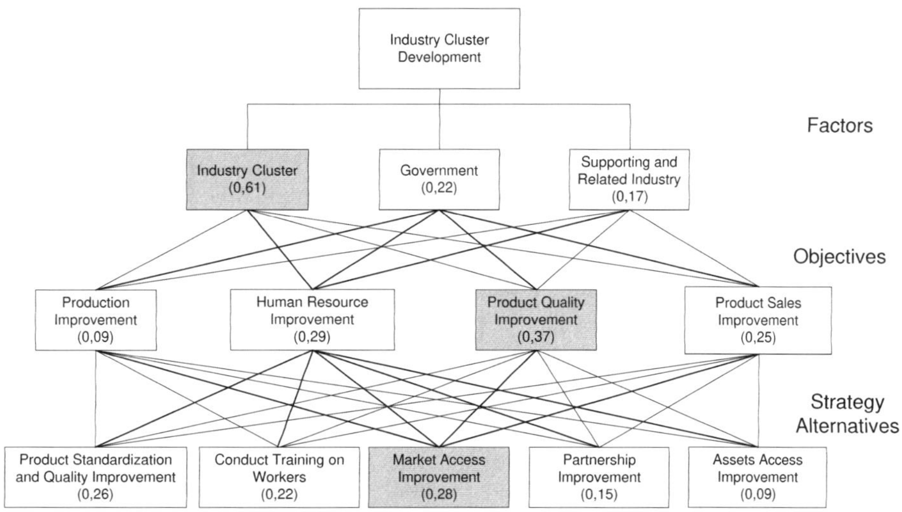

Figure 2. Hierarchy of apple dodol industrial cluster improvement for 2nd cluster

The results show that, in the 2nd cluster, the highest priority for improving the industry lies on the strategy of the product quality improvement. The study further confirmed that the quality of apple dodol products produced by the 2nd cluster SMEs was well maintained. It is obvious that, in the market, a selling price of a product is influenced by their quality. Therefore, having a good quality product can increase the selling price as well as the consumer acceptance or satisfaction. The customers always place a priority on the product's quality over the price, to fulfill their satisfaction level. Quality is conformance to requirement, or corresponds to the standard [15]. A product can be considered to have good quality if it corresponds to the approved quality standard. This study also suggested that the market access enhancement strategy need to be applied by the 2nd cluster. Marketing can be defined as the highest level in fulfilling human needs and desires. Market channel is a series of organization or activity necessary to transfering the product ownership from production process of a product or service to the consumption [18]. Market channel does two important activities, such as product grouping and product distribution [19]. Therefore, to increase the market access, it is crucial for the 2nd cluster of apple dodol SMEs to building a strong market channel.

4. Conclusion

The findings in this study confirmed that *K-means* clustering method combined with AHP approach is a powerfull tool to determine the SMEs clusters and their development strategy priority. Several considerations are required to implement the *K-means* method, include the monthly production capacity, operating period, investment value, average sale per month, total asset possession and total labor. Using those considerations, apple dodol SMEs respondents were classified into two clusters categorised as small-scale (i.e. 1st cluster) and medium-scale (i.e. 2nd cluster). This study further suggested the following priority strategy to be applied by each cluster, include quality improvement and product standarisation for the 1st cluster and market access improvement for the 2nd cluster.

References

[1] Hidayat A 2009 Perbaikan Mutu dan Penerapan Mesin Pengolahan Jenang Apel Secara Mekanis Sebagai Upaya Peningkatan Nilai Tambah Apel Kualitas *Subgrade* (The Quality Improvement and Mechanical Machinery Implementation *Jenang* Apple Production to Increase Added Value of Sub-grade Quality Apple). Universitas Brawijaya Malang Indonesia [In Indonesian]
[2] Partiwi S G 2007 Perancangan Model Pengukuran Kinerja Komprehensif Pada Sistem Agroindustri Hasil Laut IPB Bogor Indonesia [In Indonesian]
[3] Kuncoro M 2008 Strategi Pengembangan UMKM di Tengah Krisis Keuangan Global (Development Strategy for SMEs in the Global Financial Crisis) Paper presented in Rapat Kerja Nasional UMKMK Kadin 21 Oktober 2008 Jakarta Indonesia [In Indonesian]
[4] Schmitz H 2004 Local Enterprises in the Global Economy: Issues of Governance and Upgrading Edward Elgar Publishing Ltd London UK 392
[5] Porter M E 1998 On Competition: A Harvard Business Review Book Harvard College Cambridge UK 544
[6] Agusta Y 2007 K-means, penerapan, permasalahan dan metode terkait (K-means, impementation, problems and related methods) *J. Sistem dan Informatika* **3** 47-60
[7] Saaty T L 1980 The Analytic Hierarchy Process McGraw-Hill Inc New York USA 287
[8] Soekartawi 1993 Manajemen Pemasaran dalam Bisnis Modern (Marketing Management in Modern Business) Pustaka Harapan Jakarta Indonesia
[9] Rachmawati I K 2008 Manajemen Sumber Daya Manusia (Human Resource Management) ANDI Publisher Yogyakarta Indonesia [In Indonesian]
[10] Kodrat D V, Indonanjaya K 2010 Manajemen Investasi (Investment Management) Graha Ilmu Yogyakarta Indonesia [In Indonesian]
[11] Arifin I 2007 Membuka Cakrawala Ekonomi (Opening the Economic Horizon). PT. Setia Purnama Inves Bandung Indonesia [In Indonesian]
[12] Evelyn L Y 2007 Analisis pengaruh kredit permodalan dan faktor-faktor internal terhadap pendapatan pengusaha kecil yang menjadi nasabah penerima kredit BPR Gunung Ringgit KKP Ranugrati Kecamatan Kedungkandang Kota Malang (Analysis

of the influence of capital credit and internal factors on the income of small entrepreuners as the borrowers of BPR Gunung Ringgit KKP Ranugrati Kedungkandang Sub-District Malang City) Undergraduate Thesis Universitas Brawijaya Malang Indonesia [In Indonesian]

[13] Pulungan M S 2004 E-Commerce for SMEs: Kita Harus Berani Introspeksi (E-Commerce for SMEs: Brave to do self-instropection). Retrieved from http://www.lkht.net/artikel_lengkap.php?id=21. [Accessed on 26 January 2015] [In Indonesian]

[14] Diskoperindag 2014 Data Diskoperindag Kota Batu (Data from Diskoperindag Batu City). Retrieved from http://www.diskoperindag-kotabatu.com/about-us/. [Accessed on 26 January 2015] [In Indonesian]

[15] Resmi N N 2011 Strategi meningkatkan kualitas produk untuk menang dalam kompetisi (The product quality improvement strategy to win the competition) *Widyatech Jurnal Sains dan Teknologi* **10** 3 132-144 [In Indonesian]

[16] Maulana M I 2005 Loyalitas konsumen (Consumer loyalty) *J. Ilmu Administrasi Bisnis 5* 1 1-11 [In Indonesian]

[17] Soekarno K 1986 Dasar-Dasar Manajemen (Introduction to Management) Miswar Jakarta Indonesia [In Indonesian]

[18] Kotler P, Amstrong G 1997 Prinsip-Prinsip Pemasaran (Marketing Principles) Erlangga Jakarta Indonesia 515 [In Indonesian]

[19] Swastha B 2005 Asas-Asas Marketing (Marketing Principles) 3rd Edition LIBERTY Yogyakarta Indonesia 352 [In Indonesian]

Spatial pattern of agricultural land conversion in West Java Province

S Maryati, A N S Humaira and F Pratiwi

Research Cluster for Regional and City Infrastructure System, School of Architecture, Planning, and Policy Development, Institut Teknologi Bandung

E-mail: smaryati@pl.itb.ac.id

Abstract. Population growth has an implication on increasing demand for land. The demand for built-up area is filled by land conversion, mostly from agricultural land. On the other hand, population growth requires an increase in food production as well as land for agriculture. Conversion of agricultural land can threaten the availability and food security. The purpose of this study is to identify the spatial pattern of changes in agricultural land in West Java Province as input to improve food security condition in this province. Descriptive statistics and spatial analysis were used to analyse the area of agricultural land, conversion of agricultural land, and spatial pattern of changes in agricultural land in West Java Province. The data used is time series data in the period of 2005-2014. The result of analysis shows that there are still areas with a high percentage of agricultural land in West Java Province. The rate of conversion of agricultural land varies widely. Cities or regions with very high land conversion rate tend to concentrate in metropolitan areas.

1. Introduction

Rapid population growth is a global issue faced in many countries, particularly in developing country. In the period of 2007-2025, the annual rate of urban population growth in developing countries is expected to be 2.27%, whilst its rate in developed countries is 0.49% [1]. This growth commonly occurs in urban areas as it is forecasted by 2050, 66% of the world's population is projected to be urban [2]. This phenomenon also happens in Indonesia with its increase of urban population growth of 7.5% in 1930 and 42% in 2000 [3].

The rise of population growth leads to the consequence of land demand in supporting the activities of inhabitant. One of its huge impacts is the land conversion specifically from agricultural to be non-agricultural land [4]. A large number of agricultural land continues to decline replaced by the function of non-agricultural land.

Population growth has a consequence of the demand growth of food production. On the other hand, the decrease of agricultural area causes the decrease of food production that can threaten the food availability and security. In the World Food Summit held in 1996, it was declared that food security is a condition when all people, at all times, have physical and economic access to sufficient, safe and nutritious food that meets their dietary needs and food preferences for an active and healthy life [5]. This widely accepted definition points to the following dimensions of food security which are food availability, food access, utilization, and stability [5]. Food production is the main key of food availability related to demand side. Sufficient food production means the needs of food production will be fulfilled.

The aim of this paper is to identify the spatial pattern of land agricultural changes in the case of West Java Province. West Java Province is one of the main rice producing provinces in Java Island which is also the main rice producing island in Indonesia. In 2004 to 2013, most of the time, West Java Province has produced the largest rice production in Java Island. Not only in producing rice, this province also has a huge potential in producing other food commodities. It is one of 6 cassava production centers in Indonesia and gives high contribution (20%) to national production of sweet potato [6]. Given those backgrounds, this paper will explore the pattern of agricultural land conversion. Agricultural land in this study will be focused only on the land of rice field area.

2. Methods

Descriptive statistic using time series data and spatial analysis are the methods applied in this study in order to determine the spatial pattern of agricultural land changes in some period of time. According to ESRI in GIS Dictionary, "*Spatial analysis is the process of examining the locations, attributes, and relationships of features in spatial data through overlay and other analytical techniques in order to address a question or gain useful knowledge. Spatial analysis extracts or creates new information from spatial data.*" [7]. In the other words, it is a tool to understand patterns and trend of the geographic and temporal relationships among data, phenomena, and issues by mapping where things are, how they relate and what it all means [8]. Moreover, this analysis helps the decision makers to be able to generate policy of what actions to take based on the issues arisen. In this context, agricultural land conversion which is in line with the vast growth of urban areas in the last decade has become a global issue.

Time series data of agricultural land area were collected from Statistics Indonesia (BPS) in the period of 2005-2014. This time series data was used to be able to observe the patterns of changes occurred during 10 years. Analyses were done in the unit of regency/city in West Java Province. Besides, to be able to demonstrate the spatial pattern, analysis was also done in the context of comparing between metropolitan vs non-metropolitan areas as well as core vs peri-urban areas within 1 metropolitan area, considering the estimation of enormous urban growth in metropolitan areas.

Agricultural land conversion was calculated by using land agricultural area in each regency/city. In this study, agricultural land was assumed from the rice field area which tend to have the biggest area among all crops in West Java Province. Statistics descriptive was used to see the proportion of agricultural land area in each regency/city and was identified using the average variable condition in the period of 2005-2014. In addition to descriptive statistic, spatial analysis was also performed to translate the results of descriptive statistics in order to enrich the findings in the term of spatial context.

3. Study Area Context

West Java Province sits within the Island of Java which now has 18 regencies and 9 cities (see Figure 1) in the area of 35.377,76 km^2. This province accommodates around 43 million populations in 2010 and is estimated to be 57 million populations in 2035. In 2015, this province is categorized as the densest province in Indonesia. Bogor Regency, Bandung Regency, and Bekasi Regency give the biggest contributions of West Java's population which respectively are 11.69%, 7.57%, and 6.95%.

The concentration of population is still clustered around Special Capital Region of Jakarta and the rest of them is mostly concentrated in the Greater Bandung (one of 3 metropolitan areas in West Java Province). Within this metropolitan area, lies the capital city of West Java Province. The Bodebek Karpur and Greater Cirebon are another two metropolitan areas in West Java Province.

IOP Conf. Series: Earth and Environmental Science **131** (2018) 012034 doi:10.1088/1755-1315/131/1/012034

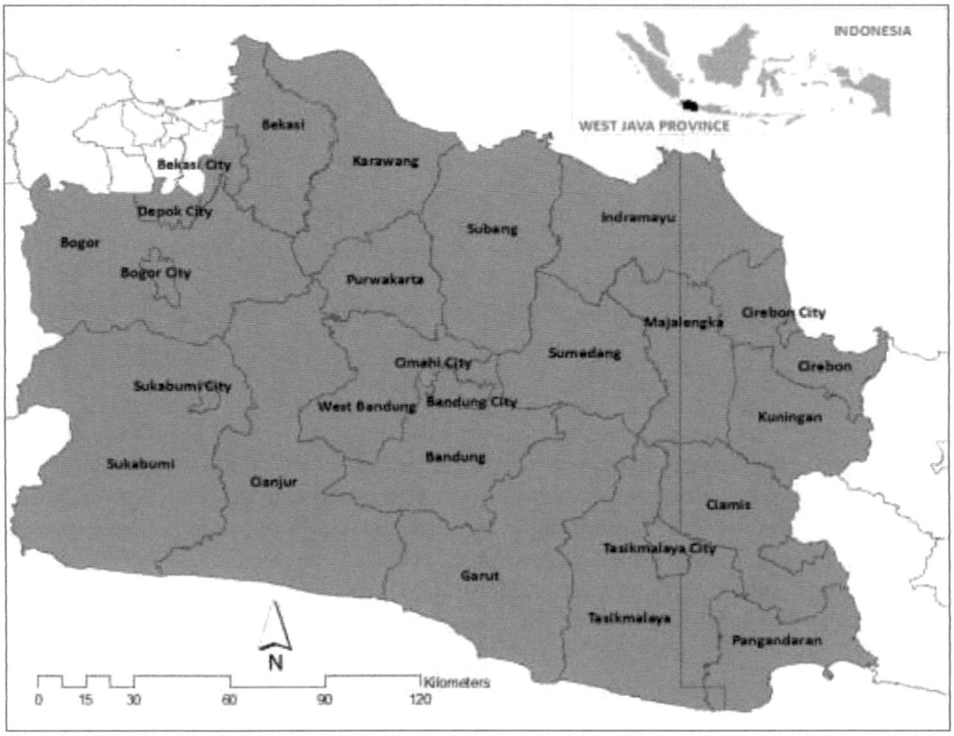

Figure 1. The map of study area

4. Results and Discussion

4.1 Proportion of agricultural land in each region

Agricultural land area in several regions of West Java Province still has a big percentage of the total area, with the percentage of more than 80% found in Bekasi Regency, Karawang Regency, Cirebon Regency, Majalengka Regency, Subang Regency, and Indramayu Regency. Some of them are the regions categorized as the peri-urban area in the metropolitan area (Bekasi, Karawang, and Cirebon Regencies) and the rest of them are not included in the metropolitan area. Indramayu is the region which has the biggest percentage of agricultural land among others and is also the only region that has legitimated LP2B in the form of its Local Regulation. LP2B (Sustainable Food Agricultural Land) is the agricultural land that has to be consistently preserved, protected, and developed in order to generate food security [9]. This policy is established by the Government of Indonesia in the Act 41/2009 regarding Protection of Sustainable Food Agricultural Land.

On the other side, the smallest percentage of agricultural land which is less than 15% is shown in all cities in West Java Province but Banjar, Tasikmalaya and Sukabumi Cities. Banjar, Tasikmalaya and Sukabumi Cities are not included in the metropolitan area, so that they still have the big percentage of its agricultural area and are still bigger compared to some regencies in the metropolitan area. Most cities in this province have been classified as the core city of its metropolitan area, therefore with its urban activity they tend to have the smallest percentage of agricultural land and also get a very small portion to pursue the mandate of LP2B policy.

Among of all, if we see it in the context of administrative region, the trend of a number of agricultural land area and its percentage to the total area always show the smaller agricultural land area in the city rather than in the regency. This proves the urban phenomenon with the nature of its activity not in the agricultural sector.

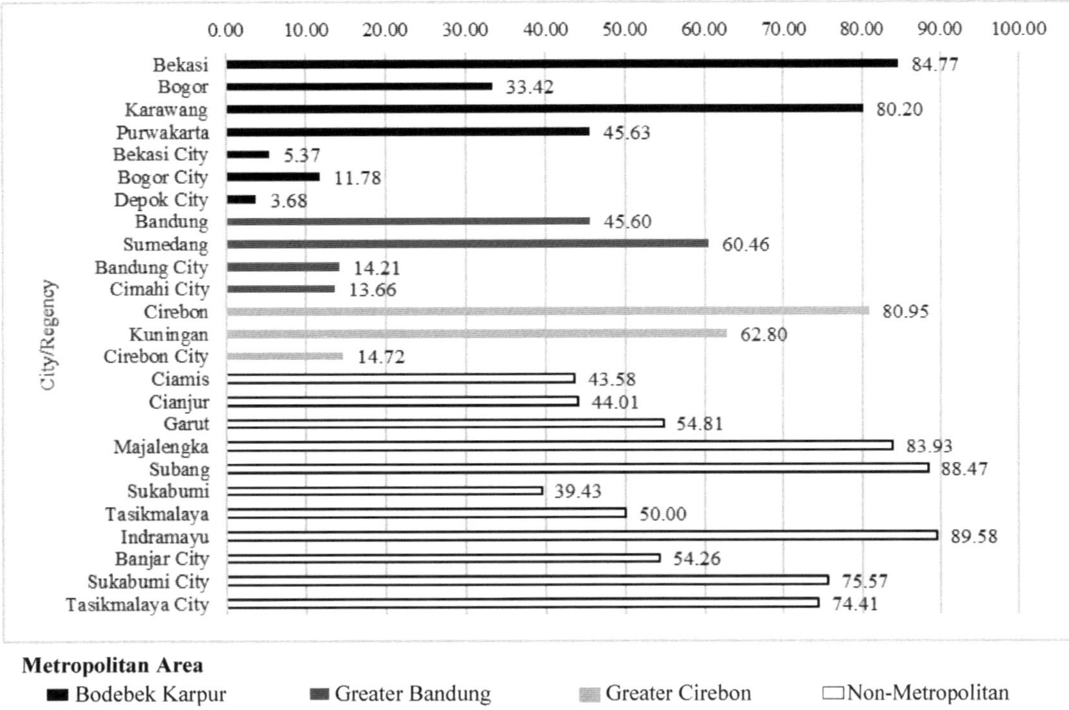

Figure 2. The average proportion of agricultural land and total area (%) in 2005-2014

4.2 The changes of agricultural land during 10 years

As shown in Figure 3, the changes rate of agricultural land can be positive and negative. Positive value means that changes direction has increased in the last 10 years, whilst negative value shows declining changes. Overall, only 9 out of 27 regions that experience the decreasing rate of agriculture land, with 6 out of those 9 are cities or the core areas in 3 metropolitan areas of West Java Province. Depok (-11.90%), Bogor (-10.04%), Bekasi (-8.66%), and Bandung (-8.36%) are the cities with the highest declining rate. Garut Regency shows the increasing tendency of 2.70% and becomes the highest position among of all, followed by Indramayu Regency (1.86%).

Even though Depok, Bogor, Bekasi, and Bandung are the cities with the highest declining rate of agricultural land, yet if it is seen from its area as shown in Figure 4, they are not the regions with the highest number of agricultural land conversion. The agricultural land conversion in all cities is found less than 3,000 Ha, meanwhile Subang and Bekasi Regencies have the highest conversion which respectively decrease of 10,476 Ha and 12,874 Ha. That can happen due to the trend of the total area of administrative region that tend to be smaller in city compared to regency so that the conversion in the regency is bigger than in the city as well.

It seems that the commitment of protecting LP2B has been implemented by most local government in many regencies. The rise of agricultural land during 10 years results significant numbers especially in several regencies, for instance Garut (33,320 Ha), Indramayu (20,392 Ha), Karawang (13,596 Ha), Majalengka (8,822 Ha), and Bogor (7,176 Ha).

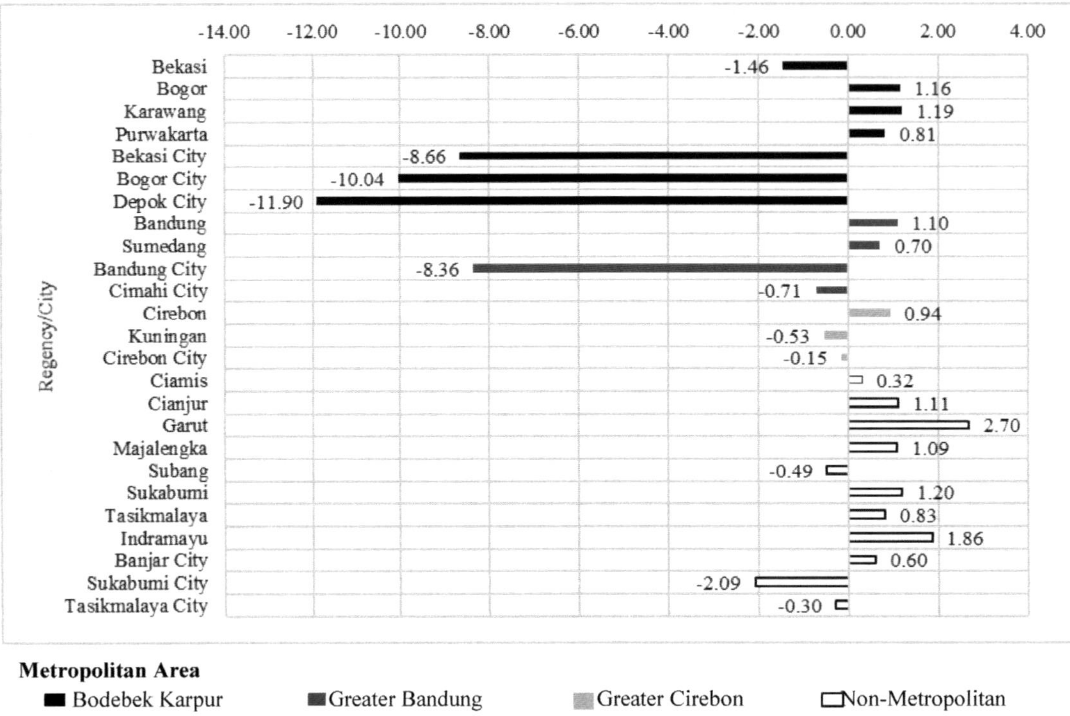

Figure 3. The average rate of agricultural land changes (%) in 2005-2014

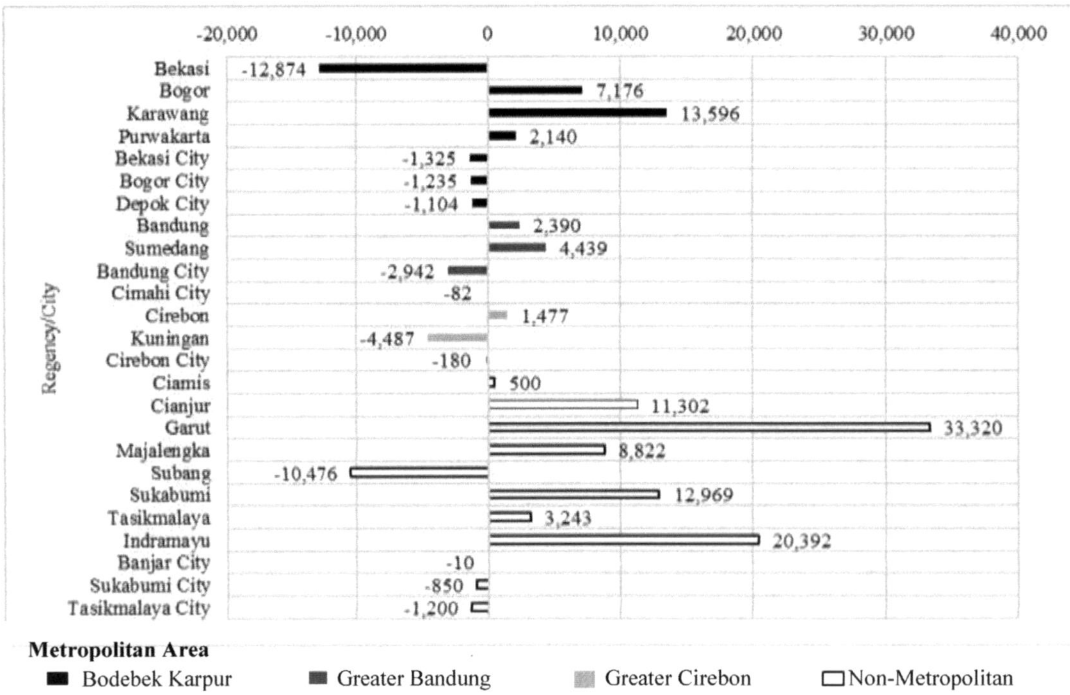

Figure 4. Changes of agricultural land area (Ha) from 2005 to 2014

4.3 Spatial pattern of changes in agricultural land

In a spatial analysis, spatial pattern is seen in the context of distribution trend based on classification made for each variable observed. The decrease rate of agricultural land that can be seen as a replaced function from agricultural land into something different mostly occur in the core areas of metropolitan area such as Depok City, Bandung City, Cimahi City, Bogor City, and Cirebon City that is indicated by the moderate and high decrease rate. Besides, this decreasing changes are also found in some regions close to the capital city of Indonesia and West Java Province (DKI Jakarta and Bandung City).

As illustrated in Figure 5, interesting finding is also shown for the south part of this province. All regions in the south part, like Sukabumi, Cianjur, Garut, Tasikmalaya, and Pangandaran Regencies, experience the increase of agricultural land. This condition in the southern part also applies for another variable of total agricultural land conversion (Ha) from 2005 to 2014 with high increase occur in Sukabumi, Cianjur, and Garut Regencies. The decreasing conversion follows the pattern of decreasing rate which mostly happen in the area close to the capital city of Indonesia and West Java Province.

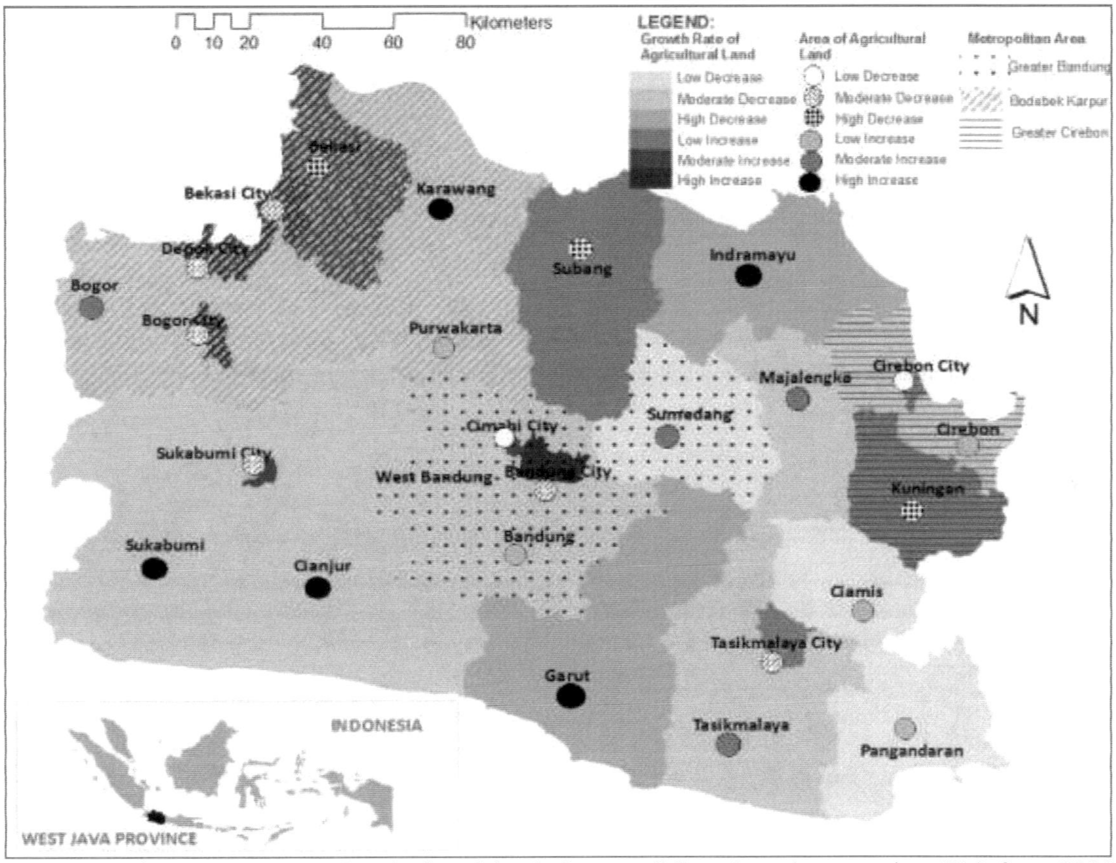

Figure 5. The average rate of agricultural land changes (%) and total conversion (Ha) from 2005 to 2014

If we go back to the previous discussion regarding the proportion of agricultural land and its total area, if it is plotted on the map in Figure 6, we can see clearly that there are some patterns in the spatial dimension. Firstly, most of the regions in the northern part of West Java Province, namely Bekasi, Karawang, Subang, Indramayu, Majalangka and Cirebon Regencies, are classified in the level of high proportion. This is in line with the government policy in the Spatial Plan of West Java Province to set the regions in the north coast (Pantura) to encourage preservation and protection of sustainable agricultural land and to prevent it from conversion of built up area. Secondly, it is obvious that low proportion of agricultural land found in the cities of metropolitan area such as Bandung and Cimahi Cities in the Greater Bandung, Cirebon City in the Greater Cirebon, and Depok City, Bogor City, and Bekasi City in the Bodebek Karpur Metropolitan Area. Thirdly, the rest of the regions are categorized in the moderate level and have no pattern.

The average rate of population shows the pattern of which high increase rate of population growth is experienced mainly by the regions close to the capital city of Indonesia as well as located in the metropolitan area of Bodebek Karpur. Besides this metropolitan, another two do not show some particular pattern and tend to be random. There are also some regencies (Sukabumi, Subang, and Kuningan Regencies) that have declining growth rate, yet until now it is assumed that it happens naturally and do not show a special pattern.

The higher the rate of population growth, the higher the urban growth. Theoretically, the urban growth is supposed to be inversely proportional with the agricultural land. Yet, as it can be seen in Figure 6, that theory does not fully apply in the context of West Java Province. Some regions meet the pattern and some do not, therefore strong correlation between the rate of population growth and the percentage of agricultural land cannot be found in this study. That means, the regions with high proportion of agricultural land does not always have to have declining rate of population growth, and vice versa. There might be other factors affecting all occurrence that are still not covered in this study.

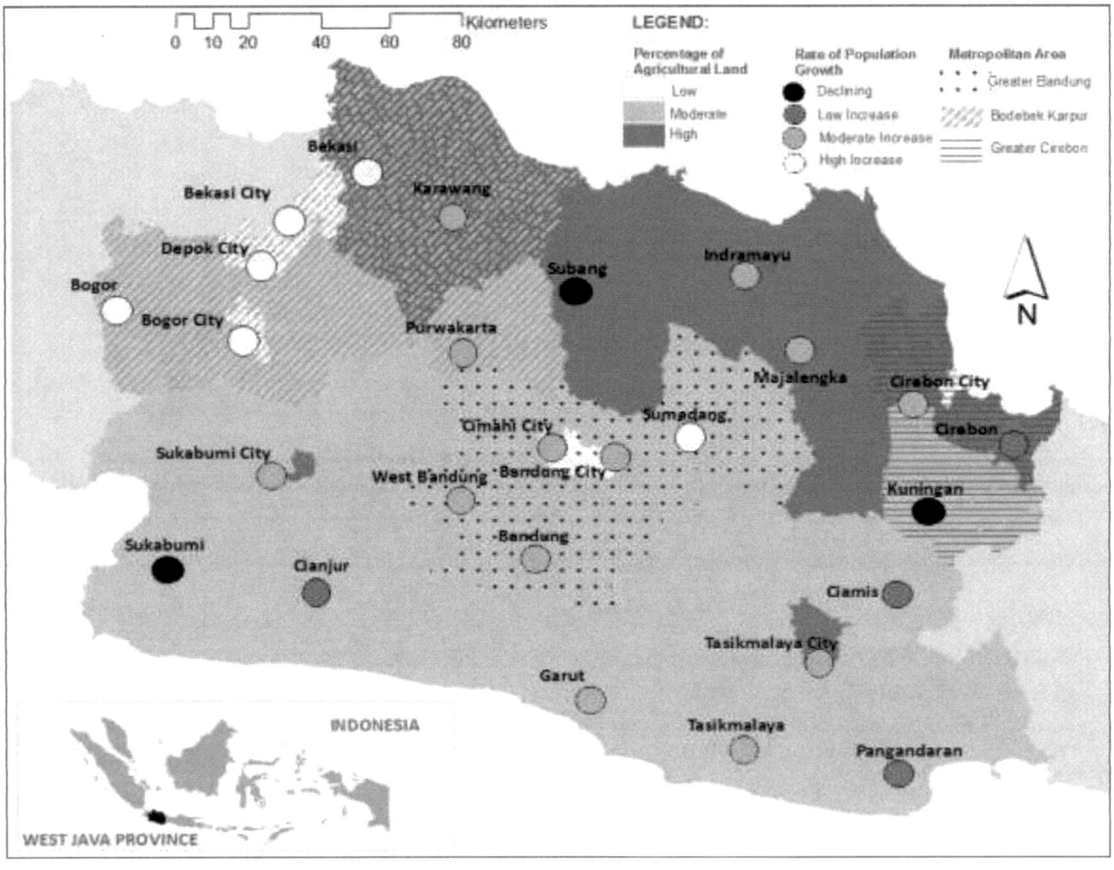

Figure 6. The average proportion of agricultural land and total area (%) and the average rate of population growth (%) in 2005 to 2014

5. Conclusion

The findings resulted from this study shows some noticeable pattern in the case of agricultural land conversion in West Java Province. First of all, agricultural land conversion in all level (varies widely) tends to be found in the core city of metropolitan area as well as around the capital city of Indonesia and West Java Province (DKI Jakarta and Bandung) such as Depok City, Bogor City, Bekasi City, Cimahi City, and Bandung City as a consequences of urban activities that also requires the land. The regions in the north coast have the highest proportion of agricultural land that is in line with the mandate of Spatial Planning of West Java Province that commit to prevent the agricultural land conversion to protect and preserve sustainable agricultural land as the rice producing centers (locally known as 'lumbung padi') of West Java Province in Pantura. In order to support LP2B policy obliged by the Act 41/2009, all regions in the southern part of West Java Province have been maintaining the agricultural land so that the area of agricultural land from 2005 to 2014 is increased significantly.

References

[1] UN Habitat (online) Planning Sustainable Cities: Fact Sheet of Global Report on Human Settlements. Retrieved from https://unhabitat.org/wp-content/uploads/2010/07/FS1.pdf. [Accessed on 11 October 2017]

[2] United Nations, Department of Economic and Social Affairs, Population Division 2015 World urbanization prospects: the 2014 revision ST/ESA/SER.A/366

[3] Pontoh N and Kustiwan I 2009 Pengantar Perencanaan Perkotaan ITB Bandung

[4] Bhatta B 2010 Analysis of Urban Growth and Sprawl from Remote Sensing Data Springer Heidelberg

[5] FAO (online) Food security (Policy Brief: June 2006 Issue 2). Retrieved from http://www.fao.org/forestry/13128-0e6f36f27e0091055bec28ebe830f46b3.pdf. [Accessed on 11 October 2017]

[6] Dewan Ketahanan Pangan, Kementerian Pertanian and World Food Programme (WFP) (online) Peta Ketahanan dan Kerentanan Pangan Indonesia 2015: Versi Rangkuman. Retrieved from documents.wfp.org/stellent/groups/public/.../ena/wfp276252.pdf [Accessed on 11 October 2017]

[7] Esri (online) GIS Dictionary: Spatial Analysis. Retrieved from https://support.esri.com/en/other-resources/gis-dictionary/term/spatial%20analysis. [Accessed on 10 December 2017]

[8] Kerski J (online) Spatial Analysis in ArcGIS Retrieved from www.josephkerski.com/wp-content/.../spatial_analysis_in_your_web_browser.pdf. [Accessed on 10 December 2017]

[9] Government of Indonesia 2009 Act Number 41 of 2009 Regarding Protection of Sustainable Food Agricultural Land The Government of Indonesia Jakarta

IOP Conf. Series: Earth and Environmental Science **131** (2018) 012035 doi:10.1088/1755-1315/131/1/012035

Applying linear programming model to aggregate production planning of coated peanut products

W G Rohmah, I Purwaningsih and EF S M Santoso

Department of Agro-Industrial Technology, Faculty of Agricultural Technology, Universitas Brawijaya, Malang, Indonesia

E-mail: wendrarohmah@ub.ac.id

Abstract. The aim of this study was to set the overall production level for each grade of coated peanut product to meet market demands with a minimum production cost. The linear programming model was applied in this study. The proposed model was used to minimize the total production cost based on the limited demand of coated peanuts. The demand values applied to the method was previously forecasted using time series method and production capacity aimed to plan the aggregate production for the next 6 month period. The results indicated that the production planning using the proposed model has resulted a better fitted pattern to the customer demands compared to that of the company policy. The production capacity of product family A, B, and C was relatively stable for the first 3 months of the planning periods, then began to fluctuate over the next 3 months. While, the production capacity of product family D and E was fluctuated over the 6-month planning periods, with the values in the range of 10,864 - 32,580 kg and 255 – 5,069 kg, respectively. The total production cost for all products was 27.06% lower than the production cost calculated using the company's policy-based method.

1. Introduction

Company X is Snacks & Food manufacturer, producing a coated peanut products. Currently, the company faced some issues in the production process, causing a failure in fulfilling the customer's demand. If such conditions remained, this will negatively influence the customer service performance of the company, as well as will reduce the company's profits [1]. Therefore, an accurate aggregate production planning is necessary to determine production capacity, helping the company to meet the market's demand.

Accurate aggregate production planning is a production planning which covers the amount of goods to be produced in a certain time unit within the overall planning horizon [2,3]. Aggregate production planning can be done using several approaches. The linear programming method introduced by Hanssmann dan Hess (1960) is one of them, which defined as a decision making technique to optimally allocate scarce resources [4]. This method was selected due to its compatibility to tackle the company's issue.

This study aimed to determine the production capacity for each product family and to compare the total production cost before and after applying the proposed aggregate production plan.

IOP Conf. Series: Earth and Environmental Science **131** (2018) 012035 doi:10.1088/1755-1315/131/1/012035

2. Materials and Methods

This study was carried out at X Snacks & Food company, Tulungagung, East Java. Data analyses were conducted at System Analysis and Computation Laboratory, Department of Agroindustrial Technology, Universitas Brawijaya. The aggregate production planning of the coated peanut products was carried out in two steps, include the demands predicting and the aggregate production planning.

2.1. Method

Aggregate production planning model formulation was developed based on the company's condition and issues. Below are several programming elements used in this study:

- Decision variable, which is the number of coated peanut product *i* family produced in period *t*, notated as X_{it}.
- Issue function, consists of:
  1) First issue is the maximum production capacity in period *t*.

     $$\sum_{i=1}^{5}\sum_{t=1}^{6} a_{it}X_{it} \le K_t \quad (1)$$

  2) Second issue is the forecast demand for coated peanut product *i* family in period *t*.

     $$X_{it} \ge D_{it} \quad (2)$$

  3) Third issue is non-negativity issue.

     $$X_{it} \ge 0 \quad (3)$$

- Purpose Function, which is the total production cost minimization.
  The calculation of the production cost for coated peanut product *i* family in period *t* is as following:
  a) Fixed cost
     - Machine depreciation cost

       $$D = \frac{P - S}{N} \quad (4)$$

     - Machine maintenance cost

       $$M = \left[H + I + J\right]m \quad (5)$$

     - Salary

       $$Gj = g \times F \times m \quad (6)$$

  b) Variable Cost
     - Main resources cost

       $$BB = \left[(Q \times q) + (R \times r) + (T \times t)\right]m \quad (7)$$

     - Packaging cost

       $$BP = W \times w \times m \quad (8)$$

     - Wage (Salary)

       $$Up = Upg \times o \times l \times m \quad (9)$$

     - Utility cost

       $$BU = \left[(E \times e) + (Y \times y) + (Z \times z)\right]m \quad (10)$$

  The production cost of coated peanut product *i* family in period *t* is the total production cost divided by production capacity, assuming that the cost is always the same throughout the planning period.

$$C_i = \frac{(D + Gj + M) + (BB + BP + Up + BU)}{K_i} \quad (11)$$

Aggregate production planning model is better described as following:

$$MinZ = \sum_{i=1}^{5} \sum_{t=1}^{6} C_{it} X_{it}$$

with issue :

$$\sum_{i=1}^{5} \sum_{t=1}^{6} a_{it} X_{it} \leq K_t$$

$$X_{it} \geq D_{it}$$

with $X_{it} \geq 0$ for i=1,2,…,5 ………………………………………………………(12)

t =1,2,...,6

The assumptions used in this study include: production activity runs smoothly; fixed number of employees; main resource, packaging material, and utility are always available; the number of production machine and their capacity are fixed; main resource, packaging material, and utility cost are fixed. There were also several parameters, such as i family in period t demand, period t production capacity, i family in period t productio`n percentage, and family i in period t production cost. Following the optimisation, sensitivity analysis on the changing objective function values and a right-hand-side value were carried out.

3. Results and Discussion

Aggregate production planning in Company X was conventionally arranged without any meticulous calculation but based on the company's policy. The company's policy required that the production capacity for the subsequent period have to be based on the product demand from the previous period with an additional 25 % was added. This addition was needed to anticipate any demand instability in the market. However, this company's policy caused issues of excess and insufficient production of the coated peanut product, resulted in an increase of the total production cost.

As shown in Table 1, the production capacity of the coated peanut product from family A, B, C, and E have relatively similar magnitude in each period. This result indicated that the production planning based on the company's policy was inappropriate. Inadequate aggregate production planning contributes to a failure to fulfill the market's demand, thus lowering both the service performance of and the financial profits of the company [5].

Table 1. Aggregate production planning based on the company's policy.

Period	Production Capacity (kg)				
	A	B	C	D	E
1	1,985	20,747	70,170	19,301	3,420
2	674	11,062	69,911	22,399	4,882
3	869	15,250	67,103	20,752	3,845
4	664	13,133	91,610	12,516	4,579
5	1,346	13,512	76,229	22,485	4,419
6	630	27,592	41,613	45,541	5,410
Total Production	6,168	101,296	416,636	142,994	26,555
Total Production Cost	IDR 4,722,816,262				

3.1. Proposed aggregate production planning

In this study, two steps were arranged in the proposed aggregate production planning. The first step was predicting the demand and followed by the aggregate production planning using the proposed linear programming. Predicting the demand was carried out to estimate the coated peanut product demand for the next period throughout the agreed planning duration. The best predicting method was chosen based on lowest error value [6] and verification process was conducted using moving distribution method [7]. Determining the best prediction method was applied to minimise the demand uncertainty occured in the next period, thus the results achieved are able to represent the actual condition [8].

The results from the best prediction method was then multiplied by business factor to obtain the prediction result that has been adapted to the future's actual condition. The demand of the coated peanut product was influenced by several factors, either internal or external factors, particulary a business factor. Therefore, this particular demand was proceeded and used to the next stage of the aggregate production planning for the coated peanut product.

The results of the aggregat production planning using the proposed method can be seen in Table 2. The table shows that the lowest production capacity was found in the product famili A, while the product family C exhibited the highest production capacity. The production capacity escalation of the coated peanut product was occurred on period 5 and 6, possibly due to an increase in demand following and during religious days. This result is in agreement with Buxey [9] who claimed that the seasonal pattern may influence the demand instability of a product. He further added that although this was happening only for a certain period, its effect on the production level is obvious.

Table 2. The results of the proposed aggregate production planning of coated peanut product.

Period	Production Capacity (kg)					Allowance (kg)
	A	B	C	D	E	
1	449	5,594	37,929	10,864	1,779	363,637
2	449	5,594	37,929	21,276	1,416	360,179
3	449	5,594	37,929	14,996	255	362,236
4	449	5,594	37,974	8,981	806	359,587
5	1,589	22,179	73,344	27,675	4,326	343,058
6	1,589	22,179	69,027	32,580	5,069	348,171
Total Production	4,974	66,734	294,132	116,372	13,651	
Total Production Cost						IDR 3,444,804,000

In linear programming method, the model parameter values are assumed as a fixed value due to its uncertainty to obtain a precise measure. Therefore, post-optimisation analysis is needed to review the influence of the model parameter value changes toward the optimal solution [10]. The sensitivity analysis result of changing objective function coefficient showed that the cost parameter value for each family for a certain period was similar, which was in the range of 0 to infinity. A decrease or an increase in the cost parameter value would not influence the optimal solution if the value is still within the tolerated limit, as stated by Wendell [11], but it may affect the Z value of the optimal solution. The sensitivity analysis results of a right-hand-side constants showed the limit values of this coefficient were varied. The right-hand-side coefficient consists of the production capacity value and the demand of the coated peanut product. If any changes occurred on those parameters values were still within the tolerated limit, this will not affect the optimal solution.

3.2. Comparisons between proposed aggregate production planning and company's policy-based method

The best aggregate production planning method was determined by comparing both aggregate production planning method. The difference between the production capacity and the total production cost from both aggregate production planning methods are shown in Table 3. The results indicated that the company policy-based aggregate production planning has a different pattern to that of the market demand. This was evidenced by the over and insufficient production occurred during the planning period. A large scale overproduction may increase both the inventory cost and the product's defect risk. Insufficient production capacity may limit the company to meet the customers demand. If the company failed to properly manage this issue, it will cost them a profit loss.

The above findings demonstrated that the proposed aggregate production planning using linear programming is more suitable to customer demand pattern compared to that of the company-based

IOP Conf. Series: Earth and Environmental Science **131** (2018) 012035 doi:10.1088/1755-1315/131/1/012035

policy method. The compatibility between the production capacity and market demand can reduce the inventory cost, as well as can provide an excellent service to satisfy the customers. Based on the average production cost, the proposed aggregate production planning decreased the total production cost with the difference value of IDR 1,278,012,262 (or 27.06 % lower). This result is supported by Fahimnia et al. [2] and Techawiboonwong et al. [3] who reported that the linear programmed-aggregate production planning can minimise total production cost and optimise the production capacity.

Table 3. Difference between production capacity and total production cost of both aggregate production planning method.

Period	Production capacity difference (kg)				
	A	B	C	D	E
1	(+)1,536	(+)15,153	(+)32,241	(+)8,437	(+)1,641
2	(+)225	(+)5,468	(+)31,982	(+)1,123	(+)3,466
3	(+)420	(-)4,069	(+)29,174	(+)5,756	(+)3,590
4	(+)215	(+)7,539	(+)53,636	(+)3,535	(+)3,773
5	(-)243	(-)8,667	(+)2,885	(-)5,190	(+)93
6	(-)959	(+)5,413	(-)27,414	(+)12,961	(+)341
Total Production Difference	(+)1,194	(+)20,837	(+)122,504	(+)26,622	(+)12,904
Total Production Cost Difference					IDR 1,278,012,262

4. Conclusion

The findings in this study confirmed that the aggregate production planning using linear programming reduces the production cost of the coated peanut product by 27.06% (or IDR 1.278 billion). A similar pattern to the customer demand was obtained when using the proposed method compared to that of by using the company's policy method, suggesting its suitability and feasibility for further application in the company. However, in-depth study is still required to include the inventory parameters into calculation, and how its impact on achieving the optimal production capacity with a low production or inventory cost.

References

[1] Herjanto E 2003 Manajemen Produksi dan Operasi (Production and Operational Management).\ 2nd Edition Grasindo Jakarta Indonesia 495 [In Indonesian]

[2] Fahimnia B, Luong L H S, Marian R M 2006 Modeling and optimization of aggregate production planning – a genetic algorithm approach *Proc. World Acad. Sci. Eng. Technol.* **15** 169-174

[3] Techawiboonwong A, Yenradee P 2002 Aggregate Production Planning Using Spreadsheet Solver: Model and Case Study *Sci Asia.* **28** 3 291-300

[4] Hsieh S, Wu M S 2000 Demand and cost forecast error sensitivity analyses in aggregate production planning by possibilistic linear programming models *J. Mech. Eng. Manufact.* **11** 4 355-364

[5] Maria A, Mattson C A, Ismail-Yahaya A, Messac A 2003 Linear physical programming for production planning optimization *Eng. Optimiz.* **35** 119-37

[6] Lapide L 2007 Don't just measure forecast errors *J. Bus. Forecasting* **26** 216-18

[7] Nahmias S 2001 Production and Operations Analysis 4th Edition Mc-Graw-Hill Book Co Singapore 674

[8] Sule D R 2008 Production Planning and Industrial Scheduling: Example, Case Studies and Applications 2nd Edition CRC Press Taylor & Francis Group USA 560

[9] Buxey G 2005 Aggregate planning for seasonal demand: reconciling theory with practice *Int. J. Oper. Prod. Manage.* **25** 11 1083-1100

[10] Hadigheh A G, Mirnia K, Terlaky T 2006 Sensitivity analysis in linear and convex quadratic optimization: invariant active constraint set and invariant set intervals *INFOR: Inform. Sys. Oper. Res.* **44** 2 129-155

[11] Wendall R E 2004 Tolerance sensitivity and optimality bounds in linear programming *Manage. Sci.* **50** 6 797-803

The influence of socio cultural dynamics on convergence communication of aquaculture agribusiness actors

Y Oktavia

Lecturer of Communication Studies, University of Andalas, West Sumatera, Indonesia

E-mail: yennyoktaviafp@gmail.com

Abstract. This research aims to: (1) Analyze the level of socio-cultural dynamics of agibusiness aquaculture actors. (2) Analyze the influence of socio-cultural dynamics on convergence communication of capacity development of aquaculture agribusiness actors.Data was collected by questionnaire and interview of group members on agribusiness. Data analyze was done by descriptive and inferential statistics with using SEM method. The result of descriptive statistics on 284 agribusiness members showed that: Socio-cultural dynamics of agibusiness aquaculture actors was in low category, as shown by lack of the role of customary institutions and quality of local leadership.The communication convergence is significantly and positively influenced by the communication behavior of agribusiness actors in access information.

1. Introduction

Globalization has provided a tremendous impact on traditional values, structure and knowledge base of society, with increasing external pressures and changing internal dynamics. Capacity development is the mainstream of development discourse, but in practice only focuses on the approach to the transfer of knowledge and technology as a solution, but it does not support communities in developing their own capacity [1]. Globalization has had negative impact on social change and economic decline in rural areas as shown by the high levels of unemployment and foreign investment [2,3].

Agribusiness of aquacultureis the primary source of income for the people in the coastal area of Padang. Besides, this business also plays an important role in supporting the fulfillment of needs for protein. But in reality, freshwater fishery agribusiness is difficult to develop. This issue is inseparable with the communication behavior of freshwater fishery agribusiness actors in accessing information to manage their business. Some of the challenges in developing capacity are: The number of people who have lack motivation and self-confidence because of their experience with the planning of development programs external top-down and cause internal conflict. Sidorenkov et al. [4] stated that conflict is based on the characteristics of each individual of group members. Socio-cultural dynamics is a symtom of changing social structure and cultural pattern in society. Socio-cultural dynamics is the most dominant factor affecting communication behavior of agribusiness in accessing information. Socio- cultural dynamics is one of the factors that need to be considered in the development of the capacity of aquaculture agribusiness actors. Based on these problems, the purposes of this study are (1) Analyze the level of socio-cultural dynamics of agibusiness aquaculture actors; and (2) Analyze the influence of socio-cultural dynamics on convergence communication of capacity development of aquaculture agribusiness actors.

2. Materials and Methods

This research uses quantitative approach. The data was collected using survey method. Based on the information from Marine Affairs and Fisheries Service/Dinas Kelautan Perikanan (DKP) Kota Padang officials, two districts were used as samples, Kecamatan Koto Tangah and Kecamatan Kuranji. Both of them are the aquaculture agribusiness development area in Padang. The number of samples was decided proportionally based on population distribution in research area, while the samples were chosen purposively.

Research was conducted in two aquaculture centers of Padang City namely Kecamatan Koto Tangah and Kecamatan Kuranji. Field research was conducted from February to March 2016. This research's population were aquaculture agribusiness actors, which include on-farm fish cultivator, cultivation business production input provider or agro industry input provider, breeder, output processor, distributor, supplier, seller, facilitator, group manager, and off-farm information provider of aquaculture agribusiness. This population was chosen based on agribusiness functions which include procurement and supply of production infrastructure, processing (agro industry) and marketing.

The number of respondents were 284 households of agribusiness actors, dispersed in Kecamatan Koto Tangah as much as 169 people and 115 people in Kecamatan Kuranji, representing 837 households of aquaculture agribusiness actors as the population. Respondents were determined using stratified random sampling. Data analysis used were quantitative analysis complemented with qualitative analysis. Explanations of social cultural dynamics were presented using descriptive statistics, while the structural equation modeling (SEM) analysis was used to analyze the dominant factor that influences communication convergence in capacity building of aquaculture agribusiness actors.

This research was done from February 2016 until March 2016. Primary data collection was done through interview to agribusiness actors and field observation, while secondary data was obtained from Marine Affairs and Fisheries Service/ Dinas Kelautan Perikanan Propinsi Sumatera Barat, Dinas Kelautan Perikanan Kota Padang, Badan Pusat Statistik Propinsi Sumatera Barat, and Balai Penyuluhan Pertanian/Perikanan in both sample districts. Data analysis was done through inferential and descriptive statistics by SEM (Structural Equation Model) analysis using Lisrel 8.7.

3. Results and Discussion

3.1. Level of socio-cultural dynamics in capacity development of aquaculture agribusiness actors

Socio-cultural dynamics are reflected by: (a) a pattern of social behavior, (b) local knowledge, (c) the role of traditional institutions, and (d) the quality of leadership. Figure 1 illustrates how the socio-cultural dynamics level in the capacity building of the aquaculture agribusiness.

Based on the distribution of socio-cultural dynamics, it appears that most of aquaculture agribusiness is in the low category, reflected by the low role of customary institutions and the low quality of leadership (Figure 1).This means that the role of the leader in managing knowledge, sharing knowledge, energizing and encouraging everyone to have the competence and tacit knowledge (ideas, creativity, experience skills to be shared together in groups are still low. From the interviews with the Wali Nagari, customary stakeholders and districts known that the customary institutions and local institutions such as KAN (Kerapatan Adat Nagari) have not been able to play a role in developing agribusiness aquaculture. Lack of local institutional and customary institution role in managing and making rules based on agribusiness fishery. This is also supported by the lack of facilitation provided by local institutions in increasing the access of farmers and freshwater fisheries agribusiness to the production facilities, capital, marketing of fishery products and processing result.

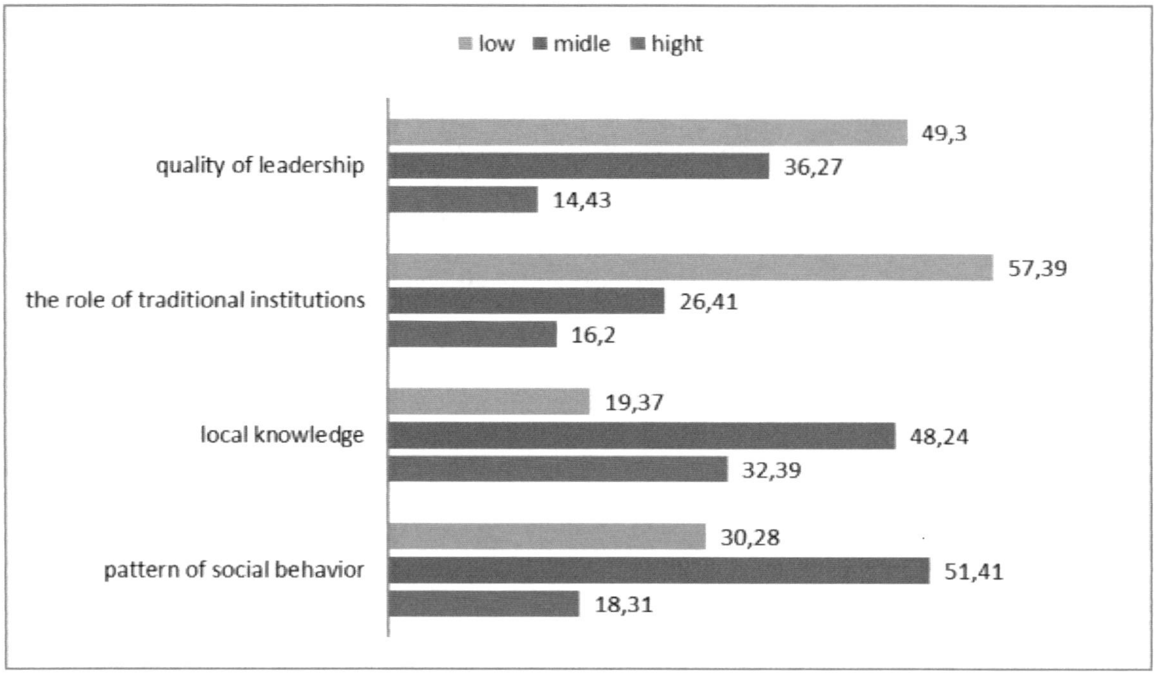

Figure 1. Socio-cultural dynamics level in the capacity building of the aquaculture agribusiness.

Until now the role of local institutions and institutional "Nagari" as one of the supporting institutions in the fishery agribusiness system, running its role so far only based on the problems of custom and custom law alone, has not developed many role in other fields such as agribusiness. Whereas local institutions have potential in developing local knowledge, which recently lively discussed because it is considered as an important part in the development program in the effort of resource management, in this case in the field of freshwater fishery agribusiness.In accordance with the research of Lubis and Zusmelia [5] said that local knowledge derived from the culture of society is considered to provide many solutions in development because it is the result of 'learning' from local communities in responding to the dynamics of their environmental change. Existing local knowledge indicates that local communities participate in exercising control over the utilization of existing resources because they have an interest in those resources.

According to Mokhtar et al. [6] and Moses et al. [7] leadership qualities correlate with the achievement of market target, organizational performance and play a role in quality management, it is imperative that a policy is formulated to support the flourishing of member's creativity. The development of science and information technology demands agribusiness actors to utilize science and technology for their business to progress, especially in the areas of product marketing, creativity post-harvest so they can have broader market channels, thus the factor of science and technology use becomes a vital concern.

3.2 *The influence of socio cultural dynamics on convergence communication of aquaculture agribusiness actors*

The convergence of communication in the capacity building of the agribusiness freshwater fisheries is a maximum common understanding between agribusiness actors, the knowledge of agribusiness that comes from the (local/implicit) and knowledge of agribusiness that comes from outside (global/ explicit), which is measured by the ability of communication effectiveness, improving quality of service information, social interaction, participation in the network and the ability of agribusiness.

IOP Conf. Series: Earth and Environmental Science **131** (2018) 012036 doi:10.1088/1755-1315/131/1/012036

Figure 2 shows the influence of socio-cultural dynamics on convergence of communication of agribusiness of aquaculture fishery

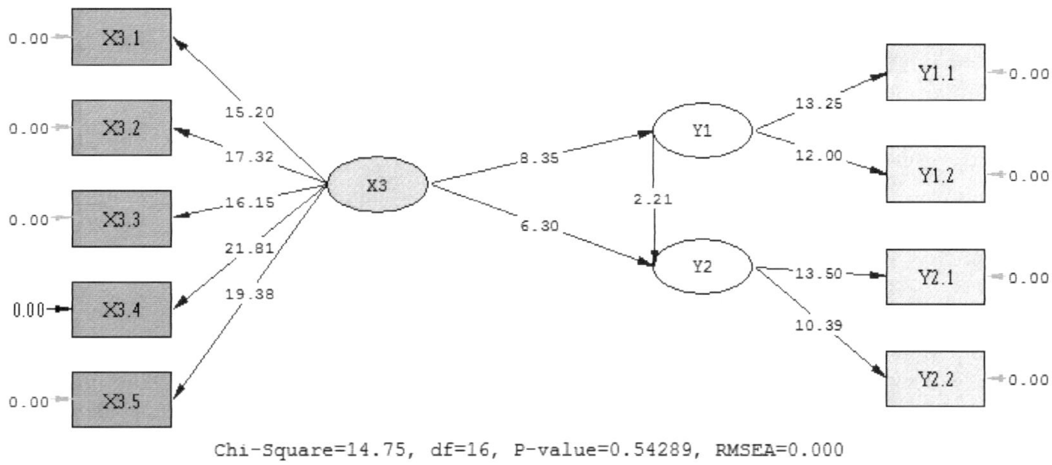

Figure 2. The influence of socio-cultural dynamics on convergence of communication of agribusiness of aquaculture fishery

$X_{3.1}$: Patterns of Social
$X_{3.2}$: Local Knowledge
$X_{3.3}$: The Role of Local Institutions
$X_{3.4}$: The Role of Traditional Institutions
$X_{3.5}$: Leadership Qualities
Y_2 : Communication Convergence of Agribusiness
$Y_{1.1}$: Contact With External
$Y_{1.2}$: Intensity of Use of Media
$Y_{2.1}$: Information Usage Level
$Y_{2.2}$: Service Quality

SEM analysis result shows that the communication behavior of agribusiness in accessing information and real positively influenced by variables socio-cultural dynamics. This means the higher the socio-cultural dynamics of freshwater fisheries agribusiness, the higher the communication behavior in accessing information.

Furthermore SEM analysis shows that all variables of socio-cultural dynamics also support in enhancing the capacity of agribusiness actors, this relationship means that the higher social behavior pattern is in the form of agribusiness support system, and the higher local wisdom owned and produced by local community the higher the communication behavior of freshwater fisheries agribusiness in accessing information. The higher role of local institutions and the role of customary institutions and the higher the role of leaders in sharing knowledge, energizing and motivating agribusiness actors, there is a tendency for better communication behavior of agribusiness actors in accessing information, reflected by increased contacts with outsiders and increased intensity of use of information media.

According to Rogers [9], in relation to the effort of disseminating innovation in a social system, the role of leaders is: (1) stimulating and inviting its followers to seek, disseminate and adopt innovation; (2) directing followers to make contact with other more advanced units of social systems; (3) creating a structure within the social system that allows for mutually motivating relationships, resulting in mutually beneficial, beneficial and mutually beneficial relationships (social learning interaction). Similar to the results reported by Marliati et al. [10] on the extent of agricultural extension.

SEM analisys shows that the variables that affect the convergence of communication is communication behavior. While communication behavior is influenced by socio-cultural dynamics, which consists of leadership qualities, patterns of social behavior, the role of traditional institutions, and local knowledge. So the increase of socio-cultural dynamics is not only needed in an effort to improve communication behavior of agribusiness actors but also to capacity development of agribusiness actors. The development of the capacity of the agribusiness, improving the socio-cultural dynamics can be done by improving the indicators that reflect the social and cultural dynamics (leadership qualities, patterns of social behavior, the role of traditional institutions, and local knowledge.

Low access to information caused by the limited availability of communication facilities and infrastructure, as well as the weakness of agribusiness institutional linkages with agribusiness freshwater fisheries. Agribusiness institutional has not been able to build a network and cooperation that can help agribusiness freshwater fisheries business development in the coastal area of Padang. In addition, local leaders may not be able to facilitate agribusiness in cooperation, help the problems that experienced by the agribusiness actors. Whereas according to Crocker et al. [11], leader's opinions are usually more trusted in conveying information in training on capacity building.

This results in a lower level of social interaction. Low levels of social interaction illustrates that the agribusiness fisheries have not been able to build a network with others, cooperate and share information with other agribusiness actors, such as from information centers or higher education institutions. As shown in the study of Sendegeyaa and Chiguvareb [12] education is important as the catalyst in capacity building and academic/higher education institutions act as the center of capacity building.

In accordance with research conducted by Sallem et al. [13], sinergy is really needed for Small and Medium Enterprises (SMEs) in managing finances and keep to attend the latest technological developments in the market, the importance of financial support from the government can also help develop SMEs.

4. Conclusion

1. The socio-cultural dynamics of agribusiness are relatively low in Padang are in the low category, especially in the quality of leadership and social behavior patterns. Increasing the role of social and cultural dynamics done by strengthening institutional support, both from the government and private sector, enhance the role of traditional institutions that are around as well as the quality of local leadership in sharing knowledge, improve their skills and are looking for ideas and new technologies
2. Socio-cultural dynamics have a real positive effect on communication behavior. All the variables of socio-cultural dynamics also support capacity building of agribusiness actors. This indicates that to improve the communication behavior of agribusiness actors in accessing information, reflected by increased contacts with outsiders and increased intensity of use of information media, it is necessary to increase the role of local institutions and the role of institutions in the form of institutional agribusiness supporters, and to increase local wisdom owned and produced by local communities, and increasing leadership roles in sharing knowledge, including increase of media information, encouraging and motivating agribusiness actors.
3. Converged communications positively and significantly influenced by agribusiness communication behavior in accessing information. The communication behavior is influenced by the social and cultural dynamics. To increase social and cultural dynamics are not only necessary for improving the communication behavior of agribusiness, but also to the development of the capacity of the agribusiness.

References

[1] Dhamotharan M 2009 *Buku Panduan Pengembangan Masyarakat Terpadu (Integrated Community Development Handbook)* Asian Productivity Organization Tokyo (In Indonesian)

[2] Fidelis OO, Olukayode EM 2016 Globalisation and economic growth in Nigeria: A Multi-Dimension Analysis *Pakistan Journal of Social Science* **9** 2 89-95

[3] Smutka L, Maitah M, Eldeeb O, Viktor V 2014 The Relationship Between The Global/Regional Rural Population Development and Agricultural Development *Research*

[4] Sidorenkov AV, Shipitko OY, Ulyanova NY, Obukhova YV 2016 Types of Interpersonal Contradictions and Conflicts in Work Group *The Social Science Medwell Journal* **11** 5 6999-7010

[5] Lubis, Zusmelia 2004 "Indigineous knowledge", Masyarakat Lokal dan Proses Pembangunan di Indonesia (Antara Harapan dan Kenyataan) (Local community and development process in Indonesia (Between expectation and reality) *Jurnal Sosial Ekonomi Pertanian* **17** 1 1-10 ((In Indonesian)

[6] Mokhtar SSM, Yusoff RZ, Yusr MM 2015 Leadership quality and the performance of market driven organization *The Social Sciences* **10** 7 1639-1644

[7] Moses CL, Ayodele OM, Mosunmola A, Gbenga AM 2016 Organization Culture and Creativity in Entrepreneurship Education: A Study of Secundary Education in Nigeria *Research Journal of Applied Sciences* **11** 8 586-591

[8] Havelock RG 1986 Modelling the Knowledge System In: G M Beal, W Dissanayake & S Konoshima (Eds), Knowledge Generation, Exchange, and Utilization, Westiew Press, Boulder pp 77-104

[9] Rogers EM 2003 *Difusion of Innovations* New York (US): The Free Press

[10] Marliati S., Pang S. A., Tjitropranoto P., Asep S 2010 Faktor-faktor yang berpengaruh terhadap kemandirian petani tanaman pangan beragribisnis di Kabupaten Kampar, Provinsi Riau. *Forum Pascasarjana* **33** 3 221-228.

[11] Crocker J, Shields KF, Venkataramanan V, Saywell D, Bartram D 2016 Building capacity for water sanitation, and hygiene programming: Training evaluation theory applied to CLTS management training in Kenya *Journal Social Science and Medicine* **166** 66-76

[12] Sendegeyaa A, Chiguvareb Z 2016 The Role Academia in Capacity Building for Sustainable Energy Development: The Case of Namibia *Journal Energy Procedia* **93** 218-222

[13] Sallem NRM, Nasir NEM, Nazif WM, Nori WM, Hisam CK 2017 Small and medium enterprisess: Critical problems and possible solution *International Business Management* 1(1): 47-52

IOP Conf. Series: Earth and Environmental Science **131** (2018) 012037 doi:10.1088/1755-1315/131/1/012037

Developing a green lending model for renewable energy project (case study electricity from biogas fuel at Palm Oil Industry)

Y A Sukirman

Sriwijaya University, Palembang, Indonesia

E-mail: yudhiapriyadi86@gmail.com

Abstract. In the last two decades, development initiatives solely aimed to generate economic growth has been placed under scrutiny, particularly amidst the rampant discussion on the quality decline of the environment, growing social divide and climate change along with its implications thereof. Considerations of the negative impacts brought about by the economic development process prompted the move to adopt the sustainable financing model that gives precedence to economic, environmental and social aspects. We introduced Green Lending Model for Renewable Energy Project (Case Study Electricity From Biogas at Palm Oil Industry) based on sustainability financing, which is used as variable to implementing financial institutions' lending policies. There are two major trends in the literature relating to sustainability and the banking industry: external and internal practices. The external practices strand analyzes the relevance of sustainability to the bank's communication with shareholders and other stakeholders, and how investors use it as a measure to help achieve optimal portfolio allocation. The internal practices literature, more relevant to the present work, studies how sustainability criteria are integrated into risk management models and lending practices. Its first implementation is in the Palm Oil industry at South Sumatera. The results explained that sustainability is not related to profit either from a short- or long-term perspective. The Sustainable Green Lending Model is related to the Equator Principles and its application is driven to project financing. It also related with short- and long-term risks and opportunities, instead of short-term sustainability impacts.

1. Introduction

In the last two decades, development initiatives solely aimed to generate economic growth has been placed under scrutiny, particularly amidst the rampant discussion on the quality decline of the environment, growing social divide and climate change along with its implications thereof. According to [1] argues that considerations of the negative impacts brought about by the economic development process prompted the move to adopt the sustainable financial model that gives precedence to economic, environmental and social aspects. Based on [2], realizing the importance of sustainable development in Taiwan, many industries have tried to amplify sustainability element in their organizations. However, in order to achieve sustainability objectives, it is crucial that the stakeholders of the industry understand the macro-environment. To analyze the macro-environment, their research employs a political, economic, social, technological (PEST) approach to develop an understanding of the big picture of the macro-environmental related to sustainable development. According to [3] argues that besides the existing wealth of proposals, mainly along the lines of better governance and co-financing strategies, need a complementary approach: parallel Quantitative Easing (QE) for Sustainable Development Goals (SDGs).

There are two major trends in the literature relating to sustainability and the banking industry: external and internal practices. The external practices strand analyzes the relevance of sustainability to the bank's communication with shareholders and other stakeholders, and how investors use it as a measure to help achieve optimal portfolio allocation [3, 4, 5, 6]. The internal practices literature, more relevant to the present work, studies how sustainability criteria are integrated into risk management models and lending practices. Weber et al. [7] use sustainability as a predictor of future financial performance and contend that bank should apply it in their credit risk models. Another example is Thoumy and Vachon [8], who show the relevance of sustainability to project finance in Canada.

We focus, on the sustainable financing model as a decision tool for financial institutions and present a case study in which a multinational bank applies the methodology to the crude palm oil and biogas industry in Indonesia. The main goal of this paper is to present a new way to incorporate sustainability into banking practices by focusing on a sustainable financing model to improve financial institutions' lending policies. Many banks now use the Equator Principles (EPs) as a framework for assessing environmental and social risks into project finance, but few integrate sustainability into more widespread credit products.

In the present paper, we analyze the case of Indonesian Government that creates a sustainable financing model to be used in loans to project renewable energy. We find no evidence of a similar approach to determine reputational risk based on sustainability issues. We contribute to the literature by discussing the implications of voluntary sustainability practices in a commercial bank's lending practices. There is a growing literature on the relationship between regulation and sustainability in the banking industry; for instance, [9] argues that regulation must target the financial sector because of its profits from unsustainable practices. However, if there is an eco-premium [10], market mechanisms may complement regulation in integrating sustainability and management practices. The aim of this study is to implemented environmental risk at financial lending practices for bank guidelines.

2. Methodology

Lending practices by commercial banks rely on credit score systems, developed to estimate the risk of default by borrowers. New regulatory standards under the heading Basel III mean that banks need to estimate the loss parameters: probability of default, loss given default, and exposure at default, Usually banks rely on credit-worthiness as the main criteria, and all information for the decision-making is related to financial information or qualitative data on how likely is that borrowers can repay their loans.

In the present case, the Sustainable Financing Model for Renewable Energy would complement regular credit rating models by yielding qualitative and quantitative information that would lead to improved decision-making. For example, it was assumed that companies would be rated in the same way as in regular credit models, with sustainability worthiness ranging from triple S (prime borrowers) to D (default). Assume a company with a low sustainability rating, below investment grade (usually in the BB range or lower). This could mean increased interest rates or denial of credit by the, even if the company is otherwise creditworthy. Conversely, companies with high sustainability ratings could get preferred treatment, either in terms of costs or access to capital.

The Indonesian Banking industry uses the same criteria as in most countries, based on Basel standards. In the present case, BNI Bank is chosen, because it has the most advanced sustainability department among the subsidiaries of the bank. The idea is that the Sustainable Financing Model for Renewable is tested in Indonesia and, if it improves the lending ability of the bank, is going to be introduced to the other banks. The methodology is based on a bottom-up approach. First is the renewable energy project industry analysis, followed by the definition of critical issues regarding the renewable energy project industry and its long-term sustainable path. then distill these issues into their environmental, economic, and social aspects and relate them to three main paths for sustainable development: Business As Usual (BAU), Sustainable Business (SB), and Future Sustainable Business (FSB).

Due to the inherent uncertainty of dealing with forecasting, adopt a generic conception of these development stages, include:

1. BAU refers to the present stage in which the industry practices are directly related to past practices, which may or may not be sustainable.
2. SB is a future stage (mid-term, around 2020), resulting from the adoption of new sustainable practices by companies. These are mainly derived from emerging technologies, new commercial practices, and evolving legislation.
3. FSB is a future stage (long-term, around 2050 marked by the foreseen role of the industry in a sustainable path that would allow continuing economic, social, and environmental development [11].

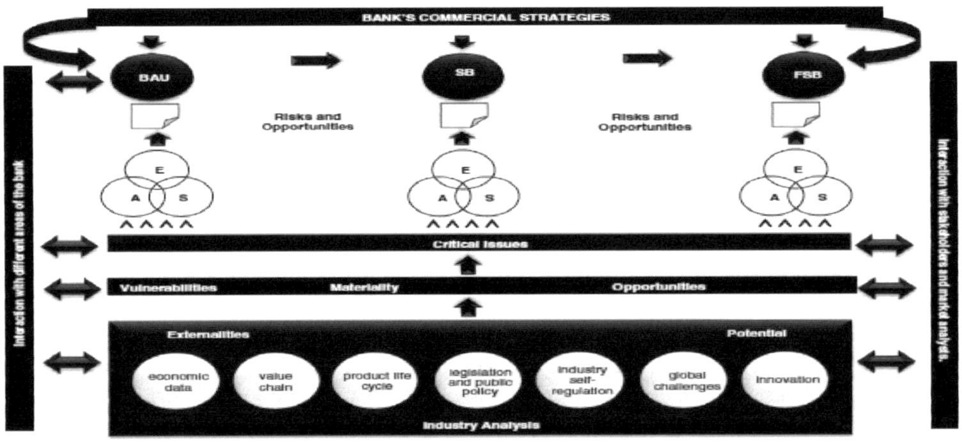

Figure 1. Analytical methodology for the green lending model

The main contribution of this methodology is that the resulting can help to develop proactive short- and long-term commercial practices. Researcher turn to this later, but a simple example of long-term commercial practices follows. By conducting a long-term analysis, researcher try to establish the possible future sustainability of the industry itself. For instance, let us assume that our analysis of the palm oil industry shows that its long-term yield productivity is going to be low and that the only way for the industry to survive is to expand geographically into protected environments and/or displace other cultures, generating negative benefits for society. In this case, the FSB analysis would show that companies would only be able to achieve FSB practices if they change their core business by focusing on more sustainability-promising products. This could result in the development of credit strategies in which loans are given to companies investing in changing their business model toward a more sustainable one. On the other hand, it may be that the FSB is one which companies are technological leaders and would be able to obtain much better yields by focusing on new crops and new techniques. In this case, directing loans to companies searching for those opportunities could be part of a proactive long-term credit policy.

3. Industry Analysis

The industry analysis methodology combines examination of the crude palm oil industry practices with an analysis of the global challenges. It forms the basis for all the subsequent analyses. Relating it to the global challenge allows us to decide on the major issues that crude palm oil companies and consequently the crude palm oil industry have to address in the coming decades.

The variables that determine environmental, social, and economic changes are many, and the relationships among them are dynamic and complex. From the countless chalenges and issues regarding sustainability [12, 13, 14, 15], among many others, thirteen are selected, either because of their global comprehensiveness (such as water and climate change issues). For each one, we identify possible sources of sustainability risks, but also opportunities for the dynamics development of the industry.

The next step is to relate these challenges to the crude palm oil industry. We achieve this by dividing the analysis into six topics: economic outlook, value chain, business life cycle, public policies and legislation, industry self-regulation and programs and innovation.

4. Critical Issues

The industry has established some important self-regulatory measures. These measures can be subdivided into programs, initiatives, and formal commitments. They attempt to mitigate some of the negative externalities generated by the industry. Some of the initiatives are related to labor practices, such as labor productivity and minimum prices paid to produce crude palm oil. There is also an initiative for an international environmental standard, called International Sustainable Palm Oil (ISPO) Certification. This attempt to gain acceptance among international market products are especially important for crude palm oil, because commodization breeds market penetration.

Innovation is a key success factor to development of the crude palm oil industry, whichever development path it should take. There are many Reserach and Development initiatives regarding the development of new palm oil varieties, but also relevant is innovation in terms of industrial flexibility in methan and crude palm oil production.

The results from the GRI G4 analyses indicate 70 issues relevant to the sustainability of the Crude Palm Oil Industry. On this set of issues, further analysis is carried out, and reseracher select 34 critical issues. Those issues are developed in eight categories. The next step is to give materiality to the groups of selected critical issues. First, researcher divide the issues and relate them to the dimensions of sustainability. The purpose is to provide guidelines for the intra-industry relationship between companies and relation to their stakeholders.

The then proceed to separate the group of critical issues into a development path represented by three categories: BAU-the average company in the industry; SB-the top tier of companies regarding sustainable practices; FSB-the average sustainable future company, in which innovations shape the industry according to sustainability requirements. There are two main paths for development: the dissemination of sustainable practices, in which the average company turns to sustainable practices (from BAU to SB); and the innovation path, which leads company from their present sustainable practices to future sustainable practices (from SB to FSB). The Researcher establish SB practices for the crude palm oil industry, which form the basis of the credit score in the next section.

The last informational step is to determine which of the some potential risks and opportunities would be relevant to the strategies of a commercial bank. Researcher filter the information through two development paths-from BAU to SB and from SB to FSB.

5. Weights, Iteration, and Measurements

For the final measurements, Weneed to assign weights to the different development paths and the different sustainability dimensions. We use subjective criteria. The subjective criteria are going to be validated through the testing of the questionnaire in the field through successive iterations of the process. After testing it with a few companies, the final measurements should be evaluated and then the criteria can be refined. There is no inherent problem with qualitative and subjective criteria in credit risk modelling in general, which has a long tradition. Agencies today are calling for more subjectivity in their credit risk models, a point that is contentious in the financial literature [16] are skeptical of allowing subjectivity in such model. Model validation is paramount to risk modelling using subjective criteria [17], and in the present case, the absence of extensive quantitative data makes the validation process even more important. The central idea behind the validation process is that it is conducted by the research department of the bank. The SCSS is an evolving tool, with changes being incorporated to reflect the feedback from the application of the questionnaires.

6. Using the Sustainable Credit Score System (SCSS)

Using the SCSS means developing a questionnaire to be completed by branch and account manager with private information regarding companies in the crude palm oil industry. There are two necessary conditions for the SCSS to be effective: information gathering and incorporation into the decission-making process. There is no hard evidence that the SCSS is actually efficient in classifying companies based on sustainability issues, and it can easily be discarded if an organization feels that it is not accomplishing its main goal. Unless it has internal ramifications, the commercial bank will be stuck in its BAU mode, even though the main objective is to move toward a future SB.

However, if used well, it can change the criteria used to make loans and to establish portfolio allocation. It fits well into the framework of [18] and can bring forth concrete information on how banks value the CSR practices of borrowers as in [19]. There is precious little information on banks' management practices regarding CSR, but if companies in the industry start adopting mechanisms similar to the SCSS, sustainability may be driven voluntarily instead of being forced by regulation; this scenario would generate positive externalities for society. The present case shows an example of a bank that is trying to act in a more socially responsible way to enhance its competitive advantage, which corroborates the result of [20].

7. Developing a green lending model for renewable energy project

Renewable energy project financing scheme generally comes from banking institutions and non-banking institutions such as multilateral institutions or financial institutions which established by the government. Some government and banking financing schemes such as: (1) Revolving Fund; (2) Investment Credit Scheme; (3) "Murabahah" Sharia Financing Scheme; (4) "Musyarakah" Sharia Financing Scheme**;** (5) "Mudharabah" Sharia Financing Scheme.

In order to accelerate the development of national infrastructure, Government attracting private funds both from local or international investor and established PT. Sarana Multi Infrastruktur (SMI). PT SMI developing various project financing at the renewable energy such as: (1) Subordinated Loan Facility Scheme**;** (2) Equity Investment Scheme.

Another alternative financing scheme for renewable energy projects such as: (1) Pooled Bond Financing; (2) Utility On-Bill Financing; (3) On-Bill Repayment; (4) Loan Purchase Program; (5) Energy Saving Insurance (ESI); (6) Linked Deposit Program**.**

8. Conclusion

Sustainability is still not a deciding factor in risk models by commercial banks. We argue that developing a SCSS can provide value creation for a bank and positive externalities to society. We show the development of such a system for a multinational bank, which is using the system for rating companies in the crude palm oil industry. The development of the SCSS consists of the following stages:

1. Selecting the industry, with corresponding analysis of its economic outlook, product life cycle, value chain, legislation and public policy, industry self-regulation, and innovation.
2. Measuring paths for the average company: BAU, SB, and FSB.
3. Defining variables related to individual companies, in each of six sustainability dimensions: economic growth (EG), environmental protection (EP), social progress (SP), socio-economic development (SD), eco-efficiency (EE), and socio-environmental development (SE).
4. Unveiling of the materiality issues pertaining to the average company, plus vulnerabilities and opportunities regarding the sustainability of the industry.
5. Combining the information to obtain questions for the composition of the score system.
6. Defining the weights for the analytical hierarchy process.

Companies that rating low on the SCSS would be denied credit, even if they are financially sound. The SCSS tries to build a competitive advantage for a financial institution by reducing the bank's long-term exposure to reputational risk. There is precious little evidence of internal management practices regarding sustainability. Financial products that are based on sustainability issues, like sustainable funds, abound, but the criteria for selecting companies are usually based on sustainability reports. Internalizing sustainability into everyday banking operations is a major challenge. Commercial bank has the ability to become proponents of societal changes. The implication for managers is clear: banks that take into account sustainability into their lending decisions would have less reputational risks and could build competitive advantage by having less default in the long run than the average bank.

There is still no hard evidence that incorporating sustainability into credit score systems leads to less defaults. Future research could show if thats is the case. There are plenty of other avenues for further research. Are market mechanisms relevant to the financial industry or should more regulation promote lending policies that are based on sustainability? Also, more direct evidence of new practices, such as the SCSS, may help us to understand how the financial industry is evolving regarding internal management practices based on sustainability.

References

[1] Casado A J, Steurer R 2014 Integrated strategies on sustainable development, climate change mitigation and adaptation in Western Europe: communication rather than coordination *J. Pub. Pol.* **34** 3 437-473
[2] Mokhtar S, Deng Y S 2015 Identification of key forces influencing sustainable development in Taiwan, *J. Sust. Dev.* **8** 2 174-186
[3] Brunhubber S 2015 How to finance our sustainable development goals (SDGs): socioecological quantitative easing (QE) as a parallel currency to make the world a better place *Soc. Sci. J.: Comp. Works* **2** 5 112-118
[4] Herzel S, Nicolosi M, Starica C 2012 The cost of sustainability in optimal portfolio decisions. *Europ. J. Finance* **18** 3–4 333–349
[5] Konar S, Cohen M A 2001 Does the market value environmental performance? *Rev. Econ.. Stat.* **83** 2 281–289
[6] Wright C 2012 Global banks, the environment, and human rights: The impact of the Equator Principles on lending policies and practices *Glob. Environ. Pol.* **12** 1 56–77
[7] Weber O, Scholz R, Fenchel M 2010 Incorporating sustainability criteria into credit risk management *Bus. Strat. Environ.* **19** 39–50
[8] Thoumy M, Vachon S 2012 Environmental projects and financial performance: Exploring the impact of project characteristics *Int. J. Prod. Econ.* **140** 1 28–34
[9] Richardson B J 2009 Keeping ethical investment ethical: Regulatory issues for investing for sustainability *J. Bus. Ethics* **87** 4 555–572
[10] Lubin D A, Esty D C 2010 The sustainability imperative *Harvard Bus. Rev.* **88** 5 42–50
[11] WBCSD (online) Vision 2050: A New Agenda for Companies. Retrieved from http://www.wbcsd.org/WEB/PROJECTS/BZROLE/VISION2050FULLREPORT_FINAL.PDF. [Accessed on 8 February 2013]
[12] KPMG (online) Expect the Unexpected: Building Business Value in a Changing World. Retrieved from http://www.kpmg.com/dutchcaribbean/en/Documents/KPMG%20Expect_the_Unexpected_ExctveSmmry_FINAL_WebAccessible.pdf. [Accessed on 13 April 2013]
[13] OECD (online) OECD Environmental Outlook to 2050 - The Consequences of Inaction Retrieved from http://www.oecd.org/env/indicators-modelling-outlooks/oecdenvironmentaloutlookto2050theconsequencesofinaction.htm. [Accessed on 2 February 2013]
[14] United Nations (online) The future we want. RIO+20 United Nations Conference on Sustainable Development Retrieved from https://rio20.un.org/sites/rio20.un.org/files/aconf.216l-1_english.pdf. [Accessed on 5 December 2012]
[15] United Nations Environment Programme (online) Keeping Track of Our Changing Environment: From Rio to Rio+20 1992–2012 Division of Early Warning and Assessment

(DEWA). Retrieved from http://www.unep.org/geo/pdfs/keeping_track.pdf. [Accessed on 23 November 2012]

[16] Griffin J M, Tang D Y 2012 Did subjectivity play a role in CDO credit ratings? *J. Financ.* **67** 4 1293–1328

[17] Jones D, Mingo J 1999 Credit risk modeling and internal capital allocation processes: Implications for a models-based regulatory bank capital standard *J. Econ. Bus.* **51** 2 79–108

[18] Weber O, Fenchel M, Scholz R 2008 Integration of environmental risks into the credit risk management process of European banks. *Bus. Strat. Environ.* **17** 149–159

[19] Ghoul S, Guedhami O, Kwok C C, Mishra D R 2011 Does corporate social responsibility affect the cost of capital? *J. Banking Financ.* **35** 9 2388–2406

[20] Chih H L, Chih H H, Chen T Y 2010 On the determinants of corporate social responsibility: International evidence on the financial industry *J. Bus. Ethics* **93** 1 115–135

IOP Conf. Series: Earth and Environmental Science **131** (2018) 012038 doi:10.1088/1755-1315/131/1/012038

Bioceramics synthesis of hydroxyapatite from red snapper fish scales biowaste using wet chemical precipitation route

D Ulfyana[1], F Anugroho[2], S H Sumarlan[3] and Y Wibisono[1]

[1]Department of Bioprocess Engineering, Faculty of Agricultural Technology, Universitas Brawijaya, Malang, Indonesia
[2]Environmental Engineering, Faculty of Agricultural Technology, Universitas Brawijaya, Malang, Indonesia
[3]Agricultural Engineering, Faculty of Agricultural Technology, Universitas Brawijaya, Malang, Indonesia

E-mail: y_wibisono@ub.ac.id

Abstract. Fish scales biowaste contain high collagens and calcium phosphates, therefore have considerable potential as raw material for value-added biomaterial such as hydroxyapatite (HAp). HAp is the main constituent component of hard tissue such as bone and teeth in the human body and is known as bioceramic materials. In this work, wet chemical precipitation method was used to syntesize HAp from Red Snapper Fish (*Lutjanus campechanus*) Scales. Two variations of calcination temperatures of 600°C (FHAp1) and 800°C (FHAp2) were conducted for 5 hours. The results showed calcium content from biowaste of red snapper fish scale was 83.62%. FTIR result shows that PO_4^{3-}, OH^-, and CO_3^{2-} functional groups presence as indicates the formation of HAp. XRD result showed the degree of crystallinity for FHAp1 and FHAp2 were 75.52% and 79.20%, respectively. The degree of crystallinity is in accordance with ISO 13779-2:2000 standard in which the minimum degree of crystallinity of hydroxyapatite used for biomedical materials is 45%. Finally, Particle Size Analyzer (PSA) results show that the particle size distribution is evenly distributed, with the size of micro-scale hydroxyapatite particles, ranging from 5.76 μm to 132.64 μm.

1. Introduction

Ceramics based biomaterial is widely used for instance for dental implant, femur bones and metal implant coating for orthopedic purposes [1-2]. Ceramic biomaterial is also called as bioceramics, especially when it derived from biological materials [3]. Bioceramics made from biological materials including biowaste can be utilized as any ceramic materials, such as membranes [4], however bioceramics has bioactive subtances which is very useful for biomedical application, for instance for repairing bones. Bioceramic made from calcium phosphate is one of bioactive ceramics widely used for repairing bones, including betatricalcium phospat (β-TCP), hydroxyapatite (HAp), and biphasic calcium phospat *(*BCP*)* that consist of HA nuclei and various HA/β-TCP [5]. Hydroxyapatite (HAp/HA) is promising bioceramics materials, is not only used for biomedical application but also for water efficient fertilizer which is potential for maintaning sustainability [6].

IOP Conf. Series: Earth and Environmental Science **131** (2018) 012038 doi:10.1088/1755-1315/131/1/012038

According to International Organization of Standardization (ISO) 13779-2 and American Society for Testing and Materials (ASTM) F1609 on the standardization of HA characteristics for coating as an application of implant for human body, the standard specification requirements for hydroxyapatite are as follows : the calcium-phosporus ratio must be in the range of 167-1.76, the crystallinity content of hydroxyapatite must be at least 45% with the crystallinity content of others maximum 5%, and the adhesive strength must be higher than 15 MPa in both metallic implant and non-metallic implant [7].

HA might be synthesized from natural materials, especially biowastes derived from cow bones waste, egg shell, and the shell of marine organisms. Hydroxyapatite derived from natural materials possesses better physicochemical attributes than synthesized hydroxyapatite [8]. Fish scale is one of the naturally hard tissues that contain organic components such as calcium [9].

Calcium phospate is the most abundant mineral contained in human body, with 99% contained in hard tissue, such as bones and tooth, especially in the form of $3Ca(PO_4)_2.Ca(OH)_2$ known as hydroxyapatite. Hydroxyapatite has synthesized from various natural materials that possess high calcium content with several methods of synthesis such as the dry method, hydrothermal method, alkoxide method, flux method, and wet method. However some of the methods above still produce hydroxyapatite with low purity and large crystal structure, therefore additional treatment is needed to minimize the particle size by conducting calcination and sonication on the sample which aim to minimize the produced particle size [10]. In this research, the synthesis of hydroxyapatite is carried out with the wet chemical precipitation route and assisted by sonication treatment and the temperatures used for calcination process are 600°C and 800°C for 5 hours.

2. Materials and Method

Materials

Red snapper fish scales waste were obtained from PT Kelola Mina Laut Gresik. Some chemicals were used for preparation and syntetis, *e.g.*: NaOH 1%, NH_3 25%, H_3PO_4 85% and aquades.

Method

2.1.1. Preparation

Fish scale samples that would be used as calcium precursor are firstly washed in order to release the impurities that could affect the final result, the samples are then boiled with boiling water for 30 minutes. After boiled, samples are washed using aquades, and then drained. Then samples are hydrolyzed using NaOH 1% for 2 hours, in order to remove fat. After that, samples are left at room temperature for 24 hours and washed using aquades and then sterilized using autoclave for 15 minutes at 121°C. After sterilization process is carried out, samples are dried using oven at 110°C for 5 hours. The hydrolyzed fish scales form $CaCO_3$ compound that needs to be converted to CaO as calcium precursor in calcination at 1000°C for 5 hours.

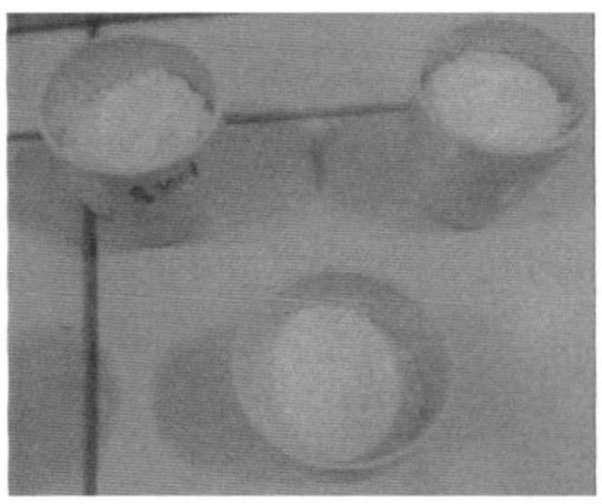

Figure 1. Calcinated hydroxyapatite.

2.1.2. Synthesis of hydroxyapatite

CaO samples obtained are then dissolved with aquades. Phosphate precursor H_3PO_4 85% was then added. $Ca(OH)_2$ and H_3PO_4 suspension are titrated with the drop rate of 1.5 ml/minutes. Then, NH_3 was added to reach pH>8. Samples are then sonicated for 1 hour at 50 Hz. After that, samples are dryed in room temperature for 24 hours. Then, the samples are washed and filtered to obtain calcium-phosphate powder. The powder is then dried using oven at 110°C for 5 hours. Subsequently, the samples are calcinated with the temperature of 600°C and 800°C, for 5 hours. The calcinated samples are then characterized by using FTIR (Fourier Transform Infra Red), XRD (X-Ray Diffraction), XRF (X-Ray Fluorescence), and PSA (Particle Size Analyzer). Calcinated HAp shown in Figure 1.

2.1.3. Characterization of hydroxyapatite

Fourier Transform Infra Red (*Shimazu IR Prestige-21*) was used to analyze chemical properties of the samples, X-Ray Fluorescence (*Philips, PANalytical Minipal 4*) was used to analyze the calcium content in the samples, and X-Ray Diffraction (*Philips, Type: PANalytical E'xpert Pro*) was used to determine the crystallinity of the samples. The formula used to calculate the crystallinity of the samples is:

$$\% = \frac{\text{Crystalline area fraction}}{\text{Crystalline area} + \text{amorphous area}}$$

Particle Size Analyzer (*Cilas 1090 Liquid*) was to analyze the particle size of the samples.

3. Results and Discussion

3.1 XRF characterization

The purposes of characterization using XRF (X-Ray Fluorescence) were to determine the content of the elements consisted in the ash of red snapper fish scale and the result of hydroxyapatite synthesis. XRF was used to detect the elements consisting in the ash of red snapper fish scale and hydroxyapatite (Table 1). Based on the characterization result using XRF, it can be seen that the Ca content within the ash of red snapper fish scale was 83.62%, and the P content was 12.7%. Then for sample FHAp1 (calcination at 600°C, for 5 hours), the Ca and P content in the sample were 82.30% and 14.9%, respectively. In sample FHAp2 (calcination at 800°C, for 5 hours), the Ca and P content in the sample were 82.09% and 15.1%, respectively. The decrease of Ca content in the sample occurred due to the calcination temperature of 600°C and 800°C.

Table 1 Percentage of elements identified using XRF.

Element	Percentage (%)		
	Red Snapper Scale Ash	HA 600°C, 5 hours (FHAp1)	HA 800°C, 5 hours (FHAp2)
P	12.7	14.9	15.1
Ca	83.62	82.30	82.09
Ti	0.05	0.077	-
Fe	0.21	0.19	0.21
Ni	1.30	0.944	0.931
Cu	0.15	0.12	0.11
Zn	0.17	0.13	0.13
Zr	0.2	-	0.2
Re	0.07	0.15	0.05
Co	-	0.04	-
Sr	-	1.1	1.1

3.2 FTIR specstroscopy

The spectroscopy resulted using FTIR for both samples showed that the main hydroxyapatite component groups are PO_4^{3-} and OH^-. This can be seen based on the peaks formed in the wave spectrum result shown in Figure 2a and 2b. As shown in Figure 2a, FHAp1 (calcination temperature at 600°C, for 5 hours), in wavenumbers 570.929536 cm^{-1} and 601.790592 cm^{-1}, PO_4^{3-} groups was observed [11]. Then for sample FHAp2 (calcination temperature at 800°C, for 5 hours), in wavenumbers 555.49908 cm^{-1} and 603.719408 cm^{-1}, PO_4^{3-} groups were observed. PO_4^{3-} groups existed at the wavenumber range of 555-602 cm^{-1}. Meanwhile, OH- groups can be seen based on the peaks observed in the wave spectrum result show also in Figure 2b. As shown, OH^- groups in sample FHAp1 and FHAp2 observed in wavenumber 3500 cm^{-1} to 3600 cm^{-1}. But, the highest peak is shown in wavenumber 3560-3580 cm^{-1}. Then, for α-TCP and β-TCP also observed in Figure 2c at wavenumber 947.048656 cm^{-1} and 981.767344 cm^{-1} as residual result in small content. This occurred due to imperfect decomposition during calcination that influenced the formation of α-TCP and β-TCP groups [12].

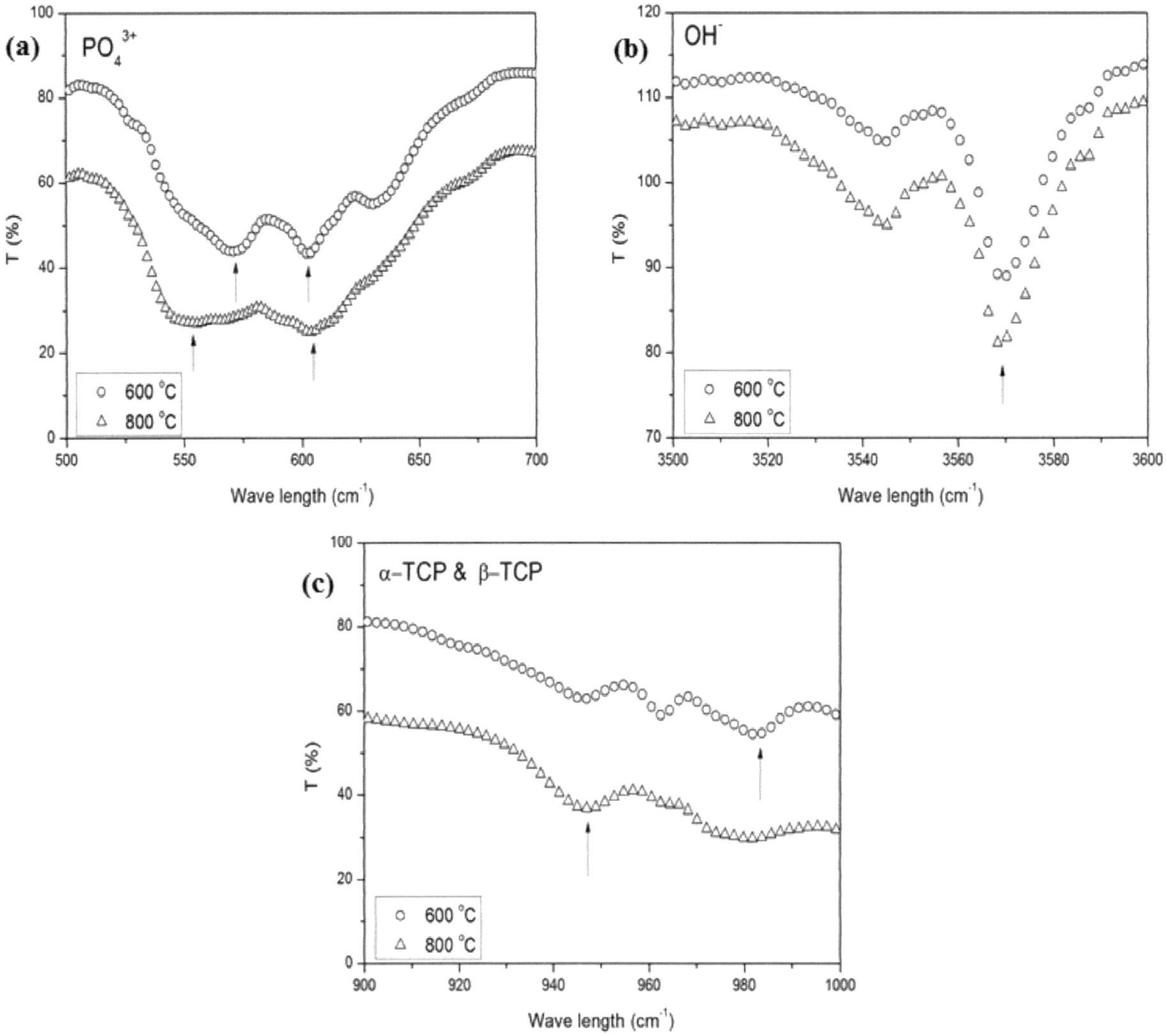

Figure 2. FTIR spectra: Phosphate groups (PO_4^{3-}) (a); Hydroxile groups (OH^-) (b); α-TCP and β-TCP groups (c).

3.3 XRD crystallinity measurement

XRD used to analyze the crystallinity of the samples, by analyzing the resulted diffractogram. The highest intensity shows the increasing HAp crystallinity. The crystallinity level can be found by using the triangle area approach. The crystalline/amorphous area fraction, in which the FWHM level is considered to be half of the base width and *height* as its height.

The diffractogram graphic which resulted from XRD is plotted in form of data as shown in Table 2 and Table 3. Data in both tables are obtained based on the highest existing intensity. Based on the data of JCPDS 00-001-1008 (*Joint Committee on Powder Diffraction Standards),* for hydroxyapatite phase the main intensity at level 2θ is characterized at 25.879; 32.196; 32.902; and 34.04. Based on the data shown, it can be determined that the crystallinity level for sample FHAp1 is 75.52%, and for sample FHAp2 the crystallinity level is 79.20%. The difference of the crystallinity levels from each sample is due to the difference of calcination temperature applied. The crystallinity percentage of a hydroxyapatite sample increases in proportion to the temperature rise during calcination process. Based on ISO standardized specification (ISO 13779-2:2000) to make hydroxyapatite coating which possesses high mechanical strength, it should have at least more than 45% crystallinity.

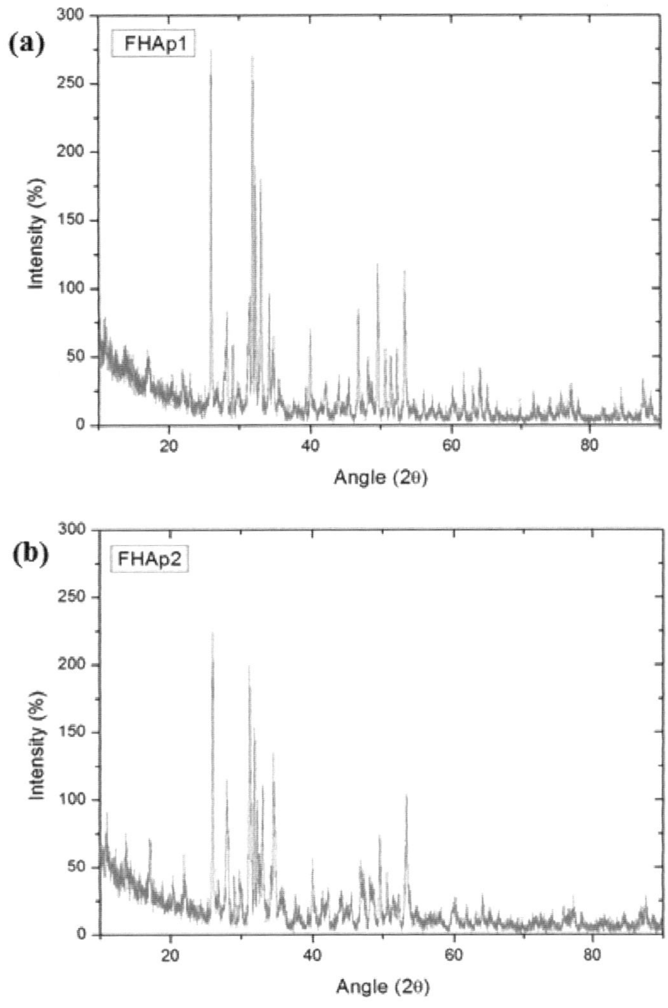

Figure 3. Diffractogram of FHAp1 (a) and FHAp2 (b).

Table 2. Crystanility of FHAp1.

	Pos (2θ)	Height [cts]	FWHM (2θ)	Fraction
	25.9894	262.71	0.1378	36.2014
	31.9081	260.35	0.1574	40.9790
Crystal			∑	77.1804
	32.3073	181.59	0.1378	25.0231
Amorph			∑	25.0231

Table 3. Crystanility of FHAp2.

	Pos (2θ)	Height [cts]	FWHM (2θ)	Fraction
	25.9972	202.49	0.1181	23.9140
	31.2274	178.89	0.2362	42.2538
Crystal			∑	66.1678
	31.9140	126.09	0.1378	17.3752
Amorph			∑	17.3752

3.4 Hydroxyapatite particle size

Crystal size of hydroxyapatite was measured using PSA. Particle size of FHAp1 sample ranged from 61.15 μm to 132.69 μm. Similar size was observed for sample FHAp2, with the size ranged from 52.33% to 132.64%. The results shown in Figure 4. Sonication affect the particle size reduction. Ultrasonic (sonicator) is mainly effective to break down the aggregates, reduce the size and reduce the polydispersity of a particle [13].

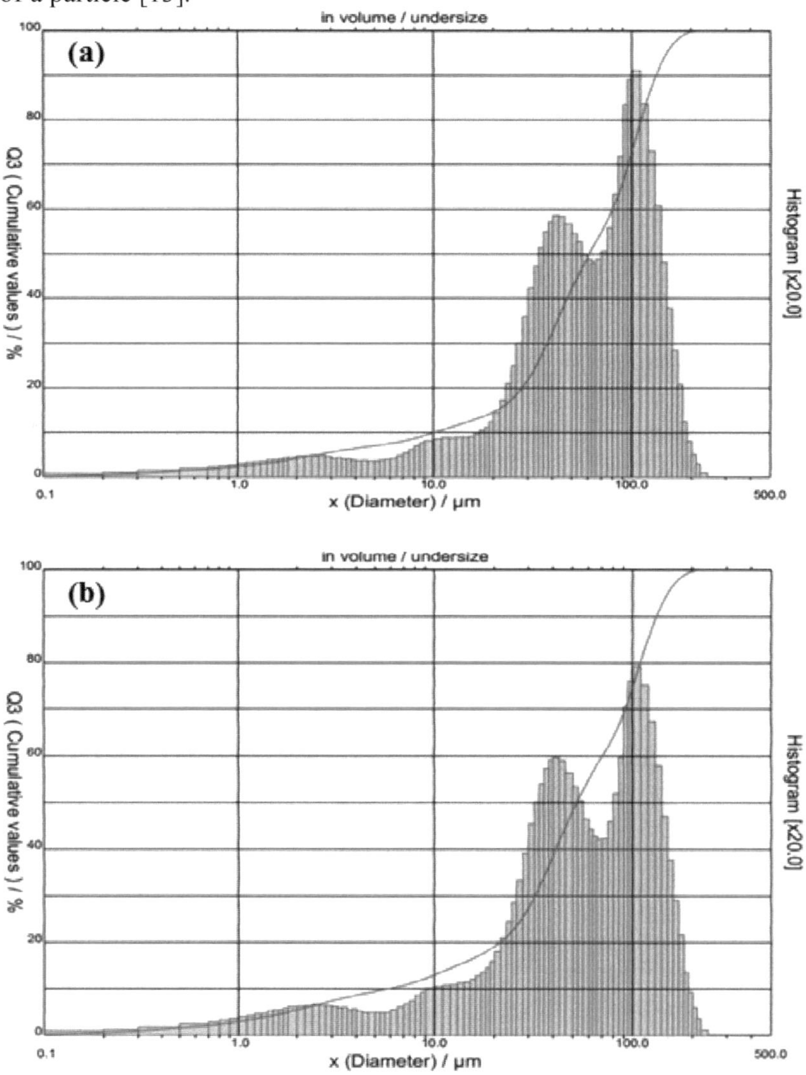

Figure 4. Particle size of FHAp1 (a) and FHAp2 (*b*).

4. Conclusion

Wet chemical precipitation method was able to syntesize hydroxyapatite from red snapper fish scales. FTIR spectra shows that PO_4^{3-}, OH^-, and CO_3^{2-} functional groups presence as indicates the formation of hydroxyapatite.The degree of crystallinity for FHAp1 and FHAp2 were 75.52% and 79.20%, respectively. Based on the ISO standard (ISO 13779-2:2000), the hydroxyapatite for medical materials must have crystallinity level of at least 45%, and the hydroxypatite resulted from this research met the those requirements. Finally, the hydroxypatite particle size distribution is evenly distributed, with the size of micro-scale hydroxyapatite particles, ranging from 5.76 μm to 132.64 μm.

Acknowledgments

The authors would like to acknowledge Mr Iqbal and Mrs Sapta from PT Kelola Mina Laut for providing red snapper fish scales as raw materials used in this study.

References

[1] Bona AD, Pecho OE and Alessandretti 2015 Zirconia as a dental biomaterial *Materials* **8** 4978-4991
[2] Piconi C and Maccauro G 1999 Zirconia as a ceramic biomaterial *Biomaterial* **20** (1-25)
[3] Wibisono Y 2017 *Biomaterial and Bioproduct* (Malang: UB Press)
[4] Wibisono Y, Ahmad F, Cornelissen E R, Kemperman A J B. and Nijmeijer K 2016 Dominant factors controlling the efficiency of two-phase flow cleaning in spiral-wound membrane elements *Desalin. Water Treat.* **57** 38: 17625-36
[5] Bayazit V, Bayazit M and Bayazit E 2010 Evaluation of bioceramics material in biology and medicine *Dig J Nanomater Biostruct* **7** 3 267-278
[6] Nugroho, W., Nugraha, R. and Wibisono, Y 2013 Autonomous Framework on Governing Water for Sustainable Food and Energy *Proceeding of Sharia Economic Conference*: 37-40
[7] Auclair-Daigle C, Bureau M N, Legoux J G and Yahia L H 2015 Bioactive hydroxyapatite coatings on polymer composites for orthopedic implants *J Biomed. Mater Res.* **73A** 4 398-408
[8] Shojai MS, Khorasani MT, Khoshdargi ED and Jamshidi A 2013 Syntesis methods for nanosized hydroxyapatite with diverse structures *Acta Biomater* **9** 8 7591-7621
[9] Torres FG, Troncoso OP, Nakamatsu J, Grande CJ and Gomez CM 2008 Characterization of nanocomposite laminate structure occuring in fish scales from Arapaima gigas *Mater Sci Eng C* **28** 8 1276-1283
[10] Poinern, G.E.J., Brundavanam R.K., Mondinos N, and Jang Z 2009 Synthesis and characterization of nano-hydroxyapatite using an ultrasound assisted method *Ultrason Sonochem* **16** 569-474
[11] Mobasherpour I., Heshajin MS, Kazemzadeh A and Zakeri M 2007 Synthesis of nanocrystalline hydroxyapatite by using precipitation method *J Alloys Compd.* **430** 1-2 330-333
[12] Wang P., Li C, Gong H, Jiang X, Wang H, Li K 2010 Effect of synthesis conditions on the morphology of hydroxyapatite nanoparticles produced by wet chemical process *Powder Technol* **203** 2 315-321
[13] Cengiz B, Gokce Y., Yildiz N, Aktas Z and Calimli A 2008 Synthesis and characterization of hydroxyapatite nanoparticles *Colloids Surf. A Physicochem. Eng. Asp.* **322** (1-3) 29-33

A sandwich-type optical immunosensor based on the alkaline phosphatase enzyme for *Salmonella thypimurium* detection

E Widyastuti[1], M F Puspitasari Schonherr[1], A Masruroh[1], R A Anggraeni[1], Y K Nisak[1] and S Mursidah[2]

[1]Dept. Agricultural Product Technology, Faculty of Agricultural Technology, Brawijaya University
[2] Dept. Agro Industrial Technology, Faculty of Agricultural Technology, Brawijaya University Veteran Street, Ketawanggede, Lowokwaru, Malang, East Java Indonesia 65145

E-mail: endrika_w@ub.ac.id

Abstract. Salmonella is pathogenic bacteria that caused foodborne diseases which being called Salmonellosis. Prevalence of Salmonellosis that being caused by *Salmonella thypimurium* in Indonesia is quite high. However, detection of Salmonella bacteria in food still limited, complicated, and required a lot time. Sensitive optical assay for *Salmonella thypimurium* paper based detection has been developed by integrating sandwich assay between antibody-antigen complex and alkaline phosphatase enzyme that produce visible bluish-purple colour with presence of NBT-BCIP substrate. The results showed that Limit of Quantitation of detection is 105 CFU mL^{-1} with detection time 15 minutes. Linearity test between Colour intensity that produced from Salmonella concentration presence on samples showed that detection has good linearity. Selectivity test exhibited excellent sensitivity with good discrimination against *Escherichia coli*.

1. Introduction

Food safety is an important factor in food requirement in order to avoid the occurrence of disease. A disease that occurred through food consumption referred as food poisoning or food-borne disease [1]. Food-borne disease is commonly caused by the presence of pathogenic microorganism such as Salmonella [2,3,4]. Salmonella does not necessarily cause change in color, smell, or taste of foods, but if there's a lot of Salmonella presence in the foods, then Salmonella will cause symptoms which is called Salmonellosis. The cases of Salmonellosis are still a major problems in food safety worldwide as reported by Pui *et al* [5,6] that there are 16 million cases of typhoid fever, 1.3 billion cases of gastroenteritis and 3 million deaths due to Salmonella. The risk of Salmonellosis is increased 5-10% each year [7,8]. Prevalence of Salmonellosis caused by *Salmonella thypimurium* in Indonesia was 1.6% in 2007 with highest cases were found in school age children [9]. However, Indonesia still use conventional methods which are limited, complicated, and needed approximately 5-7 days to complete the analysis to determine *Salmonella thypimurium* presence in foods[10,11,12]. This condition made Salmonellosis became more complicated as we hardly can detect the presence of *Salmonella thypimurium* in our food. In order to solve that problem, immunosensor based Biosensor can be an alternative of a rapid, practical and accurate *Salmonella thypimurium* detection.

Salmonella thypimurium detection is fabricated using colorimetric biosensor principle. The test result can be observed directly due to color change as indicator of positive testing result, while negative testing result will not develop any color change [13,14]. In this paper, a paper based detection and sensitive optical assay for *Salmonella thypimurium* has been developed by integrating sandwich assay between antibody-antigen complex and alkaline phosphatase enzyme that produce visible bluish-purple color with presence of NBT-BCIP substrate. This immunosensor developed are expected to be an alternative method for Salmonella detection.

The aim of this research is to developed sandwich type-optical immunosensor which is can be an alternative of Salmonella detection in Indonesia. The relevant experimental variables, including the LoQ (Limit of Quantitation), detection time, linearity and selectivity is examined and optimized both qualitative and quantitative.

2. Materials and Methods

2.1. Bacteria and Media

Salmonella choleraesuis serotype *typhimurium (S. Typhimurium)* and *Escherichia coli* was used for the experiments (obtained from Food Microbiology Laboratory, Faculty of Agriculture Technology, Brawijaya University) Indonesia. Culture of *Salmonella typhimurium* were prepared by incubating the cultures in Natrium Agar (NA) slant at 37°C for 24 h and culture of *Escherichia coli* were prepared by incubating the cultures in Lactose Broth (LB) at 37°C for 24 h.

2.2 Reagent and Antibody

Salmonella typhimurium polyclonal antibody was purchased from Bioss. Purified rabbit anti-Salmonella igG-AP (Chemicon), BSA (Nacalai), Gluteraldehyde (Sigma), and Tween 20 (Nacalai) were purchased from Bioscience Laboratory, Brawijaya University. Buffer Saline Phospate (Merck) and NBT-BCIP (Thermo Scientific) were purchased from Biomedical Laboratory, Brawijaya University.

2.3 Preparation Immobilization of Antibody

Preparation steps are done by sterilizing the Whatman paper #1 that is used as platform (110°C, 60 minutes). In spite of that, reagents required also been made, such as 1% PBST (15μL Tween 20 in 148 μL Buffer Phosphate Saline/BPS), 0.1% BSA (0.1 mg BSA in 100μL PBST), 5% Gluteraldehyde e (10μL Gluteraldehyde in 190μL PBST) and secondary antibody (combination of *Salmonella typhimurium* antibody and anti-rabbit AP-labelled antibody) as well as 0.1 mg/ml primary antibody (20 μL 1 mg/ml *Salmonella typhimurium* antibody in 180μL BPS).

2.4 Fabrication of Biosensor

Sandwich type-optical immunosensor is fabricated using modified Zourob[15]and Cao[16]steps of fabrication. Fabrication was done by the following steps: 4μl secondary antibody is dripped onto the reaction zone and left it until dried. Then, 4μL Gluteraldehyde is dropped to result zone and left to stand until it's completely dry prior to 4μl primary antibody which is added onto the result zone. After the paper is completely dry, the result zone is washed using PBST to discard unimmobilized primary antibody. After washing step, 4μL of NBT-BCIP substrate is added to the result zone and left to stand until it's dry. Lastly, 4μL BSA is added as blocking agent to block areas which did not bind to any antibody.

Figure 1 shows the fabrication steps. The detector is then incubated for 4°C, 24 hours before it is ready to use.

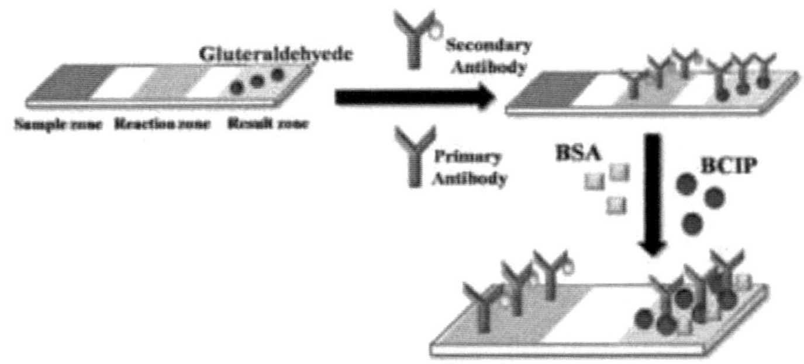

Figure 1. Schematic representation of sandwich type-optical immunosensor based *alkaline phosphatase* enzyme construction

2.5 Experiment

The experimental was conducted to determine Limit of Quantification (LoQ), linearity, detection time, selectivity and storage stability of the detector. Limit of Quantitation of the immunosensor is determined by modification of method published by Aloraefy[17]. Six different Salmonella concentrations (0, 10^4; 10^5; 10^6 ; 10^7 ; and 10^8 CFU mL^{-1}) are used as samples and the result of the tests is analyzed using Hex Color Finder to obtain RGB value of each concentrations. Then, RGB-*Salmonella* concentration standard curve is made using Ms. Excel. Limit of Quantitation (LoQ) is calculated from equation line of standard curve with the value of Y-axis is the RGB value of control/ background (0 CFU mL^{-1}). Linearity of sandwich type-optical immunosensor is determined by R2 value of RGB-Salmonella Concentration standard curve of LoQ variable test.

Detection time is determined by measure the time duration when sample is dropped into sample zone until color is developed and stable at result zone, indicated that alkaline phosphatase enzyme react with NBT-BCIP substrate to produce blue/violet colored products.

Selectivity is determined using modified method from Bhalla[18].The detectors were tested using control (BPS); *Salmonella typhimurium* and *E.coli* culture. Last experimental variable is storage stability which was determined by modified method from Rathee[19]. Three fabricated immunosensors that had been made at the same time were tested using *Salmonella typhimurium* culture at different time. The platform stability was tested for 0, 2 and 4 weeks after incubation time.

2.6 Apparatus

The test results was documented using scanner (Canon MP250 Series) and analyzed using Hex Color Finder in order to obtain the RGB value of each test results. Hex Color Finder (HFC) is an application for Windows which created by NZ Works and help users to get RGB colors to detected hexadecimal value (e.g. #2E505C). Then, the data acquired was analyzed to determine overall performances of the detector, both quantitative and qualitative.

3. Results and Discussion

3.1 Limit of Quantitation (LoQ)

Limit of Quantitation is the lowest analyte concentration that still can be detected [20]. RGB value that obtained from each concentration is 135.33(0 CFU mL^{-1}),137(10^4 CFU mL^{-1}), 135.67 (10^5 CFU mL^{-1}), 134.67 (10^6 CFU mL^{-1}), 130 (10^7 CFU mL^{-1}), and 97 (10^8 CFU mL^{-1}), after obtained these RGB value, Standard Curve RGB-Concentration is being made (Figure 2). The LoQ value of the immunosensor is 10^5 CFU mL^{-1} , this value is obtained from calculating X-axis value from Standard curve RGB equation. According to NCDC[1] and Yousef and Carlstrom[21] infection dose of *Salmonella* that can caused food poisoning is varied up to 10^9 CFU mL^{-1}. Based on NCDC [1], the immunosensor that can be developed in this paper can detect the presence of *Salmonella typhimurium* at lower concentration (10^5 CFU mL^{-1}) than the maximum concentration (10^8 CFU mL^{-1}) that can caused food poisoning. Therefore this immosensor performances is good for *Salmonella typhimurium* detection. The color that being developed from each concentration test (Figure 3).

Linearity is linear change from measurement results because of analytes (presence in a certain) concentration range [18] Standard curve had R^2value 0.642 which can be interpreted that the linearity of the standard curve still quite good. That results mean RGB values that being obtain from each tests is quite proportional to tested *Salmonella* concentrations.

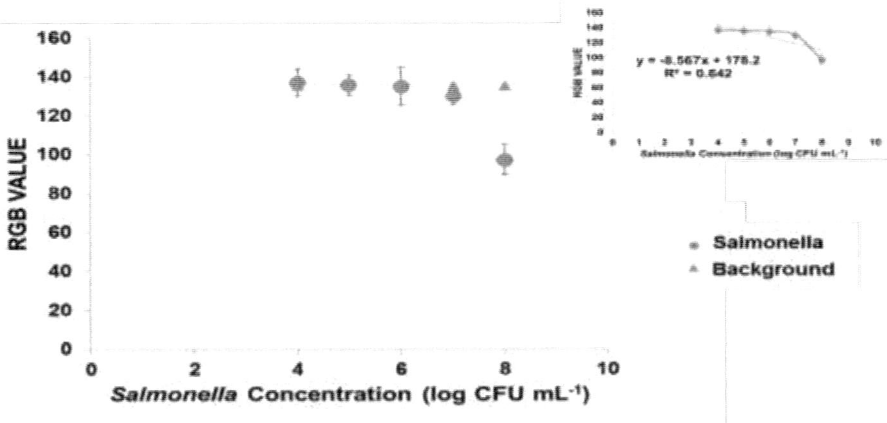

Figure 2. Calibration of *Salmonela thypimurium* concentration by sandwich hybridization using Hex Color Finder (HFC)

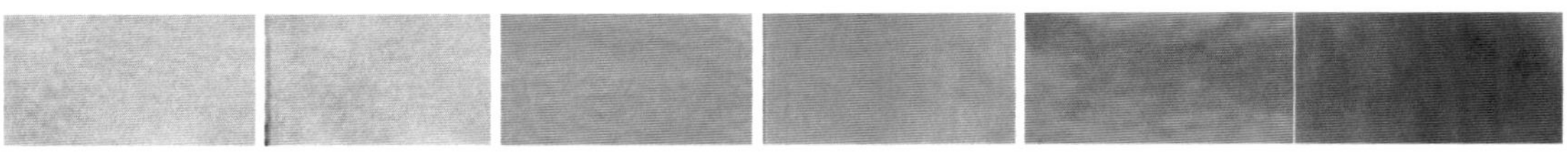

A B C D E F

Figure 3. Color developed from different tested *Salmonella thypimurium* concentrations (A) 0, (B) 10^4, (C)10^5, (D)10^6, (E)10^7 and 10^8 log CFU mL^{-1}

3.2 Detection Time

Detection time analysis of the immunosensor resulted the visible and stable colour change occurs after 15 minutes of sample addition (Figure 4). Conventional detection method of Salmonella required more than 5 days for complete isolation and confirmation processes [10,11, 12]. Therefore, immunosensor that was successfully developed in this paper has less detection time compared to conventional detection method of Salmonella.

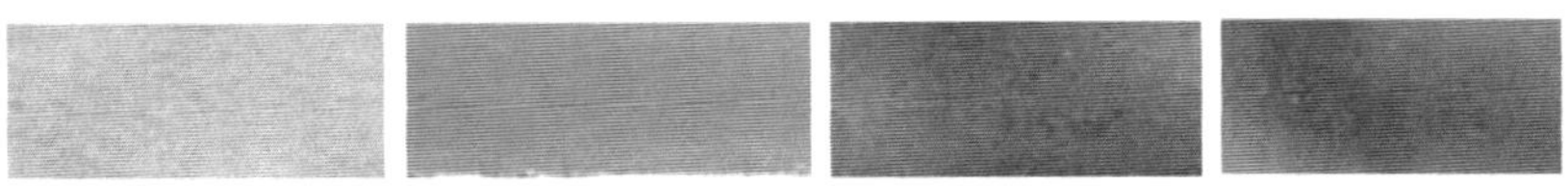

A B C D

Figure 4. Color developed from 10^5log CFU mL^{-1} *Salmonella* concentrations with different time incubation (A) 0, (B) 5, (C) 10 and (D) 15 minute.

3.3 Selectivity

Selectivity is the ability of a molecule to discriminate interaction partners. A selective binder shows little cross-reactivity[22] means that it recognizes a given partner with much higher affinity than other partners. Selectivity test results showed that the detector can only developing color when *Salmonella* is present, indicated by RGB value obtained is lower than the control while *E.coli* testing result showed that there was no significant differences with the control (Figure 5).

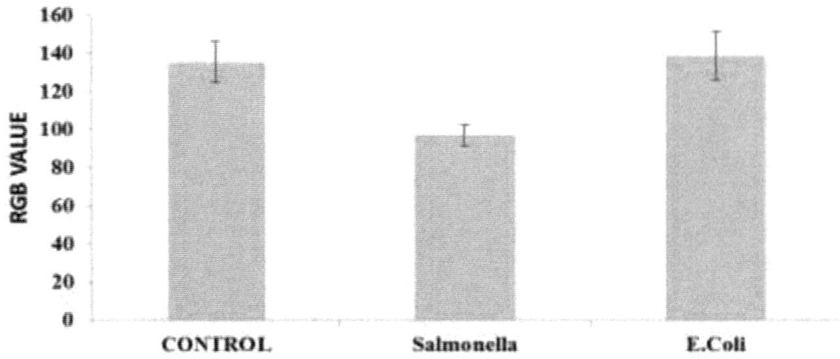

Figure 5. Use of sandwich-type optical immunosensor to measure *E.Coli*, strain BL2(DE3). The response is shown to *E.Coli* (1x10^5 CFU mL^{-1}), Salmonella (1x10^5 CFU mL^{-1}) and control (without bacteria)

3.4 Storage Stability

RGB values of three immunosensors that had been made and tested in different weeks using Salmonella culture was 116.67 for week 0, 113 for week 2 and 136.67 for week 4. This immunosensor performed good stability for 4 weeks ,indicated that there is no color development at week 4 as can be seen in Figure 6. The addition of stabilizer and encapsulation of enzyme might extend the stability of platform [23].

A B C

Figure 6. Stability sandwich-type optical immunosensor based on the alkaline phosphatase enzyme (A) 0, (B) 2 and 4 weeks

4. Conclusion

A sandwich-type optical immunosensor that had been developed in this study can be used for selective detection of *Salmonella thypimurium.* The biosensor presented here provides real time detection (15 minutes) with high sensitivity, quite good color linearity and excellent selectivity against *E.coli.* This result has shown promising application in the future as alternative *Salmonella thypimurium* detection method. However, further research about paper types, concentration of antibody and enzyme that being used and also immobilization method of antibody onto the paper should be conducted in order to optimize overall performances of detection.

Acknowledgments

We acknowledge the Ministry of Research, Technology and Higher Education of Indonesia and Universitas Brawijaya for supporting of this research.

References

[1] National Centre for Disease Control (NCDC) 2009 Food-Borne Diseases CD Alert1**3** 4 1-12
[2] Teunis P, Kasuga F, Fazil A, Ogden I, Rotariu O Strachan N 2010 Dose-Response Modeling of Salmonella Outbreak Data *International Journal of Food Microbiology* **144** 2 243-249
[3] Todd E C 1997 Epidemiology of Food-Borne Diseases: A Worldwide Review World Health Stat Q **50** 30–50
[4] Galanis E, Lo Fo Wong, D M Patrick, ME Binsztein, N Cieslik,A Chalermchikit 2006 Web-Based Surveillance and Global Salmonella Distribution 2000-2002 *Emerg.Infect.Dis.* **12**, 381–388
[5] Pui CF, Wong WC, Chai LC, Tunung R, Jeyaletchumi P, Noor Hidayah MS 2011 Salmonella: A foodborne pathogen. *International Food Research Journal* **18** 465–73
[6] Cary J W,DBhatnagar2000 Microbial foodborne diseases: Mechanisms of Pathogenesis and Toxin Synthesis p. 3-34. United States of America: Technomic Publishing Company, Inc.
[7] Kendrovski V,GDragan 2012 Climate Change: Implication for Food-Borne Diseases (Salmonella and Food Poisoning Among Human in R. Macedonia) InTech
[8] Zhang Y, Hiller JE 2010 Climate Variations and Salmonella Infection in Australian Subtropical and Tropical Regions *Sci. Total Environ.* 408 524–30
[9] National Institute of Health Research and Development 2007 National Report on Basic Health Research Ministry of Health Republic of Indonesia
[10] Naravaneni R, Jamil K 2005 Rapid Detection of Food-Borne Pathogens by Using Molecular Techniques. *Journal of Medical Microbiology* 54 1 51-4
[11] Manafi M 2000 New Developments in Chromogenic and Fluorogenic Culture Media, *International Journal of Food microbiology* **60** 2-3 205-18
[12] DownesFP, Ito K2001 Compendium of Methods for The Microbiological Examination of Foods 4ed Washington: American Public Health Association p 357-380
[13] Corcuera J, Cavalieri R 2003 Biosensor New York: Marcel Dekker
[14] Kishor K, John C, Deepalekshmi P, Mariam A, S A Al-Maadeed, Jaehwan K 2016 Biopolymer Composites in Electronics p 355. Elsevier
[15] Zourob M 2008 Principles of Bacterial Detection: Biosensors, Recognition Receptors and Microsystem New York: Springer Science and Business Media
[16] Cao R 2015 A Zero-Step Functionalization on Paper-Based Biosensing Platform for Covalent Biomolecule ImmobilizationSensing and Bio- Sensing Research **6** 13–18
[17] Aloraefy M, Pfefer J 2014 In Vitro Evaluation of Fluorescence Glucose *Biosensor Response Sensor Journal* **14** 7 12127-48
[18] Bhalla N, and Jolly P 2016 Introduction to Biosensors Assays *Biochem.* **60** 1 1–8
[19] Rathee K, Dhull V, Dhull R, Singh S 2016 Biosensors Based on Electrochemical Lactate Detection: A Comprehensive Review Biochemistry and Biophysics Reports
[20] Armbuster D,T Pry 2008 Limit of Blank, Limit of Detection and Limit of Quantitation Clinic Biochemistry Reviews **29** 49-52
[21] YousefA E,Carlstrom C 2003 Food Microbiology: A Laboratory Manual p.167-205 New Jersey: John Wiley and Sons, Inc
[22] Nomine Y, Choulier L, Trave G, Vernet T, Altschuh D 2015 Antibody Binding Selectivity: Alternative Sets of Antigen Residues Entail High-Affinity Recognition.PLoS ONE **10** 12 1-1
[23] Chen W, Yang W, Lu Y, Zhu W, Chen X 2017 Encapsulation of Enzyme into Mesoporous Cages of Metal-Organic Frameworks for The Development of Highly Stable Electrochemical *Biosensor Analytical Methods* **9** 3213-3220

IOP Conf. Series: Earth and Environmental Science **131** (2018) 012040 doi:10.1088/1755-1315/131/1/012040

Characteristic of physical, chemical, and microbiological kombucha from various varieties of apples

E Zubaidah, S Yurista, N R Rahmadani

Department of Food Science and Technology, Universitas Brawijaya, Malang, Indonesia

Email: elzoeba@yahoo.com, elok@ub.ac.id

Abstract. Kombucha is a fermented tea beverage with the addition of kombucha starter also called SCOBY (Symbiotic Culture Of Bacteria and Yeast). The purpose of this research was to know the physical, chemical and microbiological characteristics of kombucha from various varieties of apple kombucha. The study used Randomized Block Design (RAK) with one factor of apple varieties (Anna, Fuji, Granny Smith, Manalagi, Red Delicious, Rome Beauty, Royal Gala). Each treatment was repeated three times. Data was analyzed with ANOVA (Analysis of Variance). The best treatment was selected using Multiple Attribute method. Data of hedonic test was analysed using Friedman Test. The best treatment was obtained on Fuji varieties of kombucha apple with the characteristics as follows: total acid 1.33%; pH 2.95; Total phenol 268.57 µg/ml GAE; Total sugar 6.74%; Antibacterial activity against *S.aureus* 21.30 mm; Antibacterial activity *E.coli* 21.20 mm; Antioxidant activity 35.62%; Organoleptic aroma 3.55, taste 3.3; Color 3.4 (on a scale of 1-5)

1. Introduction

Kombucha is a functional beverages produced by fermenting tea using a SCOBY (Symbiotic Culture of Bacteria and Yeast). SCOBY consists of symbiotic interaction between Bacteria (*Acetobacter xylinum*) and Yeasts (*Brettanomyces, Zygosaccharomyces, Saccharomyces,* and *Pichia*) [1]. The symbiotic interaction during tea fermentation converts sugar into short-chained organic acid, vitamins, and alcohols [2].

Kombucha contains some minerals, vitamins, enzymes, and organic acids which is considered as functional beverages. Kombucha is effective to treat constipation, increasing microflora inside the guts, increasing immune system, reducing blood pressure and cholesterol, eliminating cancer cell, has an antidote effect, antioxidant activity, and also antibacterial effect [3]. Symbiotic interaction in kombucha breaks down the substrate in aerobic environment, for seven to ten days, to produce fresh, acidic carbonated drinks [4].

In general, kombucha is made from black tea and sugar. Many researches have been conducted to increase its functional value of kombucha by substituting the black tea with other materials that contain high amount of natural antioxidant compounds. Apple is one of the fruit that is highly consumed in Indonesia. Three of the most common apple variety cultivated in Indonesia are the Rome Beauty, Manalagi, and Anna. Another varieties of Apples like Fuji, Red Delicious, Roya Gala, and

Granny Smith are largely imported from various countries. Apples are often processed into apple-based products such as apple cider, apple chips, apple toffee, apple vinegar, and many more. Malang, East Java is one of the biggest apple supplier in Indonesia with the apple cropping area reaching 2.500 m^2 which produces up to 10 tons of apples. These apples have not been entirely utilized, therefore more apple-based products innovation is necessary to be developed.

Apple-based kombucha fermentation is expected to increase the functional value and increase the organoleptic value of kombucha. Apple is used to substitute tea as the main ingredients to produce kombucha. Fermentation will increase the bioavailability of several nutritions from raw material such as antioxidant and antibacteria which will make it easier for the substance to be absorbed into digestive system. Therefore, a study is necessary to determine the characteristics of kombucha made from various kinds of apple to obtain kombucha with high quality.

2.Materials and Methods

2.1 Experimental design

This research used Randomized Block Design using single factor with three duplicates of each treatment. The factor were apples varieties (Anna, Manalagi, Fuji, Granny Smith, Red Delicious, Rome beauty and Royal Gala). Black Tea was used as the comparative control. The data were analyzed using Analysis of Variance (ANOVA) and Duncan's Multiple Range Test (DMRT) with 5% of confidence range. Multiple Attribute Method was used to determine the best treatment.

2.2. Apple juice preparation and Kombucha fermentation

Apple was peeled, washed and cut into small size. Each apples were carefully weighed using the analytical balance until 300 g and added with water at a ratio of 1 :2. The mixture was blended for 30 seconds and then filtered with filtering cloth to obtain the apple juice. Apple juice was added with 10% cane sugar and then pasteurized at 65 ^oC for 30 minutes and cooled at room temperature. The pasteurized apple juice was put into sterilized glass jar. Kombucha starter were added for 10% (v/v) and then covering with cloth. The fermentation process was maintained for 14 days at room temperature. Microbial growth and chemical changes were monitored during the fermentation.

2.3. Data Analysis

The post-fermentation characteristics determination of kombucha included analysis of total microbial cells (TPC method) [5], pH[6], total acids content [7], total sugars content [7], total phenols content [8], antioxidant activity (DPPH method) [9], antibacterial activity analysis [10] and organoleptic analysis [11].

3.Result and discussion

3.1 Total bacteria

Figure 1 shows that total microbial cells in apple kombucha increase from day 0 to 7 days of fermentation, and then decreased after 14 days fermentation. Microbes utilize glucose as the energy source which is abundantly present in the fermentation medium from added sugar and apple itself. The increasing number of total microbes occurred because microbes utilize sugars to replicate cells and grow. A part of the sugars are also converted into metabolites such as alcohols. The number of microbes increased until 7 days of fermentation then reaching stationary phase [12]. Total microbial decreased after 14 days due to the production of various metabolite compounds that inhibit microbial growth. The optimum pH for *Acetobacter sp.* is between 5.4 - 6.3. While, yeast has an optimum pH at 4.0-5.0. In the acid state, the growth of microbes is inhibited.

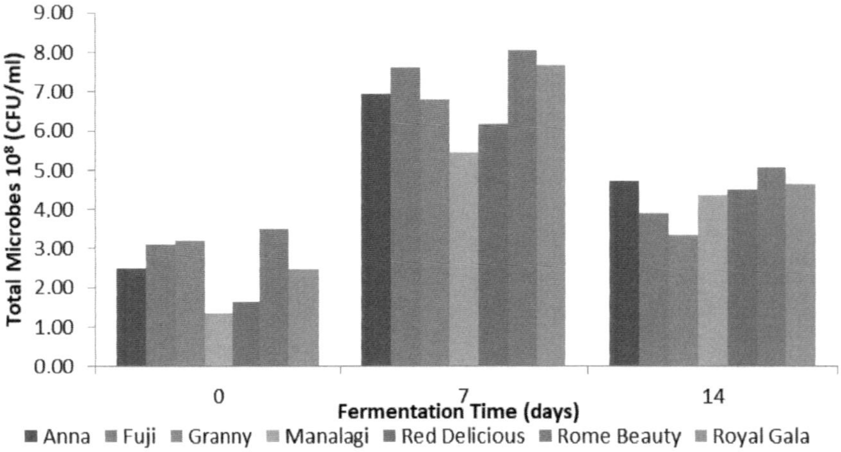

Figure 1. The growth of microbes from various apples varieties during kombucha fermentation

3.2 Total acid and pH kombucha

Figure 2 and Figure 3 show the total acid and pH of kombucha during fermentation, respectively. Total acids increased during fermentation. During fermentation, glucose is metabolized by microbes to produce organic acids. Therefore, the longer the fermentation, the more organic acids are being produced. Kombucha fermentation is initiated by the enzimatic hydrolisis of sucrose resulting glucose and fructose. The yeasts (*Saccharomyces cereviseae*) metabolize the glucose resulting ethanol and carbon dioxide. The ethanol is then oxidized into acetaldehyde by the acetic bacteria (*Acetobacter*) which is later converted into acetic acid. The acetic acid concentration in kombucha increases during fermentation. *Acetobacter* also breaks down glucose into various low molecular weight organic acids. Fermentation of kombucha also produce the other kind of acids such as glucanoic acid, glucoronic acid, and lactic acid. Invertase enzyme activity from the yeasts allows sucrose breakdown, while the acetic acid bacteria do not have the similar ability because of its lack of hydrolase and kinase enzyme [3]. The result of variance analysis showed that the treatment of apple variety was not significantly different ($\alpha = 0.05$). The result of raw material analysis showed that there was no significant difference between apple varieties to sugar content, total acid, pH, and total phenol.

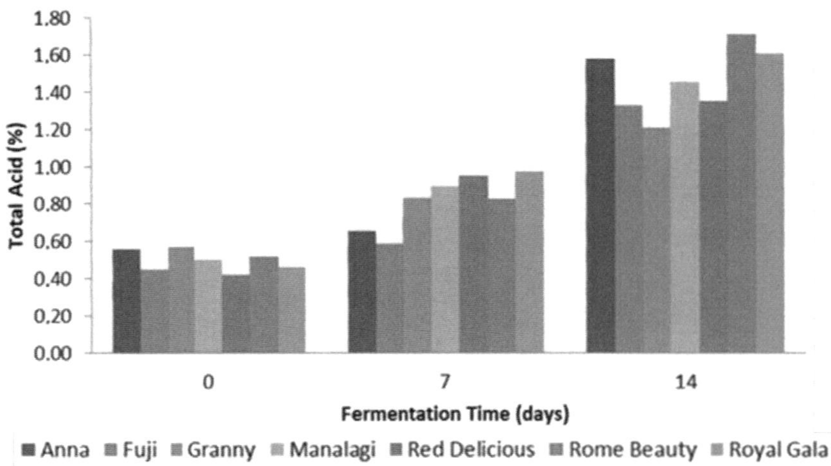

Figure 2. Total Acids of kombucha during fermentation

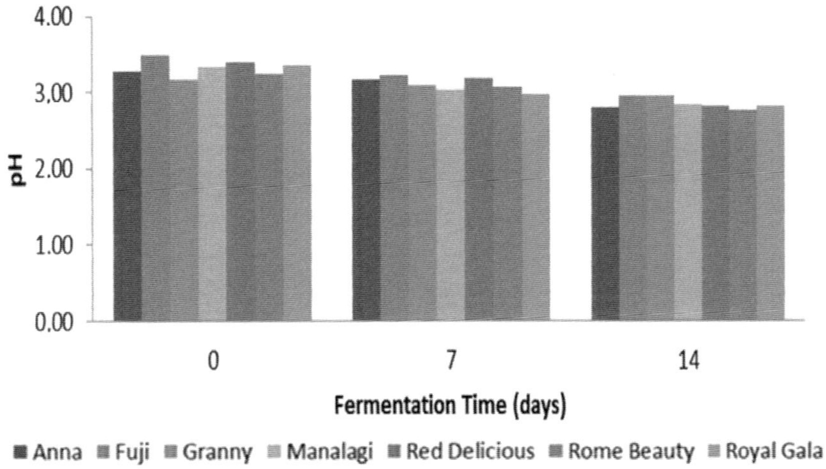

Figure 3. pH of kombucha during fermentation

3.3 Total sugar

Figure 4 shows that total sugars of each of apple kombucha varieties decreased during fermentation. Microbes utilizes glucose to produce new cells during fermentation that produce organic acids as the metabolites. *Saccharomyces cereviseae* metabolizes glucose as its energy sources. Yeasts utilize glucose and fructose during fermentation as their energy sources although they mostly break down glucose to produce ethanol and carbon dioxide, while fructose is metabolized by another minor microorganisms. Sugars are the source of glucose which can be metabolized by the microorganisms to grow their cells and to produce acetic acid. During the fermentation the yeasts metabolize glucose in the media to produce alcohols and other substances simultanously, continued by alcohol oxidation by the *Acetobacter xylinum* producing acetic acid. According to Aditiwati and Kusnadi [3], *Acetobacter xylinum* also oxidize glucose into glucoronic acids and other organic acids at the same time.

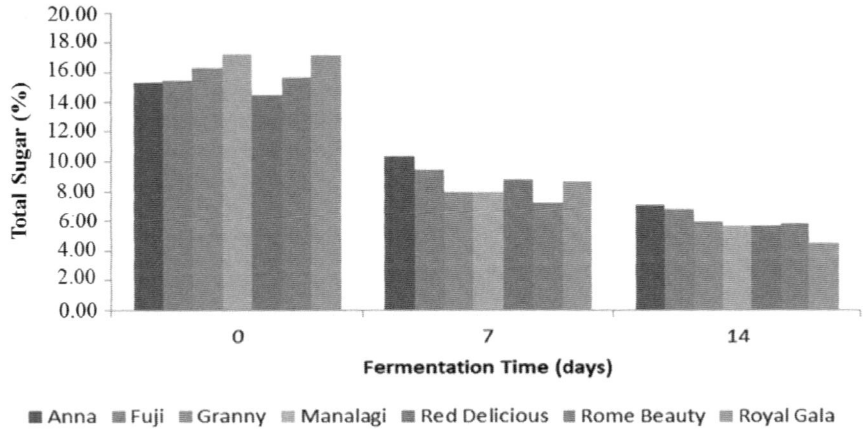

Figure 4. Total sugar content during kombucha fermentaion

3.4 Total phenols and antioxidant activity

Figure 5 and Figure 6 show total phenol and antioxidant activity during kombucha fermentation, respectively. Total phenols content and antioxidant activity each varieties of apple kombucha increased during fermentation. Yeasts and bacteria both have the ability to produces the enzyme that convert polyphenolic complex into its less complex phenolic components during fermentation. During fermentation, the phenols content increases because of biotransformation. Biotransformation is a process that changes particular functional group into its composing substances mediated by enzymes [12]. Biotransformation utilizes enzymes from a certain plant to increase particular biological activity [13]. According to Bhanja et al. [14], phenol is one of many organic compounds that have antioxidant activity, therefore when more phenols are produced as metabolites, the higher the antioxidant activity.

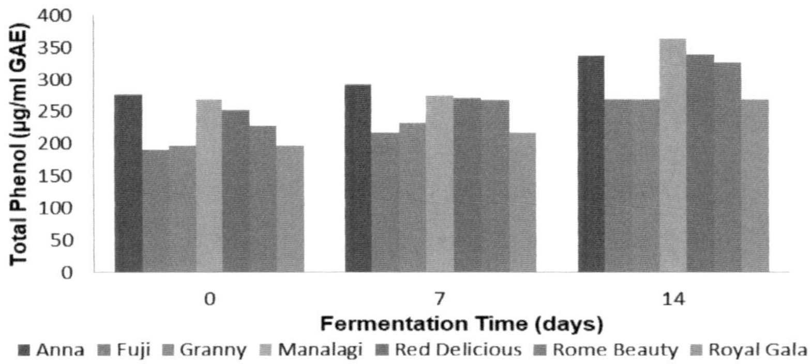

Figure 5. Total Phenols content during kombucha fermentation

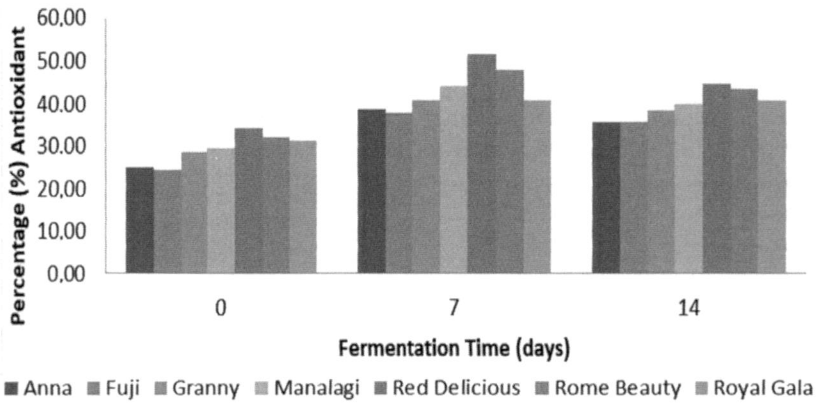

Figure 6. Antioxidant activity during apple kombucha fermentation

3.5 Antibacterial activity against Staphylococcus aureus and E.coli

Figure 7 and Figure 8 show antibacterial activity against *Staphylococcus aureus and E.coli* during kombucha fermentation, respectively. During fermentation, the antibacterial activities of apple kombucha against *Staphylococcus aureus* increased. The bigger inhibition zone is likely caused by the organic acids produced during kombucha fermentation. Kombucha itself has already demonstrated its antibacterial activity against pathogenic microorganism [15]. Acetic acid is lipophilic that makes it penetrates easily into the gram positive bacteria cell compared to gram negative ones. Antimicrobial activity is also determined by the value of pKa which defined as the amount of un-dissociated acid molecule in water. The lower the pKa, the higher the antimicrobial activity. Naidu and Clemens [16] studied that acetic acid dissosiates and releases protons resulting lower the pH value. These protons released by the acetic acid disturb the membrane cell function of the bacteria. Acids cause enzyme denaturation and interfere the membrane cell's permeability. The antibacterial activity of kombucha is highly caused by the presence of acetic acid produced from the glucose metabolism by bacteria and yeast, and also polyphenols that did not undergo biotransformation during the fermentation [15].

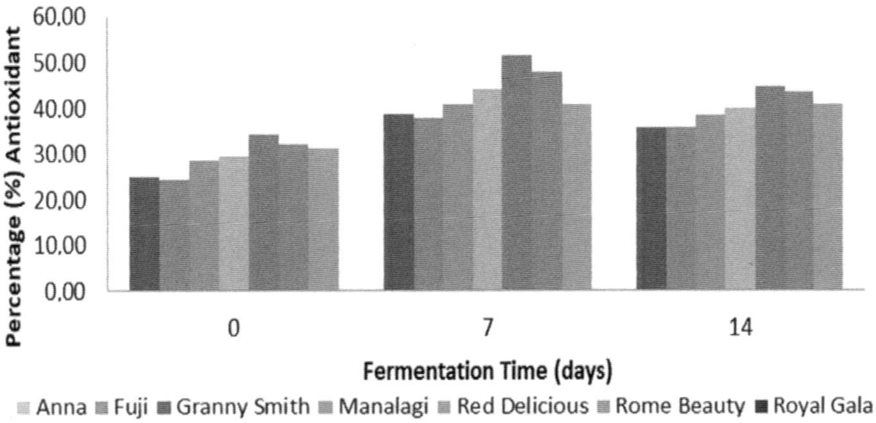

Figure 7. Antibacterial activity of apple kombucha against *Staphylococcus aureus*

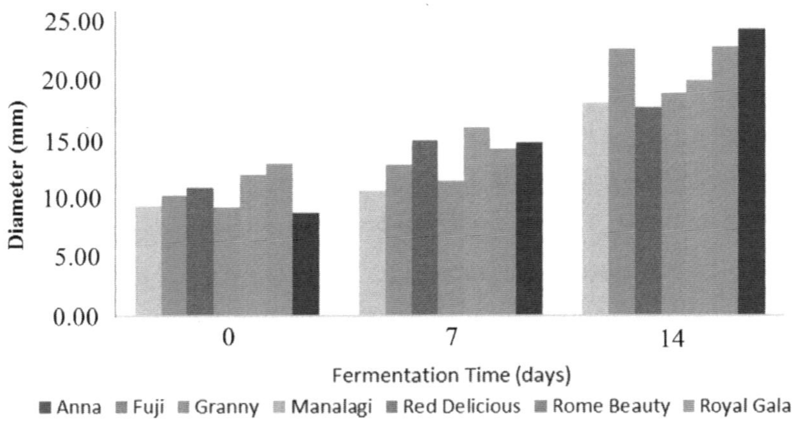

Figure 8. Antibacterial activity of apple kombucha against *E.coli* during kombucha fermentation

3.6 Taste and aroma

The acceptance on the taste and aroma of all the apple kombucha from various types of apples was different (Figure 9 and Figure 10). The result from statistical analysis using Friedman's test showed that the variety of apples gave significant effect (α=0.05) on the degree of likeness in respect to the taste of apple kombucha. The Non-Parametric test showed significant difference between the following varieties; anna-manalagi, anna-royal gala, fuji-manalagi, granny smith-manalagi, granny smith-red delicious, granny smith-royal gala, and granny smith-rome beauty. According to Anugrah [17], kombucha became more acidic as the longer the fermentation. This is most likely caused by the sucrose metabolism which produces organic acids such as acetic acid, glucoronic acid, and gluconic acid by the bacteria and yeasts. The taste of kombucha depends on the ratio of sugar and acids, the variety and amount of flavoring agent, and also vitamins. Friedman's test showed that the variety of apples gave significant effect (α=0.05) on the degree of likeness in respect to the aroma of apple kombucha. Short chain organic acids are aroma source of the apple kombucha. The longer the fermentation, the more organic acids are being produced. More organic acids enhance the aroma of kombucha, yet decrease the acceptance of the taste. According to Wood [18], khamir has the ability to produce components that are responsible for aroma and taste.

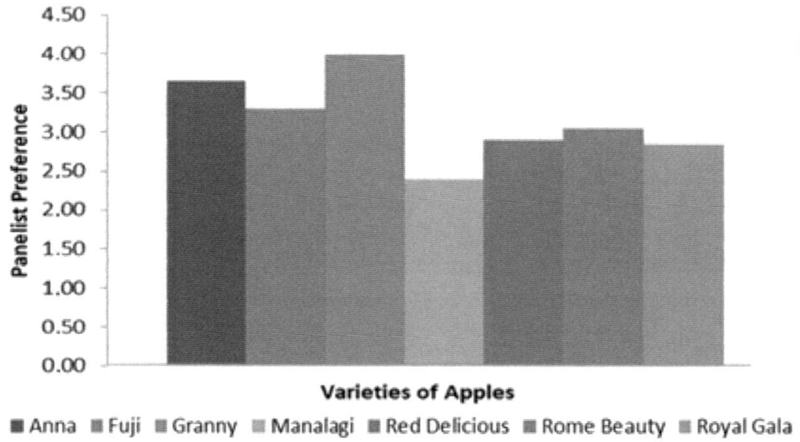

Figure 9. Panel's preference score of the taste of apple kombucha

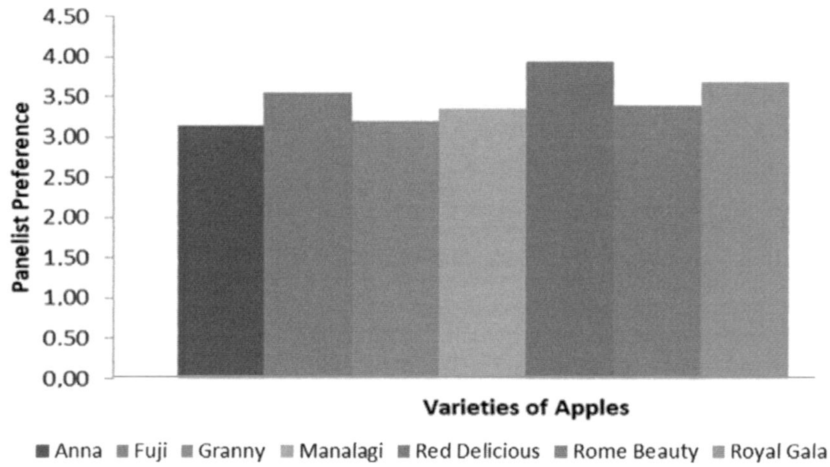

Figure 10. Panel's preference score of aroma of apple kombucha

4. Conclusion

Based on Multiple Attribute Test, Fuji apple variety is the best apple to be processed to kombucha. This apple kombucha has characteristics as follows: degree of likeness in terms of aroma 3.55, taste 3.3, total acid content of 1.33%, pH value of 2.95, antioxidant activity of 35.62 %, total sugars content of 6.74 %, total phenols content of 268. 57 µg/ml GAE, antibacterial activity against *S.aureus* of 21.30 mm, antibacterial activity against 21.20 mm.

References

[1] Watawana I M, N Jayawardena, C B Gunawardhana, dan V Y Waisundara 2015 Health, Wellness, and safety aspects of the consumption of kombucha Journal of Chemistry ID 591869
[2] Naland, H 2008 Kombucha Teh dengan Seribu Khasiat PT Agromedia Pustaka Jakarta

[3] Aditiwati and Kusnadi 2003 Kultur Campuran dan Faktor Lingkungan Mikroorganisme Yang Berperan dalam Fermentasi Tea Cider Jurnal ITB Sains dan Teknologi **35** 2 147-162
[4] Malbasa, R, Loncar, E, Vitas, J, & Canadanovic-Brunet, J M 2011 Influence of starter cultures on the antioxidant activity of kombucha beverage *Food Chem.* **127** 4 1727-1731
[5] Fardiaz 1988 Fisiologi fermentasi PAU IPB Bogor Indonesia
[6] AOAC (Association of Official Analytical Chemist) 1999 Official Methods of Analysis Washington DC
[7] Apriyanto, AD, Fardiaz, Puspitasari, NL, Sedarnawati dan Budiyanto, S 1989 Analisis pangan PAU Pangan dan Gizi IPB Bogor
[8] Strycharz, S dan K Shetty 2002 Effect of agrobacterium rhizogenes on phenolic content of menthapulegium alite clonal line for phytomedition application *Journal of Process Biochemistry* **38** 287-293
[9] Hatano, T, Edamatsu, R, Hiramitsu, M, Mori, A, Fujita, Y, Yasuhara, T, Yoshida, T and T Okuda 1989 Effect of the interaction of tannins with co-existing substances vi. effect of tannins dan related polyphenols on superoxide anion radical dan on 1,1-Diphenyl-2-Picrylhydrazyl Radical *Chem. Pharm. Bull.* **37** 2016-2021
[10] Wolf, CE and WR Gibbons 1996 Improved method of quantification of bacteriosin 1 nisin *J. App. Bacteriol.* **80** 453
[11] Rahayu, W P 2001 Penuntun praktikum penilaian organoleptik jurusan teknologi pangan dan gizi Fakultas Teknologi Pangan IPB Bogor
[12] Goh, WNA, A Rosma, B Kaur, A Fazilah, AA Karim, and R Bhat 2012 Fermentation of Black Tea Broth (Kombucha): I. Effects of Sucrose Concentration and Fermentation Time on Yield of Microbial Cellulose *International Food Research Journal* **19** 1 109-117
[13] Jayabalan, R, Subathradevi, P, Marimuthu, S, Sathishkumar, M, & Swaminathan, K 2008 Changes in free-radical scavenging ability of kombucha tea during Fermentation *Food Chemistry*. **109** 227- 234
[14] Bhanja T, Kumari A, Banerjee R 2009 Enrichment of phenolics and free radical scavenging property of wheat koji prepared with two filamentous fungi *Bioresour Technol.* **100** 112861-6
[15] Greenwalt, CJ, Ledford, RA. and Steinkraus, KH 1998 Determination and characterization of the anti-microbial activity of the fermented tea Kombucha *Lebensm Wiss Technol.* 1998; 13: 291-296
[16] Naidu, A.S and Clemens, R.A 2000 Natural Food Antimicrobial System:Probiotik CRC press. New York.
[17] Anugrah, S.T 2005 Pengembangan produk Kombucha Probiotik Berbahan Baku teh hitam (*Camelia sinensis*). Skipsi. Fakultas Teknologi Pertanian, IPB: Bogor.
[18] Wood B.J.B 1998 Microbiologi of Fermented Food 2nd ed Blackie Academy

IOP Conf. Series: Earth and Environmental Science **131** (2018) 012041 doi:10.1088/1755-1315/131/1/012041

Purification and characterization of a fibrinolytic enzyme from tempeh *bongkrek* as an alternative of thrombolytic agents

I R A Sasmita, A Sutrisno, E Zubaidah and A K Wardani

Department of Food Technology and Science, Faculty of Agricultural Technology,Universitas Brawijaya, Malang, Indonesia

E-mail: ireneratri91@gmail.com

Abstract. Tempeh is one of Indonesia's traditional foods that contain fibrinolytic enzymes. Tempeh *bongkrek* shows very strong activity among various tempeh. The fibrinolytic enzymes of *bongkrek* tempeh are obtained by steps of purification i.e, ammonium sulphate precipitation, ion exchange chromatography and gel filtration chromatography. The fibrinolytic enzymes has been successfully purified with a yield of 4.37%, specific activity of 3,361 U / mg and purification fold of 44.02. SDS PAGE analysis showed that the enzyme was purified in to single band with estimated molecular mass of 75.82 kDa. The purified enzyme has optimum pH of 7 and optimum temperature of 50°C and pH stability between pH 4 - 7 with temperature stability from 30°-50°C. The fibrinolytic activity is increased with addition of $CaCl_2$ but inhibited with $CuSO_4$, phenylmethylsulfonyl fluoride (PMSF), sodium dodecyl sulfate (SDS), and ethylenediaminetetraacetic acid (EDTA).

1. Introduction

Cardiovascular disease is a disease caused by heart and blood vessels disorders and is one of the most deadly diseases in the world. Based on the results of the report published by the World Health Organization (WHO) in 2011, an estimated 17.3 million people or 30% of the world's deaths caused by cardiovascular disease [1]. In poor and developing countries, cardiovascular disease accounts for 80% of deaths in both women and men caused by heart attacks and strokes [2]. One of the things that can lead to heart disease and stroke are the occurrence of thrombosis. Thrombosis is a disorder of blood vessels that lead to the blockage of blood vessels [3]. Abnormalities caused by thrombosis can be treated with thrombolytic therapies that can be performed via injection and orally using agents containing urokinase, tissue plasminogen activator (TPA), and streptokinase. Although widely used, these agents have the effect of immunogenic [4], a short half-life, the price is relatively very expensive[5], specificity to fibrin is reocclusion, and requires a large of therapeutic dose [6].

In addition to the above thrombolytic agents, there is an enzyme that can break up blood clots that is fibrinolytic enzyme [7]. Fibrinolytic enzyme is a protease which has fibrinolytic activity that is able to degrade fibrin clot [8]. Based on several previous studies, the activity of fibrinolytic enzyme can be found in fermented food such as *nattokinase* in *natto*, *katsuwokinase* on *Skipjack shiokara* [8];

Substilisin DJ-4 on *Doen-Jang* [9]; *Douchifibrinolytic enzyme* (DFE) on *Douchi* [5] and fibrinolytic enzyme on *tempe gembus* [10].

Since there are many variants of tempeh in Indonesia, we previously investigated the fibrinolytic activity of various tempeh that are produced in Malang and Batu, East Java using in vitro blood degradation methods (unpublished data). Among various tempeh i.e. *tempeh bungkil*, *tempeh menjes* and *tempeh bongkrek, the tempeh bongkrek* showed the highest fibrinolytic activity (2.11 FU/ml). Therefore, a more in-depth study of the purification and characterization of fibrinolytic enzymes derived from *tempeh bongkrek* is necessary. The understanding of the properties of the enzyme will open the possibilities of the enzyme to be an alternative for thrombolytic therapy for cardiovascular diseases and increase the economic value of tempeh.

2. Materials and Methods

2.1 Extraction of Enzyme from Tempeh Bongkrek

Five hundred (500) grams of tempeh sample was used and put into sterile plastic containing 500 mL of PBS buffer pH 7.4. Then the sample was homogenized using stomacher tool with medium speed for 3 min. The homogenate is centrifuged at 8000 rpm for 10 min at 4°C. The supernatant was taken as a crude enzyme extract [12].

2.2 Ammonium Sulfate Precipitation

Ammonium sulphate was then added with 80% saturation level and stirred overnight at temperature of 0-4°C. After stirred overnight, the mixture was centrifuged at a rate of 10.000 rpm for 20 min at temperature of 4°C. The precipitate was resuspended in buffer Tris-CL 50mM pH 7 and then dialyzed against 20 volumes of the same buffer for 24 h at 4°C with four buffer changes [10,13].

2.3 Ion Exchange Chromatography

The dialyzed enzyme was chromatographed on a column of DEAE-Sepharose. The sample was loaded on to a column of DEAE-Sepharose (10 cm x 1 cm) equilibrated with 50 mMTris-Cl pH 7.0. The absorbed protein solution was eluted at a flow rate of 1 ml/min with a discontinuous gradient ranging 0.1 to 1 M NaCl dissolved in same buffer. The eluted fraction was collected in 4 ml per tube. The fraction measured at λ 280 nm and fibrinolytic activity of each fraction was measured with the fibrinolytic enzyme assay using spectrophotometry method [14].

2.4 Gel Filtration Chromatography

Gel filtration chromatography was carried out using a Sephadex G-100 Column (10 cm x 1 cm). The column was equilibrated with 50mM Tris-Cl pH 7.0. The flow rate was 1 ml / min. The eluted fraction was collected in tubes with a volume of 4 ml each. The fraction was measured at λ 280 nm and fibrinolytic activity of each fraction was measured.

2.5 Fibrinolytic Activity Assay

Fibrinolytic activity was carried out according to the method described by Chang et al. [15]. Briefly, 1.4 mL of 50mMTris-Cl pH 8.0 and 0.4 mL of 0.72% (W/V) fibrinogen solution were mixed and incubated at 37°C for 5 min. Then, 0.1 mL (20 U /mL) of thrombin was added and incubated at 37° C for 10 min. Then, 0.1 mL of sample was added and incubated for an additional 60 minutes. The reaction was stopped by addition of 0.2 M trichloroacetic acid (TCA). After centrifugation (8000 rpm for 10 min), the absorbance of supernatant at 275 nm was read and recorded. One unit of fibrinolytic activity (FU) is defined as the amount of enzyme required to produce an increase in absorbance equal to 0.01 in 1 min at 275 nm [15].

2.6 Characterization of The Fibrinolytic Enzyme

To determine the effect of pH on fibrinolytic enzyme activity with 0.1 ml fibrinolytic enzyme sample incubated at 50 mM acetate buffer (pH 4.0 - 5.0), 50 mM phosphate buffer (pH 6.0 - 7.0), 50mMTris-Clbuffer (pH 8.0 - 9.0) for 1 h at 37°C. Then the enzyme activity was determined by the fibrinolytic activity assay [12].

For thermal stability measurements, 0.1 ml of fibrinolytic enzymes were suspended in 50 mMTris-Cl (pH 7.0) and incubated in a water bath for 1 h at various temperatures (30–70 °C). After incubation at each temperature, the residual fibrinolytic activity was measured [12].

To determine the effect of metal ions and inhibitors on the activity of fibrinolytic enzyme, 0.1 ml samples were incubated in Tris-Cl (pH 7.0) containing 1mM metal ions (NaCl, $MgCl_2$, $CaCl_2$, $CuSO_4$) and 1mM inhibitor (phenylmethylsulfonyl fluoride (PMSF), SDS, ethylenediaminetetraacetic acid (EDTA), and $HgCl_2$) for 1 hour at temperature of 37°C. Then the residual enzyme activity was determined by the fibrinolytic activity assay [12].

2.7 Sodium dodecyl sulfate polyacrylamide gel electrophoresis (SDS-PAGE) and zymogram analysis

SDS-PAGE analysis was carried out using 12% polyacrylamide gel. Zymogram analysis was performed as described by [16], using 12% of the SDS-PAGE gel containing 0.12% fibrinogen and 10NIH U/ ml thrombin. After electrophoresis, the gel was washed with 0.05M Tris-Cl pH 7.8 and then immersed in the same buffer for overnight at temperature of 37° C. The enzyme activity was revealed as a colorless band with a blue background, after washed with 0.11M trichloroacetic acid and staining with Coomassie blue.

3. Results and Discussion

3.1 Purification of Fibrinolytic Enzyme from Tempeh Bongkrek

The purification of fibrinolytic enzyme from tempeh *bongkrek* is summarized in Table 1. The enzyme was purified to homogeneity by a three-step procedure, resulting in a yield of 4.37%., specific activity of 3.361 U / mg and purification fold of 44.02. The SDS-PAGE and zymography analysis showed that the enzyme is purified in to a single band of fibrinolytic enzyme corresponding to a molecular mass of about 75.82 kDa (Figure 1). Previous studies have reported different molecular mass for different fibrinolytic enzymes: 70 kDa and 140 kDa for *Bacillus lichenniformis* RO3 [11], 29.93 kDa for *Bacillus subtilis* [15], and 20 kDa for *Bacillus pumilus* 2 g [10].

Table 1. Purification of Fibrinolytic enzymes.

	Volume (ml)	Total Activity (U)	Total Protein(mg)	Specific Activity (U/mg)	Yield (%)	Purification Fold
Crude Extract	500	1190	14425	0.082	100	1
Ammonium Sulfate	75	442.5	803.25	0.551	37.18	6.68
DEAE - Sepharose	28	175	127.60	1.372	14.71	16.63
Sephadex G-100	8	45.36	14.32	3.631	4.37	44.02

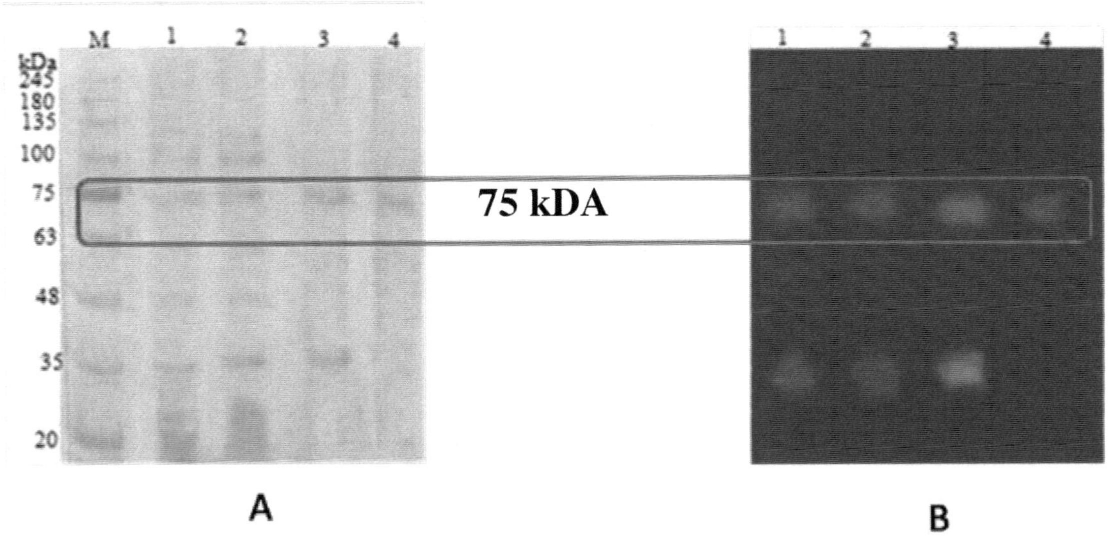

Figure 1. The SDS Page and Zymogram of fibrinolytic enzyme produced by *Tempe Bongkrek* (M) Maker; (A) SDS-PAGE; (B) Zymography result; (1) Crude extract; (2) ammonium sulfate; (3) DEAE-Sepharose; (4) Sephadex. A fibrinolytic enzyme with an estimated molecular mass of 75.82 kDa was successfully purified in to a single band of protein.

3.2 Effect of Temperature on Enzyme Activity and Stability

The effect of temperature on enzyme activity was studied by incubating the enzyme in temperature of 10 to 70°C. The enzyme has optimum temperature of 50°C and stable at temperature up to 50°C. At temperature of 60°C, its activity is reduced as much as 40%. The enzyme is stable in temperature range of 30-50°C and rapid decrease in activity was observed at temperature above 60°C (Figure 2). Previous studies have reported in natto from *Bacillus substilis* RJAS19 was optimal at 50°[15], in douchi from *Bacillus substilis* DC33 was optimal at 40°C [5], in *Gembus* from *Bacillus pumilus* was stable in temperature of less than 60°C [10].

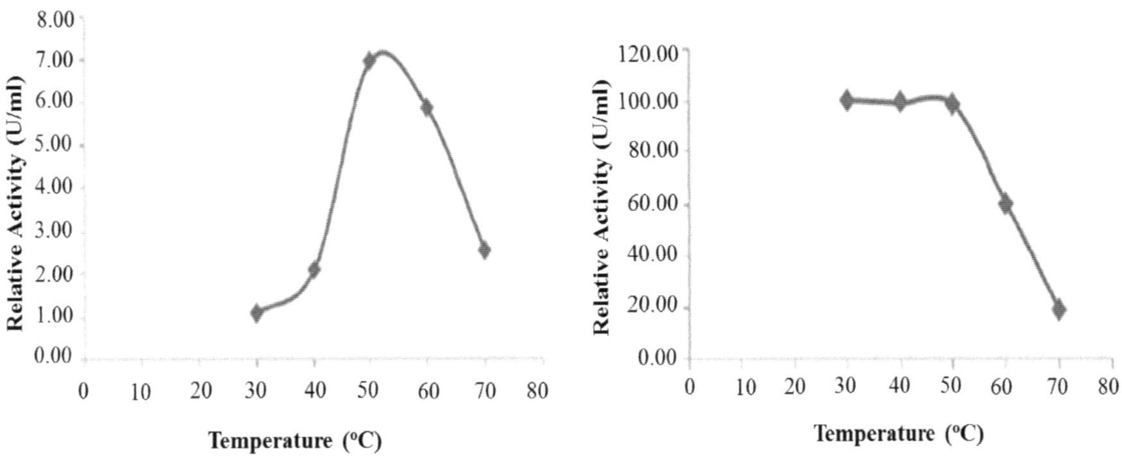

Figure 2. Effect of temperature on activity and stability of fibrinolytic enzyme from tempeh bongkrek. The enzyme has optimum temperature of 50°C and stable at temperature up to 50°C. At temperature of 60°C almost half of its activity is reduced.

3.3 Effect pH on Enzyme Activity and Stability

The effect of pH on the activity and stability of the enzyme was determined using buffers at various pH values. Fibrinolytic enzyme from *tempeh bongkrek* is optimal in pH 7 and stable in pH range of 4–7 (Figure 3). These results are similar with the results from previous reports. A fibrinolytic enzyme of *meju* fermented using *Bacillus amyloliquefaciens* MJ5-41and the one from *Bacillus pumilus isolated from* Tempeh g*embus* [10] are optimum at pH 7. The pH stability is slightly different, other enzymes were reported have stability in pH range of 5 – 9 [9, 10, 16], but the purified enzyme from this work have more acidic resistance.

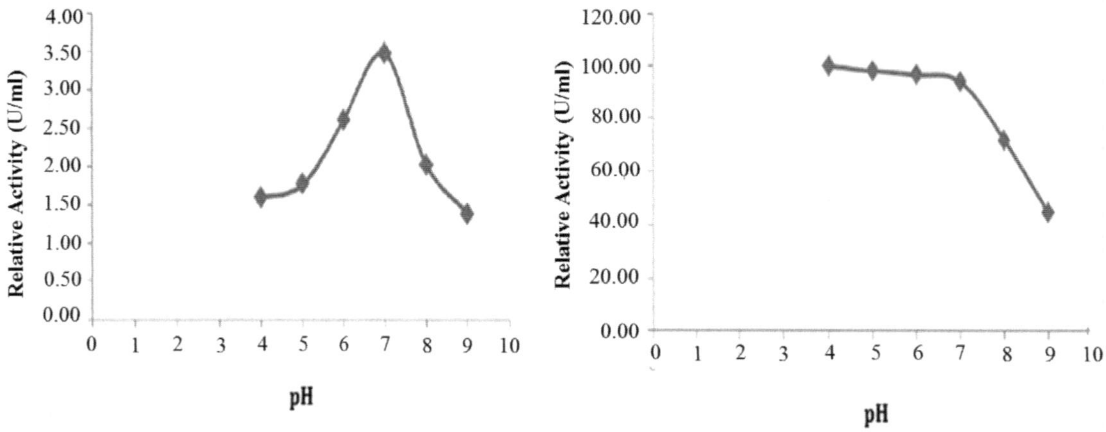

Figure 3. Effect pH on stability fibrinolytic enzymes

3.4. Effect of metal ion on the activity of fibrinolytic enzyme

Effect of various metal ions was observed (Figure 4). Calcium ion of $CaCl_2$ could increase the fibrinolytic enzyme activity. It showed that calcium ion has significant role to the enzyme. On the other hand, the Cu decreased the enzyme activity. It was reported that Cu can bind to sulfhydryl group which caused the fibrinolytic enzyme inactive [10]. In this research, the addition of PMSF decreased the activity of fibrinolytic enzyme > 60%. PMSF is specific inhibitor for serine proteases, so it is possible that fibrinolytic enzyme of tempeh *bongkrek* is also a serine protease [17], but this prediction needs further study. Serine Protease is a group of proteolytic enzymes which had active group (OH) from serine amino acid. PMSF which contained sulfonyl group would hammer on irreversible competitive basis toward this enzyme with the reaction on serine OH group in the active side [16].

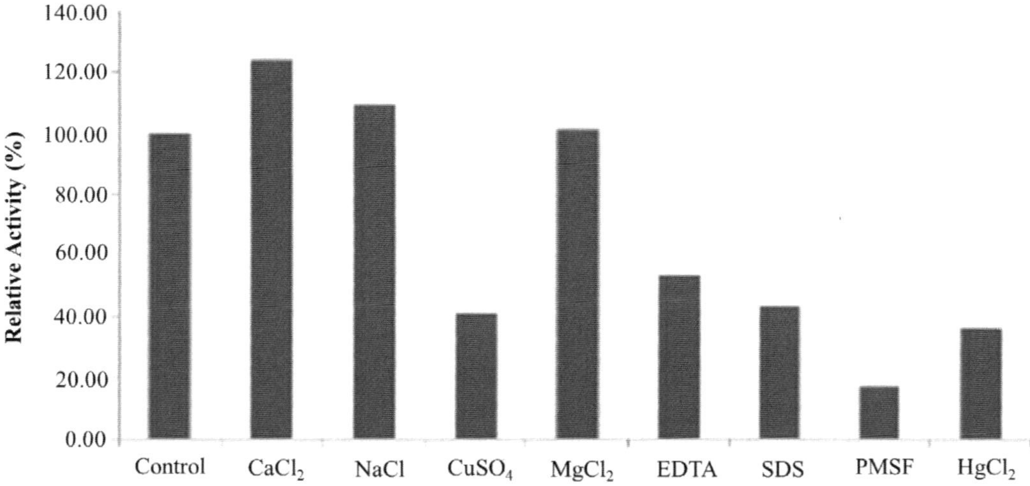

Figure 4. Effect of metal ion on fibrinolyticenzymes

4. Conclusion

The fibrinolytic enzymes have been successfully purified with a yield of 4.37%, specific activity of 3.361 U / mg and purification fold of 44.02. SDS PAGE combined with zymography analysis showed that the enzyme was purified in to single band with an estimated molecular mass of 75.82 kDa. The purified enzyme showed optimum pH of 7 and optimum temperature of 50°C and showed its pH stability between pH 4-7 with temperature stability from 30-50°C. The fibrinolytic activity is increased with addition of $CaCl_2$ but inhibited with CuSO4, phenylmethylsulfonyl fluoride (PMSF), sodium dodecyl sulfate (SDS), and ethylenediaminetetraacetic acid (EDTA). Effect of EDTA showed that metal ion is important for the activity. The fibrinolytic enzyme isolated and purified from this work are able to degrade blood clod that indicated the enzyme can be applied for thrombosis therapy related to cardiovascular disease.

Acknowledgments

The author would like to thank Fund Managers and Institutions of Education, Ministry of Finance for the financial support and Universitas Brawijaya for facilities provided throughout the duration of the study.

References

[1] Oley J S, Grace D K, Iyone E T S, Ronald I O, Lilian R, Pandelaki A J, Wulan P J K 2014 The description of the eating habits of the minahasa ethnic foods are at risk of coronary heart disease in the environment chapter 5 the village of paal 2 manado city. *J. Kedok. Kom. dan Trop.***2** 14 – 19

[2] Raju E V N and Divakar G 2013 Purification and biochemical characterization of novel fibrinolytic protease produced by mesophillic bacillus circulans gd25. *IJPSR* **4** 12 4559-4566

[3] Choi J-H, Sapkota K, Park S E, Kim S, Kim S J 2013Thrombolytic, anticoagulant and antiplatelet activities of codiase a bi-functional fibrinolytic enzyme from *codium fragile. Biochimie.* **95** 1266 – 1277

[4] Banerjee A and Chisti Y 2003 Streptokinase-A Clinicslly Useful Trombolytic Agent *J. Biotech.* 2 287 307

[5] Yuan J, Jing Y, Zhenhong Z, Yanling Y, Ling L, Shihua W 2012 Thrombolytic effect of douchi fibrinolytic enzyme from bacillus subtilis ld-8547 in vitro and in vivo *BMC Biotech.* **12** 36

[6] Madaniyah 2013 Screening of bacterial origin of the fibrinolytic soil at waste disposal idea Thesis Jember University

[7] Sumi H, Hamada H, Tsushima H, Mihara H, Muraki, H 1990 Enhancement of the fibrinolytic activity in plasma by oral administration of nattokinase. Acta. *Haematol.* **84** 3 139-43

[8] Mine Y, Wong, Ada H K, Jiang B 2005 Fibrinolytic enzymes in asia traditional fermented foods. *Food Research International* **38** 243 – 250

[9] Kim S H and Choi N S 2000 Purification and characterization of subtilisin dj-4 secreted by *bacillus* sp. strain dj-4 screened from *doen-jang Biosci. Biotechnol. Biochem*. **64**. 1722-1725.

[10] Afifah D N, Muhammad S, Dahrul S, Yanti, Maggy T S, Jeong HK 2014 Purification and characterization of a fibrinolytic enzyme from bacillus pumilus 2.g isolated from gembus, an indonesian fermented food. *Prev Nutr Food Sci*. **19** 3 213-219

[11] Sugimoto S, Fujii T, Morimiya T, Morimiya, Johdo O, Nakamura T 2007 The fibrinolytic activity of a novel protease derived from a tempeh producing fungus, *fusarium* sp. blb.*J. Biosc. Biotech. Biochem.***71** 9 2184- 2189

[12] Singh, Thangjam A K R D, Giasuddin A, Kumaraswamy J 2014 Microbial and endogenous orign of fibrinolytic activity in traditional fermented foods of northeast india. *Food Research Intern.* **55** 356-362

[13] Rhasmi B and Liny P 2013 Production and characterization of novel fibrinolytic enzyme from different soil fungal sp. *Intern. J. of Pharma and Bio Science.* **4** 3 454-463

[14] Chen Y C, Chiang Y C, Hsu F Y, Tsai L C, Cheng H L 2012 Structural modeling and further improvement in ph stability and activity of a highlyactive xylanase from an uncultured rumen fungus.*Bioresource Tech.*123 125-134

[15] Chang C T, Wang P M, HungY F, Chung Y C 2012 Purification and biochemical properties af a fibrinolytic enzyme from *Bacillus subtilis*-fermented red bean *Food Chem.* **133** 1611–1617

[16] Sajuthi, Dondin I S, Yanti, Willy P 2010 Purification and characterization of protease fibrinolytic enzymes from merang mushroom extracts. IPB Bogor Indonesia

[17] Jo H D, Lee H A, Jeong S J, Kim J H 2011Purification and characterization of a major fibrinolytic enzyme from bacillus amyloliquefaciens mj5-41 isolated from meju *Microb. Biotech.* **21** 11 1166-1167

IOP Conf. Series: Earth and Environmental Science **131** (2018) 012042 doi:10.1088/1755-1315/131/1/012042

Marine microalgae *Nannochloropsis oculata* biomass harvesting using ultrafiltration in cross-flow mode

L A Devianto[1], D N Aprilia[2], D W Indriani[2], S Sukarni[3], S H Sumarlan[4] and Y Wibisono[2,5]

[1]Department of Environmental Engineering, Faculty of Agricultural Technology, Universitas Brawijaya, Malang, Indonesia
[2]Department of Bioprocess Technology, Faculty of Agricultural Technology, Universitas Brawijaya, Malang, Indonesia
[3]Department of Mechanical Engineering, State University of Malang, Malang, Indonesia
[4]Department of Agricultural Engineering, Faculty of Agricultural Technology, Universitas Brawijaya, Malang, Indonesia
[5]MILI Water Research Institute, Malang, Indonesia

E-mail: y_wibisono@ub.ac.id

Abstract. Microalgae is a potential bioenergy source. It can grows rapidly, even it could be harvested within 7 days. Harvesting is an important part of microalgae cultivation due to the method used. It should be undamaging toward essential content of microalgae and should produces high yields of biomass. In this study, the harvesting of *Nannochloropsis oculata* was carried out using capillary ultrafiltration in cross flow mode. This study aims to test ultrafiltration membrane performance in *Nannochloropsis oculata* harvesting accompanied by *Backwash* and *Non-Backwash* modes and to analyse its total lipid content. The harvest was done under 1; 1.5; and 2 bar of trans membrane pressure. Some observed parameters were permeate flux, cell density, biomass recovery, microalgae's dry weight, yield, and total lipid content. The application of high pressure and backwashed treatment have boosted slurry production time which lead to microalgae's biomass abundance. The result showed that the best treatment of *Nannochloropsis oculata* harvesting using capillary ultrafiltration membrane in cross flow mode is under 2 bar of pressure with backwashed treatment. This is the fastest condition to produce slurry within 1800 s with the highest recovery percentage 79.50%, 16.05 x 10^6 cell/ml of post-treatment cell density, 6.8 grams of biomass' dry weight, 22.66 % of yield, and 2.52 % of total lipid content.

1. Introduction
Microalgae has been considered as one of biodiesel production raw material due to its high content of lipid and rapid growth compared to other biodiesel raw material crops [1]. Microalgae has to undergo additional treatment prior to biomass acquirement; harvesting. It is a separation process between microalgae and its medium through liquid-solid separation.

Harvesting is an essential process in microalgae cultivation due to the method used. It should not damage essential content of microalgae and should produces high yields of biomass. It is difficult to choose suitable separation method considering microalgae's tiny size of 3 to 30 µm. Membrane technology is promising methods for separation of microsize particles in aqueous solution [2], including separation of microalgae biomass from its liquid growth medium [3].

Despite its potential, membrane technology is suffered by fouling [4,5]. Although chemical approach such as modification of surface properties promotes low fouling [6], yet the physical cleaning is still effective [7]. In this work therefore, a utilization of ultrafiltration (UF) membrane accompanied by backwash treatment as physical cleaning method have been conducted in marine microalgae harvesting process. The use of UF membrane is hoped that it could produce high lipid content within the acquired dry biomass without damaging essential substances of microalgae.

2. Materials and Method

2.1. Materials

Marine microalgae species *Nannochloropsis oculata* obtained from Centre for Development of Brackish Water Aquaculture at Situbondo, cultivated in sea water sequencially in glass carboy of 3-4 L, jars of 10 L and fiber container of 500 L. Prior to microalgae cultivation, the growth medium sterilized to prevent any contaminants. Natrium thiosulfat ($Na_2S_2O_3$) was used for deactivation of microalgae solution after sterilization. Afterwards, the water was enriched to provide the culture with nutrients. In this step, aeration applied and pro analyse-grade diatomic fertilizer and Technical Growth (TG), were added. Furthermore, mass culture was conducted in a open pond.

Polysulfone (PS) UF membrane with pore size of 0.01 µm of was used for microalgae harvesting, driven by water pump Shimizu PN-125 BIT to pressurize the microalgae solution prior to the membrane modules. The membranes were used for another experiment run after splased with 0.1% of Natrium hydroxide (NaOH). For calculating cell density, a centrifuge Nesco 80-2 (8 *hole*), microscope and haemocytometer were used.

2.2. Method

2.2.1. Observed factors

There were two observed factors in filtration process: operation mode and operation pressure. Operation mode was divided by two treatments: non-backwash and backwash treatment. And three variations were used in pressure: 1 bar, 1.5 bar, and 2 bar. Backwash process was done for 10 seconds in every 10 minutes interval at 2 bar pressure.

2.2.2. Microalgae harvesting.

Readily harvested Nannochloropsis oculata was distinguished by rather deep brown appearance, 1-2 million cell/ml of cell density reached, and was cultivated for 5-6 days. Nannochloropsis oculata was harvested while in stationary phase due to higher lipid total content compared to exponential phase. In the beginning of 9 days of Nannochloropsis oculata feed fetching in the mass culture pool, it was known that the average cell density was $\pm$ 3.38 x 10^6 cell/ml.

Thirty liters of 5-6 days-growth microalgae were harvested on each treatment variation. UF membrane used had inside-out system of hollow fiber module. A cross flow mode was used instead of dead end mode due to low fouling potential. Liquid feed was pushed entering the membrane channel by using a gear pump, while the trans membrane pressure set at 1, 1.5 or 2 bar. Permeate would pass through membrane while concentrate would be pushed back to reservoir by feed flow. Filtration process produced microalgae biomass concentrate slurry and sea water permeate which then were collected using measuring cup. The volume were measured in every minute.

IOP Conf. Series: Earth and Environmental Science **131** (2018) 012042 doi:10.1088/1755-1315/131/1/012042

2.2.3. Membrane washing
UF membrane washing was done in every end of harvesting process using ± 20 liters of 0.1% NaOH which was continually circulated every 10 minutes at 2 bar. It then rinsed using distilled water for 10 minutes.

2.2.4. Membrane performance testing
Membrane performance was tested using flux by measuring permeate volume every minute per unit area of membrane.

2.3. Data Analysis
Data analysis was conducted by measuring cell density, biomass recovery percentage, dry weight, yield, and fatty acid content.

3. Results and Discussion

3.1. Water flux test
One micron-filtered tap water was used in this test, to check the performance of newly used PS membranes. The test was carried out every minute for 30 minutes with three pressure variations: 1 bar, 1.5 bar, and 2 bar. The water flux test result showed a stable flux flow which meant there was no fouling found. At 1 bar, 1.5 bar, and 2 bar of pressure, the flux value were stable on 0.018, 0.030, and 0.037 L/m^{2}s.

3.2. Backwash effects
Flux value test was carried out in order to determine the best treatment to generate cross flow permeate flux on UF membrane. *Nannochloropsis oculata* backwash harvesting was done first followed by non-backwash one. Backwash was done to reduce biomass fouling effect which might decrease cross flow permeate flux. Besides, flux value showed membrane capability to flow permeate through UF membrane in certain time unit. The tight membrane pores were supposed to optimally hold microalgae biomass. Backwash efficiency in minimizing flux decline caused by fouling were affected by several factors such as, pressure, interval, and backwash duration.

Different operation pressure used in microalgae harvesting has caused filtration time difference, 2,880 seconds, 2,040 seconds, and 1,800 seconds at 1 bar, 1.5 bar, and 2 bar respectively. Given that condition, backwashing was done differently for each pressure, five times of backwashing at 1 bar and three times for 1.5 and 2 bar.

The result showed that the fastest and most stable process was occurred at 2 bar of pressure. Meanwhile, at 1 bar, the slowest one has occurred, additionally, at the end of process the flux was sharply down. Overall, fluctuated fluxes were always found during all three treatments. Nevertheless, at 2 bar, the process occurred faster and more efficient. The flux value at 2 bar also indicated that the best filtration process has occurred among three variations. Backwashing effect can be seen on Figure 1.

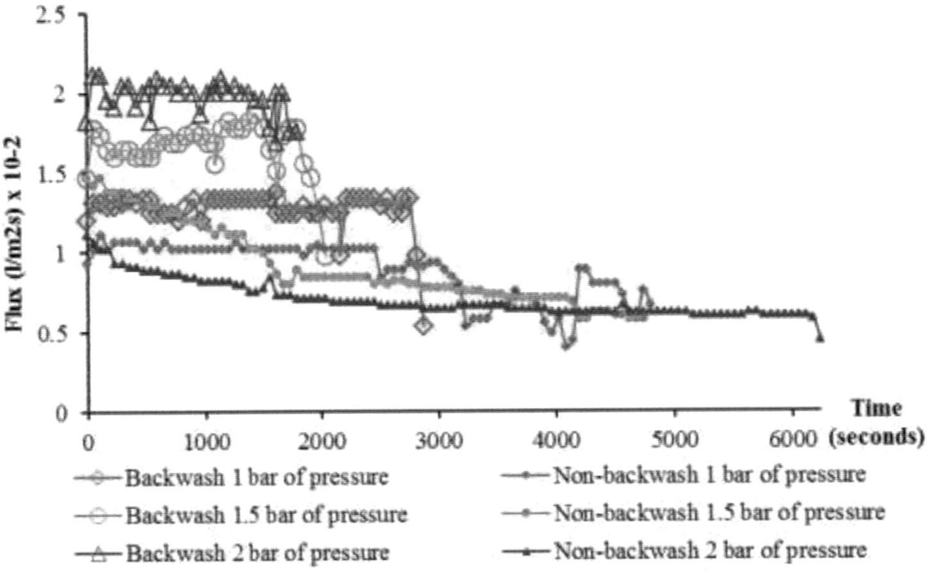

Figure 1. Effect of backwashing toward flux behaviour

It was known that backwashed process at 2 bar of pressure considered to be the best treatment, followed by 1.5 bar and 1 bar respectively using the same method. Besides offered faster process, backwashed operation also showed steady flux compared to non-backwashed one which showed declined trend caused by accumulated fouling inside the membrane. As for non-backwashed harvesting, the highest flux value was found at 1.5 bar of pressure. Unfortunately, it could not keep the stability. Nevertheless, 1.5 bar of pressure was proved to be the fastest harvesting pressure among others.

Short backwash interval has caused higher flux and flow in every filtration cycle. This indicated that formed fouling was better treated frequently by backwash cycle [8]. Ninety percent of permeate flux could be kept by backwashing for 10 seconds at 1 bar of pressure and 20 minutes interval. The result was still lower than was by 10 minutes backwashing, which was 95%.

However, Bhave et al. [9] shows that harvesting with two-step membrane filtration of *Nannochloropsis sp.* 2 g/L-150 g/L in batch system with cross flow using microfiltration have constant TMP at 2.3-25.6 Psi at 2.5 hour/cycle with flux 35-684 L/m^{2}h . Zhang et al. [10] using membrane ultrafiltration in batch system with cross flow mode harvest *Scenedesmus quadricauda* with constant TMP at 34.5 kPa at 30 min/cycle with flux 45 L/m^{2}h.

3.3. *Cell density during backwash and non-backwash modes*

In backwashed treatment (Figure 2), cell density measurement was done once at the start of feeding and five times at post-backwash. On the 1 bar pressure harvesting, the initial cell density was 3.31 x 10^6 cell/ml and 7.37 x 10^6 cell/ml; 8.46 x 10^6 cell /ml; 8.99 x 10^6 cell /ml; 10.7 x 10^6 cell /ml; and 14.31 x 10^6 cell /ml at the post-backwash. Backwashing was occurred five times at the 1 bar pressure due to the longer harvesting time. At 1.5 bar, backwashing was done three times and resulted: 8.22 x 10^6 cell/ml; 11.59 x 10^6 cell/ml; and 14.93 x 10^6 cell/ml accompanied by initial measurement which was 3.56 x 10^6 cell/ml. While at 2 bar, backwashing was also done three times and resulted: 11.25 x 10^6 cell /ml; 13.52 x 10^6 cell /ml; 16.05 x 10^6 cell /ml, followed by the initial cell density: 3.29 x 10^6 cell/ml.

Harvesting process at 1.5 bar and 2 bar was completed faster compared to 1 bar process so that there were only three backwashing. At the 2 bar process, cells would become dense faster than those at 1 and 1.5 bar. At the first backwashing, the cell density was 11.25 x 10^6 cell/ml and sharply increased at the last which reached 16.05 x 10^6 cell/ml. Cell density at the 1.5 bar process was also relatively

IOP Conf. Series: Earth and Environmental Science **131** (2018) 012042 doi:10.1088/1755-1315/131/1/012042

high, the first one was 8.22 x 10^6 cell/ml and the last was 14.93 x 10^6 cell/ml. The correlation among flux, backwashing, and cell density at the single pressure can be observed in figure 2.

In non-backwashed treatment cell density measurement was done before and after harvesting. Initial cell density at 1; 1.5; and 2 bar were 2.74 x 10^6 cell/ml; 2.72x 10^6 cell /ml; 2.41 x 10^6 cell /ml respectively and the post-harvest one were 7.924 x 10^6 cell/ml; 8.521 x 10^6 cell/ml; 9.43 x 10^6 cell/m also respectively.

Non-backwashed process cell density was concluded to be relatively low. This was caused by the microalgae had started to enter their death phase considering the fact that the harvesting process took a lot of time. Backwashed harvesting was done before the non-backwashed one. High temperature on the tank 35°C caused by improper water circulation, killed the microalgae faster.

Although the cell density at 2 bar treatment was higher, it was feared that microalgae cell would be broken in enormous number due to the high pressure. One of the aim of this research is to achieve maximum yield of total lipid extracted from dried microalgae. In this study, the number of broken cell analysis was not carried out. The amount of broken cells were rising as the increasing TMP. Therefore, maximum TMP 1.5 bar will be used in future study. High temperature would inhibit microalgae's growth and eventually leads to their death.

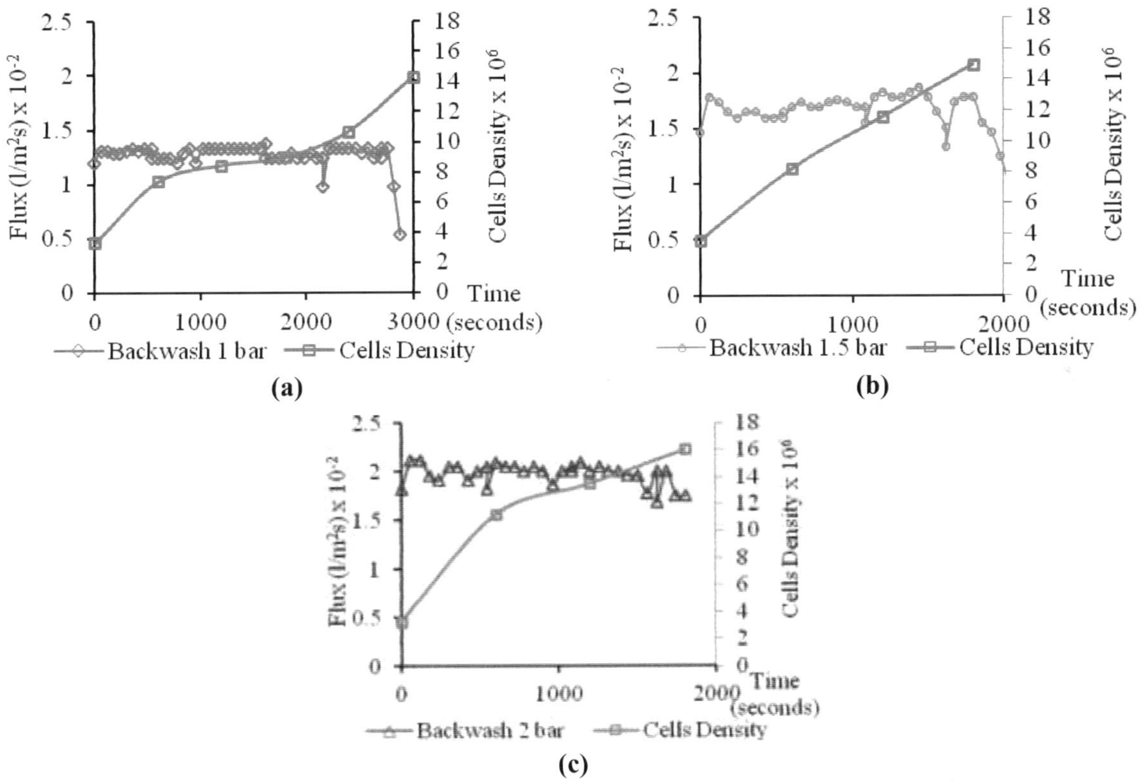

Figure 2. Flux, backwashing, and cell density at different TMP: (a) 1 bar; (b) 1.5 bar; (c) 2 bar.

3.4. Biomass recovery, dry weight and harvesting yield

Biomass recovery percentage describes the amount of biomass which has been successfully acquired at the end of harvesting process. At 1; 1.5; and 2 bar backwashed treatment, 76.18%; 76.15%; and 79.5% of pre-harvest product has been successfully retrieved, respectively. Highest recovery percentage at 2 bar process resembles the high cell density which was 16.05 x 10^6 cell/ml. On the other hand, non-backwashed treatment generated lower recovery percentage compared to backwashed one. In consecutive order, the recovery percentage at 1; 1.5; and 2 bar non-backwashed treatment were

65.42 %; 68.16%; and 74.44%. The recovery percentage difference between backwashed and non-backwashed process is proportional to cell density (slurry). However, Zhang et al. [10] in the previous study shows results that harvesting can be achieve up to 98% recovery of *Scenedesmus quadricauda* with backwash and membrane cleaning using 2% NaOH, 0.5% citric acid and 200 mg/L NaOCl, for 1 h at 20 °C [10].

Dry weight described as the final steady biomass weight after being put in the 80° C oven for 24 hours. At 1; 1.5; and 2 bar backwashed treatment, the dry weight were: 5.78 gram; 6.33 gram; and 6.8 gram respectively. Highest dry weight at 2 bar process resembles the recovery percentage which was 79.50 %. Biomass' dry weight at backwashed process was affected by the pressure that was used which might affect slurry's weight at the end of harvesting.

Non-backwashed treatment resulted lower dry weight compared to backwashed one. In consecutive order, the recovery percentage at 1; 1.5; and 2 bar non-backwashed treatment was 3.83 gram; 4.15 gram; and 4.79 gram. This relative low dry weight might have been caused by the microalgae had started to enter their death phase so that there was not much slurry to be produced which might affect the recovery percentage that was going to be used in dry weight calculation later on.

Harvest yield defines the amount of acquired dry weight each liter of harvested microalgae. Backwashed treatment yield was higher than non-backwashed one, they were: 19.26% at 1 bar pressure; 21.1 % at 1.5 bar; and 22.66% at 2 bar. %. Yield value at backwashed process was affected by the pressure that was used which might affect slurry's weight and might affect its dry weight as well. As for non-backwashed treatment, in consecutive order, the yield value at 1; 1.5; and 2 bar non-backwashed treatment 12.76 %; 13.83 %; 15.96 %. The lower yield might have been caused by the microalgae had started to enter their death phase so that there was not much slurry to be produced.

3.5. Lipid content

Total lipid content resembles the amount of fatty acid content in the microalgae as biofuel's raw material. The sampling for this test was done at the end of harvesting process by extracting 2 grams of *Nannochloropsis oculata* biomass using soxhlet.

Extracting result showed that total lipid content was 2.52%. This percentage is quite low since according to previous study that was done by another researcher that *Nannochloropsis sp* is one species of microalgae which has considerable high total lipid content ranging from 37% to 60% of its dry weight, higher than other microbial strain [11]. This phenomenon might have been caused by lipid content dependence on harvesting time and extraction method. Besides, 2 bar of pressure seemed too high for microalgae so that they would be collapse and leak their cytoplasm which contains the fatty acid itself. Therefore, there was only little amount of available lipid to be extracted.

4. Conclusion

Nannochloropsis oculata harvesting using UF membrane in cross flow mode shows that backwashed treatment under 2 bar of pressure would do the fastest harvesting process which is 1,800 seconds. Backwashing was done three times for 10 seconds every 10 minutes. At this particular condition, 79.50% of recovery percentage, 6.8 grams of biomass' dry weight, and 22.66 % of yield were successfully acquired. But there was only 2.52 % of total lipid content which is relatively low.

References

[1] Wibisono Y 2017 Biomaterial and Bioproduct UB Press Malang

[2] Wibisono Y 2014 Two-phase flow for fouling control in membranes University of Twente Enschede Twente

[3] Bilad M R, Arafat H A, Vankelecom I F J 2014 Membrane technology in microalgae cultivation and harvesting: A review *Biotechnol. Adv.* **32** 7 1283-1300

[4] Wibisono Y, Nugroho W A, Chung T W 2014 Dry degumming of corn-oil for biodiesel using a tubular ceramic membrane *Procedia Chemistry* **9** 210-219

[5] Wibisono Y, Widodo S 2015 Concentration boundary layer in membrane degumming: a CFD

model and neural network approach *Procedia Environ. Sci.* **28** 224-33

[6] Wibisono Y, Yandi W, Golabi M, Nugraha R, Cornelissen E R, Kemperman A J B, Ederth T, Nijmeijer K 2015 Hydrogel-coated feed spacers in two-phase flow cleaning in spiral wound membrane elements: a novel platform for eco-friendly biofouling mitigation *Water Res.* **71** 171-86

[7] Wibisono Y, Ahmad F, Cornelissen E R, Kemperman A J B, Nijmeijer K 2016 Dominant factors controlling the efficiency of two-phase flow cleaning in spiral-wound membrane elements *Desalin. Water Treat.* **57** 38 17625-36

[8] Mulder M 1996 *Basic Principles of Membrane Technology* 2nd Dd. Kluwer Academic Publisher Dordrecht

[9] Bhave R, Kuritz T, Powell L, Adcock D 2012 Membrane-based energy efficient dewatering of microalgae in biofuels production and recovery of value added co-product *Environ. Sci. Technol.* **46** 10 5599-5606

[10] Zhang X, Hu Q, Sommerfeld M, Puruhito E, Chen Y 2010 Harvesting algal biomass for biofuels using ultrafiltration membranes *Bioresour Technol* **101** 14 5297-5304

[11] Xiao-Nian M, Tian-Peng C, Bo Y, Jin L, Feng C 2016 Lipid production from *Nannochloropsis Mar. Drugs* **14** 4 61

Characterization of thermostable cellulase produced by *Bacillus* strains isolated from solid waste of carrageenan

N P Listyaningrum[1,2], A Sutrisno[1] and A K Wardani[1]

[1]Department of Biotechnology, Faculty of Agricultural Technology, Universitas Brawijaya, Malang, Indonesia
[2]Fisheries Product Processing Technique, Marine and Fisheries Polytechnic, Sidoarjo, Indonesia

E-mail: agustinwardani@ub.ac.id, nikenprawesti@apsidoarjo.ac.id

Abstract. Cellulase-producing bacteria was isolated from solid waste of carrageenan and identified as *Bacillus licheniformis* C55 by 16S rRNA sequencing. The optimum condition for cellulase production was obtained at pH and temperature of 8.0 and 50°C, respectively in a medium containing glucose as carbon source and 1.0% carboxymethyl cellulose (CMC) to stimulate the cellulase production. Most remarkably, the enzyme retained its relative activity over 50% after incubation at 50°C for 90 minutes. Substrate specificity suggested that the enzyme is an endoglucanase. The molecular mass of *Bacillus licheniformis* C55 crude cellulase was found about 18 kDa by SDS-PAGE analysis. This thermostable enzyme would facilitate development of more efficient and cost-effective forms of the process to convert lignocellulosic biomass into high-value products.

1. Introduction

Cellulose is the most abundant compound in the world and is a renewable source that can be easily utilized by humans. Cellulase is an enzyme that hydrolyzes the cellulose bond β-1,4-glycosidic and is generally divided into endoglucanase and selobiohydrolase. Cellulase has potential in several industries such as food industry, textile and laundry industry, pulp and paper industry, as well as agricultural industy [1]. Thermostable cellulase produced by thermophilic microorganisms are useful for improvement the efficiency of laundry detergents, paper removal and improving pulp drainage, because most of these processes operate at high temperatures.

Cellulose is also found in solid waste of carrageenan processing. Seaweed from *Euchemacottonii* types is a raw material for the manufacture of carrageenan flour. Based on data of Directorate General of Aquaculture Ministry of Marine Affairs and Fisheries (MMAF) Indonesia, seaweed production in Indonesia in 2010 reached 3.91 million tons and increased to 10.23 million tons in 2014. In Indonesia, carrageenan processing industries are only able to process 20% from the total production of seaweed (540,000 dry tons per year), it will generate solid waste of 10,800 tons per year. This leads to very serious pollution problems because the wastes contain cellulose which is very difficult to be degraded. Cellulose is commonly degraded by cellulose which is produced by several microorganisms, commonly by bacteria and fungi. Bacteria which have high growth rate as compared to fungi have good potential to be used in cellulase production. For this application, the utilization of thermophilic cellulolytic microbes is more advantageous than the mesophilic one because the cellulase enzyme produced remains stable at high temperature and still has optimum activity [2]. Thermostable cellulase enzymes with high stability can be obtained by isolating from thermophilic bacteria. Most of these bacteria are isolated from soil, compost, animal waste or haystack and sugar cane. Isolation of thermophilic cellulase-producing bacteria isolated from waste of carrageenan processing has not been studied further. This aim of the present study is to isolate and identify cellulase-producing bacteria from solid waste of carrageenan processing and to characterize the thermostable cellulose produced by bacteria.

2. Materials and Methods

2.1. Materials

Waste of carrageenan processing was obtained from PT. Kappa Carrageenan Nusantara, Pasuruan, East Java. Waste was obtained aseptically using a sterile spatula, to prevent exogenous microbial contamination during sampling. Samples were brought to the laboratory in polypropylene tubes and stored at temperature of -20°C until analyzed. Carboxy Methyl Cellulose (CMC), Tryptone Soya Agar (TSA), Tryptone Soya Broth (TSB) and glucose were used as growth media.

2.2. Enrichment and isolation of thermophilic cellulose-degrading bacteria from carrageenan waste

The growth media used for enrichment of cellulose-producing bacteria from waste of carrageenan processing were TSB and 1% CMC sodium salt as the carbon source. Before autoclaved, the pH of growth medium was adjusted to 8.0 using NaOH 1 M and then 100 g of culture was inoculated and incubated at 50°C for 72 hours in a shaker waterbath. After enrichment, the mixture of culture was used to isolate thermophilic cellulose-producing bacteria by dilution method using sterile NaCl (0.85%). One hundred (100) µL of each dilution was streaked on media TSA containing 1% CMC sodium salt pH 8.0 and incubated at 50°C for 24 hours. Screening of cellulase-producing bacteria was performed on TSA pH 8.0 containing 1% CMC sodium salt and incubated at 50°C for overnight, then flooded with congo red, continued by washing with 0.1% NaCl. Isolates that produce the largest clear zone were considered as potential isolates.

2.3. Molecular identification

Potential cellulase-producing isolates were identified morphologically, physiologically and biochemically and confirmed of gene sequencing 16S rRNA. Extraction and purification of DNA was performed using bacteria kit *(The Presto™ Mini kit gDNA Bacteria geneaid).* Extraction and purification of DNA in accordance with manufacturer protocol. Primer used for sequencing is a universal primer of 16S rRNA27F/1492R as follows: forward primer 27F:AGAGTTTGATCMTGGCTCAG and reverse primer 1492R:TACGGYTACCTTGTTAACGACTT. The PCR condition of 35 cycles covering of denaturation at 95°C for 10 seconds and annealing at 54°C for 30 seconds then elongates for 1 minute at 72°C. The similarity search of sequence was carried out using BLAST on the site http://blast.ncbi.nlm.nih.gov/Blast.cgi, to determine the level of homology species of bacteria selected by bacteria in *Genebank.* The phylogenetic tree was designed using the sequences ratio of 16S-rRNA of other bacteria in the database tracking program *Basic Local Alignment Search Tool* (BLAST) on the site http://blast.ncbi.nlm.nih.gov/Blast.cgi. Alignment was carried out using ClustalW program. The phylogenetic tree was constructed using neighbor-joining method by MEGA7 program.

2.4. Enzyme production

Isolates C55 were grown in a liquid medium TSB contain 1% CMC sodium salt. Isolates that were inoculated as much as two-ose into 10 ml of medium of liquid TSB-CMC and incubated at 50°C in *waterbath shaker*, 120 rpm, for 24 hours. After a phase of exponential growth, the cells from cultures were inoculated back into 90 mL of liquid medium TSB contain 1% CMC sodium salt, then incubated in waterbath shaker at 50° with 120 rpm agitation. Cellulase activity and *Optical Density* ($OD_{600\ nm}$) were measured every 3 hours for 24 hours incubation. The cellulase was harvested at the optimum

IOP Conf. Series: Earth and Environmental Science **131** (2018) 012043 doi:10.1088/1755-1315/131/1/012043

incubation time with the value of the highest cellulase activity. The crude extract of the cellulase enzyme was separated from bacterial pellets with 10000 × g for 15 minutes at 4°C.

2.5. Cellulase activity assay

The measurement of cellulase activity was performed by incubating the mixture of 100 μL enzyme and 100 μL of 1% (b/v) CMC in 50 mM Tris-HCl (pH 8.0) at 50°C for 30 minutes. The reaction was stopped by adding 3,5-dinitrosalicylic acid solution, boiled for 15 minutes, then cooled in ice for color stabilization. Optical absorbance was measured at 540 nm and the amount of reducing free sugar (glucose equivalent) was estimated on the glucose standard curve. One unit (U) of enzyme activity was defined as the amount of enzyme that releases 1 μmol glucose per minute under the test conditions. The control of enzyme and appropriate substrate were included in all tests. Each experiment was repeated three times and the mean value plotted for each experiment.

2.6. Substrate specificity

The specificity of the crude cellulase substrate was determined by testing different substrates, namely Carboxy Methyl Cellulose (CMC), filter paper No.1 (Whatman), carrageenan flour, carrageenan waste. Cellulase activity was measured by incubating 200 μL diluted enzyme solution (100 μL enzyme + 100 μL substrates 1% in 50 mM Tris-HCl, pH 9.0) at 50°C for 30 minutes. The reaction was stopped by adding 3,5-dinitrosalicylic acid solution, boiled for 15 minutes, then cooled in ice for color stabilization. The amount of reducing sugar produced was measured by the method of DNS [3].

2.7. Effect of temperature on activity and stability of crude enzyme

The optimum temperature of the crude enzyme activity was determined by incubating the mixture of enzyme reactions as much as 100 μL added with 100 mL of 1% (b/v) CMC for 30 minutes at various temperatures between 30°C to 90°C. The stability of temperature was done by incubating crude enzyme on buffer Tris-HCl 50 mM (pH 8) at the optimum temperature for 15, 30, 45, 60, 90, 120 and 240 minutes and residual activity was assayed by standard DNS method [3].

2.8. Effect of pH on activity and stability of crude enzyme

The optimum pH of crude enzymes secreted by isolates was estimated range from pH 4.0 - 10.0 by using a variety of different solvents. Enzyme activity test performed by incubating the reaction mixture at a temperature of 50°C for 30 minutes. As many as 100 μL of enzyme was added with 100 μL 1% (b/v) CMC dissolved in 50 mM acetate buffer (pH 4, 5, 6); 50 mM phosphate buffer (pH 6, 7, 8); 50 mM Tris-HCl buffer (pH 7 and 8); 50 mM glycine buffer-NaOH (pH 8, 9, 10). The stability of the pH of enzyme were determined by incubating enzyme as much as 100 μL added with 100 μL 1% (b/v) CMC dissolved in buffer mentioned above for one hour at the temperature of 50°C and residual activity was assayed by the DNS standard methods [3].

2.9. SDS-PAGE and zymogram analysis

The molecular weight of the crude cellulase was determined by Sodium Dodecyl Sulfate Polyacrylamide Gel Electrophoresis (SDS-PAGE) with Bio-Rad system using 4% stacking and 12% resolving gel. After electrophoresis, the gels were stained with 0.1% (w/v) Coomassie blue, 30% (v/v) methanol, and 10% (v/v) acetic acid. Zymograms of cellulase activity performed in 12% SDS-polyacrylamide gels were polymerized with 0.2% (w/v) carboxymethyl cellulose according to a previously described protocol. Renaturation of proteins following electrophoresis was performed by incubating the gels in 10 mM Tris–HCl buffer (pH 8.0), containing 1% (w/v) Triton X-100, and the incubation period was 2 h at 50°C. To identify protein bands with cellulase activity, the gels were stained with 0.2% (w/v) congo red for 30 minutes and washed with 1 M NaCl until cellulase bands became identified.

3. Results and Discussion

3.1 Microorganisms

The cellulase-producing isolate was designated as isolate C55. On the basis of the morphological, physiological, and biochemical characteristics, C55 is gram-positive and endospore-forming *Bacillus* with catalase, which grows in both aerobic and anaerobic environments. The 16S rRNA gene sequencing analysis evidenced that C55 showed high homology (100%) with strain of *Bacillus licheniformis* (Figure 1). Based on the evolution range and phylogenetic tree, this C55 strain was identified as *Bacillus licheniformis* and designated as *Bacillus licheniformis* C55 (accession of GenBank number KY392939.1).

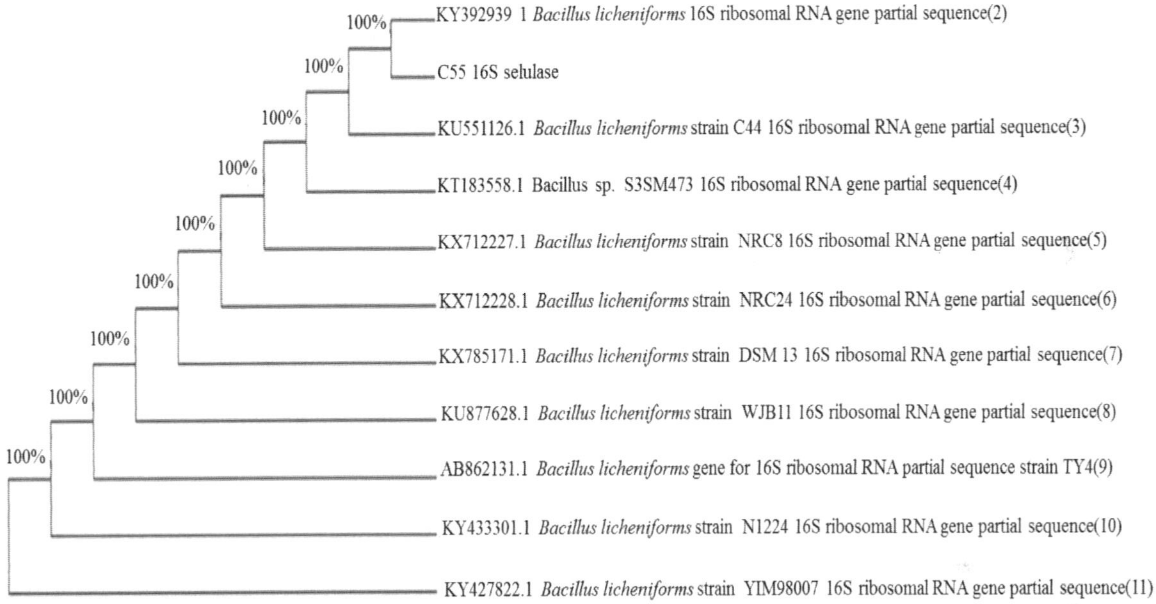

Figure 1. Phylogenetic tree of isolate C55 16S rRNA gene sequence with other *Bacillus* species/strain

3.2 The effect of temperature and pH on enzyme activity and stability

The effect of temperature on the activity and stability of cellulase was determined on various temperature ranges from 30 and 70°C. The optimum temperature for cellulase activity of isolate C55 were found at 50°C in pH 9.0 (Figure 2A). Thermostability studies revealed that crude enzyme was 100% stable in optimum temperature at 50°C for 30 minutes and it retained relative activity more than 50% for 90 minutes (Figure 2B). The optimal cellulase temperature and thermostability which was obtained by isolate C55 was the same as obtained by *B. licheniformis* WBS1 and *Bacillus* sp. WBS2 [4], *B. cereus* JD0404 [5], *B. amyloliquefaciens* DL-3 [6] and *B. subtilis* subsp. *Subtilis* A-53 [7] and *Bacillus* sp. HSH-810 [8].

The effect of pH on crude cellulase activity produced by isolate C55 was examined at various pHs ranged from pH 4.0 until pH 10.0. The optimal pH for cellulase activity was 8.0 and it exhibited more than 60% relative activity between pH 7.0 and 8.0 (Figure 2C). The Optimum pH of crude cellulase from isolate C55 was higher than cellulase produced by *B. amyloliquefaciens* DL-3 [6], *Bacillus sphaericus* JSI [9] and *B. subtilis* subsp. *Subtilis* A-53 [7] which exhibited optimum activity between pH 6.0 and 7.5. The pH stability studies showed that crude cellulase from isolate C55 was active in the range of pH between 7.0 and 8.0. Crude cellulase was 100% stable in pH 8.0 and it retained 50% of its original activity even at pH 7.0 (Figure 2D). The pH stability of crude enzyme of cellulase from isolate C55 was the same with cellulase produced by *Bacillus spherical* JS1 [9] and *Bacillus* sp. HSH-810 [7] which were stable at pH 8.0.

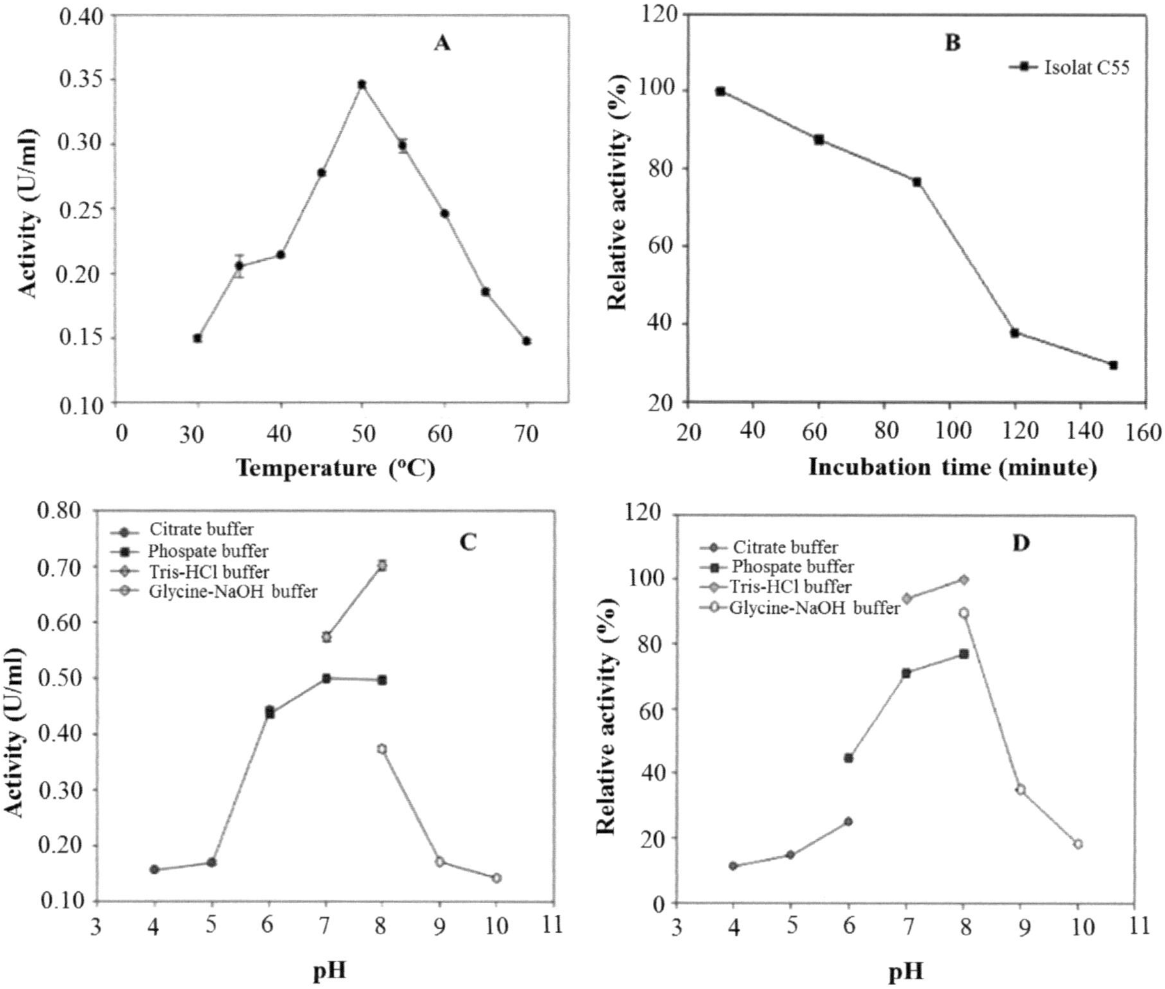

Figure 2. The effect of temperature and pH on activity of crude cellulase of isolate C55. (A). Effect of temperature on activity of crude cellulose; (B). Effect of temperature on stability relative activity of crude cellulose; (C). Effect of pH on activity of crude cellulose; (D). Effect of pH on stability relative activity of crude cellulase

3.3 Substrate specificity

Substrate specificity assayed revealed that crude cellulase which was produced by *Bacillus licheniformis* C55 is shown in Figure 3. This result indicated that the enzyme is an endoglucanase. It displayed high activity against CMC, low activity against filter paper, carrageenan and waste of carrageenan. The most *Bacillus* has high enzyme activity in hydrolyzing carboxymethyl cellulose [10].

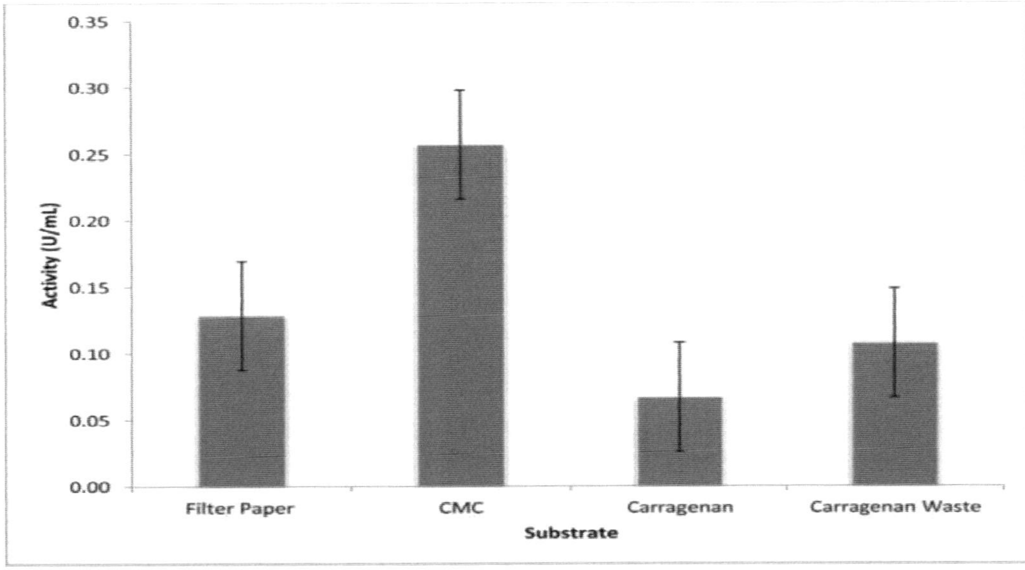

Figure 3. Effect of substrate specificity on crude cellulase of isolate C55

3.4 *SDS-PAGE and zymogram analysis*

The crude cellulase protein was observed as a single band with a molecular weight of about 18 kDa on SDS-PAGE gel, and the protein band was visualized by colloidal Coomasie blue staining (Figure 4). In zymogram gel, the appearance of proteins with carboxymethylcellulase activity was observed with supplementation of 1 % carboxymethyl cellulose (Figure 4). The molecular mass of *Bacillus licheniformis* C55 crude cellulase was found about 18 kDa by SDS-PAGE analyses, which was similar size to *Chryseobacteriumindologenes* McR-1.

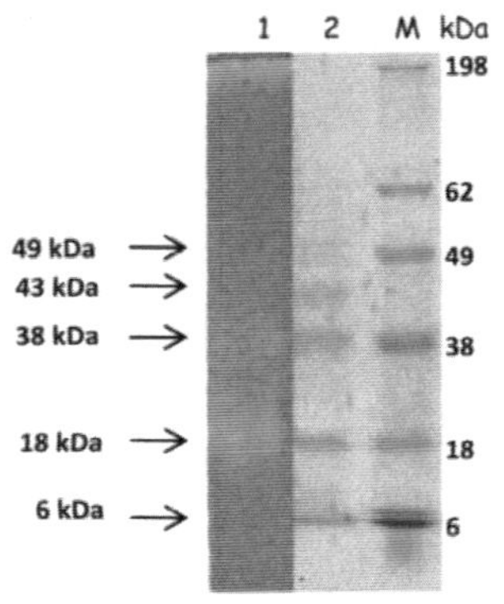

Notes: 1. Zymogram, 2. Crude protein, M. Molecular weight standards)

Figure 4. SDS-PAGE and zymogram analysis of the crude cellulose from *Bacillus licheniformis* C55

4. Conclusion

In this research, isolate C55 has the similarity 100% with *Bacillus licheniformis*. Enzyme that was produced by isolate C55 worked optimum in pH 8.0 and at 50°C. The enzyme can retain relative activity over 50 % at 50°C for 90 minutes. The molecular mass of *Bacillus licheniformis* C55 crude cellulase was found about 18 kDa. Substrate specificity indicated that cellulase enzyme which is produced by isolate C55 is endoglucanase.

References

[1] Bhat M K 2000 Cellulases and related enzymes in biotechnology *Biotechnol Adv.* **18** 355–383

[2] Baharuddin A S, Razak M N A, HockL S, Ahmad M N, Aziz S A, Rahman N A A, Shah U K M 2010 Isolation and Characterization of Thermophilic Cellulase Producing Bacteria from Empty Fruit Bunches-Palm Oil Mill Effluent Compost *Am J Appl Sci.* **7** 1 56-62

[3] Miller G L 1959 Use of dinitrosalicyclic acid reagent for determination of reducing sugar *Anal. Chem.* **31** 426-428

[4] Acharya S and Anita C 2011 Effect of nutritional and environmental factors on cellulase activity by thermophilic bacteria isolated from hot spring *J. Sci. and Ind Res.* **70** 142-148

[5] Chantarasiri A 2015 Aquatic *Bacillus cereus* JD0404 isolated from the muddy sediments of mangrove swamps in Thailand and characterization of its cellulolytic activity *Egypt J Aquat res.* **41** 257-264

[6] Lee Y J, Kim B K, Lee, B H, Jo K I, Lee N K, Chung C H, Lee Y C, Lee J W 2008 Purification and characterization of cellulase produced by *Bacillus amyloliquefaciens* DL-3 utilizing rice hull *Bioresour Technol.* **99** 378–386

[7] Kim B K, Lee B H, Lee Y J, Jin I H, Chung C H, Lee J W 2009 Purification and characterization of carboxymethylcellulase isolated from a marine bacterium, *Bacillus subtilis* subsp. subtilis A-53 *Enzyme Microb Technol.* **44** 411–416

[8] Kim J Y, Hur S H, Hong J H 2005 Purification and characterization of alkaline cellulase from a newly isolated alkalophilic *Bacillus sp.* HSH-810 *Biotechnol Lett.* **27** 313–316

[9] Singh J, Batra N, Sobti R C 2004 Purification and characterization of alkaline cellulose produced by a novel isolate *Bacillus sphaericus* JS1 *J Ind Microbiol Biotechnol.* **31** 51–56

[10] Ladeira S A, Cruz E, Delatorre A B, Barbosa J B, Martins M L L 2015 Cellulase production by thermophilic *Bacillus* sp. SMIA-2 and its detergent compability *Electron J Biotechn.* **18** 110-115

IOP Conf. Series: Earth and Environmental Science **131** (2018) 012044 doi:10.1088/1755-1315/131/1/012044

Characterization of pterin deaminase from *Mucor indicus* MTCC 3513

M Thandeeswaran[1], P Karthika[1], R Mahendran[1], M Palaniswamy[2] and J Angayarkanni[1]

[1]Cancer Therapeutics Lab, Department of Microbial Biotechnology, Bharathiar University, Coimbatore, Tamil Nadu, India.
[2]Department of Microbiology, School of Life Sciences, Karpagam University, Coimbatore, Tamil Nadu, India

E-mail: angaibiotech@buc.edu.in

Abstract Pterin deaminase is an amidohydrolase enzyme which hydrolyses pteridines to produce lumazine derivatives and ammonia. Even though the enzyme was shown as early as 1959 for its anticancer efficacy there was a long gap in the communique after that which was in 2013. In our study we have chosen *Mucor indicus* MTCC 3513 which was a promising strain for production of different industrial products.The pterin deaminase enzyme was harvested and extracellular from *M. indicus*.The extracellular sample was partially purified by using ethanol precipitation and ion exchange column (Hi-Trap QFF) in Fast Protein Liquid Chromatography. The molecular weight of the purified pterin deaminase enzyme was apparently determined by SDS-PAGE. The purified enzyme was further biochemically characterized. Molecular docking studies with the predicted sequence showed higher binding affinity towards folic acid interaction. The structure of this protein may open the windows for new drug targets for cancer therapy.

1. Introduction

Having about 1.5 million species, fungi have the second highest number of species after insects [1] with development of biotechnology, fungi have gained a crucial importance in the production of a wide range of products such as unique foods, biofuels, enzymes, antibiotics, and chemicals. Zygomycota is the most varied and least studied phylum of fungi that are divided into two subphyla, trichomycetes and zygomycetes. Being a source of human food for several hundred years together with ensuring performance in production of different products in the current decades has put zygomycetes at the center of a large number of recent research activities [2]. Among different zygomycetes, Mucoris a genus of about 3,000 members that are widely disseminated in the soil, digestive systems, plant surfaces, and decomposed vegetables. Indeed, *Mucor indicus* were one of the earliest discovered and most studied species of this genus. Followed by discovery in 1860s, different features of biology including morphology, biosynthesis, and metabolism as well as applications of this fungus have been extensively studied [3-10]. However, despite the broad and long term studies, several aspects of the fungus especially morphology, cell characteristics, and new applications continue to be unknown. A new wave of research on *M. indicus* has been initiated since the beginning of the 21st century, when *M. indicus* was introduced as a promising microorganism with several potential industrial importance such as co-production of ethanol and chitosan from lignocellulosic waste materials [11-16].Enzyme therapy involves taking enzyme supplements as an alternative form of cancer treatment. They are normal proteins that stimulate and accelerate many biological reactions in the body. Metabolic enzymes build new cells and repair damaged ones in the blood, tissues and organs. Enzymes have specific affinities for binding and catalytic properties towards specific targets with low toxicity make them as ideal drug. Cancer cells are more susceptible to enzymes than a normal cell because enzymes dissolve fibrous coating on cancers cells, allowing the immune system to work [17]. Pterin deaminase is an amidohydrolase enzyme which hydrolyses pteridines to produce lumazine derivatives and ammonia. The enzyme has been reported in several fungi viz., *Aspergillus oryzae* [18], *Aspergillus* Y-8-5 (ATCC NO. 20413), *Mucor lamprosporus* IAM 6114 (ATCC No.20410), *Rhizopus arrhizus* IAM 6052 [19] *Aspergillus tamarii* IFO 4287, *Aspergillus oryzae* var.no 13, *Aspergillus gymnosardae* IAM 2149, *Mucor alboater* IAM 6141, *Rhizopus japonicus* IAM 6002 (ATCC NO. 20409), *Aspergillus* Y-43-3 (ATCC NO. 20414), *Penicillium* Y-70-2 (ATCC No.20411) and *Penicillium* Y-110-2(ATCC No.20412) [20]. The enzyme substrates are involved an important role in the biological activity [21]. This study was conducted to produce and purify enzyme pterin deaminase from *M. indicus* MTCC 3513. Along with this, the study involved the structure prediction of pterin deaminase through bioinformatic tools.

2. Materials and methods

2.1 Collection of strain and the Preparation of extracellular enzyme from M. indicus

Lyophilized culture of *M. indicus* (MTCC 3513) was procured from Microbial Type Culture Collection and Gene Bank (IMTECH). The Fungal cells were grown in modified PDB (Potato dextrose broth) for 7 days under aerobic conditions. After incubation period the cells were harvested by centrifugation at 10,000 rpm at 4°C for 30 min. The supernatant was collected from the broth culture and served as crude enzyme. It was followed by enzyme precipitation using ethanol. The enzyme activities (IU ml^{-1}) against the substrate 0.01M folic acid were estimated in triplicates in extracellular culture filtrates by Nesslerization [22].

2.2 Protein precipitation

An initial step in purification of protein is concentration of protein sample which may be achieved by precipitation. Required volume of ethanol was cooled at -20°C. The protein sample was placed in ethanol compatible tubes. Different concentration of ethanol ranging from 3 to 9 volumes was added to the tubes. The tubes were vortexed and incubated at 4°C overnight. After incubation the tubes were centrifuged at 10,000rpm for 10 min at 4°C. The supernatant was properly discarded without disturbing the protein pellet. The ethanol was allowed to evaporate from the uncapped tube at room temperature for 30 min. The protein pellet was suspended with 0.01M potassium phosphate buffer (pH-7) for further analysis. The ethanol concentration which yielded higher protein and enzyme activity was chosen for further precipitation studies.

2.3 Ion exchange chromatography (FPLC)

Chromatographic process was performed at 4°C.A pre-packed Hi Trap Q FF (1ml) column a strong anion Exchanger, which was connected to the ÄKTA FPLC system. The column was equilibrated with 10 column volumes (CV) of binding buffer (buffer A: 0.01M PBS pH 7). The precipitated sample was injected to the column through a sample loop. Then the column was washed with 10 CV of buffer A to remove unbound or weakly bound particles. The adsorbed particles were eluted with 70 CV of final elution buffer with 0 %-100% NaCl (buffer B: 1M NaCl) in gradient mode. Flow through and elute were collected with a fraction collector included in the system (Frac - 920) at fixed volumes (1ml/min) [23].

The peak fractions were collected and assayed for enzyme activity and protein concentration was estimated. The selected fractions were analysed by SDS-PAGE [24]. Silver staining was done for the chosen fraction by the method of Switzer.1979 [25].

2.4 Enzyme characterization

Optimum pH was determined by incubating the reaction mixture in various pH ranging from 6-8 and optimum temperature for enzyme activity was ascertained by incubating the reaction mixture in various temperatures ranging from 10ºC-70ºC. Ammonia liberation in all the reactions were calculated by Nessler's Reagent for calculating enzyme activity.

2.5 Inhibition assay

The effect of inhibitors on pterin deaminase was determined by using different inhibitors such as Ethylene diamine tetra acetic acid (EDTA), Dithiothreitol (DTT), Phenyl methyl sulfonyl fluoride (PMSF), SDS and Triton X100. The enzyme was incubated for 30min with various concentrations of inhibitors like 10mM, 5mM and 1mM. After incubation, the enzyme activity was estimated by assaying the ammonia released by enzyme by deamination the substrate folic acid by a modified method of Wurster and Butz. 1980 [26].

2.6 *In silico* analysis of pterin deaminase

A complete genome of one of the zygomyceta family species *Malassezia sympodialis* and a sequence of Pterin deaminase (WP_012652141.1) were retrieved from NCBI through Z28140.1 identifier (NCBI-Gene sequence). Pterin deaminase perfectly matched with a small part of the sequence present in the *Malassezia sympodialis.* Further folic acid structure was retrieved from Pubchem database with ID of CID_6037. Using Prime modules of Schrodinger suit 2016, a model protein structure of pterin deaminase was predicted and validated. The modelled structure of pterin deaminase and folic acid from Pubchem database was used for interaction studies by glide modelling followed by extra precision method and rigid docking parameter.

3. Results

3.1 Extraction of extracellular enzyme

M. indicus MTCC 3513 used for this study was procured from Microbial Type Collection Centre (MTCC), Chandigarh and maintained in PDA medium. The extracellular enzyme was extracted from the fungal cells through centrifugation and the crude enzyme retained in the supernatant was used as crude enzyme extract. The protein content of the extracellular enzyme was found to be 3.3 mg/ml and the enzyme activity using folic acid as substrate was found to be 9 U/ml.

Maximum extracellular enzyme was precipitated (15.36 U/mL) with 3 volumes of ethanol. The highest concentration of protein being precipitated at 1:3 ratio (Crude Extract: Ethanol) validated 3 volumes as the preferred concentration of ethanol.

Table 1. Purification of pterin deaminase.

Purification Steps	Volume (ml)	Enzyme activity (U/ml)	Protein Concentration (mg)	Specific activity (U/mg)	Purification (fold)
Crude	50	9	3.3	2.72	1
Ethanol precipitation	30	15.36	2.35	6.56	2.411
Hi Trap Q FF (Ion-Exchange Column)	2	47.84	0.40	119.6	18.231

The ethanol precipitated enzyme, after suspension with potassium phosphate buffer was estimated to have 15.36 U/ml of pterin deaminase activity and protein content was measured to be 2.35 mg/ml. The specific activity was increased from 2.72 U/mg to 6.56 U/mg and the purification fold was increased to 2.411 (Table 1).

The ion exchange column purification process increased the purity of the enzyme which was revealed by the increase in enzyme activity to 47.84 U/ml with the specific activity of 119.6 U/mg. The purifications fold was increased to 18.231 fold. Even though the chromatogram revealed 5 active peaks, the fractions from 39 to 43 under the peak alone exhibited enzyme activity corresponding to 60% NaCl gradient solution (Table 1; Figure 1).

The protein profile of the crude enzyme showed more than 4 bands and appeared very thin due to very low concentration of loaded protein. Single band was visualized upon PAGE analysis of the active fractions (pooled) of ion exchange chromatography and apparent molecular weight of the enzyme was approximately 60 kDa (Figure 2).

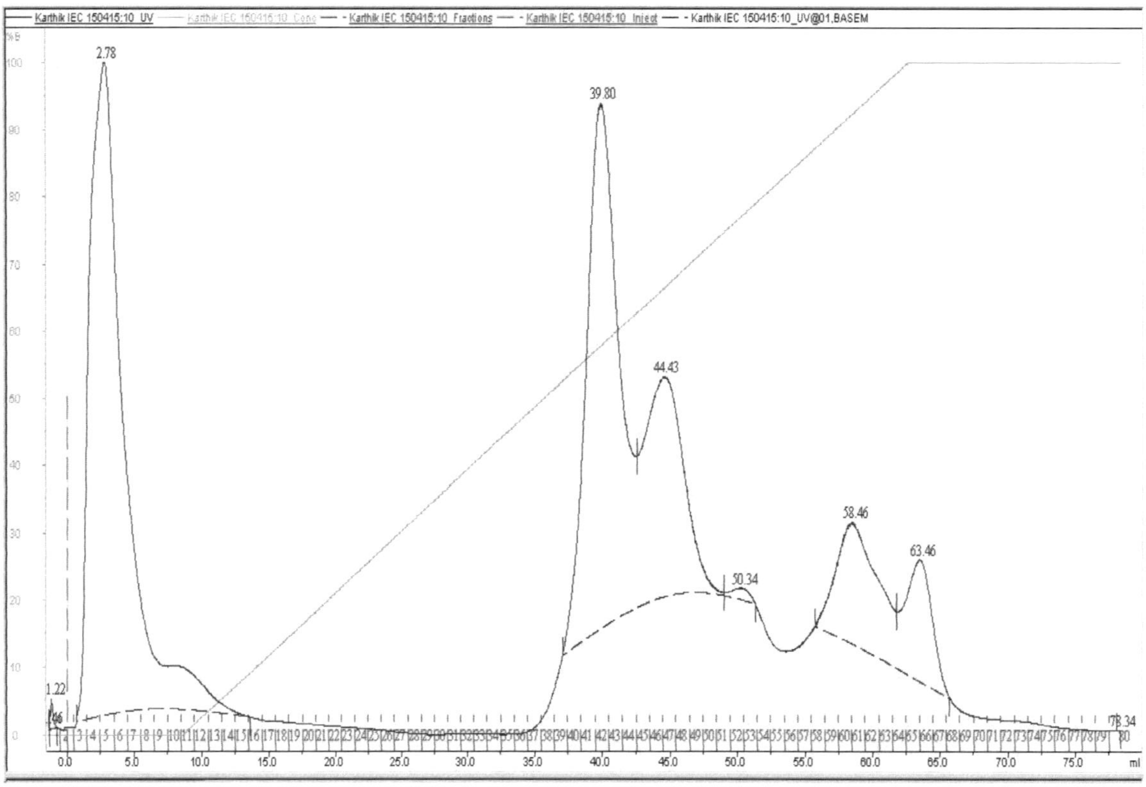

Figure 1. Ion Exchange chromatogram of the purified pterin deaminase

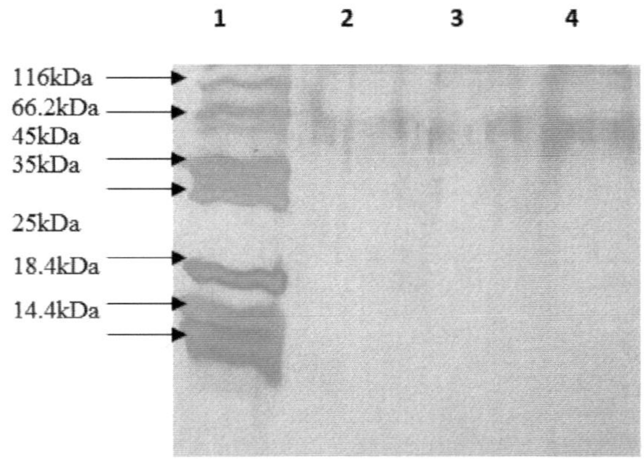

Figure 2. SDS-PAGE (Silver staining)

Lane 1- Marker, Lane 2- Purified and concentrated Fraction (15µl), Lane 3- Purified and concentrated Fraction (20µl), Lane 3- Purified and concentrated Fraction (30µl)

3.2 Optimization of purified pterin deaminase

The optimum temperature was found to be 30°C which corresponded to the maximum enzyme activity of 51 U/ml (Figure 3) whereas alkaline pH 7 was optimized condition for pterin deaminase activity (Figure 4).The molecular weight of the purified protein 60kDa determined in silver staining.

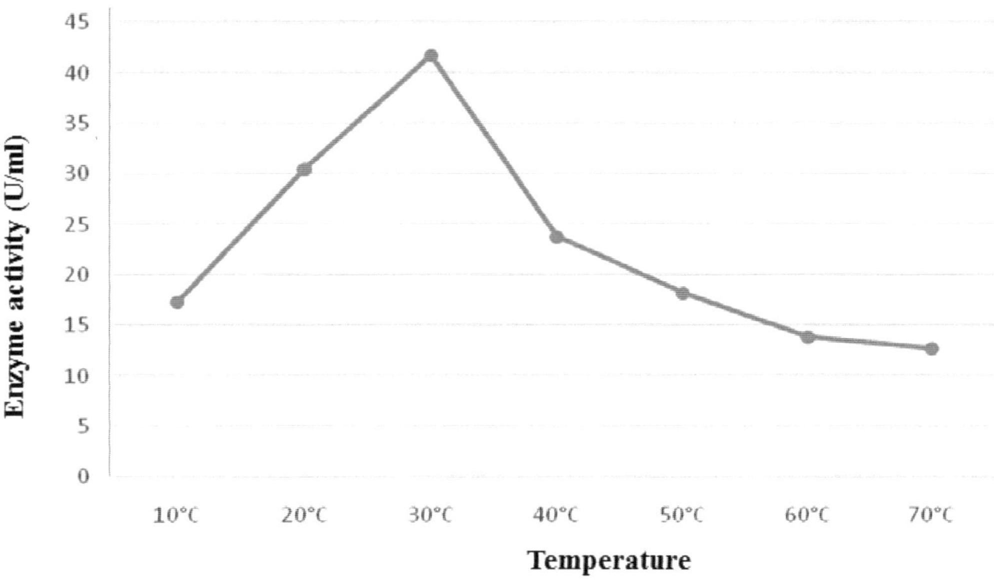

Figure 3. Effect of temperature on enzyme activity

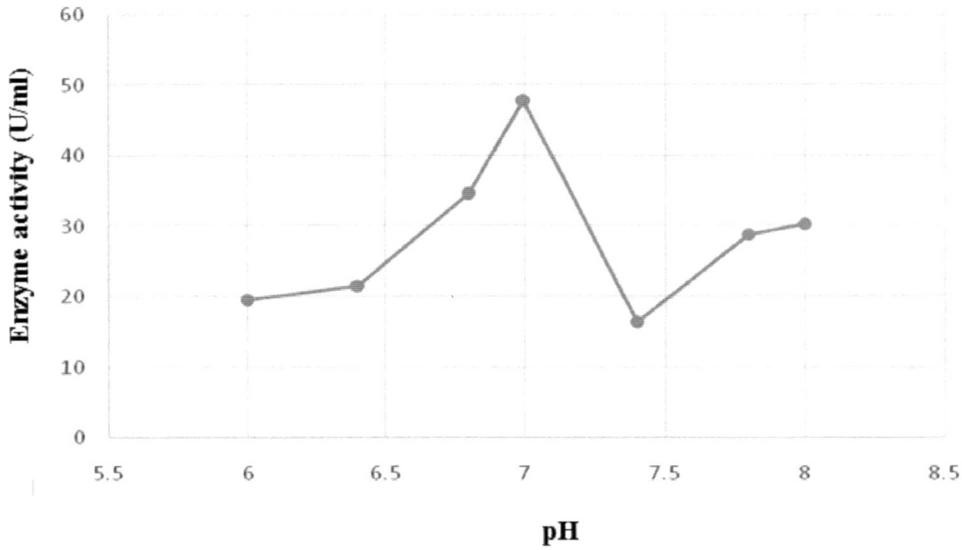

Figure 4. Effect of pH on enzyme activity

3.3 Effect of inhibitors

To predict the nature of the protein, inhibitor studies were carried out with pure protein. The enzyme was inhibited by EDTA at 10mM, 5mM and 1mM concentrations (Figure: 5). Triton X 100 also partially inhibited the enzyme activity at 10mM concentration. EDTA disodium salt inhibited the enzyme activity which indicated that the enzyme may be a metalloenzyme [27].

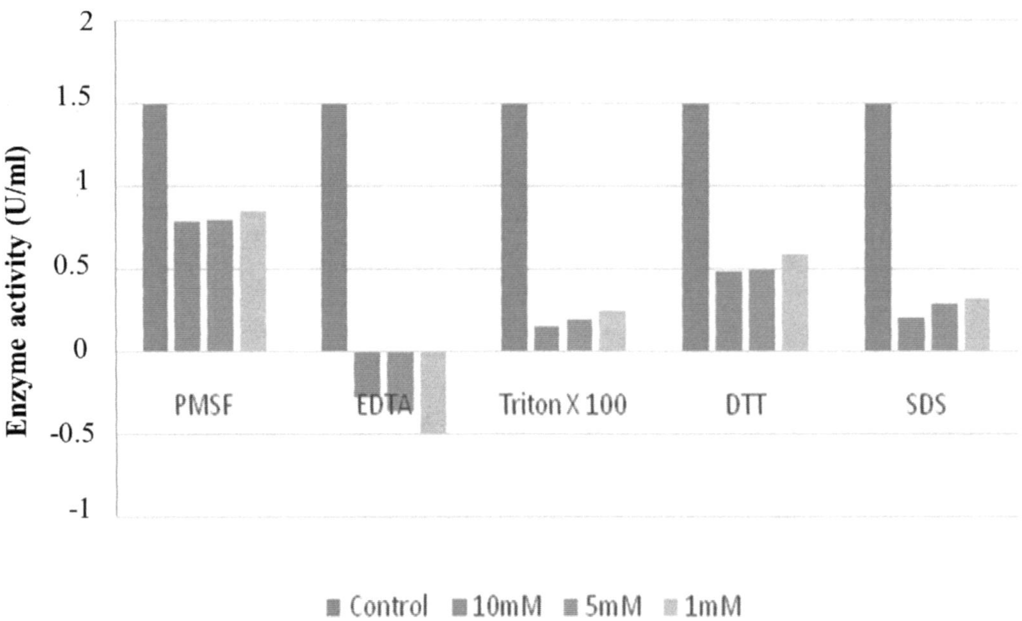

Figure 5. Effect of inhibitors on enzyme activity

3.4 Insilco analysis of Pterin deaminase (Homology Modelling of Protein Sequence Retrieved)
Retrieved sequence was then subjected to perform Homology modelling against Protein Databank which has only structural conformation of each atom's configuration. 3-Dimensional structure was predicted for translated protein sequence by homology modelling technique. The template structure was retrieved from crystally studied protein in protein data bank. The template ID is 4R7Y_A. By translating six frame translation method. It was taken first frame which is higher in number of protein sequence.

3.5 Model protein Vs Folic acid interaction
Interaction profile of Model protein with folic acid showed that the ligand interacted at 6 sites of the Protein having residual atom types ARG 626 (H), ARG 626 (H), ARG 624 (H), PHE 371 (H), ARG 624 (H), HIS 370 (H), ARG 126 (H), ASP 369 (O), ASP 369 (O) & ASP369(O) receptor via H (Hydrogen) and O (Oxygen) and atom types forming bond distance 1.715Å, 2.482Å, 2.325Å, 2.196Å, 1.653Å, 2.307Å, 2.072Å, 2.309Å, 2.309Å & 1.72Å respectively (Table 2, Figure 6 and 7). The higher interaction of Modelled protein with folic acid can be noticed from the Glide score -6.74 kJ/mol as higher interaction profile (Table 2).

Table 2. Ligand (folic acid) and protein interaction position.

Protein	Ligand	Bond	Protein atom	Ligand atom	Protein Residue name	Residue No	Bond Length
Pterin Deaminase Modelled	Folic Acid	1	Hydrogen	Oxygen	ARG	626	1.715
		2	Hydrogen	Oxygen	ARG	626	2.482
		3	Hydrogen	Oxygen	ARG	624	2.325
		4	Hydrogen	Oxygen	PHE	371	2.196
		5	Hydrogen	Oxygen	ARG	624	1.653
		6	Hydrogen	Oxygen	HIS	370	2.307
		7	Hydrogen	Oxygen	ARG	126	2.072
		8	Oxygen	Hydrogen	ASP	369	2.309
		9	Oxygen	Hydrogen	ASP	369	2.309
		10	Oxygen	Hydrogen	ASP	369	1.72

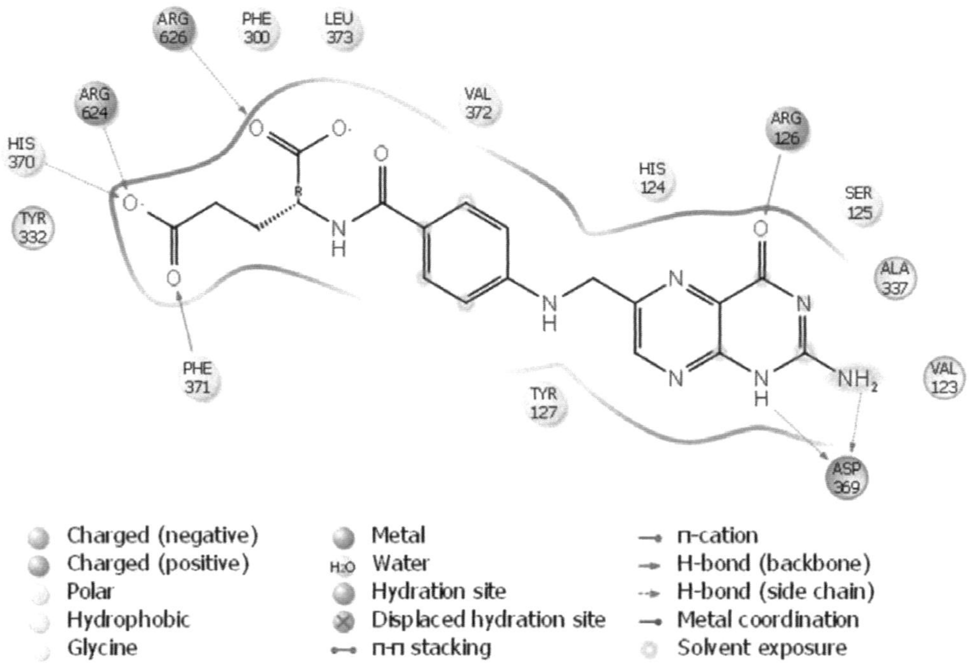

Figure 6. Ligand (folic acid) and Protein 2D- interaction

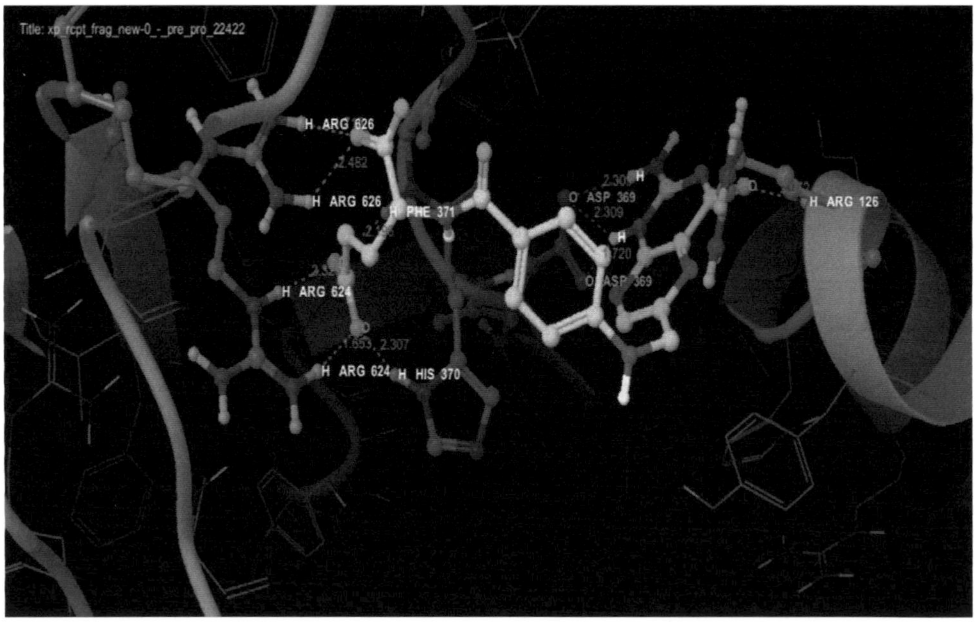

Figure 7. Ligand (folic acid) and Protein 2D- interaction

The docking studies revealed that the protein sequence taken for study had folic acid binding capacity with highest Glide score. Hence using this as base sequence the genes responsible for pterin deaminase in *M. indicus* may be identified by further recombination and expression studies.

4. Discussion

The potential of pterin deaminase to possess antitumor activity has been recorded since early 1980's [28].In spite of its immense biological significance, there is but little information available about the enzyme and its biological role. The present study was designed to purify and characterize pterin deaminase from *M. indicus* MTCC 3513.The extracellular enzyme was extracted from the fungal cells by centrifugation at high speed. Maximum extracellular enzyme was precipitated with 3 volumes of ethanol, Ethanol is the most commonly used alcohol compound for precipitation of proteins and nucleic acid. Also ethanol induces conformations changes which results in the formation of inter helical interaction allowing the proteins to aggregate [20].

Protein purification is essential for characterization of protein of interest. The purification process may separate proteins based on molecular weight or interaction with the charged resins. Ion exchange chromatography separates compounds according to their nature and degree of ionic charge. Anion exchange resins packed in the column have a positive charge and are used to retain and separate negatively charged compounds. Literally, there were no reports based on the purification of pterin deaminase enzymes in fungi whereas Kusakabe, 1976 [20] reported the enzyme was purified from fungi Aspergillus Y8-5 (ATCC 20413) by DEAE cellulose column chromatography and obtained 84,975 units of pterin deaminase with the enzyme activity 825 units/mg of powder (1428 units/mg of protein). In this study, ethanol precipitated protein sample was subjected to purification by ion exchange column chromatography. The chromatogram revealed 5 active peaks, the fractions from 39 to 43 under the peak alone exhibited enzyme activity corresponding to 60% NaCl gradient solution.

SDS-PAGE analysis was utilized to check the purity of protein by separating the protein according to their sizes based on their electrophoresis mobility. SDS-PAGE was made according to the method proposed by Laemmli.1970 [24]. Silver staining was performed for protein staining as it is more susceptible and is capable of staining less than 1 ng of protein. Through the purity of protein was not confirmed the single band formation around 60 kDa, denoted that the molecular weight of the enzyme may be apparently 60 kDa. The enzyme molecular weight reported by Takikawa, 1979 [29] was considered to be around 110kDa, which was higher than that obtained in the present study.

Previous reports have expressed the enzyme activity in terms of micromoles of substrate such as pterin carboxylic acid [30] and 2- amino -4-hydroxypteridine [31], deaminated in one min. The reduction in absorbance at 490nm being the absorbance peak for the substrate was taken for calculating the deaminated product. As per Takikawa [29] for each mole of substrate deaminated, equivalent moles of ammonia are released. Hence in the present study, the ammonia liberated during the deamination process was evaluated and the enzyme activity was expressed as micromoles of ammonia released/ml/min.

The temperature of a system is tantamount to some extend a measure of the kinetic energy of the molecules in the system. Increase in temperature of a system results in the simultaneous increase in the kinetics of the enzyme reaction due to energetic collision between the enzyme and substrate molecules. As the temperature of the system is raised, the internal energy of the molecules of the system will increase. The internal energy includes transitional energy, vibration energy and rotational energy of the molecules. The energy involved in chemical bonding of the molecule as well as the energy involved in nonbonding interactions. If the chemical potential energy increases greater enough, some of the weak bonds that determine the 3D shape of the active protein may be broken. This could result in a thermal denaturation and inactivation of protein. Thus, too much heat can cause the enzyme catalyzed reaction to reduce as the enzyme or substrate becomes denatured and inactive. In the present study, the optimum temperature was found to be 30°C which corresponded to the maximum enzyme activity of 51 U/ml.

Optimum pH values for enzyme activity reflects the point at which bonds within them are influenced by H+ and OH- ions in such a way that the shape of their active site is the most complementary to the shape of their substrate i.e. at optimum pH the rate of reaction is at an optimum. Any change in pH above or below the optimum will quickly cause a decrease in the rate of the reaction, since more of the enzyme molecules will have operational sites whose shape are not complementary to the shape of their substrate. Slight changes in pH above or below the optimum do not cause a permanent change to enzyme, since the bonds can be reformed. However, extreme changes in the pH can cause enzymes to denature and permanently lose their function. In the present study, the optimum pH of 7 was recorded with maximum enzyme activity of 52 U/ml. Pterin deaminase from *Rattus norvergicus* liver was reported to have an optimum pH of 7. In *Bacillus megaterium* and *Bacillus subtilis*, the pterin deaminase enzyme was observed to exhibit an optimum pH of 7.3 [32]. The result of the present study is in accordance with earlier reports wherein alkaline pH of around 7 was optimum.

To predict the nature of the protein, inhibitor studies were performed with pure protein. The enzyme suffered as a result by EDTA at 10mM, 5mM and 1mM concentrations. Triton X 100 also partially inhibited the enzyme activity at 10mM concentration. EDTA disodium salt inhibited the enzyme activity which indicated that the enzyme may be a metalloenzyme [27]. Besides, this study added a further significant unique result of the protein structure of pterin deaminase in eukaryotic organisms *M. indicus* MTCC 3513. Sequence from reference (ID: gil222086854) was taken and blast against nucleotide sequence database (non-reductant) to find out similar region from zygomycetes. The sequence was modeled and submitted to interact with folic acid. Glide score obtained from the interaction studies revealed that the folic acid has study interaction towards modelled pterin deaminase structure. More recently, one of the authors Fan [33] determined a multidisciplinary approach that integrated homology modeling, molecular docking screens of a metabolite library and physical library screening by kinetic assays in Arad3529 from *Agrobacterium radiobacter* K84 regarding pterin deaminase.

5. Conclusions

The study highlighted the possibility of extracting pterin deaminase enzyme from *Mucor indicus*. The enzyme activity and specific activity of the partially purified enzyme has clearly reflected the role of chromatographic separation in protein purification. The current research also throws light on the influence of temperature and pH on the enzyme activity. Molecular docking studies with the predicted sequence showed higher binding affinity towards folic acid interaction. The structure of this protein may open the windows for new drug targets for cancer therapy.

References

[1] Feofilova E P 2001 *Applied Biochemistry and Microbiology* **37** 124-137
[2] Zamani A, Edebo L, Niklasson C, Taherzadeh JM 2010 *Int J Mol Sci* **8** 2976-2987.
[3] Bartnicki-Garcia S 1963 *Bacteriol Rev* **27** 293–304
[4] Bartnicki-Garcia S, Nickerson WJ 1961 *J Bacteriol* **82** 142–8.
[5] Bartnicki-Garcia S, Nickerson WJ 1962 *Biochim Biophys Acta* **64** 548–51.
[6] Bartnicki-Garcia S, and Reyes E 1964 *Arch. Biochem. Biophys.* **108** 125-133
[7] Bettinger P 1917 *Bull Assoc Chim Sucrerie Distillerie France Colonies* **35** 129-33.
[8] Bettinger P 1920 *Bull Assoc Chim Sucrerie Distillerie France Colonies* **37** 427-33
[9] Gordon PA, Stewart PR, Clark-Walker GD 1971 *J Bacteriol* **107** 114–20.
[10] Haidle CW, Stock R 1966 *J Bacteriol* **92** 1236–44.
[11] Abedinifar S, Karimi K, Khanahmadi M, and Taherzadeh M J 2009 *Biomass and Bioenergy* **33** 828-833
[12] Heidary Vinche M, Asachi R, Zamani A and Karimi K 2013 *J Chem Tech Biotech* **88** 255-260
[13] Javanbakht V, Zilouei H, Karimi K 2011 *Int Biodeterior Biodegradation*, **65** 294-300

[14] Jeihanipour A, and Taherzadeh M J 2009 *Bioresour technol* **100** 1007-1010
[15] Millati R, Edebo L, Taherzadeh M J 2005 *Enzyme Microb Technol* **36** 297-300
[16] Millati R, Karmi K, Edebo L, Niklasson C and Taherzadeh M J 2008 *BioResources* **3** 1020-9
[17] Sabu A 2003 *Indian J Biotechnol* **2** 334-341
[18] Nawa S 1958 Annual Report. *National Institute of Genetics*. **9** 97.
[19] Kusakabe H, Kodama K, Midorika Y, Machida H, Kuninaka A, Yoshine H 1974 *Agric Biol Chem* **38** 1753–1754.
[20] Kusakabe H, Kodama K, Midorikawa Y, Machida H, Kuninaka A, Yoshino H 1976 US patent.
[21] Jayaraman A, Thandeeswaran M, Priyadarsini U, Sabarathinam S, Ayub Nawaz KA, Palaniswamy M 2016 *Appl Microbial Biotechnol* **100** 4779-4789
[22] Imada A, Igarasi S, Nakahama K, Isono M 1973 *J General Microbiol* **76** 85–99.
[23] Murugesan T, Velliayadevar K, Easwaran M, Kiran KG, Ayub Nawaz KA, Ramasamy M, Muthusamy P, Jayaraman A 2017 *Pteridines* doi:10.1515/pterid-2017-0011
[24] Laemmli U K 1970 *Nature* **227** 680–5
[25] Switzer RC, Merril CR and Shifrin S Anal 1979 *Biochem* **98** 231–7
[26] Wurster B. Butz U Eur 1980 *J Biochem* **109** 613–8
[27] Koprowski H J, Abramczuk D Solter 1977 *Dev Biol* **61** 378–383
[28] Eickbush,T H, Moudrianakis EN 1978 *Cell* **13** 295–306
[29] Takikawa S, Kitayamayokokawa C, Tsusue M 1979 *J. Biochem* **85** 785-790
[30] Levenberg B, Hayaishi O 1959 *J Biol Chem* **234** 955-961
[31] Klein R, Tatischeff I, Tham G, Mano N 1994 *Chirality* **6** 564–571.
[32] Tsusue M, Takikawa S, Yokokawa C K 1978 *Dev Biochem* **4** 153–8.
[33] Fan H, Hitchcock DS, Seidel RD, Hillerich B, Lin H, Almo SC, Sali A, Shoichet BK, Raushel FM 2013 *J Am Chem Soc* **2** 795–803

IOP Conf. Series: Earth and Environmental Science **131** (2018) 012045 doi:10.1088/1755-1315/131/1/012045

Optimization of sol-gel medium for entrapment of acetylcholinesterase enzyme in biosensor for pesticide detection

S D Wijayanti, F S Rahayu and T D Widyaningsih

Department of Agricultural Product Technology, Faculty of Agricultural Technology, Universitas Brawijaya, Malang, Indonesia

E-mail : dee_ta2002@yahoo.com

Abstract. Pesticides are chemical substances used to kill and control pests or diseases that can damage crops. The use of pesticides should be done precisely because the accumulation of chemicals contained in pesticides can cause various health effects. Therefore, detection of pesticide residues on plants is important to reduce the risk of poisoning due to pesticide residues. Some of the conventional methods that have been done to detect pesticide residues have weaknesses among expensive tools, takes a long time, and are generally performed by trained laboratory technicians. Biosensors are analytical devices that can measure the quantitative or semi-quantitative targets of analyte by utilizing a bioreceptor such as enzyme. Several studies have shown that enzyme-based acetylcholinesterase-based biosensors can be used to detect pesticide residues in vegetable samples. The objective of this research was to get a proper silica based sol-gel formulation with molar ratio of H_2O:TEOS and NaOH concentration as immobilization medium of acetylcholinesterase enzyme for biosensor application. Response Surface Methodology (RSM) was used in order to determine the interaction between the parameters studied and resulting responses which were amount and activity of acetylcholinesterase enzyme. Based on the research, the best result for immobilized enzyme activity was shown by molar ratio (H_2O: TEOS) 1: 8 and 4 mM NaOH treatment.

1. Introduction

Generally, horticultural crop farmers tend to use excessive pesticides to improve the quality of their products. The use of pesticides should be done precisely because the accumulation of chemicals contained in pesticides can lead to increased nerve activity such as headache, vomiting, shortness of breath, muscle spasms and paralysis [1] Therefore, detection of pesticide residues in plants is important to reduce the risk of poisoning due to pesticide residues in plants. Biosensors are analytical devices that can measure analyte target quantitatively or semi-quantitatively by utilizing bioreseptor such as enzymes, antibodies, DNA, nucleic acids and whole cells [2]. Enzyme-based biosensor is mostly applied because of its high specificity to the target analyte. Several studies have shown that acetylcholinesterase (AChE) enzyme-based biosensor can be used to detect pesticide residues in vegetable samples.

In the fabrication of biosensors, enzymes are immobilized in a solid matrix in order to maintain their catalytic activity. Entrapment is one of the enzyme immobilization techniques conducted by trapping enzymes in micro-polymer matrices such as silica-based sol-gel [3]. Immobilization of enzymes by using sol-gel media has several advantages such as large enzyme capture, stable to high temperature, and flexible or controllable pore diameter [4]. However, the use of sol-gel for immobilized enzyme media also has a disadvantage as it is porous that may lead the enzymes to seep out. Modifying the molar ratio between silica (Tetraortosilicate/ TEOS) with H_2O can be an alternative to overcome this weakness as H_2O is used in the TEOS hydrolysis process so that the molar ratio used can affect the TEOS structure in the formation of sol-gel. Furthermore, other factors that can be used to regulate the size of the sol-gel pores can be done by modifying the concentration of NaOH as a catalyst. NaOH can accelerate the process of hydrolysis of TEOS so that the pore size of the sol-gel becomes larger and the enzyme is more easily trapped [6]

The aim of this research is to get a proper silica-based sol-gel formulation with TEOS and H2O molar ratio and NaOH (mM) concentration as immobilization medium of acetylcholinesterase enzyme. Using the statistical method of Response Surface Methodology (RSM) it is expected that the interaction between the parameters studied and the resulting response can be known. The minimum data used In the TEOS and H_2O molar ratio factors is 1: 4, the central data is 1: 8, and the maximum unit data is 1:12. Meanwhile, the concentration factor of NaOH on the minimum data used is 3mM, central data is 4mM, and data unit maximum of 5 mM. Analysis on the number of immobilized enzyme is investigated as the response.

2. Materials and Method

The optimization of silica-based sol-gel was done using Response Surface Methodology (RSM) method with the factor studied is the molar ratio between TEOS with H_2O and NaOH concentration which is a sol-gel catalyst. The research is divided into several stages including:

a. Sol-gel Formulation with Software Response Surface Methodology (RSM)

Response Surface Methodology (RSM) method was conducted using Design Expert 7.0 software. The initial stage is designing the desired experimental variables. The factor used in the optimization is the consideration of the molar ratio between TEOS and H_2O. Another factor is the optimization of the concentration of NaOH (mM) while the response is the amount of immobilized enzymes. Table 1 presents the experimental design of Response Surface Methodology (RSM) with the Central Composite Design (CCD) model. About 13 designs are obtained from the RSM method.

Table 1. Design of Experiment by RSM Method

Design	Factor	
	Factor 1 Molar Ratio (H_2O:TEOS)	Factor 2 Concentration of NaOH (mM)
1	1:4	3
2	1:12	3
3	1:4	5
4	1:12	5
5	1:2.36	4
6	1:13.66	4
7	1:8	2.59
8	1:8	5.41
9	1:8	4
10	1:8	4
11	1:8	4
12	1:8	4
13	1:8	4

b. Response Determination

Number of immobilized enzyme was determined as a response (result) of optimization carried out using RSM method. The number of immobilized enzyme analysis was conducted by calculating the amount of lysis enzyme (ΔA1) after immobilization process was done. 5 µl of acetylcholinesterase enzyme (ΔA0) was first immobilized on each 5 µl sol-gel formulation. A mixture of enzyme and sol-gel was then dropped onto Whatman no. 1 (size of 0.8 cm x 3 cm). The Whatman paper was left to stand for 15 minutes and then immersed in 250 µl Tris-HCl buffer pH 7.5 for 20 minutes at room temperature. The solution was assumed as lysis enzyme (ΔA1). As much as 195 µl of the solution was placed on the microplate and added by 5 µl indoxyl acetate (5 mg / ml). Furthermore, a sol-gel absorbance measurement was performed using a microplate reader every 2 minutes at a wavelength of 630 nm for 20 min. ΔA0 is free enzyme absorbance, while ΔA1 is the lysis enzyme absorbance at 630 nm

$$Amount\ of\ Immobilized\ Enzyme\ (\%) = \frac{\Delta A0 - \Delta A1}{\Delta A0} \times 100\%$$

c. Model Selection and Model Analysis with ANOVA

Selection of appropriate model (Linear, Quadratic, or Cubic) was determined in this study. Furthermore, after obtaining the appropriate model, an analysis of variance (ANOVA) was performed on the selected model which aims to determine the interaction between two factors, i.e the TEOS and H_2O molar ratio and NaOH concentration towards the number of immobilized acetylcholinesterase enzyme.

d. Determination of Optimum Condition

From the results of the response analysis, the Expert Design software will also suggest optimal sol-gel formulation results along with an approximate response value of the greatest number of immobilized enzyme. In addition to providing predictive response values, the Expert Design program also provides prediction interval (PI). PI is a range that shows the results of the next response measurements with the same condition at a significance level of 5%.

e. Verification of Optimization Result

Verification was carried out by analyzing the number of immobilized enzyme based on the optimum formulation obtained with two replications. The purpose of verification is to confirm whether the response is generated in accordance with the approximate response generated by the Expert Design software. If the difference of predicted value and a research result is no more than 5%, it indicates that the model is sufficiently precise and thus the solution of the given free variable is acceptable.

3. Results and Discussion

3.1 Response Analysis Results

Figure 1 shows the effect of TEOS : H_2O molar ratio and NaOH concentration to amount of immobilized enzyme, as a response based on the 13 designs provided. The central data is obtained from TEOS: H_2O molar ratio of 1: 8 and NaOH concentration of 4 mM. The central data is repeated 5 times from 13 RSM designs. The central data tends to show higher results than other designs. The amount of immobilized acetylcholinesterase enzyme in the central data ranged from 72.73% to 86.36%. The number of enzymes that are immobilized is higher when compared to the literature of research conducted by [4] which immobilized trypsin enzyme in sol-gel with a precursor of 0.1 ml TEOS silica; Deionized H_2O; PEG 6000 0.1 ml; and H_3PO_4 1 M as much as 2µl. The result of the number of immobilized trypsin enzymes in the study ranged from 59.9-67.7%. This may be due to the addition of PEG 6000 which acts as a sol-gel stabilizer will have an effect on the sol-gel that makes the particle size of the silica larger because PEG will envelop the silica particles so the size is larger [5]. After obtaining the data, we select appropriate model to determine the optimum response based on sum of squared sequence model, statistical model summary (Model Summary Statistic), and Lack of Fit Test.

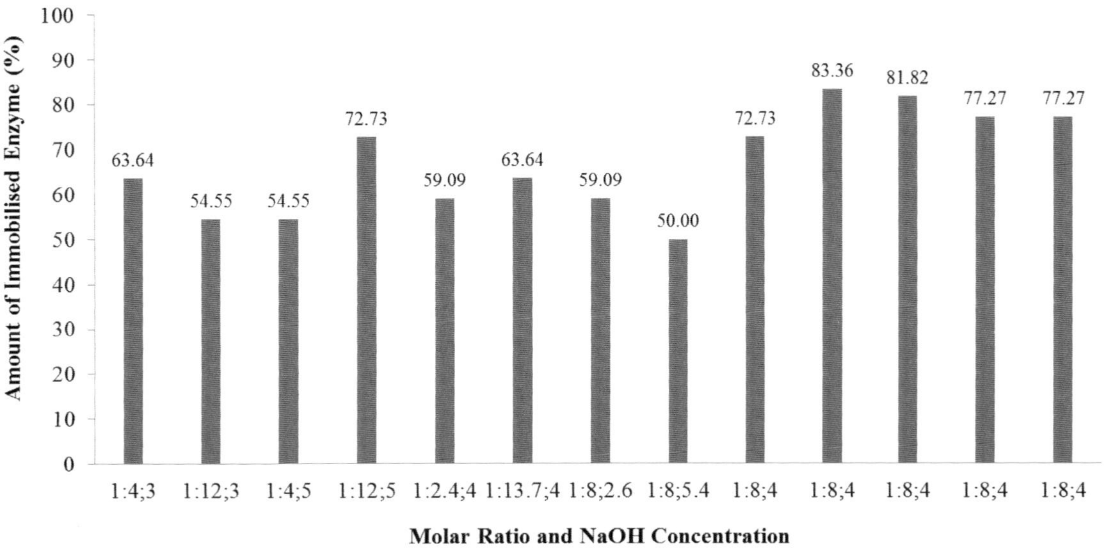

Figure 1. Amount of Immobilized Enzyme Response Result

3.2 Result of Optimization Analysis on the Number of Immobilized Enzyme

3.2.1. Analysis of Selection of Statistical Model

Analysis of statistical model selection was conducted to determine the appropriate model in describing the significance of the research results obtained. Some types of models that can be used in Design Expert program include: linear model, 2 factor interaction (2FI), Quadratic and Cubic models. The model selection analysis is based on the sum of quadratic (Sequential Model Sum of Squares), Lack of Fit Test and Summary of Statistic statistics. Based on these three methods, the Quadratic model has a significant effect on the response so that the Quadratic model is the best and suggested design.

3.2.2 Analysis of Variance (ANOVA) of RSM on the Number of Immobilized Enzyme Responses

After getting the selected model that is Quadratic model, then the analysis of variance (ANOVA) is used to analyze the accuracy level of the selected model based on the analysis value of variance to the model and the value of Lack of Fit presented in Table 2.

Table 2. Result of Analysis of Variance (ANOVA) from *quadratic* model.

Source of Variance	Number of Squares	Db	Mean Squared	F Value	P Value Prob>F	Description
Model	1426.15	5	285.23	10.410	0.0038	*significant*
A	30.11	1	30.11	1.100	0.3294	*not significant*
B	1.77	1	1.77	0.065	0.8065	*not significant*
AB	185.95	1	185.95	6.790	0.0352	*significant*
A^2	446.48	1	446.48	16.300	0.0050	*significant*
B^2	907.32	1	907.32	28.110	0.0007	*significant*
Residual	191.78	7	27.40			
Lack of Fit	84.34	3	28.11	1.05	0.4633	*not significant*
Pure Error	107.44	4	26.86			
Cor Total	1617.93	12				

Description: A = Variabel X_1 (Molar Ratio of TEOS and H_2O)
B = Variabel X_2 (Concentration of NaOH (mM))
AB, A^2, B^2 = Interactions between treatment

In this research,the analysis of model is shown with P value 0.0038 so that the model is stated significant. The value of Lack of Fit shown is 0.4633. Therefore, based on the analysis value of the model and Lack of Fit it is concluded that the Quadratic model corresponds to the whole design value. The interaction between the TEOS and H_2O (squared) molar ratios and NaOH concentration has a positive and significant effect on the number of mobilized enzymes, so that it can be expressed when the interaction between molar ratio (squared) and NaOH concentration increases, the number of immobilized enzyme will be higher

3.3 The Effect of Molar Ratio and NaOH Concentration on Response of Number of Immobilized Enzymes

Figure 2 illustrates the 3-dimensional curves and plot contours for sol-gel media optimization for immobilization of acetylchololesterester enzymes. The image illustrates the effect of two parameters on the number of mobilized enzymes. The values listed on the box on the plot contour indicate the number of enzymes mobilized under various optimization process conditions studied.

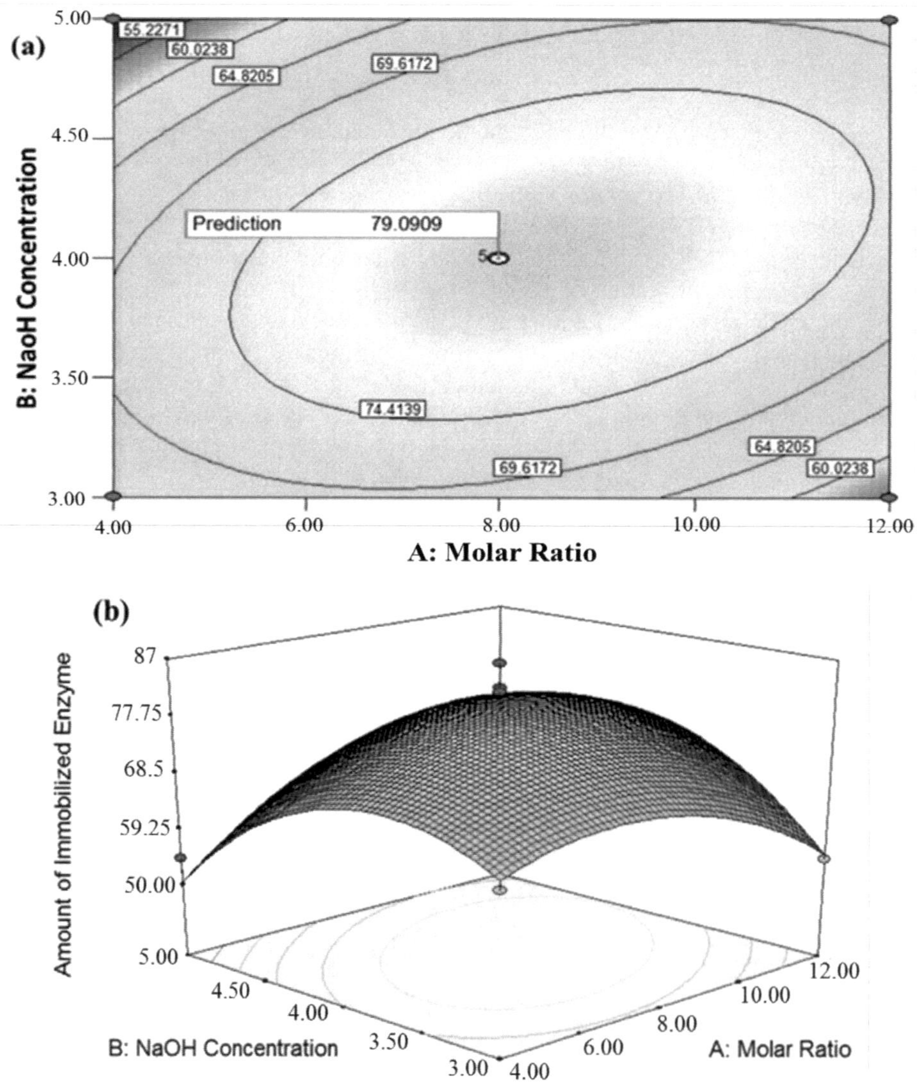

Figure 2. (a) Contour plot (b) surface response (3D) curve of molar ratio and NaOH concentration against the amount of immobilized enzyme response

Figure 2 shows that the curve shape of the saddle represents the possibility of the variable at maximum and minimum points. Such contour systems are called saddle or minimax system. The x and y axes in Figure 2 (a) show the optimized variables. The x-axis shows the variables of the TEOS and H_2O molar ratios, whereas the y-axis shows the NaOH concentration variable. The circular lines indicate the response. Optimal response is indicated by the flag in the center of the contour indicating the optimal point information located at the point (node) shown in the flag. On the plot contour, the number of optimized mobilized enzymes is shown at the point (node) of 79.0909%.

3.4 The Determination of Amount of Immobilized Enzymes Optimum Response
Software Design Expert 7.1.5 is used to identify the best combination of sol-gel formulation parameters used to optimize the number of immobilized enzymes and the relative activity of immobilized enzymes. Desirability is a method used to explain how well the optimum solution is offered to match the objectives of the response . The desirability value 1 indicates the perfect case, but the desirability value 0 indicates the response that should be discarded. In this study, the optimal solution offered by the model is the molar ratio of 1: 8.58 (TEOS: H_2O) and the NaOH concentration of 4.09 (mM) for a response prediction of 79.155%, with a desirability value of 0.77. The optimum point of each variable is the stationary point which is supposed to be the optimum response.

3.5 The verification on optimization result
The verification of optimization results needs to be done to prove the predicted results and response values of the optimum formula solution suggested by the Design Expert program. Verification is done by comparing the value of response analysis on the research with the response value of the results of Design Expert software calculations. Verification and prediction results are shown in Table 3.

Based on the data, the comparison of predicted results of Design Expert program with actual calculation on all responses shows that the value of verification is in accordance with the prediction value, because the response is still around the range of PI.

Table 3. Comparison of Verification Result and Predictions.

	Molar Ratio TEOS:H_2O	NaOH Concentration (mM)	Amount of Immobilized Enzyme (%)
Prediction	1:8.58	4.09	79.15
Verification	1:8.58	4.09	84.09 ± 3.21
95% PI low	1:8.58	4.09	65.61
95% PI high	1:8.58	4.09	92.70

4. Conclusions
Based on this research, it is found that the result of sol-gel optimization for TEOS molar ratio factor: H_2O is 1: 8.58 while for NaOH concentration factor is 4.09 mM. From the sol-gel optimization, the verification result of immobilized enzyme is 84.09 ± 3.21%.

References
[1] Alegantina, S , Raini, M, Lestari,P 2005 *Penelitian Kandungan Organofosfat dalam Tomat dan Slada yang Beredar di Beberapa Jenis Pasar di DKI Jakarta* Media Litbang Kesehatan XV **1** 2005

[2] Guo, Xishan, Xueyin Zhang, Qiang Cai, Tao Shen, dan Songming Zhu 2013 *Developing a Novel Sensitive Visual Screening Card for Rapid Detection of Pesticide Residues in Food. Food Control* **30** 1 15–23

[3] Souza, Ranyere L, Emanuelle L P Faria, Renan T Figueiredo, Alini T Fricks, Gisella M Zanin, Onélia A A Santos, Álvaro S Lima, dan Cleide M F Soares 2014 *Use of Polyethylene Glycol in the Process of Sol–gel Encapsulation of Burkholderia Cepacia Lipase. Journal of Thermal Analysis and Calorimetry* **117** 1 301–6

[4] Lan, Wensheng, Guoping Chen, Feng Cui, Feng Tan, Ran Liu, dan Maolidan Yushupujiang 2012. *Development of a Novel Optical Biosensor for Detection of Organophoshorus Pesticides Based on Methyl Parathion Hydrolase Immobilized by Metal-Chelate Affinity*. Sensors **12** 12 8477–90
[5] Sakka, S 2005 *Handbook of Sol-gel Science and Technology Vol III: Processing, Characterizing, and Applications.* Kluwer Academic Publisher Boston
[6] Sotiropolou, S, Nikos, A, dan Chaniotakis 2005 *Tuning the Sol-gel Microenvironment for Acetylcholinesterase Encapsulation Biomaterials* **26** 6771-6779

Electricity generation from wetlands with activated carbon bioanode

E. Sudirjo[1, 2], C J N Buisman[1] and D P B T B Strik[1]

[1]Sub-Department of Environmental Technology ,Wageningen University & Research, The Netherlands
[2]Mining and Energy Agency, Government of Landak Regency, West Kalimantan Province, Indonesia

Email: emilius1.sudirjo@wur.nl / emiliuss@gmail.com; david.strik@wur.nl

Abstract. Paddy fields are potential non-tidal wetlands to apply Plant Microbial Fuel Cell (PMFC) technology. World widely they cover about 160 million ha of which 13.3 million ha is located in Indonesia. With the PMFC, in-situ electricity is generated by a bioanode with electrochemically active bacteria which use primary the organic matter supplied by the plant (e.g. as rhizodeposits and plant residues). One of limitations when installing a PMFC in a non-tidal wetland is the usage of "expensive" large amounts of electrodes to overcome the poor conductivity of wet soils. However, in a cultivated wetland such as rice paddy field, it is possible to alter soil composition. Adding a conductive carbon material such as activated carbon is believed to improve soil conductivity with minimum impact on plant vitality. The objective of this research was to study the effect of activated carbon as an alternative bioanode material on the electricity output and plants vitality. Lab result shows that activated carbon can be a potential alternative for bioanode material. It can continuously deliver current on average 1.54 A/m3 anode (0.26 A/m2 PGA or 66 mW/m2 PGA) for 98 days. Based on this result the next step is to do a test of this technology in the real paddy fields.

1. Introduction

Plant Microbial Fuel Cell (PMFC) is considered as a new source for sustainable bioelectricity production which does not compete with food production. It produces sustainable *in-situ* energy, without harvesting the biomass, by conversion of solar energy into electricity via plants and electrochemically active bacteria [1]. Hereby plant residues like rhizodeposits are used as fuel to produce electricity. For large scale electricity generation, PMFC should be applied in wetlands because they need waterlogged condition to maintain anaerobic condition and to facilitate ion transport [1]. Paddy fields are potential wetlands to apply PMFC technology (Figure 1). World wide they cover about 160 million ha of which 13.3 million ha of it is located in Indonesia [2].

It is known that in a non-tidal wetland such as rice paddy field, soil conductivity is relatively low compared to tidal wetlands [3]. Some researchers have tried to install PMFC system in a rice paddy field without any modification on the soil composition. The maximum power output of PMFC in the paddy field is about 140 mW/m^2 projected anode surface area [4]. This power output is half of the best performance of PMFC in the lab, which is 240 mW/m^2 [5].

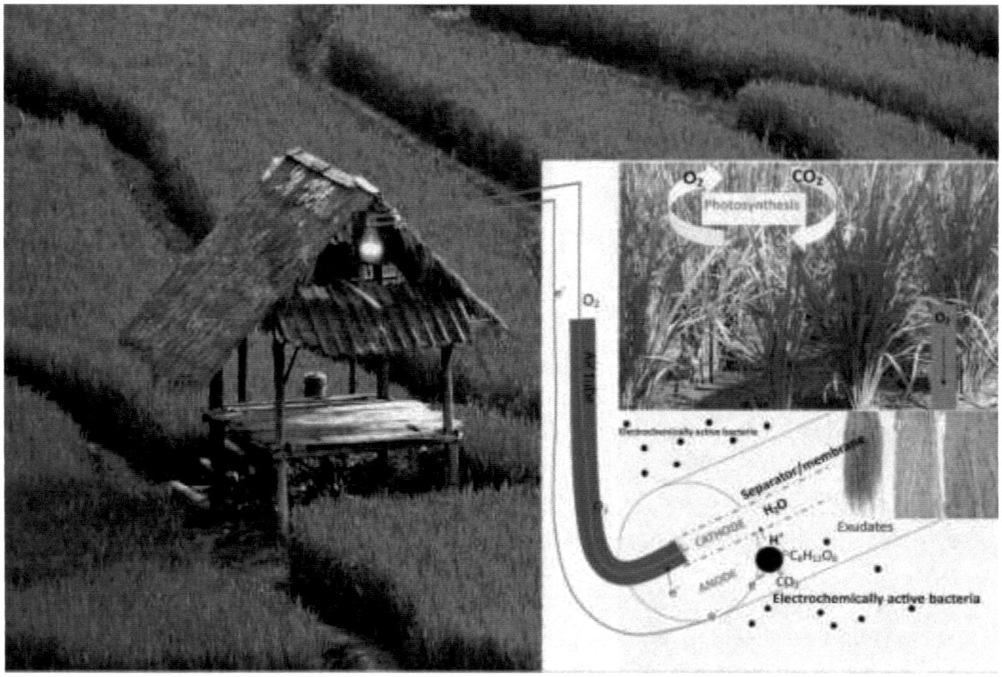

Figure 1. An illustration how to install a PMFC into a paddy field

In a cultivated wetland, most of the time before the growing season, soil composition of the paddy field is altered. The top layer soil and the lower layer is mixed while adding fertilizer. Considering this activity some inert conductive material such as carbon based material may be applied in the rice paddy fields. Researches have shown that adding biochar is not only increasing rice paddy soil fertility and the rice yield but it also reduces total methane emission [6, 7]. As biochar electrical conductivity (2-4.4 uS/cm) are typically about 1000 times lower than activated carbon [8], using activated carbon (AC) in a PMFC can allow electricity generation and potentially profiting on the additional advantages of biochar.

Based on this concept, this research aims to investigate the effect activated carbon as an alternative bioanode material on the electricity generation and plants vitality within PMFCs at laboratory condition.

2. Materials and Methods

This research was conducted with 2 identical flat plate PMFC reactors in a climate chamber (10 hours dark, 14 hours light, 25^{0}C, 70% humidity). The flat plate reactor was made from plexiglas material. The reactors consist of 2 chambers (anode and cathode chamber) separated by a Cation Exchange membrane (CEM). Two graphite rods connected with titanium wire (18x1x0.2 cm) were place in both side of the anode chamber with a distance of 18 cm. The anode chamber had total volume 722 ml (19x19x2 cm) but only 650 ml were used. The anode chambers were filled with activated carbon Norit PK 1-3. Nitrate-less, ammonium-rich plant growth medium [9] was continuously pumped into anode chamber using minipuls-3 Gilson pump at a constant speed of 15 rpm. Anode plant growth area (PGA) was 19 cm x 2 cm. Cathode electrode was made from graphite felt (thickness: 3mm), FMI composites Ltd., Galashiels, Scotland. This electrode was woven with titanium wire (1 mm diameter) as a current collector. The same plant growth medium was used as catholyte and aerated with ambient air using aquarium pump.

There were two phase of operation of the experiment. The first phase was blank characterization without plants for 159 days (day 1 - 159) and the second one was with plant for 129 days (day 160 – 288). In the first 96 days, both reactors were operated only with plant growth media without bacteria inoculum. On day 97, 20 mL inoculum from previous PMFC's research [5] was added in both anode reactor. One day later another 10 mL fresh inoculum from running MFC's anode was also added into both anodes. Anode, cathode, cell and membrane potential were recorded every 60 minutes from day 23 until the end of experiment (day 288). All data were logged in a computer using field point data logger and LabVIEW software from National Instrument, The Netherlands. Anode and cathode potential were measured against Ag/AgCl reference electrode. Both PMFC reactors were controlled either with 1 kohm external load or at -100 mv anode potential using a potentiostat (16 channel Ivium BV, The Netherlands).

3. Results and Discussion

After 288 days of operation, AC bioanode PMFSs were successfully operated. Number of stems and total plant's height were growing in both reactors.

3.1. Activated carbon (AC) bioanode can function as an alternative bioanode

When the experiment was started, AC anode was operated only with plant growth media without inoculation for 96 days. During this period both control modes did not show any current production. This result was predicted before due to absence of electrochemically active bacteria (EAB). The similar trend was also observed (data not shown) after adding bacteria inoculum on day 97 and 98 until day 160.

On day 160, plants *(Spartina anglica)* were planted into the anode PMFCs. Reactors were kept at open cell until day 163. From day 163 until 260, both anodes were controlled at -100 mV. During this period the PMFC reactors was starting to deliver current as can be seen on

Figure *2*. In the first 12 days after starting generating current, both PMFC delivered current of about 1 mA (equal to 0.25 mW). However, from day 175 current generation in PMFC 2 was gradually decreasing and reaching 0 on day 187. On day 260 PMFC 2 current was about –0.5 mA, meaning it functions as a cathode instead of an anode. At the same time, PMFC 1 continuously generated current on more stable condition around 1 mA (1.54 A/m3 anode or 0.26 A/m PGA) until day 260.

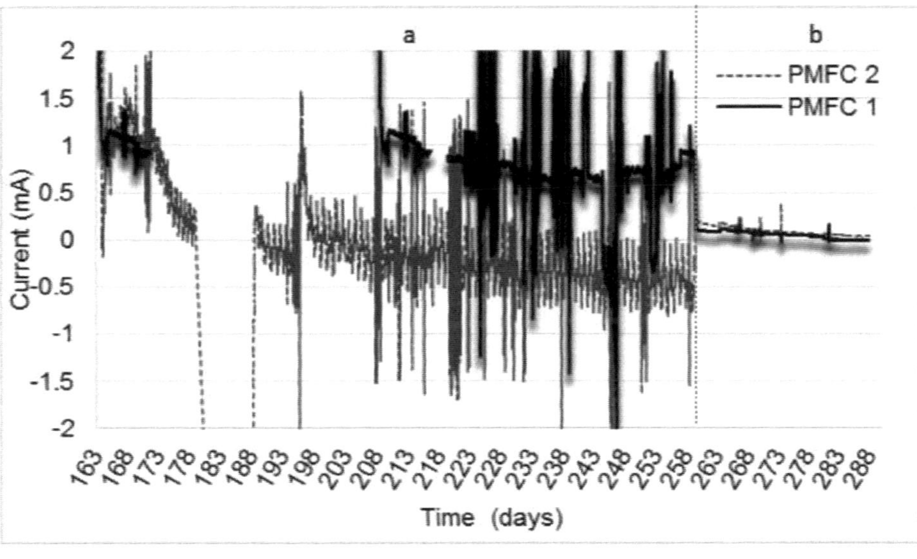

Figure 2. Current production of PMFC 1 and PMFC 2: (a) from day 163 until day 260 at -100 mV anode control voltage; (b) from day 260 until day 288 at 1000ohm external load.

The drop on current generation of PMFC 2 was more likely caused by oxygen present in the anode. This oxygen can be (bio)electrochemically reduced to water [10]. As plant can deliver oxygen via its root [11], it was confirmed that in the first 100 days after planting more plant's stems and leaves were growing in PMFC 2 than PMFC 1 (Figure 3 and Figure 4). This effect was also indicated by clear day and night pattern on PMFC 2 during this period. The current production was decreasing during the day (light regime) and increasing during the night (dark regime). When reactors were controlled with external load, this effect was not observed. However due to cathode limitation, the current production both PMFCs were gradually decrease from about 0.1 mA (0.01 mW) to 0.02 mA(0.002 mW).

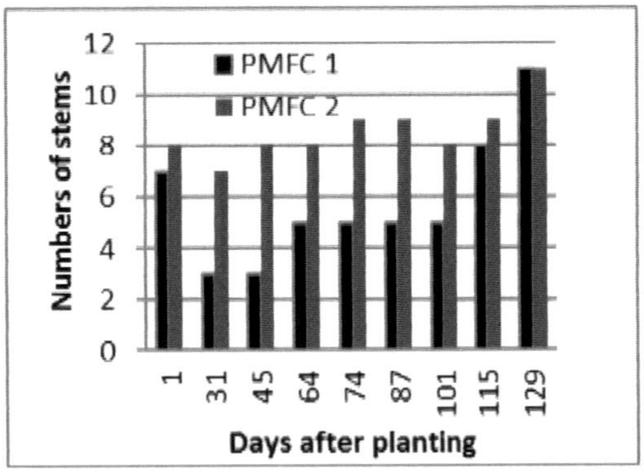

Figure 3. Number of living stems

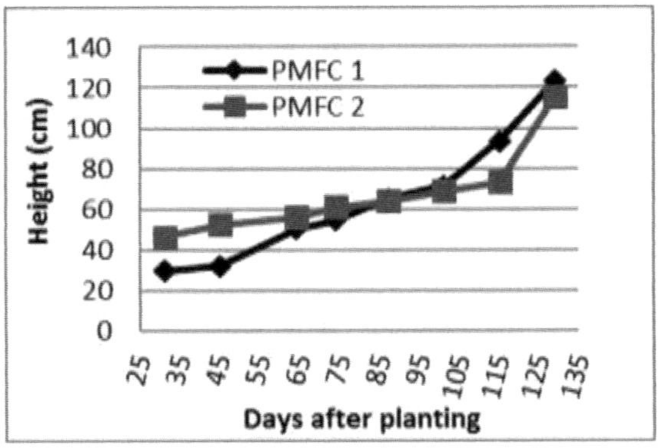

Figure 4. Total stems' height

In either case, AC bioanodes were successfully operated in PMFC and delivering total electric charge as much as 4791 C for PMFC 1 and 4338 C for PMFC 2 for 129 days operation with plants (***Table 1***). Moreover, the activated carbon anode can also support plant growth. Plants on both reactors were growing (Figure 3 and Figure 4).

Table 1. Charge production (Coulomb).

	PMFC 1	PMFC 2
Day 163-260 (98 days potentiostat control)	4655	4103
Day 260-288 (29 days 1000ohm ex load)	136	235
Total	4791	4338

3.2. AC PMFC generating cost is cheaper than other PMFCs

Comparing costs among PMFC system is not always feasible because most of current production was reported based on PGA instead of anode material volume. As a normalise parameter, the PGA cannot give us a real amount of used electrode. The most fair comparison to calculate the effect of an anode material and its ability to deliver such amount of current in a PMFC system should be based on the number of used material itself. Therefore, a unit of electricity generating cost based on utilised anode material per unit power is calculated (Table 2). For this calculation the prices for graphite felt, graphite granule (high purity) and activated carbon are based on these data: 1. Graphite felt PGF grade, CGT carbon GmBh (0.12 g/cm^3), thickness 3mm, €62/m^2 or equal to €172/kg [12] ; 2. Huaming graphite granule (0.798 g/cm^3); €0.76/kg[13] ; 3. Taoshi activated carbon (0.55 g/cm^3); €1.68/kg [14]. For comparison with other studies, average current production of 0.5 mA (0.125 mW) and 1 mA (0.25 mW) were used.

Based on our calculation we found that activated carbon is a promising material for anode on a PMFC system. Investment cost on bioanode electrode material per installed mW is about €2.4 – €4.8 (Table 2).

Table 2. Power density and electricity generating cost based on anode material price from various PMFC studies

No	System	plant species	Lab /field	average power density (mW/m2 PGA)	Power (mW)	Anode material	Anode material weight (kg)	Anode cost (€)	Investment cost based on anode (€/mW)	Ref
1	2 chambers flat-plate PMFC with CEM	Spartina anglica	lab	66	0.25	activated carbon	0.3575	0.60	**2.40**	This study
				33	0.125		0.3575	0.60	**4.80**	
2	2 chamber pot PMFC with CEM	Reed mannagrass (Glyceria maxima)	lab	67*	0.258	graphite felt+graphite granule	0.0138 felt & 0.8288 graphite granules	3.01	11.68	[1]
3	without membrane	Paddy	paddy field	6*	0.810	graphite felt	0.0486	8.37	10.33	[15]
4	container with CEM	Paddy	lab	26	0.708	graphite mat+graphite rod; in soil	0.0165	0.88	1.25[a]	[16]
				15.8	0.246	graphite mat+graphite rod; in graphite granules	5.6479	4.27	17.33[b]	
5	without membrane	Paddy	paddy field	14.44*	0.907	graphite felt	0.0226	3.89	4.29	[17]
6	2 chamber pot PMFC with CEM	Spartina anglica	lab	50	0.1	graphite granules	0.1650	0.12	1.25	[18]
7	PVC tubular PMFC, 2 chamber with ultra filtration membrane	Reed mannagrass (Glyceria maxima)	lab	10	0.020	graphite felt	0.0426	7.34	366.83	[19]
				12	0.024	graphite granules	0.7500	0.57	23.63	
8	2 chamber Cylindrical PMFC with CEM	Spartina anglica	lab	21	0.646	graphite grain	4.6661	3.53	5.46	[20]
9	2 chambers flat-plate PMFC with CEM	Spartina anglica	lab	440*	0.376	graphite felt	0.0194	3.35	8.91	[9]
10	3 chambers: cathode-anode-cathode with BPM	Spartina anglica	lab	240	0.6528	graphite felt	0.1299	22.38	34.29	[5]
11	Floating cathote without membrane; anode below	Paddy	field	80*	0.64	graphite felt	0.00288	0.5	0.78	[21]
12	Floating larger cathote than anode without membrane; anode below	Paddy	field	140*	0.7	graphite felt	0.0018	0.31	0.44	[4]

Notes: *maximum; [a]graphite rod as a curent collector is not considered; [b]calculated only based on graphite granule

4. Conclusion

Based on the above result it can be concluded that activated carbon can be used as an alternative anode material for PMFC system. Plant vitality was shown with growing plants under laboratory studies. Investment costs on bioanode electrode materials per installed mW power are between €2.4 and €4.8. This is relatively cheaper compare than other PMFC anode materials.

Acknowledgements

This research is funded by Government of Landak Regency, West Kalimantan Province, Republic of Indonesia.

References

[1] Strik D P B T B, Hamelers H V M, Snel J F H, Buisman C J N 2008 Green electricity production with living plants and bacteria in a fuel cell *Int J Energy Res* **32** 870–876. doi:10.1002/er.1397

[2] FAO of the UN. FAO Statistical Year Book 2013 World Food and Agriculture Food and Agriculture of the United Nation Rome

[3] Anderson C J, Lockaby B G 2011 Foliar nutrient dynamics in tidal and non-tidal freshwater forested wetlands. *Aquat. Bot.* **95** 153–60. doi:10.1016/j.aquabot.2011.05.010

[4] Ueoka N, Sese N, Sue M, Kouzuma A, Watanabe K 2016 Sizes of anode and cathode affect electricity generation in rice paddy-field microbial fuel cells *J Sustain Bioenergy Syst.* **6** 1 10–15. doi:10.4236/jsbs.2016.61002

[5] Wetser K, Sudirjo E, Buisman C J N, Strik D P B T B 2015 Electricity generation by a plant microbial fuel cell with an integrated oxygen reducing biocathode *Appl. Energy* **137** 151–157. doi:10.1016/j.apenergy.2014.10.006

[6] Thammasom N, Vityakon P, Lawongsa P, Saenjan P 2016 Biochar and rice straw have different effects on soil productivity, greenhouse gas emission and carbon sequestration in Northeast Thailand paddy soil *Agric Nat Resour.* **50** 192–198. doi:10.1016/j.anres.2016.01.003

[7] Koyama S, Inazaki F, Minamikawa K, Kato M, Hayashi H 2015 Increase in soil carbon sequestration using rice husk charcoal without stimulating CH_4 and N_2O emissions in an andosol paddy field in Japan *Soil Sci Plant Nutr.* **61** 873–884. doi:10.1080/00380768.2015.1065511

[8] Chen S, Rotaru A-E, Shrestha P M, Malvankar N S, Liu F, Fan W, et al 2014 Promoting interspecies electron transfer with *Biochar*. **4** 5019

[9] Helder M, Strik D P, Hamelers H V, Buisman C J 2012 The flat-plate plant-microbial fuel cell: the effect of a new design on internal resistances *Biotechnol. Biofuels* **5** 70. doi:10.1186/1754-6834-5-70

[10] Chen Z, Huang Y, Liang J, Zhao F, Zhu Y 2012 A novel sediment microbial fuel cell with a biocathode in the rice rhizosphere. *Bioresour Technol* **108** 55–59. doi:10.1016/j.biortech.2011.10.040.

[11] Koop-Jakobsen K, Wenzhöfer F 2015 The dynamics of plant-mediated sediment oxygenation in *Spartina anglica* Rhizospheres—a planar optode study. *Estuaries Coasts* **38** 951–963. doi:10.1007/s12237-014-9861-y

[12] Wetser K 2016 Electricity from wetlands : technology assessment of the tubular plant microbial fuel cell with an integrated biocathode Wageningen University Netherlands

[13] Huaming (online) Graphite Granule. Retrieved from https://www.alibaba.com/product-detail/Graphite-granules-made-in-China_60601895228.html?spm=a2700.7735675.35.1. 50d7548c8Xv66U. [Accessed on 27 September 2017]

[14] Taoshi (online) Activated Carcbon. Retrieved from https://www.alibaba.com/product-detail/Factory-price-granule-coconut-shell-activated_60547953161.html?spm=a2700.7735675. 30.1.6168a84fLppqRR&s=p. [Accessed on 27 September 2017]

[15] Kaku N, Yonezawa N, Kodama Y, Watanabe K 2008 Plant/microbe cooperation for electricity generation in a rice paddy field *Appl Microbiol Biotechnol.* **79** 43–49 doi:10.1007/s00253-008-1410-9

[16] Schamphelaire L D, Bossche L V den, Dang H S, Höfte M, Boon N, Rabaey K, Verstraete W 2008 Microbial fuel cells generating electricity from *Rhizodeposits* of rice plants *Environ Sci Technol.* **42** 3053–3058. doi:10.1021/es071938w

[17] Takanezawa K, Nishio K, Kato S, Hashimoto K, Watanabe K 2010 Factors affecting electric output from rice-paddy microbial fuel cells *Biosci Biotechnol Biochem.* **74** 1271–1273. doi:10.1271/bbb.90852

[18] Timmers R A, Strik D P B T B, Hamelers H V M, Buisman C J N 2010 Long-term performance of a plant microbial fuel cell with *Spartina anglica Appl Microbiol Biotechnol.* **86** 973–981. doi:10.1007/s00253-010-2440-7

[19] Timmers R A, Strik D P B T B, Hamelers H V M, Buisman C J N 2013 Electricity generation by a novel design tubular plant microbial fuel cell *Biomass Bioenergy* **51** 60–67. doi:10.1016/j.biombioe.2013.01.002

[20] Helder M, Strik D P B T B, Hamelers H V M, Kuhn A J, Blok C, Buisman C J N 2010 Concurrent bio-electricity and biomass production in three plant-microbial fuel cells using *Spartina anglica*, *Arundinella anomala* and *Arundo donax Bioresour Technol.* **101** 3541–3547. doi:10.1016/j.biortech.2009.12.124

[21] Kouzuma A, Kaku N, Watanabe K 2014 Microbial electricity generation in rice paddy fields: recent advances and perspectives in rhizosphere microbial fuel cells *Appl. Microbiol. Biotechnol.* **98** 9521–9526. doi:10.1007/s00253-014-6138-0.

Performance of microbial fuel cell double chamber using mozzarella cheese whey substrate

M D Darmawan, L C Hawa and B D Argo

Department of Agricultural Engineering, Faculty of Agricultural Technology, Universitas Brawijaya, Malang, Indonesia

E-mail : doddydarmawan0495@gmail.com

Abstract. Nowadays the availability of electric energy is decreasing, hence there is a need for innovation of electric energy producer alternative; one of them is microbial fuel cell (MFC). MFC is a bioelectrochemical system generated by bacterial metabolism that utilizes organic substrate. One of the substrates that can be used is whey, a waste generated from cheese production. Therefore, this study aimed to determine the power of potential current and voltage generated from the use of whey cheese as a substrate for bacterial metabolism. In this research, double chamber system was used in microbial fuel cell reactor by using cheese whey as substrate at anode and potassium permanganate as cathode and utilizing membrane nafion 212 as membrane of proton exchange. The variable of experiment was bacteria type. The types of bacteria used in this study were *Lactobacillus bulgaricus*, *Streptococcus thermophillus* and *Lactobacillus casei*. While the operating time used was 100 hours. The highest current produced was 74.6 μA and the highest voltage was 529.3 mV produced by *Lactobacillus bulgaricus* bacteria. In this study, it was also found that the death phase of the three bacteria was at 70-80 hours.

1. Introduction

Electrical energy is a basic requirement in the community today, where almost all human activities require electrical energy. The vital importance of electricity for humankind, in case of Indonesia increase electricity demand is estimated to continue growing by 4.6% annually and will be tripled in 2030. If this is not accompanied by efforts to increase energy production, it is alarming that Indonesia will experience an energy crisis. Therefore, an effort should be made to generate sustainable alternative energy sources. Some innovations in renewable energy are available today; one of them is microbial fuel cell or MFC. Microbial Fuel Cell (MFC) is one of the prospective technological alternatives to be developed. MFC is a system that uses bacteria to produce electricity from organic or non-organic compounds. Various types of microbes have been used in the MFC, among others *Shewanella oneidensis* [1], *Saccharomyces cerevisiae* [2], *Lactococcus lactis* [4], *Geobacter sulfurreducens* [6], and *Escherichia coli* [7], As well as some existing studies utilizing a variety of substrates such as tofu and other substrates that can be substrate for bacterial metabolism. One of the substrates that have the potential to be developed with MFC technology is whey, a waste generated from cheese production. Currently, cheese whey has not been further utilized. Cheese industries are beginning to grow along with the increasing interest of the community in processing food by adding cheese. The existing cheese industry only discharges whey through waste channels. Based on the potential of MFC as an electricity supplier and the potential of whey for bacterial substrate, the present study aimed to utilize whey as a substrate in MFC reactors to generate electrical energy. In this study, two variables were used to determine the effectiveness of cheese whey as a substrate in generating electricity through the MFC reactor. The first variable observed was bacterial type, because bacteria are the main factor in electricity production using MFC reactor. It is important to know the most optimal bacteria for electricity production. Another variable tested was MFC reactor operation time to determine the optimum time for electricity production in MFC reactor.

2. Materials and Method

Whey in this research was obtained from CV. Brawijaya Dairy Industry, Batu. The cultures of *L. bulgaricus, S. thermophillus* and *L. casei* were obtained from Microbia Laboratory, Jakarta. MFC reactors were constructed in Mechatronics and Agroindustrial Machine Laboratory University of Brawijaya. The voltage and current were measured using digital multimeter (Sanwa Electric Instrument Co., Ltd., Japan).

2.1. MFC configuration

The set up comprised of a dual- chambered MFC consisted of an anode and a cathode with a reactor volume of 700 mL in each chamber and the reactors were connected in series. The two compartments were connected with a proton exchange membrane (PEM; Nafion 212, Lyntech, USA) and joined tightly. The reactor that used in this research is shown in Figure 1.

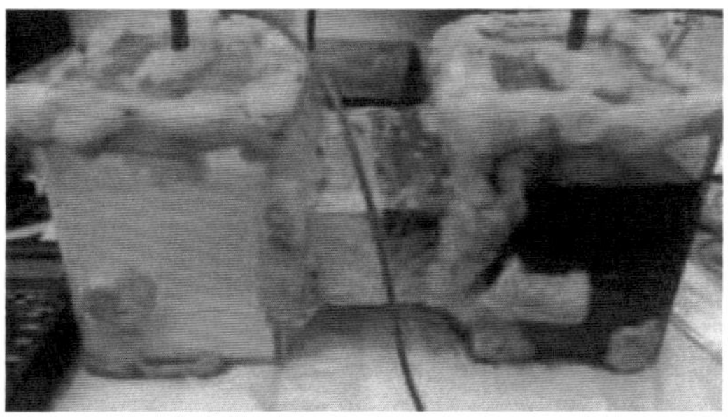

Figure 1. MFC Double Chamber Reactor

The electrodes were carbon graphite purchased from CV. Cahaya Sejati – Surabaya, with an effective surface area of 7,598 cm^2. The electrodes were soaked in 1M HCl for one day, followed by immersion in 1M NaOH for one day. The electrodes were completely immersed in the anode and cathode solutions and connected using a wire. Before use, the electrodes and the Nafion membrane were pre-treated to remove any contaminants.

2.2 MFC experiments

The anode chamber was filled with three different inoculums that were *L. bulgaricus, S. thermophillus,* and *L. casei.* As well as added glucose, aquadest, and phosphate buffer (pH 7). The number of initial bacteria was 5,7 x 10^4 CFU/ml. The cathode chamber was filled with potassium permanganate and a phosphate buffer (pH 7). The MFC was operated for 3 hours, 30 hours and 100 hours.

3. Results and Discussion

The performance of microbial fuel cell double chamber reactor was evaluated by current strength and voltage. The main goal of this research was to find out the electrical output generated by microbial fuel cell double chamber reactor using mozzarella cheese whey substrate. Figure 2 shows current produced and Figure 3 shows voltage output from 100 hours of reaction time.

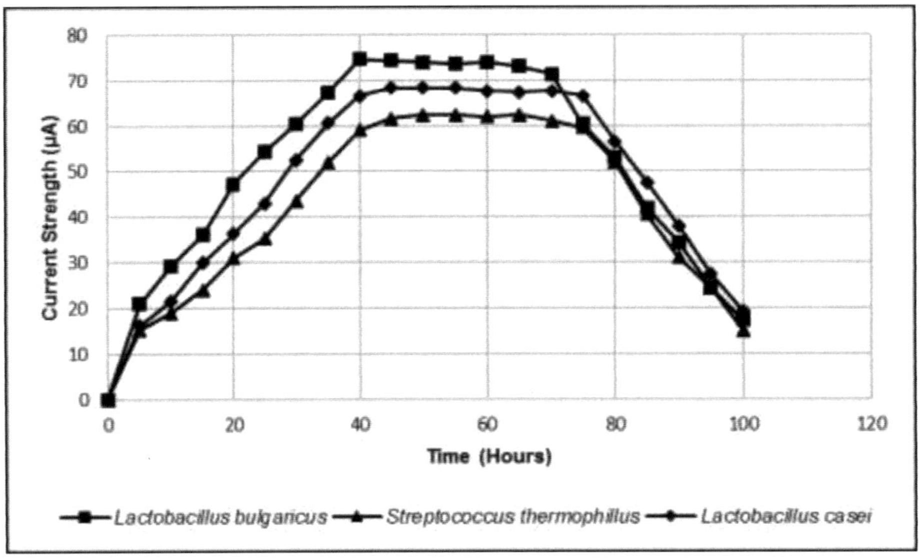

Figure 2. Current strength of 100 hours reaction time

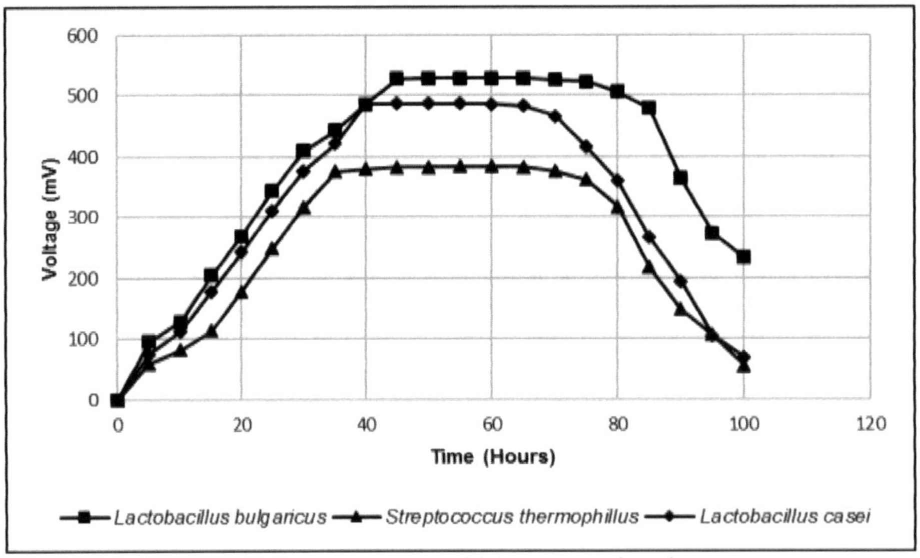

Figure 3. Voltage of 100 hours reaction time

Figure 2 shows the results of the current produced by each type of bacteria with reaction time of 100 hours. Based on the measurements, the three types of bacteria showed that current strength was increasing until 30 hours but afterwards undergoing stationary phase until 70 hours, after 70 hours the current strength was decreasing. The highest current strength value was generated by *Lactobacillus bulgaricus* bacteria with an average maximum current of 74.6 µA. The second highest current strength was produced by *Lactobacillus casei* with average maximum current strength at this treatment was 69 µA. While the lowest current strength produced by *Streptococcus thermophillus* with average maximum current strength at this treatment was 62.3 μA. It is shown at Figure 2- that the three bacteria had maximum current strength at 40 hours and then undergo stationary phase but current value fluctuates about 1-3 μA. Then the three bacteria entering the death phase when reaction time was reaching around 70-80 hours where the value of the voltage decreased significantly even when it reached 100 hours there was still a voltage generated.

Figure 3 shows the voltage generated by each type of bacteria used in this study with a variation of reaction time 100 hours. The measurement indicated that the three types of bacteria were able to generate an increased current and voltage until 30 hours, next the bacteria undergoing stationary phase until 70-80 hours then it was dropped afterwards. The highest voltage value was generated by *Lactobacillus bulgaricus* with a maximum voltage of 529.3 mV. The second highest voltage was generated by *Lactobacillus casei* with a maximum voltage at this treatment was 486.6 mV. While the lowest voltage producer was *Streptococcus thermophillus* with a maximum voltage generated of 383,6 mV. The three bacteria were experiencing maximum current strength at 35-40 hours, then undergoing stationary phase until 70-80 hours and the bacteria entering death phase where the value decreased significantly. Interestingly, at 100 hours there was still a low voltage generated. These phenomena were in accordance with the bacterial growth phase consisted of four phases of growth namely, lag phase, log phase or exponential, stationary phase and phase of death. Lag phase is a phase of bacterial adaptation to the environment. Log or exponential phase is the phase of exponential increase in the number of bacteria. Increasing the number of these bacteria occur due to binary division in bacteria up to two cells. Then bacterial division occurs again in each cell and so on [3]. Increasing electric voltage and electric current strength at 0-40 hours was the result of bacterial log phase. Then at 40-70 hours the bacteria entering stationary phase i.e., a phase in which the growth rate equals to the death rate consequently the cell number was constant. The last phase undergone by bacteria was the death phase or declining phase where the amount of substrate was no longer meets the needs of bacteria resulting in bacterial death. In this case declining phase began to occur from 70 hours until the end of the observation at 100 hours but still a low voltage was generated.

Based on the results of the present study, the growth phase is the phase when *Lactobacillus bulgaricus* , *Streptococcus thermophillus* and *Lactobacillus casei* convert most lactose into electrical energy in microbial fuel cell. In addition, the results of this study were in accordance with other studies on growth of lactic acid bacteria using whey from milk fermentation using *Lactobacillus bulgaricus* [8]. The declining in the currents strength may be due to the decline in the number of bacteria as a consequence of the reduced availability of substrate. Another factor that occurred the decline phase was biofouling on proton exchange membrane. According to other studies biofilm enrichment during long-operating times may pollute the membrane fouling which decreases some factors such as concentration, ion transfer capacity and diffusion coefficient [5].

4. Conclusion

Whey, a waste that generated from cheese factory can be used as a substrate of lactose for bacteria in the microbial fuel cell double chamber reactor. The maximum voltage and current strength obtained from the use of *Lactobacillus bulgaricus* was 529.3 mV and 74.6 μA respectively. *Lactobacillus bulgaricus* maximum electricity at 40 hours and entering stationary phase with fluctuating values until 70 hours. While *Streptococcus thermophillus* peak point was reached at 50 hours and entering stationary phase until 80 hours, *Lactobacillus casei* reached the peak electricity at 45 hours.

5. References

[1] Biffinger, Justin C 2008 *J. Biotechnol* 2008 **2** 3 150-5
[2] Catal T L, Kaichang, Bermek H and Liu H 2007 *J. Power Sources* **175** 1 196-200
[3] Chae K J, et al 2008 *J. Energy & Fuels* **22** 169-176
[4] Masuda, M 2010 Flavins Contained In Yeast Extract are Exploited For Anodic Electron Transfer by Lactobacillus Lactis *Division of Applied Life Science* Kyoto University

[5] Rahimnejad M J, Haghparast T F, Najafpour G D and Ghoreyshi A A 2010 *J. Engineering Enviromental Science* **34** 289-292
[6] Richter H, Covalla S F, Johnson J P, Woodard T L, Orloff A L, Jia H, Zhang M and Lovley D R 2009 *J. Compilation Society for Applied Microbiology and Blackwell Publishing Ltd, Environmental Microbiology*
[7] Scott K and Murano C 2007 *J. Chemical Technology and Biotechnology* **82** 92-100
[8] Silva J C, et al 2011 Effect of The pH Growth on The Survival of Lactobacillus delbrueckii subsp. Bulgaricus to Stress Conditions During Spray-Drying Universidade de Aveiro Portugal

IOP Conf. Series: Earth and Environmental Science **131** (2018) 012048 doi:10.1088/1755-1315/131/1/012048

Can anaerobic digestion of sugar beet pulp support the circular economy? a study of biogas and nutrient potential

S Suhartini[1], S Heaven[2] and C J Banks[2]

[1] Department of Agro-industrial Technology, Faculty of Agricultural Technology, Universitas Brawijaya, Malang, Indonesia
[2] Faculty of Engineering and the Environment, University of Southampton, Southampton, United Kingdom

E-mail: ssuhartini@ub.ac.id

Abstract. Anaerobic digestion (AD), known as a biological process without oxygen to convert complex organic materials into biogas, is capable of processing large tonnage quantities of biomass, such as sugar beet pulp (SBP). In addition to biogas production, its use allows nutrients and organic carbon recycle back to agriculture through the spreading of digestate. Digestate still contains high amount of nutrients (N, P, K) for use as biofertiliser. The aims of this research were to determine biogas/methane potential as a baseline for comparison with performance in semi-continuous digestion, and to determine nutrient and potentially toxic elements (PTE) of digestate fractions with respect to their potential for utilisation in agriculture. The Biochemical Methane Potential (BMP) test was performed in triplicate against blank and positive controls over a period of 28 days with gas measured at regular intervals. Semi-continuous AD of SBP was operated under mesophilic and thermophilic condition for 206 and 165 days. The results indicated that SBP is a very promising feedstock for AD, with the average BMP of 0.321 l CH_4 g-1 VS and biogas potential of 0.605 l g-1 VS. Under semi-continuous operation, SBP also demonstrated positive results. Digestates from mesophilic and thermophilic AD of SBP contained useful quantities of N, P and K, with an acceptable Ni concentration in accordance to limits for PTE. These results suggest that digestate has the potential to be utilised on agricultural and arable land. This study illustrated the positive effects of applying AD to the achievement of economic savings and environmental-friendly performance.

1. Introduction

Sugar Beet Pulp (SBP) is the solid organic residue that arises from sugar beet processing and is generated worldwide, in particular in Europe and the USA. In the UK, which has an annual SBP production of around of 500,000 tones year^{-1} [1], it is mainly used as cattle feed as it contains all the basic constituents of forage including fibre and amino acids.

Anaerobic digestion (AD) plays an important role in providing a source of renewable energy. In addition to biogas production, AD as one of the promising technology for treating biomass produces digestate, which still rich in organic matter and macro nutrients (such as N, P dan K), thus suitable for biofertiliser or soil conditioner [2]. Also, AD is capable of processing large tonnage quantities, as well as the nutrient and organic carbon can be recycled back to agriculture through the spreading of digestate. The process also gives a reduction in biological risks and the potential for malodour [3]. AD also offers economic benefits through reduced material handling and transport costs and better revenue returns from energy production compared to the animal feed market [4-5].

The use of digestate as alternative soil conditioner or biofertiliser is likely to make a beneficial contribution in reducing negative impact on the environment as opposed to the use of synthetic fertilisers. It is because digestate usage can reduce nitrogen losses to groundwater, surface water and the atmosphere and/or minimise the carbon footprint through a reduction in green house gasses (GHG) emissions [6]. Although digestate may contain useful nutrients, it can also contain potentially toxic elements (PTE) (e.g. Cd, Ni, etc.) or pathogens [7]. Therefore, application of digestate to agricultural land is normally regulated to fulfil the requirements of both agricultural best practice and environmental protection

Before implementing AD at continuous or semi-continuous condition, studies under batch condition is necessary. This is known as the Biochemical Methane Potential (BMP) test which is used to determine the biodegradability of a substrate under anaerobic conditions by monitoring the cumulative methane production during the test period [8]. This measurement can provide important information including the anaerobic digestibility and potential biogas (methane) production from substrates which is useful for evaluating, designing, and optimising the AD process [9]. Besides measuring the conversion of organic matter to methane, the BMP test can also be used to determine the residual organic material amenable to further anaerobic treatment, the non-biodegradable fraction remaining after treatment, and the potential efficiency of AD for a particular substrate [10].

Therefore, this study was aimed at determining biogas/methane potential from SBP as a baseline for comparison with performance in semi-continuous digestion, as well as determining nutrient and PTE of digestate fractions with respect to their potential for utilisation in agriculture.

2. Materials and Methods

2.1. Feedstocks and inoculums

SBP was collected fresh from British Sugar's Wissington Factory, Kings Lynn, UK. The inoculum for BMP test was collected from a mesophilic digester treating municipal wastewater biosolids (Millbrook Wastewater Treatment Works, Southampton, UK). The inoculum for semi-continuous trials under mesophilic condition was prepared by mixing 1 part of digestate taken from an anaerobic digester treating sugar beet pulp (British Sugar, Wissington, UK) with 1 part of digestate taken from Millbrook Wastewater Treatment Works. While, the thermophilic digesters used an inoculum taken from this mesophilic municipal wastewater biosolids digester, which was then acclimated to thermophilic conditions.

2.2. BMP test set-up

The BMP test in this work was performed in 550 ml sealed bottles placed in a temperature controlled water bath at 37 ^{o}C. The inoculum-to-substrate (I/S) ratio used was 4:1 on a VS basis and the test was run over a period of 28 days. No supplements were added, as the inoculum used was known to be sufficiently rich in the required nutrients. Biogas was collected in perspex cylinders by displacement of a 75% saturated sodium chloride solution acidified to pH 2, in order to reduce losses of methane by dissolution. The height of the solution in the collection cylinder was recorded manually for a certain interval on a daily basis. Vapour pressure and salt solution density were taken into account in correction of gas volumes to a standard temperature and pressure (STP) of 0 ^{o}C and 101.325 kPa [11]. Samples for gas composition analysis were taken from the cylinders each time they were refilled, at intervals of no more than 7 days to avoid the risk of overfilling or losses of methane. The bottles were shaken each day before the gas level measurement was taken to provide mixing. Samples were run alongside blanks (inoculums only) and positive controls (cellulose powder from Sigma-Aldrich, Dorset-UK), all in triplicate.

2.3. *Semi-continuous trials set-up*
The experiments were carried out in four mesophilic (37 °C) and four thermophilic (55 °C) digesters with a 4-litre working volume. These were all seeded with an inoculum as described in section 2.1. Duplicate digesters were run at each temperature at organic loading rate (OLR) of 4 and 5 g VS l^{-1} day^{-1} for 3 hydraulic retention time (HRT), equivalent to 206 and 165 days. Each digester received trace element (TE) supplementation based on the amount of feedstock added.

2.4. *Analysis*
TS and VS determination was based on Standard Method 2540 G [12]. pH was measured using a Jenway 3010 meter (Bibby Scientific Ltd, UK) with a combination glass electrode, calibrated in buffers at pH 7 and 9.2. Total ammonia nitrogen (TAN) analysis was based on Standard Method 4500-NH_3 B and C [12]. Alkalinity measurement was based on Standard Method 2320B [12]. Biogas composition was quantified using a Varian Star 3400 CX gas chromatograph (GC), (Varian Ltd, Oxford, UK). Biogas volume was measured using a weight-type water displacement gasometer [11]. Nitrogen (N) was measured using Total Kjeldhal Nitrogen (TKN) analysis. Phosphorus (P) content of the digested sample was determined using a UV-Visible scanning spectrophotometer (Cecil 3000 series, Cecil Instruments). Potassium (K) was measured using a Spectr AA-200 Atomic Absorption Spectrometer (Varian, Australia). Trace element was analysed using ICP-MS at Severn Trent Laboratory Limited, UK. This was measured according to Standard Method 2510 [12] using a LF330 conductivity meter (WTW, Germany). COD was measured by the closed tube reflux method [12]. Turbidity was measured according to Bruus et al. [13] at a wavelength of 650 nm.

3. Results and Discussion
The characterisation on SBP in this study has been reported in Suhartini et al [14]. The results indicated a high potential for conversion to biogas due to its high VS content at value of 225.5 g kg^{-1} WW (or 93.14 % of TS) and high calorific value of 16.86 MJ kg^{-1} TS.

3.1. *BMP test results*
Figure 1 shows that the specific biogas and methane production of the SBP and the cellulose positive controls against the blank samples. The blank samples reached a stable value after approximately 14 days with 28-day specific biogas productions of 0.110, 0.092, and 0.092 l g^{-1} VS, giving an average value of 0.098 l g^{-1} VS. The positive controls had a rapid biogas production after a short lag time and reached a plateau after 10 days: the specific biogas yields were 0.566, 0.568 and 0.581 l g^{-1} VS with an average value of 0.572 l g^{-1} VS (Figure 1a). SBP with fresh inoculum demonstrated the same trend as the positive control, showing a short lag time and reaching a plateau after 10 days with net specific biogas productions of 0.577, 0.584, and 0.653 l g^{-1} VS, respectively, thus giving an average value of 0.605 l g^{-1} VS. This value was slightly lower than the values of 0.664 l g^{-1} VS reported in a previous study [15] for a different batch of material from the same source.

Figure 1b shows the specific methane production of the SBP and the positive controls against the blank sample. The blank samples had specific methane productions of 0.084, 0.071, and 0.071 l CH_4 g^{-1} VS, respectively; with the average value 0.075 l CH_4 g^{-1} VS. The positive controls had a specific methane yield of 0.296, 0.295 and 0.303 l CH_4 g^{-1} VS with an average of 0.298 l CH_4 g^{-1} VS. The positive controls showed reasonable agreement and the average specific methane yields were typical of values obtained for these materials, indicating the suitability of the assay conditions. Using fresh inoculum gave values of 0.302, 0.307 and 0.355 l CH_4 g^{-1} VS with an average BMP of 0.321 l CH_4 g^{-1} VS. This BMP value is fairly typical for crop and agro-waste substrates which have high cellulose content and relatively low lignin [16].

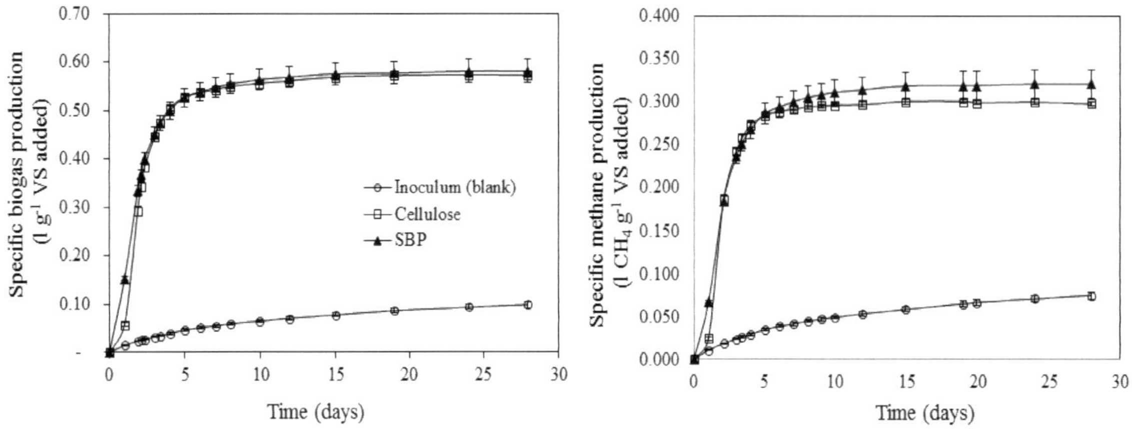

(a) **(b)**

Figure 1. Specific biogas (a) and methane (b) production from the BMP test: the blank samples, positive control, and SBP. Data are expressed as means of triplicate samples and bars represent range

3.2. Semi-continuous digestion performance

Although SBP was shown to be easily degraded under ideal conditions in the BMP test, with a good gas yield and VS destruction, in practice semi-continuous digestion under mesophilic conditions proved to be much more difficult, particularly at high OLR. Details findings on semi-continuous digestion performance has been reported in our previous study [14]. This study reported that thermophilic digestion was superior and better suited to the digestion of SBP when compared to mesophilic digestion. It could operate at higher organic loading rates, produced a higher specific methane yield, and offered better solids destruction. The mesophilic process when operated at a high organic loading rate suffered further disadvantages in that it showed a relatively high residual methane production, indicating a potential for methane emissions during any digestate storage phase. The major finding, however, was that thermophilic digestion was unlikely to result in foaming problems in the digester. The results for SBP therefore support those from other studies that have reported a better performance from thermophilic digestion than mesophilic in terms of: organic matter destruction, process stability, and specific biogas and methane production.

In this regards, it seems that in the case of SBP, operating AD at thermophilic condition is more preferable and stable.

3.3. Digestate characteristics

The characteristics of the whole digestate, solid and liquid fractions are presented in Table 1. Digestates from mesophilic and thermophilic AD of SBP contained useful quantities of N, P and K, with an acceptable Ni concentration with respect to limits for PTE. After separation, both liquid and solid fractions of digestates still have a high amount of nutrient N. These results confirm that the digestate has potential for application on agricultural land to substitute the use of mineral N, P and K fertiliser both as whole digestate, solid and/or liquid fraction.

In addition, the concentrations of TKN and ammonia N in the supernatant were in the range of 1 - 2 g l^{-1} (assumed density of supernatant = 1 kg l^{-1}), very high for directly discharge to watercourses. Furthermore, the removal of ammonia N in wastewater treatment is energy intensive as it requires ~4.3 g O_2 per g N while typical transfer efficiencies in aeration plants range from 0.9 to 3.1 kg O_2 kWh^{-1} [17]. There is considerable interest in development of technologies to recover N from liquid supernatants though struvite precipitation etc. [18], but these are not considered further here.

Furthermore, this study has also determined the system boundary for carbon, energy and nutrient footprint from AD of SBP (Figure 2), in both mesophilic and thermophilic conditions. The system boundary includes the anaerobic digestion process, feedstock and digestate transportation, digestate dewatering and application, and biogas utilisation/upgrade. The calculation showed that the potential mineral N fertilizer replacement was in the range of 14-18% of the net savings in GHG emissions, from 8 diferent scenarios (Figure 3). This indicates that the use of digestate as a biofertiliser has a useful role to play in terms of carbon balance or GHG emissions or environmental contribution as it makes up about 1/3 of the emissions saving in these cases.

Table 1. Nutrient and trace elemental content of mesophilic and thermophilic digestates.

Parameter	Mesophilic						Thermophilic					
	OLR 4			OLR 5			OLR 4			OLR 5		
	WD	SL	SF	WD	SL	SF	WD	SL	SF	WD	SL	SF
Macro nutrient (g kg^{-1} WW)												
TKN (N)	4.04	1.54	8.15	3.86	1.75	8.09	4.22	1.81	8.19	4.34	1.60	8.47
P	0.61	-	-	0.47	-	-	0.80	-	-	0.77	-	-
K	0.72	-	-	0.80	-	-	0.80	-	-	0.94	-	-
Elemental analysis (mg kg^{-1} WW)												
Co	0.27	-	-	0.23	-	-	0.34	-	-	0.30	-	-
Fe	66.16	-	-	39.73	-	-	15.32	-	-	12.96	-	-
Mo	0.14	-	-	0.19	-	-	0.12	-	-	0.11	-	-
Ni	0.74	-	-	0.94	-	-	0.68	-	-	0.57	-	-
Se	0.11	-	-	0.07	-	-	0.12	-	-	0.11	-	-
Other parameters												
Crude Protein (g kg^{-1} WW)*	25.23	18.58	9.63	24.12	20.81	10.92	26.35	23.94	11.29	27.09	37.94	9.98
Conductivity (mS cm^{-1})	9.7	-	-	1.3	-	-	14.1	-	-	14.6	-	-
Total dissolved salts (g l^{-1})**	7.76	-	-	1.04	-	-	11.28	-	-	11.68	-	-
COD (g l^{-1})	-	96.48	-	-	197.93	-	-	66.09	-	-	68.67	-
Turbidity (NTU)	-	2435	-	-	8015	-	-	1255	-	-	1235	-
Alkalinity (g $CaCO_3$ kg^{-1} WW)	16.1	8.53	39.41	11.6	5.61	16.99	16.9	8.65	34.99	16.7	9.08	39.39
TAN (g NH_3-N kg^{-1} WW)	1.40	1.25	1.85	0.80	0.55	0.85	2.09	1.56	1.90	1.98	1.63	2.04

Note: WD = whole digestate, SL = separated liquid, SF = separated fibre,
* = TKN x 6.25, OLR is expressed in g VS l^{-1} day^{-1},
** TDS = k_eEC, where k_e is a conductivity factor ranging from 0.55 - 0.8 (0.55 for natural water, 0.65 for hard/alkaline water, and 0.8 for inorganic nutrients), while EC is the electrical conductivity in $\mu S\ cm^{-1}$ at 25 °C. For the purposes of the calculation, a factor of 0.8 is selected.

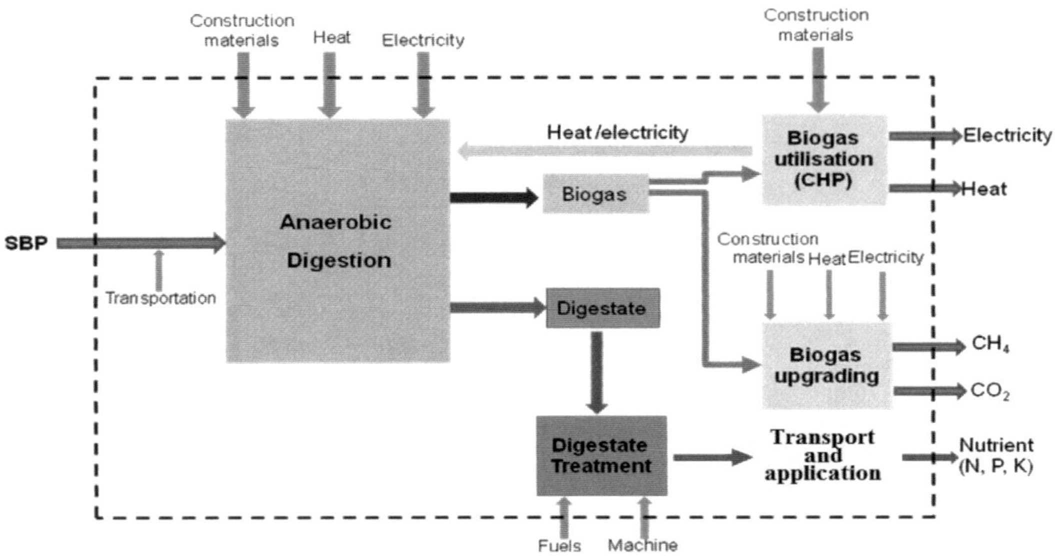

Figure 2. System boundary for the energy balance analysis on AD of SBP

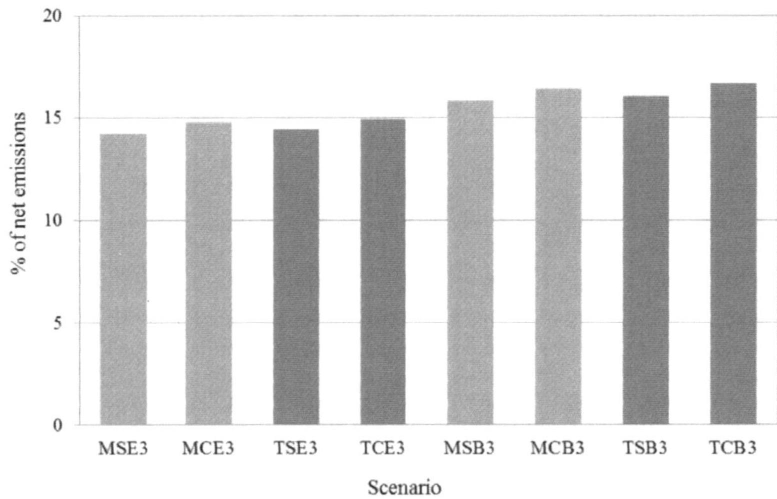

Figure 3. Mineral N fertilizer replacement as % of net GHG emissions savings. The scenarios were identifed in the following codes: M= mesophilic, T= thermophilic, S= simple system, C= complex system, E= biogas used in CHP unit for electricity and heat generation, B= biogas upgrading to biomethane. Unless specified, all loading rates are 3 kg VS m^{-3} day^{-1}

4. Conclusion

SBP is a very promising feedstock for AD either at batch and semi-continuous condition. Digestates from mesophilic and thermophilic AD of SBP are potential to be utilised on agricultural and arable land due to its high nutrient and low PTE. Implementation of AD treating SBP can provide economic savings and environmental-friendly performance. Therefore, AD can be used for supporting the circular economy through energy, nutrient, and biofertiliser production from biomass and organic waste.

Acknowledgments
Thanks are due to the Indonesian Government - Directorate General of Higher Education (DGHE) and Universitas Brawijaya - Indonesia for scholarship funding to support this research; and to British Sugar PLC for the supply of materials and good advice.

References
[1] British Sugar 2011 UK Beet Sugar Industry: Sustainability Report 2011. British Sugar plc and NFU Sugar. Retrieved from http://www.britishsugar.co.uk/Files/Beet_Sugar_Industry_Sustainability_Report_2011.aspx. [Accessed on 9 October 2017].
[2] Holm-Nielsen J B, Seadi T A, Oleskowicz-Popiel P 2009 The future of anaerobic digestion and biogas utilization *Bioresour. Technol.* **100** 5478-5484
[3] Berglund M, Börjesson P 2006 Assessment of energy performance in the life-cycle of biogas production *Biomass Bioenerg.* **30** 254-266
[4] Mata-Alvarez J, Macé S, Llabrés P 2000 Anaerobic digestion of organic solid wastes: an overview of research achievements and perspectives *Bioresour. Technol.* **74** 3-16
[5] Parawira W, Read J S, Mattiasson B, Björnsson L 2008 Energy production from agricultural residues: high methane yields in pilot-scale two-stage anaerobic digestion *Biomass Bioenerg.* **32** 44-50
[6] Chen S, Chen B, Song D 2012 Life-cycle energy production and emissions mitigation by comprehensive biogas–digestate utilization *Bioresour. Technol.* **114** 357-364
[7] Andreoli C V, Pegorini E S, Fernande F, dos Santos H F 2005 Land application of sewage sludge. In: von Sperling M, Chernicharo C A D L (eds.) *Biological Wastewater Treatment in Warm Climate Regions* IWA Publishing London UK
[8] Jensen P D, Ge H, Batstone D J 2011 Assessing the role of biochemical methane potential tests in determining anaerobic degradability rate and extent *Water Sci. Technol.* **64** 880-886
[9] Lim S J, Fox P 2013 Biochemical methane potential (BMP) test for thickened sludge using anaerobic granular sludge at different inoculum/substrate ratios *Biotechnol. Bioprocess Eng.* **18** 306-312
[10] Speece R E 1996 Anaerobic Biotechnology for Industrial Wastewaters Archae Press Tennessee USA
[11] Walker M, Zhang Y, Heaven S, Banks C 2009 Potential errors in the quantitative evaluation of biogas production an anaerobic digestion processes *Bioresour. Technol.* **100** 6339-6346
[12] APHA 2005 Standard Methods for the Examination of Water and Wastewater American Public Health Association, American Water Works Association Water Environment Federation Washington USA
[13] Bruus J H, Nielsen P H, Keiding K 1992 On the stability of activated sludge flocs with implication to dewatering. *Water Res.* **26** 1597-1694
[14] Suhartini S, Heaven S, Banks C J 2014 Comparison of mesophilic and thermophilic anaerobic digestion of sugar beet pulp: performance, dewaterability and foam control *Bioresour. Technol.* **152** 202 – 211
[15] Banks C J 2009 Additional methane potential from pressurised carbon dioxide treatment of sugar beet digestate and pulp Interim Report University of Southampton Southampton UK
[16] Brulé M, Bolduan R, Seidelt S, Schlagermann P, Bott A 2013 Modified batch anaerobic digestion assay for testing efficiencies of trace metal additives to enhance methane production of energy crops *Environ. Technol.* 1-12
[17] Tchobanoglous G, Burton F L, Stensel H D 2003 Wastewater Engineering (Treatment and Reuse) McGraw-Hil, Inc New York USA
[18] Rahman M M, Liu Y H, Kwag J-H, Ra C S 2011 Recovery of struvite from animal wastewater and its nutrient leaching loss in soil *J. Hazard. Mater.* **186** 2026-2030

IOP Conf. Series: Earth and Environmental Science **131** (2018) 012049 doi:10.1088/1755-1315/131/1/012049

Comparative study between Federer and Gomez method for number of replication in complete randomized design using simulation: study of Areca Palm (*Areca catechu*) as organic waste for producing handicraft paper

A Ihwah, P Deoranto, S Wijana and I A Dewi

Department of Agro-industrial Technology, Faculty of Agricultural Technology, Universitas Brawijaya, Malang, Indonesia

E-mail: azimmatulihwah@gmail.com

Abstract. The part of Areca Palm (*Areca catechu*) that economical is the seed. It is commercially available in dried, cured and fresh forms, while the fibre is usually thrown away. Cellulose fibers from agricultural waste can be utilized as raw material for handicraft paper. Laboratory research showed that Areca palm fibre contained 70.2% of cellulose, 10.92% of water, and 6.02% of ash. This indicated that Areca palm fibre is very potential to be processed as handicraft paper. Handicraft paper is made of wastepaper or plants which cointain celluloce to produce rough-textured paper. In order to obtain preferred sensory quality of handicraft paper such as color, fiber appearance and texture as well as good physical quantity such as tensile strength, tear resistance and grammage, the addition of wastepaper to provide secondary fibre and sometimes adhesive are needed in making handicraft paper. Handicraft paper making was one alternative to treat the solid waste and to reduce the use of wood fiber as paper raw material. The aim of this study is to compare the two most famous method, i.e. Federer and Gomez Method, for calculate the number of replications. This study is preliminary research before do the research in order to get the best treatment to produce handicraft paper. The Gomez method calculates fewer replications than the Federer method. Based on data simulation the error generated using 3 replicates of 0.0876 while using 2 replicates of 0.1032.

1. Introduction

The areca palm is one of plants that has not received serious attention, compared to other palm trees. In Indonesia, this plant is widely found in Sumatra (Aceh, North Sumatra and West Sumatra), Kalimantan (South Kalimantan and West Kalimantan), Sulawesi (South Sulawesi and North Sulawesi) and Nusa Tenggara (West Nusa Tenggara and East Nusa Tenggara) [1]. Areca palm is still limitedly utilized. This can be seen in the use of its seeds which is only eaten with betel and it is only used in some traditional events served as an ingredient which become a habit for generations in some specific areas in Indonesia. In other words, parts of areca palm which are frequently used so far are the leaves, stems and seeds [2].

Meanwhile, a part of areca palm that is areca coat nut is often wasted and its utilization is still low. Areca coat nut (Areca catechu L.) is included as waste containing cellulose. Laboratory analysis showed that the areca coat nut contains 70.2% cellulose content, 10.92% water, 6.02% ash. Given that fact that the current utilization of areca coat nut is still limited in number, so innovations are needed to take advantage of areca coat nut such as utilize it as an art paper material.

The potential pulp market in Indonesia is quite high and this can be seen from the pulp and paper industry which numbers keep increasing by time. Indonesia is in the 9th rank having 3.6% global capacity of pulp production after China, USA, Japan, Germany, Canada, Finland, Sweden, and South Korea. It can be seen from the number of pulp exports in 2010 which reached 2.6 million tons [3]. The pulp and paper industry in Indonesia is still far from the expectation, and the need of pulp for art paper in the country reaches 4 million tons/year [4].

Pulp is a paper slurry derived from the separation of cellulose fibers from lignin that comes from fibrous materials both wood and non-wood through some processing [5]. The purpose of the pulping process is to degrade lignin and to separate it from cellulose by minimizing the occurrence of cellulose damage. It can be concluded that the quality of the pulp is affected by the small amount of lignin present in the pulp, the smaller the lignin content in the pulp, the better the quality. The statement is also supported by some research on pulp stating that good pulp is the one that has the least lignin content because lignin will affect the physical properties and quality of the produced pulp [6].

In the pulping process, there are several methods of manufacture which are mechanically, chemically and semi-chemically. The optimal pulping process for non-timber plant fibers is the alkaline process using NaOH, NaOH solution can solubilize the lignin in the fibers. The chemical pulping process is carried out by developing a soda process, an alkaline cooking process with NaOH solution. NaOH solution can degrade fibrous bonds therefore the formation of pulp is accelerated [7]. In general, using NaOH results dark fiber, bleaching (brightening) is necessary to be done using H2O2 solution.

The presence of NaOH, both in the pulp and paper industry is one of the most important components. From the research conducted by Saenah [4], the best conditions of NaOH use of pulp and kenaf characteristics are 12% and 14%. In addition, a research by Onggo and Tri Astuti[8] stated that the effect of pulping process on the color of art paper from reedstakes 10% NaOH as optimal result. Based on the above research, appropriate use of NaOH may vary depending on the type of material. So in this study we will look for proper NaOH concentration to produce quality pulp with lignin content, expected cellulose from midrib and nipah leaf.

Research on areca fiber used as art paper is conducted in two stages. The first stage was to find out the proper concentration of NaOH and the quality of the pulp produced from the pin fiber which in the process is done chemically by the process of soda using NaOH solution. The best treatment in the first stage is used in the second stage of research, which was the manufacture of art paper from pulp of areca coat nut.

One of the interesting physical properties of paper is its tensile strength. The tensile resistance is defined as the paper's resistance to the tensile force acting on both ends of the path to break, expressed in units of force per unit of test width (kN/m)[9]. Tensile strength is important for assessing the strength of paper when stretched and subjected to force on each side other than that tensile strength plays a significant role in the formation of pulp into sheets of paper[10].

Research that involves treatment to produce the response requires a proper experimental design to produce research with reliable conclusions. There are three important elements in the Experiment Design that are treatment, environmental control and replication[11]. The repeat function in an experiment is to provide a means of estimating experimental errors (error = random = residual = error), reducing experimental or research errors and allowing for a more rigorous and precise estimation [12].

There are several formulas in calculating the number of replicates in the experimental design. This paper describes the comparative method of Federer and Gomez Methods in counting replications in the experimental treatment of adding NaOH concentration in pulp fiber manufacturing based on the results of the tests that have been done. The purpose of this comparative study was to find the number of replications with the smallest error for designing advanced research which is planned to use 8 treatments ie Newspapers: (Pulp of areca coat nut + NaOH) = 90: 10, Newspapers: (Pulp of areca coat nut + NaOH) = 80: 20, Newsprint paper: (Pulp of areca coat nut + NaOH) = 70: 30, Newsprint: (Pulp of areca coat nut + NaOH) = 60: 40, Cardboard Paper: (Pulp of areca coat nut + NaOH) = 90: 10, Cardboard: (Pulp of areca coat nut + + NaOH) = 80: 20, Cardboard: (Pulp Shell + Semi + NaOH) = 70: 30, Cardboard Paper: (Pulp of areca coat nut + NaOH) = 60:40. Before that, the data simulation using the result test of tensile strength from eight materials was processed (Table 1). Data simulation is done to estimate what method that produce smaller error.

2. Materials and Methods

This research was a preliminary research prior to the research on paper tensile strength test with 8 treatments. This study uses comparative studies to compare two methods of counting the number of replications. The bootstrap resampling method is used to simulate the error comparison obtained using Federer Method and Gomez Method. The data used for resampling is a paper tensile strength test data from paper with 6 treatments and 2 types of paper circulating on the market used as control.

2.1. The Federer Method

The number of samples according to Federer's formula[13] follows the following formula:

$$t(r-1) > 15 \quad (1)$$

where t = number of treatments and r = number of replications

2.2. The Gomez Method

According to Gomez[14], Test F should be counted only if the degrees of error free are six or more. Where the degree of error in Complete Randomized Design is calculated by $t(r-1)$.

3. Results and Discussion

The data used for the simulation is the data of tensile strength test of paper using from 6 types of pulp maker and two types of art paper as control. In tensile strength test of art paper, there are 6 samples tested that are, first, areca paper without additional filler with bleaching treatment using H_2O_2. Second, areca paper with 24 hour drying treatment. Third, betel paper plus re-use paper filler with 50:50 ratio. Fourth, areca paper with boiling material for 1 hour. Fifth, betel paper plus used cardboard filler with 50:50 ratio. Sixth, betel paper plus filler used newspaper with 50:50 ratio. All samples have been through cooking process for 120 minutes using 20 % NaOH. The sample used as a control is a white art paper from a straw obtained from one of the paper stores sold in Malang.

The tensile strength test process was conducted using Brazillian Test instrument. The testing process is carried out in the following steps: first, all the samples tested along with the control sample being tested are measured regarding its width respectively. Based on these measurements, the results obtained is that each paper has a length of 8 cm with a width of 4 cm, so a paper area was 32 cm^2 each. The second step was the measurement of the applied tensile force carried out using a micrometer tool. The tensile force was the maximum value on the micrometer reading of the dial when the paper is pulled until the sample is broken or torn at the time of the test in kilogram force. In Apriani study (2016) it is stated that the average yield of tensile strength of corn stem fiber art paper ranged between 0,213 - 0,554 (kgf/mm^2). The mean value of the highest tensile strength of 0.554 kgf/cm^2 was obtained from the composition of corn stalk raw materials and scrap paper with ratio : 25%:75% and cooking time of 180 minutes. The average value of the lowest tensile strength is 0.213 kgf/mm^2 on the composition of the raw material of corn and sracp paper with ratio 100%:0%, with a cooking time of 60 minutes. The higher the composition of corn stalks used in making art paper, the tensile strength of the resulting paper tends to decrease. It is suspected that the lignin content of corn stalk pulp is higher compared to waste paper, resulting in the tensile strength of the paper being low. The data on paper tensile strength test in the preliminary study is presented in Table 1, where each ingredient is added NaOH with a concentration of 20% based on material weight.

Table 1. Paper Tensile Strength Testing.

Material	Cross-sectional area (cm^2)	F (kgf)	P (F/A) ($kgf.cm^{-2}$)
Bleached areca nut paper	32	2.2	0.069
Boiled dry areca nut without glue	32	0.4	0.013
Material	Cross-sectional area (cm^2)	F (kgf)	P (F/A) ($kgf.cm^{-2}$)
50% re-use: 50% areca nut	32	1.3	0.041
Boiled areca nut (for 1 hour)	32	1.4	0.044
50% Cardboard : 50% areca nut	32	3.7	0.116
50% newspaper : 50% areca nut	32	1.2	0.038
White paper	32	1.3	0.041
Red paper	32	12.3	0.384

Data P in the column used for simulation. Data on resampling results with eight treatments are presented in Table 2 and Table 3.

Table 2. Result of resampling using 3 replications.

Replication	Treatment							
	T1	T2	T3	T4	T5	T6	T7	T8
1	0.013	0.013	0.041	0.116	0.038	0.384	0.013	0.069
2	0.013	0.038	0.038	0.038	0.069	0.044	0.044	0.038
3	0.044	0.041	0.044	0.116	0.044	0.044	0.041	0.116

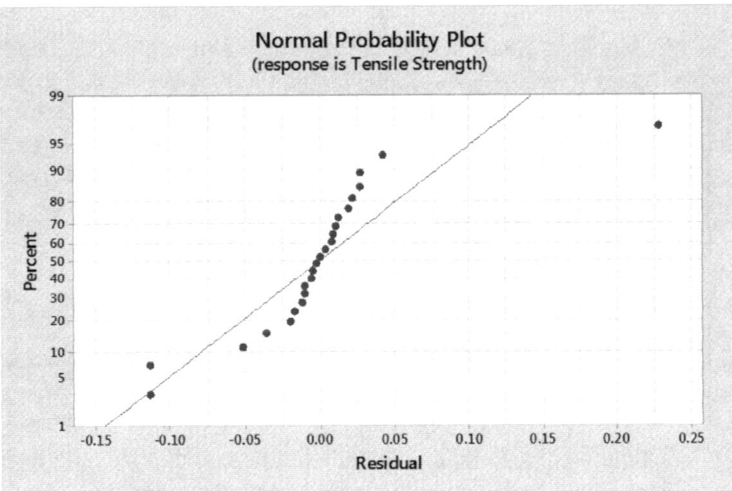

Figure 1. Normal Probability Plot of Residual Using 3 Replication

Table 3. Result of resampling using 2 replications

Replication	Treatment							
	T1	T2	T3	T4	T5	T6	T7	T8
1	0.116	0.041	0.013	0.038	0.384	0.069	0.384	0.116
2	0.069	0.041	0.116	0.041	0.384	0.013	0.041	0.384

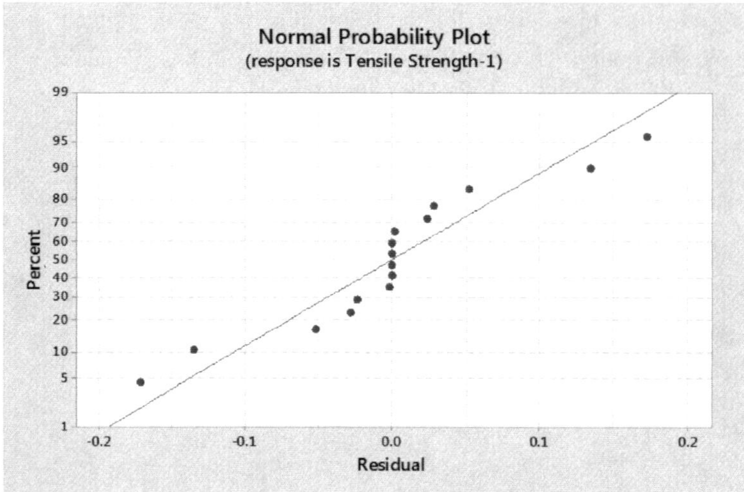

Figure 2. Normal Probability Plot of Residual Using 2 Replication

3.1 The Federer Method

In the follow-up study, there are 8 treatments planned so that according to the recalculation formula using Federer method, at least 3 replications are obtained. Analysis of variance shows that sum square error of 0.08678 was obtained.

3.2 Gomez Method

Using the Gomez method, the replication can be reduced in number because the recalculating formula in the Gomez method requires $t(r-1) \geq 6$, so it can be determined that using 2 replications on each treatment is still valid to do. Analysis of Variance shows that the sum square error of 0.1032 was obtained.

From the results it is obtained that the higher the number of replications then the error becomes smaller. This is consistent as stated by Montgomery that the replication is used to minimize the error [2]. The researcher can adjust the recalculation method using various considerations. For example in terms of price of raw materials in making the sample, the length of the production process, and so forth. But the researchers have to note that when replication is taken with the extra-large number, the error will bigger than the reasonable of replication [15].

4. Conclusion

The Gomez method calculates fewer replications than the Federer method. Error produced using 3 and 2 replicates were 0.0876 and 0.1032, respectively. The next study is focusing on the amount of replication to minimize the error.

References

[1] Pertanian dan Perkebunan Indonesia. Retrieved from https://id.wikipedia.org/wiki/Pertanian_dan_Perkebunan_di_Indonesia. [Accessed on 6 December 2017]

[2] Balai Penelitian Tanaman Palma 2010 Prospek Pengembangan Tanaman Pinang Sulawesi

[3] Pusat Kebijakan Perdagangan Luar Negeri Badan Pengkajian dan Pengembangan Kebijakan Perdagangan 2015 Indonesia Export Target Analysis (In Indonesia) Kementrian Perdagangan. Jakarta

[4] Endah A 2016 The Influence of Raw Material Composition and Length of Cooking Time on Drag Strength on Art Paper Making from Corn Stover and Used Paper Waste (In Indonesia)

J.Mekanika dan Sist.Termal **1** **2** 38-42

[5] Saenah E 2002 Effect of Soda Dose on Abaca Pulp and Pulp Kenaf Pulping Soda-Antaquinon Characteristics (In Indonesia) Skripsi Jurusan Kimia FMIPA Universitas Brawijaya Malang

[6] Fengel D dan Wegener 1995 *Wood, Ultrastructure Chemistry and Reactio*ns (In Indonesia) Penerjemah H. Sastrohamidjojo Gajah Mada University Press. Yogyakarta

[7] Malo B A 2004 *Membuat Kertas Seni dari Pelepah Pisang* (Making Art Paper from Banana Bark) (In Indonesia) Kanisius Yogyakarta

[8] Onggo H dan Tri Astuti J 2004 Effects of Sodium Hydroxide and Hydrogen Peroxide Against Rendeman and Pulp Color from Leaf Fiber (In Indonesia) *J. Tolok Ukur Pemanfaatan Serat Alam Bagian Proyek Penelitian dan Pengembangan Otomotif, Transportasi, dan Energi LIPI*

[9] Giancoli D 2000 *Physics for Scientists & Engineers Third Edition* Upper Saddle River: Prentice Hall

[10] E J Pavlina and C J Van Tyne 2008 Correlation of Yield Strength and Tensile Strength with Hardness for Steels *J. of Mat. Eng. and Perform* **17** 6

[11] Component of an Experimental Study Design. Retrieved from https://stats.libretexts.org/Core/Experimental_Design/Analysis_of_Variance/Components_of_an_experimental_study_design. [Accessed on 6 December 2017]

[12] Montgomery C D 2002 *Design and Analysis of Experiment* John Wiley and Son, Inc United State of America

[13] Federer W T 1967 *Experimental Design, Theory and Application* Oxford and IBH Publ. Co New Delhi

[14] Gomez and Gomez 1995 *Statistical Procedures for Agricultural Research* John Wiley and Son, Inc United State of America

[15] Ihwah A 2012 Bootstrap in Censored Data to Estimate The Regression Coefficient on The Cox Regression Models (In Indonesia) Thesis of Postgraduate in Mathematics Gadjah Mada University Yogyakarta

IOP Conf. Series: Earth and Environmental Science **131** (2018) 012050 doi:10.1088/1755-1315/131/1/012050

The existence of microplastic in Asian green mussels

A Khoironi[1], S Anggoro[1,2] and Sudarno[1,3]

[1] Doctoral Program Environmental Sciences of Diponegoro University, Semarang, Indonesia
[2] Faculty of Fisheris and Marine Science of Diponegoro University, Semarang, Indonesia
[3] Environmental Engineering Departement of Diponegoro University, Semarang, Indonesia

E-mail: batik.ayacantik@gmail.com

Abstract. Due to resistance of polymer as basic properties of plastic, several studies have been conducted to understand the fate of plastic debris in the marine environment. Degradation is the most important process to control distribution of plastic debris a long the marine environment until the existence of plastic in the food chain. The physical and chemical changes of plastic because of degradation process will lead to the release of polluted substances which eventually more toxic for the environment. Furthermore, when plastic degraded become a microplastic will lead to easy ingested by biotic such as mussell which commonly consumed by humans. The aim of this research was to investigate the concentration of microplastic adsorbed and ingested by mussels considering characteristic of sea water. About 30 samples have been collected from 3 different locations that is brackish water (31 ppb), high salinity (36 ppb) and low salinity (33 ppb) for measuring a number of microplastic in mussels on three different salinity. The result of microstructure analysis by microscope showed that mussel evaluated from the marine environment contaminated by microplastic with average size of 211.163 μm. In high salinity sea water, microplastic found in mussel was greater than low salinity and brackish water. The SEM/EDX analysis showed the presence of SiO_2 0.14 % (w/w), Na_2O 24.27 %(w/w) and Al_2O_3 0.27 % (w/w) in the microplastic obtained in the mussel indicating the components which are mostly found in the plastic industries. The amount of microplastic in mussell could be used as pollution indicator in the marine environmental by plastic waste.

1. Introduction

Since plastics debris has been found as a major component of the mix of debris in the oceans [1] and also contributed to 80% of all anthropogenic debris in the oceans [2], many studies had been conducted to investigate the fate of the plastic debris. One of the most important phenomena occurs in the marine environment is plastic degradation due to chemical, physical and biological processes. The degradation will lead to bond scission of chemical chains and then change its molecule structure [3].

However, when the plastic debris discarged to the marine environment, they are eventually exposed to a combination of chemical and physical components as well as subtances produced during degradation process [4]. The plastic degradation is also producing chemicals which is potentially harmful and toxic. The toxicity is not only because the release of organic pollutants from plastic surface and chemicals additives leaching out of the plastic, but it is also due to the release of chemical produced by degradation of the plastic itself such as microplastic.

Law and Thompson [5] categorized microplastic as a plastic materials or fragments in diameter less than 5 mm. Microplastic originated from various sources that dirrectly discarged into the oceans as small particles is defined as primary microplastic i.e. powder and pellets, and microplastic as results of the degradation processes of macroplastic into smaller fragments is categorized as a secondary microplastics [6-7].

Microplastic has been found in almost every marine species due to plastic composition and marine environmental conditions which significantly affecting their distributions [8]. Pollution by microplastic have a negative effect related to entanglement and ingestion of lower trophic level prey [9]. Furthermore, Lambert et al,[4] has found that non-selective and filter-feeding consumers like fish, shell and some sea species could be susceptible to ingest microplastic which lead to the potential passage up the food chain to secondary and tertiary consumers.

Asian green mussels (*Perna viridis*) is one of the very common bivalves belonging to the Mytilidae family that have high economic value [10]. As a highly nutritious food, green mussels is very popular to people in Indonesia and it is cultivation extensively in Tambak Lorok coastal area. The cultivation of green mussels in Tambak lorok is done by installing bamboo rumpon as a place to stick and growth of the green mussels. Green mussels is considered as feeder filter that is able to filter organic particles, vegetable plankton, animal and microorganisms in water. The activity of green mussels is influenced by water temperature, salinity, and concentration of food particles in water. The growth is attached to various substrates in water using a tool of fibers called byssu. The ability of green mussels to accumulate various types of pollutants making green mussels is used as bioindicators for pollution of an environment in the waters of the sea. This research has purposed to investigate the existence of microplastic in mussels that considering characteristic of sea water, to prove the assumption that the product of primary plastic degradation involved in the food chain.

2. Materials and Methods

2.1. Sample collection

Location of sampling is along the Java Sea in Tambak Lorok coastal area. Sample were acquired directly from 3 locations in September 2017. Sampling locations are based on salinity levels of sea water including high salinity (36 ppb) is about 5 km from coastal area , low salinity (33 ppb) is about 0,5 km from coastal area and brackish water (31 ppb). 10 mussels are collected from each location. The level of contamination from seawater also becomes one of the considerations in choosing the sampling location. All mussels acquired are adult mussels with size 4 - 5 cm that cultivated and stick on bamboo *rumpon* at a depth of 2 meters. Fresh mussels are then stored in ice boxes to be transfered to the laboratory and kept in refrigerators for analysis preparation.

Sumber: Bappeda Kota Semarang, 2012

Figure 1. Sampling location in Tambak Lorok Coastal area, Java Sea.

2.2. Preparation sample

The initial step was to ensure that H_2O_2 as a solvent has capability in separating organic compounds from soft tissue [11]. To make sure that H_2O_2 only leach the organic compounds but will not disolved microplastic, the method was done by dissolving Polyethylene Terephtalate (PET) plastic that has been cut at the size of 1 mm x 0.5 mm into the glass container containing 30% H_2O_2 solution, seawater and distilled water. Furthermore, mussels were carefully removed from their skin and fed into 200 mL of 30% H_2O_2 to digest the organic matter of the soft tissue. The saturated saline solution (1.2 g/mL) was added to dissolve liquid of the soft tissue to separate the microplastic. The solution was then mixed and the bottle was stood for some times. The overlying water was filtered with filter paper and then the filter was placed in the petri-disc for next analysis step.

2.3. Identification and validation of microplastic

Prior to identifying the samples, samples containing microplastic with size of 1 mm x 0.5 mm was used as control in determining microstructural visualization of microplastic through microscope. Filter were visually analyzed using a microscope (Olympus Model MD 50 at magnification 10 x). Microplastic were classified into four morphotypes i.e. fiber, sphere, flake and fragment [11]. After obtaining a filter containing microplastic, the filter was then dried using a freeze dryer at a temperature of 0 - 5°C for 3 days. The dried sample was used for morphology examination using Scanning Electron Microscopy (SEM) as well as confirmation of the element composition of the particles. The result of both SEM and Electron Dispersive X-Ray (EDX) were used to verify that the items were microplastics.

IOP Conf. Series: Earth and Environmental Science **131** (2018) 012050 doi:10.1088/1755-1315/131/1/012050

3. Results and Discussion

From the results of visual analysis using microscope (Figure 1), it can be seen that there was a significant variation of both microplastic size and shape based on green mussel location during the sampling. The morphological form of microplastic particles inside mussels in high salinity can be seen in Fig.1. A total of 30 green mussels taken from different locations gave different results that were influenced by the location of the breeding mussels. In mussels that breed in high saline waters (36 ppb) where the location is located about 5 km from the coastline, microplastic particles size of 51.31 to 87.43 μm are found and the number of plastic particles was in average of 5 particles per 0.25 gram of sample weight.

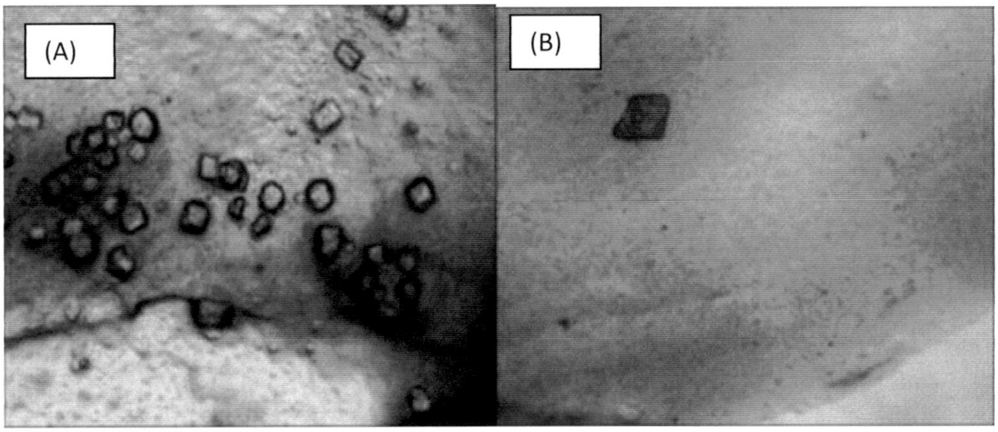

Figure 2. (A). Microplastics detected in green mussels grown in the high salinity sea water, (B). In the low salinity sea waters.(at magnification of 10x)

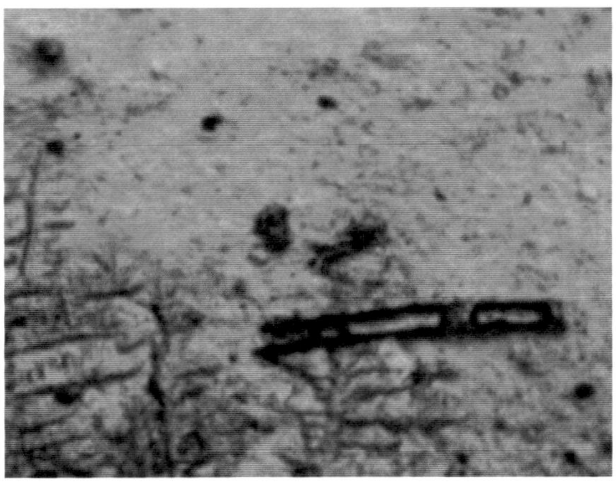

Figure 3. Microplastics detected in the brackish water green mussels. (at Magnification of 10x)

The microplastic found in mussels living in the sea waters at lower salinity (33 ppb) was fewer than microplastic particles found in high salinity sea water mussel. The microplastic sizes found in the low salinity water was in the range of 45.33-124.74 μm and has concentration of 2 particles per 0.25 gram. On other hand, at brackish environment (31 ppb), the microplastic sizes showed larger than other environments (154.76-232 μm). The average concentration of microplastic in brackish water was 1 particle per 0.25 gr sample. Figure 2 shows the microplastic detected in brackish mussel. Furthermore, in order to support the microscope analysis, the SEM analysis was employed to evaluate microplastic morphology form in mussels. From the SEM/EDX result, the identified tissue showed components of Silica (SiO_2), Alumina (Al_2O_3) and Sodium Oxide (Na_2O) which are mainly used in plastics industry. This also strengthen the previous observation by microscopic analysis.

Table 1. Size range and average concentration of microplastics in green mussels tissue.

Location by degree of salinity	Microplastic size (μm)	Exposure concentration (particles/0.25 g)
High Salinity	51.31 – 87.42	5
Low Salinity	45.33 – 124.74	2
Brackish	154.76 – 232.00	1

In overall, this study (Table 1) shows that location of sampling greatly affected the number of microplastic in the mussels. At longer distance from the costal line (high salinity) there was less sizes of microplastic founded in the mussel, however the concentration showed higher. This is probably due to high salinity water will degrade more plastic than the lower salinity. Moreover, at high salinity water, the accumulation of plastic debris will also larger which consequently increase the number of microplastic particles in the mussels. The chemical, physical and biological degradation processes in higher saline water will greatly contribute to the increasing number of microplastic outcomes from primary plastic degradation [4, 7]. Salinity levels of seawater also play an important role in the accumulated microplastic concentrations in mussels. This is because the ingested microplastic has the potential to be taken by epithelial cells from the intestinal tract and even translate through the intestinal wall to the open mussels system [12]. All the energy used for the digestive system metabolism in mussels is strongly influenced by the salinity of the waters where mussels grow. The salinity of the waters is also one of the physiological factors that affect the utilization of feed and growth of mussels [13] and this becomes the main cause of the amount of accumulated microplastic in higher salinity mussels than lower salinity mussels.

4. Conclusion

This study discovered the presence of microplastic inside the mussels grown in Tambak Lorok coastal area. It also showed that the level of pollution by microplastic in marine waters has exceeded the limit of tolerance of inside living biota. Furthermore, this study also depicts that green mussels become potential medium for microplastic transfer from seawater to the food chain and lead mussel can be considered as one of the bioindicator of pollution level of microplastic in the aquatic environment.

References

[1]. Galgani, F., Hanke, G and Maes, T. 2015. In M.Bergmann, L.Gutow & M.Klages (Eds), Marine anthropogenic litter (pp.29 – 56). Berlin: Springer.
[2]. Derraik, J.G.B., 2002. *Mar.Pollution Bulletin*.**44(9)**, 842 - 852
[3]. Shah, AA., Hasan, F., Hameed, A., Ahmed, S., 2008. *Biotechnol. Adv.* **26(30)**, 246-265
[4]. Lambert, S, Sinclair, J.C. & Boxall, A. 2014. *Environment Contamination and Toxicology*, **227**,1-53
[5]. Law, KL and Thompson, R C. 2014. *Science*,**345**,144-145
[6]. Andrady, A. L .2011. *Mar Pollut Bull.* **62(8)**, 1596-1605
[7]. Cole M, Lindeque, P, Halsband, C, Galloway ,T. S. 2011. *Mar Poll Bull.* **62(12)**, 2588-2597
[8]. Luser, A L *et al.* 2015. *Environ Pollution.* **199**, 185-91
[9]. Farrel, P.,Nelson, K., 2013. *Environ. Pollut.***177**, 1-3
[10]. Normah, I & Asmah, A. H. 2016. *International Food Research Journal.* **23(4)**, 1409 - 1417
[11]. Li, J. et al 2016. *Environmental Pollution* **214**. 177 – 184
[12]. Browne, M.A *et al* 2007. *Integr. Environ.Assess.Manag.* **3**, 559 - 561
[13]. Anggoro, S et al. 2008. *Laporan Penelitian RISTEK.LPPM, Undip, Semarang* . 147

IOP Conf. Series: Earth and Environmental Science **131** (2018) 012051 doi:10.1088/1755-1315/131/1/012051

Optimization of free radical scavenging capacity and pH of *Hylocereus polyrhizus* peel by Response Surface Methodology

A W Putranto[1], S R Dewi[1], Y Puspitasari[2] and F A Nuriah[2]

[1] Department of Agricultural Engineering, University of Brawijaya, Malang, ZIP 65145, Indonesia
[2] Department of Food Science and Technology, University of Brawijaya, Malang, ZIP 65145, Indonesia

E-mail : angkywahyu@ub.ac.id

Abstract. Red dragon fruit (*Hylocereus polyrhizus*) peel, a by-product of juice processing, contains a high antioxidant that can be used for nutraceuticals. Hence, it is important to extract and investigate its antioxidant stability. The aim of this study was to optimize the free radical scavenging capacity and pH of *H. polyrhizus* peel extract using Central Composite Design (CCD) under Response Surface Methodology (RSM). The extraction of *H. polyrhizus* peel was done by using green-Pulsed Electric Field (PEF)-assisted extraction method. Factors optimized were electric field strength (kV/cm) and extraction time (seconds). The result showed that the correlation between responses (free radical-scavenging capacity and pH) and two factors was quadratic model. The optimum conditions was obtained at the electric field strength of 3.96 kV/cm, and treatment time of 31.9 seconds. Under these conditions, the actual free radical-scavenging capacity and pH were 75.86 ± 0.2 % and 4.8, respectively. The verification model showed that the actual values are in accordance with the predicted values, and have error rate values of free radical-scavenging capacity and pH responses were 0.1% and 3.98%, respectively. We suggest to extract the *H. polyrhizus* peel using a green and non-thermal extraction technology, PEF-assisted extraction, for research, food applications and nutraceuticals industry.

1. Introduction

The anthocyanin pigment is one of the water-soluble pigments that is widely used as a natural ingredient for consumption because it is easily absorbed by the body. In addition, it has the capability of capturing free radicals and high antioxidant capacity and shows the inhibitory effect on the growth of some cancer cells [1]. Recently, the research focussed on utilization of waste material from the industrial processing of fruits and vegetables as a good source of bioactive compounds. The types and amounts of phytochemicals retained in the discarded material are important to be considered for deriving value-added ingredients [2]. Red dragon fruit (*Hylocereus polyrhizus*) peel was potentially to be extracted since it has a high anthocyanin pigment that can provide natural dyes to the food, high antioxidant activity 10 times more than flesh [2], high flavonoids and betacyanins content, also a strong inhibitor of the growth of melanoma cells than the flesh [3].

The extraction technology to extract *H. polyrhizus* peel that are commonly used are batch maceration and Microvawe Assisted Extraction (MAE). However, the technology still has many

IOP Conf. Series: Earth and Environmental Science **131** (2018) 012051 doi:10.1088/1755-1315/131/1/012051

shortcomings including: long extraction process, more solvents needed [4], MAE needs heat. The heating process can cause degraded pigments during the process, decreasing color quality, and affecting antioxidant activity [5].

In recent years, non-thermal PEF-assisted extraction with low electric field strength has been reported in several studies to improve the high extraction rate of some antioxidants such as total carotenoid from carrot pomace (2.5 kV/ cm) [6], anthocyanin from purple fleshed potatoes and grape pomace using 1.5 kV/cm and 3 kV/cm of electric field strength respectively [7]. The non thermal extraction using PEF technology has many advantages such as short extraction time consequently the energy used is very low [8], maintaining antioxidant qualities in heat sensitive materials [9], as well as high extraction yield due to damage to cell membrane and improved extraction characteristics [10].

PEF on food processing cannot be separated from the cell electroporation that inhibit the synthesis or formation on the membrane phospholipids. The formation of these pores depends on the electric field strength and the treatment time [11]. The studies of cell division to retrieve the valuable substances in waste product have not been determined, moreover the extraction of *H. polyrhizus* peel using non-thermal method, PEF-assisted extraction has not been fully explored. The present study is expected to explore the *H. polyrhizus* peel as a by-product juice processing to be a source of antioxidants and a natural colorant substitute for synthetic dyes. Therefore, the aim of the present study was to optimize the free radical scavenging capacity and pH of *H. polyrhizus* peel extract using Central Composite Design (CCD) under Response Surface Methodology (RSM).

2. Materials and Methods

2.1. Plant material, chemical and reagent

Fresh red dragon fruit (*Hylocereus polyrhizus*), a local type, were purchased in a market in Malang City, East Java, Indonesia. The chemicals used such as potassium chloride, hydrochloric acid 0.2 N, Na-acetate, distilled water and 2,2'-diphenyl-b-picrylhydrazyl (DPPH) reagent 0.2 mM were purchased from Sigma-Aldrich Chemicals.

2.2. Apparatus and instruments

Pulsed Electric Field (PEF) batch system used in this study was the main apparatus. In present study, the PEF was applied in green extraction of *H. polyrhizus* using low electric field strength. The treatment chamber (capacity of 9.1 L) was made of stainless steel as well as food grade material and configuration of electrodes inside adopts from coaxial models. PEF is also being equipped with control systems such as input voltage regulator, frequency regulator and timer. The PEF apparatus is specifically designed for the improvement of mass transfer, with electric field ranged from 1.3 to 6.45 kV/cm. The frequency from 0.5 to 31 kHz, the timer (Omron AT-8N) ranged from 0.05 second to 100 hours and pulse width produced of 78 μs. While the other instruments which were used in this study such as pH meter, Spectrophotometer UV-Visible (Genesys 10), analytical scale (Denver Instrument M-310), vortex, aluminum foil, filter paper and cuvette.

2.3. Green-PEF-assisted extraction of H. polyrhizus

The fresh red dragon fruit (*H. polyrhizus*) separated from the flesh and from the green fruit petals. Then, the sample were cut into dimension of 2x12x1 mm to enlarge the surface area of *H. polyrhizus* peel. In line with a green extraction concept, PEF assisted extraction using distilled-water as a solvent, since it was a green and safe solvent for food and also was a polar solvent to dissolve the anthocyanin. The sample and solvent with ratio of 1: 5 (w/v) were blended and filling into treatment chamber. Then set the PEF frequency of 8 kHz. The electric field strength and extraction time was adjusted by following the actual variables on run order of central composite design matrix (Table 1). After the extraction, the *H. polyrhizus* peel extract was placed in a dark bottle to prevent degradation of the compounds contained in the extract due to exposure from the outside before analysis.

IOP Conf. Series: Earth and Environmental Science **131** (2018) 012051 doi:10.1088/1755-1315/131/1/012051

2.4. DPPH radical-scavenging measurement
The ability of anthocyanin in extracts and fractions from *H.Polyrhizus* peel to scavenge the 2,2'-diphenyl-b-picrylhydrazyl (DPPH) radical was measured using the method from [12]. Thus, an aliquot of extract (0.1 mL) was added to 3.9 mL of ethanolic DPPH (60 mM). The mixture was shaken vigorously and left to stand at room temperature for 30 min in the dark and absorbance was measured at 517 nm. The free radical scavenging capacity was calculated as follows:

$$\%\text{ Free radical-scavenging capacity}=\left[\frac{(A_{blank}-A_{sample}}{A_{blank}}\right]x100\% \quad (1)$$

where A_{blank} was the absorbance of the control reaction (containing all reagents except the test compound), and A_{sample} was the absorbance of the test compound.

2.5. pH measurement
The *H. polyrhizus* peel extract measured by using pH meter. Measurements were performed by calibration stage of the apparatus, by using a pH 4 distilled water then continued with pH 7 distilled water. Furthermore, pH measurement were taken on *H. polyrhizus* peel extract.

2.6. Experimental design
The experimental design in this study was experimental design using central composite design (CCD). In previous study consisted of two factors: 1). electric field strength (three levels): 3.5; 4.0 and 4.5 kV/cm and treatment time (five levels): 15, 30, 45 seconds. The best results of the previous study will be used as a central point in the present study. The optimization of free radical-scavenging capacity and pH in from *H. polyrhizus* peel extract used by Response Surface Methodology (RSM) in Design Expert (DX) 7.1.6 software. A two-factor central composite design was adopted to optimize the total carotenoid in carrot juice. The second order second-order polynomial equation shown in Eq.2:

$$Y = \beta_0 + \sum_{i=1}^{k}\beta_i X_i + \sum_{i=1}^{k}\beta_{ii} X_iX_i^2 + \cdots + \sum_{i<j}\sum \beta_{ij} X_iX_j + \varepsilon \quad (2)$$

Where Y for the reponses, β0 denoted the model intercept, i and j were the linear and quadratic coefficients, respectively, βi, βii and βij were the regression coefficient, k was the number of factors studied and optimized in the experiment and ε was the random error. The goodness of fit of the model was evaluated by the coefficient of determination (R^2) and the analysis of variance (ANOVA).

2.7. Verification of model
Verification of the model is needed for testing the accuracy of the model in describing the actual conditions. Verification was done by comparing the results of optimal treatment (electric field strength and extraction time) based on predicted value from RSM with the results of actual or experimental study.

3. Results and Discussion
3.1. Optimization by response surface method
Central composite design (CCD) was used to further optimize the extraction conditions with respect to the free radical-scavenging capacity and pH of *H. polyrhizus* peel extracts. An electric field strength of 4 kV/cm and extraction time of 30 second, were chosen from previous study. The responses values free radical-scavenging capacity and pH of extracts obtained under various experimental conditions are shown in Table 1.

Based on Table 1, the maximum free radical-scavenging capacity (76.54%) and pH (5.1) was recorded during Run no. 5, while the lowest antioxidant capacity (71.89%) in Run no. 11 and the lowest pH (4.6) was detected in Run no. 4. The response surface analysis using CCD, has several statistical models offered for analyzing research data to obtain the most optimal response. Some of the statistical models offered include quadratic models, linear models, 2FI models or also called two-factor interactions, and cubic models. The four models was analyzed each other based on Sequential Model Sum of Squares, and supported by Lack of Fit Tests, as well as Model Summary Statistics to determine the model corresponding to optimum response.

Table 1. Actual and Coded Variables Combinations uisng Central Composite Design (CCD) matrix

Run Order	Actual Variables		Coded Variables		Responses	
	Electric Field Strength (kV/cm)	Extraction Time (seconds)	X_1	X_2	Free Radical-Scavenging capacity (%)	pH
1	4.50	45.00	1.000	1.000	72.99	4.9
2	4.00	51.21	0.000	1.414	76.04	4.7
3	4.00	8.79	0.000	-1.414	75.79	5.0
4	4.50	15.00	1.000	-1.000	72.49	4.6
5	4.00	30.00	0.000	0.000	76.54	5.1
6	4.00	30.00	0.000	0.000	75.64	5.0
7	3.50	15.00	-1.000	-1.000	72.09	4.9
8	4.00	30.00	0.000	0.000	77.44	5.1
9	4.00	30.00	0.000	0.000	75.19	4.9
10	3.50	45.00	-1.000	1.000	73.19	5.0
11	4.71	30.00	1.414	0.000	71.89	4.8
12	3.29	30.00	-1.414	0.000	73.39	4.7
13	4.00	30.00	0.000	0.000	73.94	4.9

Based on the three statistical models analysis, the chosen (suggested) model to explain the relationship between electric field strength variable (X_1) and extraction time (X_2) to free radical-scavenging capacity and pH responses was quadratic model. This is clarified by the analysis of variance (F-test) for quadratic model. The DX 7.1.6 modeling results obtained second order polynomial models. The three-dimensional plots and contour plots of optimal free radical-scavenging capacity and pH of *H. polyrhizus* peel extract was presented in Figure 1. It indicates an increase in free radical-scavenging capacity and pH with increasing electric field strength variable (X_1) and extraction time variable (X_2), but when these variables is increased even further the both responses were decreased.

The electric field strength variable is a factor that strongly influences the extraction using PEF since it will determine the pore formation rate in the electro-permeabilization process. However, the increase of electric field strength in this study will actually decrease the antioxidant activity and pH (Figure 1a and 1b). This is expected due to the higher electric field strength, more pores are formed on the membrane cell since it exceeds the critical field of the cell ($Ef > Ec$), hence the pores are irreversible [13]. When the cell membrane is irreversible, mass transfer from within the cell will occur rapidly. Displacement of these compounds will not only dissolve the antioxidants into the solvent, but also other compounds contained in cells that can actually reduce the percentage inhibitor antioxidants capacity. The decreased of antioxidants with increasing electric field strength also reported by [14] study that the significant decrease in total carotenoid in orange juice using bipolar pulses of 30 kV/cm, and significant decrease in carotene content in the PEF-treated (35 kV/cm, 4 μs bipolar pulses at 100 Hz) in tomato juices [15]. Therefore, in this study, the suggested optimum condition of electric field strength is 3.96 kV/cm which can be extract the anthocyanin optimally.

On the other hand, the anthocyanin stability is influenced by several factors, one of which is pH. The pH is very important for the color of anthocyanins. Some anthocyanins are red in acid solutions, violet or purple in neutral solutions, and blue in alkaline pH [16]. In this study, the higher of electric field strength will actually decrease the pH to reach 4.6 (Figure 1b). The pH change was expected due to other compounds that followed extractable anthocyanin pigments thus making it pigment color changes.

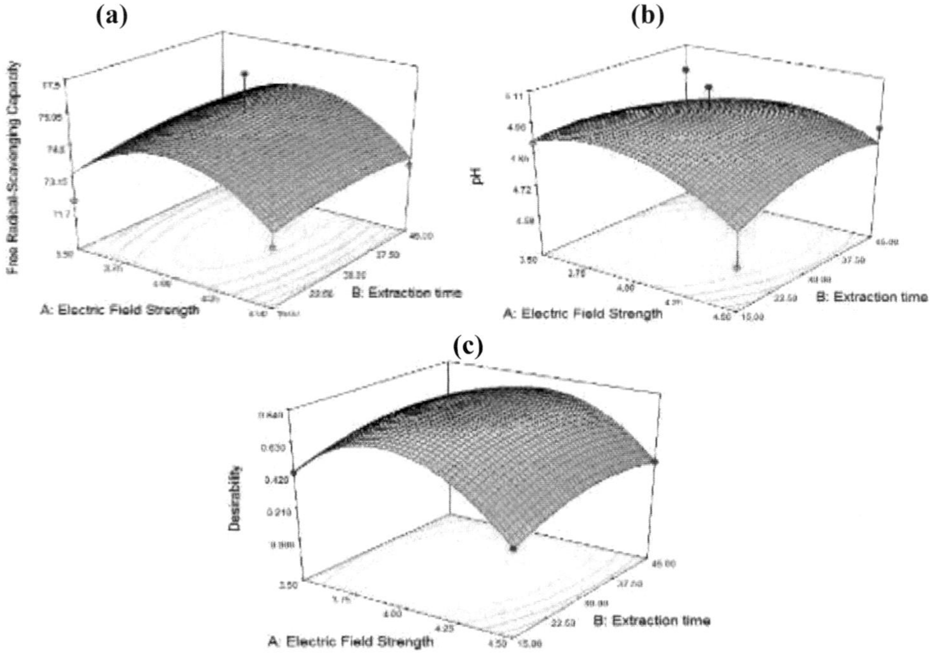

Figure. 1 Three dimensional graph between combination of electric field strength and extraction time on optimal responses. (a) free radical-scavenging capacity response, (b) pH response, (c) desirability value of optimum condition

The optimum treatment is based on suggested model are shown in Figure 1c, where the optimum point is indicated by the desirability value of 0.84. The desirability value (ranged from 0 to 1.0) is the function value of optimization objective which indicates the program's ability to satisfy the desire based on the criteria specified on the final product. The desirability value closer to the 1.0 value, indicates the ability of the program to produce the desired product more perfect. The results of computerized prediction DX 7.1.6 software, the optimum point of electric field strength variable was 3.96 kV/cm, whereas the optimum extraction time variable was 31.9 second. Under these condition, the predicted optimum response values of free radical-scavenging capacity and pH of *H. polyrhizus* peel extract was 75.784% and 4.9, respectively.

3.2. Verification of optimal conditions

Verification of the model were conducted under optimal conditions compared with actual values of dependent variables. The comparison between predicted value and actual values or experimental study are shown in Table 2.

Table 2. Predicted and Actual Values of Response Under Optimal Conditions

Responses	Model equations	Predicted value	Actual value (experimental)[a]	Error rate value
Free radical-scavenging Capacity (%)	$Y = -51.39457 + 62.57967\ X_1 + 0.18011\ X_2 - 0.02\ X_1X_2 - 7.8075\ X_1^2 - 0.0014\ X_2^2$	75.784	75.86 ± 0.2	0.10 %
pH	$Y = -1.338536 + 3.33536\ X_1 - 0.010202\ X_2 + 0.0067\ X_1X_2 - 0.45\ X_1^2 - 0.00027\ X_2^2$	4.9	4.8	3.98%

Notes: [a] experimental using electric field strength : 3.96 kV/cm, extraction time : 31.9 second

The actual value or experimental results were in good agreement with the predicted values. The difference of free radical-scavenging capacity and pH between the predicted value and actual value has an error rate of 0.1% and 3.98%, respectively. The confidence level of the predicted values with the actual value more than 95%, which indicates that the estimation models are reasonable and accurately in predicting an optimal response. In this study, the *H. polyrhizus* peel extraction with PEF using optimal conditions have an antioxidant capacity of 75.86 ± 0.2% and higher than extraction by using maceration with a ratio of 1:50 (w/v) using a water solvent, which is having an antioxidant capacity of 51.35 ± 0.87% [17]. Whereas the experimental value based on optimal conditions obtained pH of 4.8 (red-purple color). The anthocyanin belongs to the group of natural dyes responsible for some colors in the red-blue range, this spectrum depends on an abundance of anthocyanin in the natural source. As a result anthocyanin can be food colorant with variation pH [16].

4. Conclusion

The green extraction technology of *H. polyrhizus* peel extract was well implemented using PEF. Factors that affect the improvement of mass transfer in *H. polyrhizus* peel extract using PEF is electric field strength and extraction time. By using RSM method, the correlation between responses and two variables was quadratic model. At this model, the optimum conditions was obtained at the electric field strength of 3.96 kV/cm, and treatment time of 31.9 seconds. Under these conditions, the actual free radical-scavenging capacity and pH were 75.86 ± 0.2 % and 4.8, respectively. The verification of the model showed that the actual value are in good agreement with the predicted values and has an error rate value of free radical-scavenging capacity and pH of 0.1% and 3.98%, respectively. The green PEF-assisted extraction method of *H. polyrhizus* peel extract not only has a sufficiently low electric field strength but also had a rapid extraction time. Therefore we suggest to extract the *H. polyrhizus* peel using a green and non-thermal extraction technology, PEF-assisted extraction, for research, food applications and nutraceuticals industry.

References

[1] Santoso W E A and Estiasih T 2014 *Jurnal Pangan dan Agroindustri* **2** 121-127
[2] Tenore G C, Novellino E and Basile A 2012 *J. Func. Foods* **4** 129 – 136
[3] Wu L C, Hsu H W, Chen Y C, Chiu C C, Lin Y I, Ho A 2006 *Food Chem.* **95** 319-327
[4] Safdar N M, Tusneem K, Saqib J, Amer M, Karam A, Ambreen A S 2017 *J. Food & Drug Analysis* **25** 488-500
[5] Patras A, Nigel P B, Colm O D, B K Tiwari 2010 *Trends. Food Sci. & Technol.* **21** 3-11
[6] Putranto A W, Argo B D, Wijana S 2014 *Indonesian Green Technol. J.* **3** 1-9
[7] Toepfl S 2006 Pulsed electric fields (PEF) for permeabilization of cell membranes in food and bioprocessing – applications, process and equipment design and cost analysis Prozesswissen-schaften der Technischen Universität Berlin Jerman
[8] Zderic A and Edwin Z 2016 *J. Chem. Eng.* **109** 586-592
[9] Zhang Z, Xin-An Z, Charles S B, Margaret B, Zhong H, Xia-Yu X 2015 *J. Mol. Sci.* **16** 24159-24173
[10] Maskooki A and Mohammad N E 2012 *J. Food & Bioprod. Proc.* **90** 377-384
[11] Raso J and Heinz V 2006 Pulsed Electric Fields Technology For The Food Industry - Fundamental and Applications Springer Science Business Media USA
[12] Jiao Y, Jiang Y, Zhai W, Yang Z 2012 *J. Biotechol.* **11** 7046-7054
[13] Shamsi K 2008 Effects of Pulsed Electric Field Processing on Microbial, Enzymatic and Physical Attributes of Milk and the Rennet-Induced Milk Gels RMIT University, Melbourne
[14] Cortés C, Torregrossa F, Esteve M J, Frígola A 2006 *J. Agri. and Food Chem.* **54** 6247-6254
[15] Odriozola-Serrano I, Aguilo-Aguayo I, Soliva-Fortuny R, Gimeno-Ano V and Martin-Belloso O 2007 *J. Agri. & Food Chem.* **55** 9036–9042
[16] Wahyuningsih S, Wulandari L, Wartono M W, Munawaroh H, Ramelan A H 2017 *IOP Conf. Ser.: Mater. Sci. Eng.* **193** 012047

[17] Manihuruk F M, Suryati T, Arief I 2016 *Media Peternakan* **40** 47-54

IOP Conf. Series: Earth and Environmental Science **131** (2018) 012052 doi:10.1088/1755-1315/131/1/012052

Single cell protein production of *Chlorella* sp. using food processing waste as a cultivation medium

D Putri, A Ulhidayati, I A Musthofa and A K Wardani

Department of Agricultural Product Technology, Faculty of Agricultural Technology, Universitas Brawijaya, Malang, Indonesia

E-mail: dininurilmi@gmail.com

Abstract. The aim of this study was to investigate the effect of various food processing wastes on the production of single cell protein by *Chlorella* sp. Three various food processing wastes i.e. tofu waste, tempeh waste and cheese whey waste were used as cultivation medium for *Chlorella* sp. growth. Sea water was used as a control of cultivation medium. The addition of waste into cultivation medium was 10%, 20%, 30%, 40%, and 50%. The result showed that the highest yield of cell mass and protein content was found in 50% tofu waste cultivation medium was 47.8 x 10^6 cell/ml with protein content was 52.24%. The 50% tofu waste medium showed improved cell yield as nearly as 30% than tempeh waste medium. The yield of biomass and protein content when 30% tempeh waste was used as cultivation medium was 37.1 x 10^6 cell/ml and 52%, respectively. Thus, food processing waste especially tofu waste would be a promising candidate for cultivation medium for single cell production from *Chlorella* sp. Moreover, the utilization of waste can reduce environmental pollution and increase protein supply for food supplement or animal feed.

1. Introduction

Single cell protein (SCP) is an interesting object in research since early 1950. Single cell protein is the dried cells of microorganism, has been explored for being an alternative protein source as a food supplement or animal feeds as the preventive protein deficiency supply [1]. Some microorganism such as yeast, fungi, bacteria, and microalgae can be used in SCP production. This protein source is named as single cell protein because it describes the protein production from biomass (cell) that different from vegetable protein and animal protein. The production of SCP from various microbes gives some different characteristics either.

The production of SCP from fungi and bacteria has been received considerable attention, in contrast, only a few studies have dealt from SCP of microalgae. Comprehensive analysis and nutritional studies have demonstrated that these algae proteins are a high quality and comparable with conventional vegetable proteins [2].

Microalgae are the unicellular photosynthetic microorganism that grows in marine. Microalgae has diameter or length of approximately 3-10 μm. Microalgae metabolism reacts to change in the external environment with changes in its intracellular environment. Thus, the manipulation of the culture condition or the presence nutrients stimulates the biosynthesis of specific compounds [3]. Microalgae are the best choice for single cell protein productions because they also have simple and inexpensive growth requirements (free sea water, inexpensive nitrogen, phosphorus, and carbon sources) and rapid growth rates with sufficient light. Single cell protein production using microalgae has been known as an effective and economical way process. Microalgae are a potential source of protein. Cultivation of microalgae is the preferred way for the development and production of these high-value products [4].

Many studies have created to develop microalgae cultivation technique that purposes for decreasing production cost and produce a better quality of cell yield. Microorganisms like algae, fungi, yeast, and bacteria utilize inexpensive feedstock and wastes as sources of carbon, nitrogen and energy for growth to produce biomass and protein concentrate. A study shows that microalgae *Chlorella vulgaris* could grow well in tofu industrial waste [5]. It shows that nitrogen source in waste from food processing can be used as microalgae growth support.

Waste from food industry is one of causing the environmental pollution. Food industries such as tofu industry, tempeh industry and cheese industry are widely spread in Indonesia's areas. Waste from food industries contains complex organic material, such as carbohydrates, protein, lipids, and several microelements. The protein contained in its raw waste occupies about 60% of all solid fraction depends on the characteristic of the waste [5]. Protein can be degraded into various organic compounds, which some of them can be used either as macronutrient or micronutrient for the autotrophic organism, such as microalgae. It is expected that waste of food industry is able to enhance the microalgae cultivation. In this study, three various food processing wastes i.e. tofu waste, tempeh waste, and cheese whey waste were utilized as a nitrogen source for microalgae growth. Among various microalgae, *Chlorella* sp. was selected in this study as single cell producer because of its high growth rate, high protein content, high chlorophyll content and low nucleic acid content.

2. Materials and methods

2.1. Microalgae strain and cultivation medium

Chlorella sp. strains were obtained from Balai Budidaya Air Payau (BBAP) Situbondo, East Java, Indonesia. Tofu waste was obtained from Dusun Gendangan, Kabupaten Malang, tempeh waste was obtained from Sanan, Kota Malang, and cheese whey waste was taken from Pujon, Kabupaten Malang. The cultivation medium consisted of control medium and modified medium growth. Control medium was contained filtered seawater and Walne as nitrogen source. The modified medium contained seawater and each various waste with 10%, 20%, 30%, 40% and 50% concentration. Seawater was sterilized at 121°C for 15 min [6]. The tofu waste and tempeh were pasteurized at 70°C for 15 min [7]. The cheese whey waste was sterilized at 121°C for 15 min. Volume of medium growth was 300 ml. *Chlorella* sp. utilized waste as a nitrogen source such as nitrate during the growth. Table 1 shows the nitrate content of each nitrogen source.

Table 1. Nitrate (NO_3) content.

Nitrogen Sources	Nitrate (ppm)
Tofu waste	9.21
Tempe waste	13.33
Cheese whey waste	14.70
Walne	1.9

The initial pH of modified medium was adjusted from 3.5 to 7.0 using NaOH 1N solution. Vitamin B12 was added as much as 0.3 ml into each cultivation medium as a supported growth of microalgae. Microalgae *Chlorella* sp. added into medium was 30 ml.

2.2 Cultivation conditions
Chlorella sp. was grown in each modified medium at 25-35°C with aeration, salinity was adjusted at 30-34 ppt and pH was adjusted at 7.5-8.5[8]. Light source of 18-watt lamp (2500 lux) was used in this study to give the optimum growth of *Chlorella* sp. The proper Dissolved Oxygen (DO) in medium was 5-7 mg/l. Cultivation of *Chlorella* sp was did in three step which are to produce stock culture, determine the effect of each modified medium, then scale up (Figure 1).

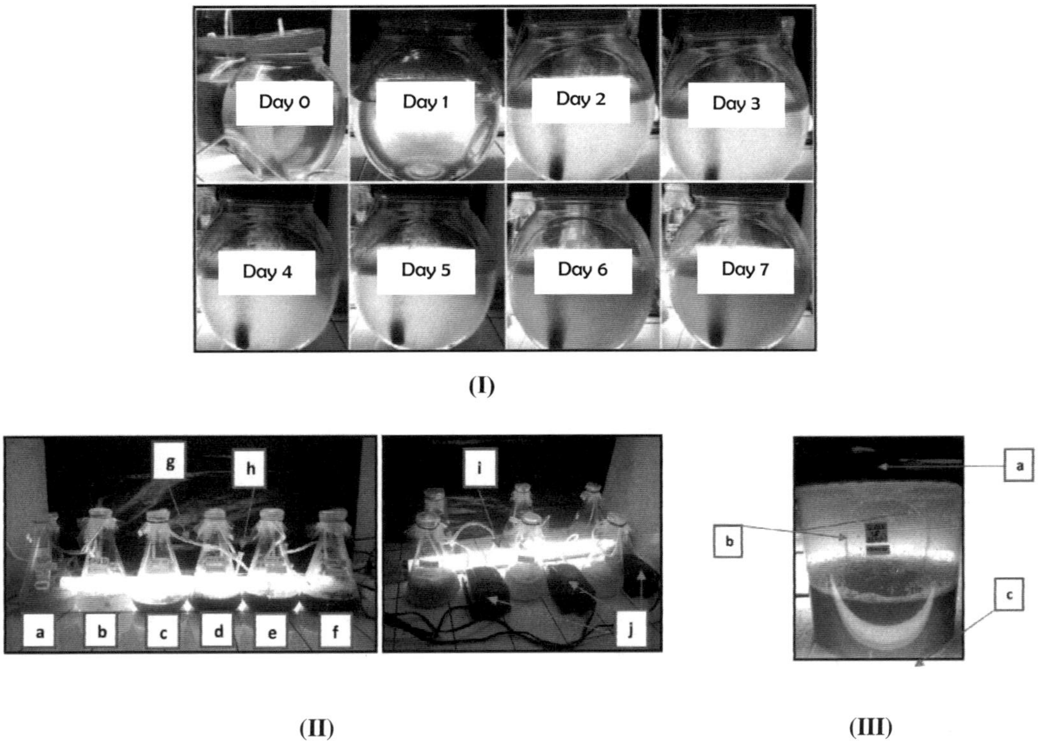

Figure 1. (I) Stock culture of *Chlorella* sp; (II) Waste medium formation of *Chdlorella* sp - (a) control medium (b) 10% waste medium (c) 20% waste medium (d) 30% waste medium (e) 40% waste medium (f) 50% waste medium (g) aeration tube (h) aeration regulator (i) 18-watt lamp (j) aerator ; (III) Scale up - (a) aerator (b) aeration tube (c) 18-watt lamp

2.3 Analysis
Aliquots of 300 ml were taken every 24 hours to determine cell concentration, temperature medium, pH medium, salinity, and Dissolve Oxygen (DO). Cell concentration was determined by microscope using haemocyotometer method [6]. Temperatur and DO were determined by DO meter, salinity was determied by handrefractometer, pH was determined by pH meter [9]. Protein content was determined every 48 hours. The protein content of *Chlorella* sp. was determined by using Biuret method [10]. Nitrat content in medium growth was determined in day 0 and day 10 using spektrofotometry method.

3. Results and discussion
Chlorella sp. cultivation media in the study included control media and waste media. Waste media included tempeh waste, tofu waste, and cheese whey waste. Each waste medium was added 10%, 20%, 30%, 40% and 50% of waste. The more concentration of waste added to the growth medium, the higher the increase of the nitrate content of the growth media. The nitrate content of the growth media with various concentrations of waste is shown in Table 2.

3.1 Protein production by Chlorella sp.

Increased protein occurs due to the presence of nitrate assimilation by *Chlorella* sp. Assimilation of nitrate during the growth period produces glutamate. Glutamate is one of the protein amino acids [11]. Glutamate will undergo metabolism to produce other amino acids to form proteins. Figure 2 presents an increase of protein content from three types of waste.

Table 2. The Nitrate (NO_3) content of the growth media.

Media	Nitrate (ppm)
Control	1.90
Tofu waste medium 10%	0,.92
Tofu waste medium 20%	1.84
Tofu waste medium 30%	2.77
Tofu waste medium 40%	3.68
Tofu waste medium 50%	4.61
Tempe waste medium 10%	1.33
Tempe waste medium 20%	2.66
Tempe waste medium 30%	3.99
Tempe waste medium 40%	5.32
Tempe waste medium 50%	6.65
Cheese waste medium 10%	1.40
Cheese waste medium 20%	2.81
Cheese waste medium 30%	4.22
Cheese waste medium 40%	5.63
Cheese waste medium 50%	7.04

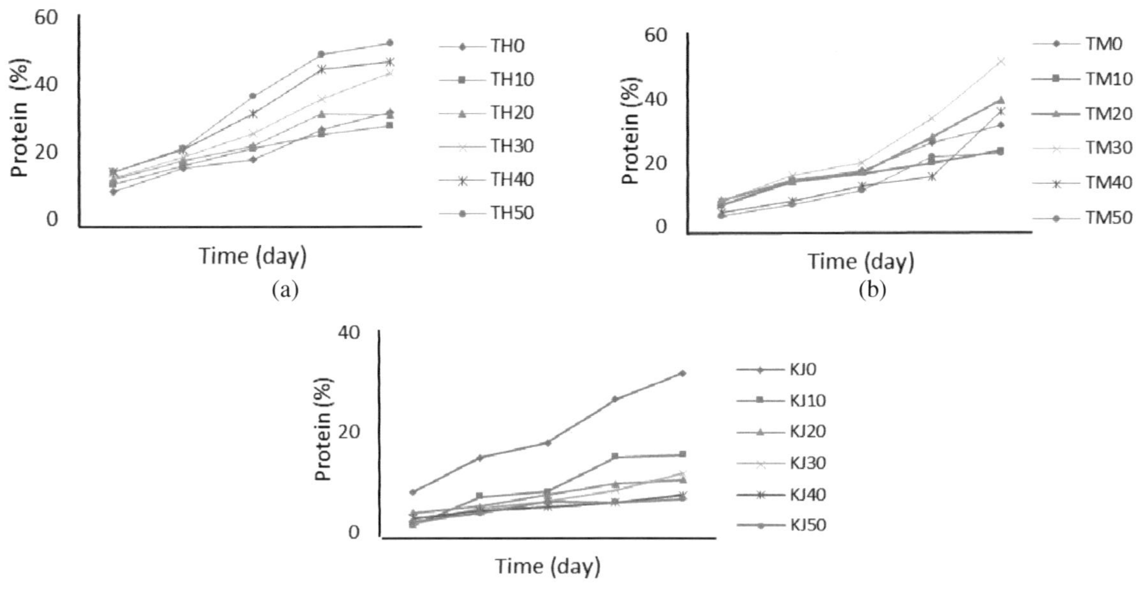

Figure 2. The effect of waste concentration on protein production (SCP) by *Chlorella* sp.
TH10 = Tofu waste 10%; TH20 = Tofu waste 20%; TH30 = Tofu waste 30%; TH40 = Tofu waste 40%; TH50 = Tofu waste 50%; TM10 = Tempeh waste 10%; TM20 = Tempeh waste 20%; TM30 = Tempeh waste 30%; TM40 = Tempeh waste 40%; TM50 = Tempeh waste 50%; KJ10 = Cheese whey waste 10%; KJ20 = Cheese whey waste 20%; KJ30 = Cheese waste 30%; KJ40 = Cheese whey waste 40%; KJ50 = Cheese whey waste 50%.

3.2 Effect of waste type as nitrogen source

Based on the results, growth media with the addition of different waste produced a different effect on the total protein of *Chlorella* sp. The highest total protein was found when 50% of tofu waste was added in the medium growth. Table 3 shows the protein content of *Chlorella* sp. in waste media and control media.

Table 3. Cell concentrate and total protein of *Chlorella* sp.

Medium	Cell concentrate (cells/ml)	Total Protein (%)
Control	$17{,}67 \times 10^6 \pm 2{,}25$	31,80% ± 3,10
50% of Tofu waste	$42{,}52 \times 10^6 \pm 2{,}80$	52,32% ± 3,31
30% of tempeh waste	$37{,}17 \times 10^6 \pm 1{,}52$	52,00% ± 1,80
10% of cheese whey waste	$8{,}5 \times 10^6 \pm 1{,}45$	15,43% ± 2,55

Table 3 shows the protein concentration of cell higher when waste media was used for protein production than that of control media. However, the protein concentration of *Chlorella* sp. in cheese whey waste media was lower than that of the control media. The main source of nutrition for the growth of *Chlorella* sp. is carbon, nitrogen, and phosphorus content [12]. The carbon source comes from CO_2 obtained from the atmosphere and organic carbon. Nitrogen sources are utilized in the formation of proteins in the form of ammonium and nitrate. The source of nitrate needed to support the growth of *Chlorella* sp. is 0.9-3.9 mg/l. Phosphorus is used as a universal energy source in the photosynthesis, respiration, and protein formation process; the ratio of nitrogen and phosphorus needs is 15:1 to 30:1 [12]. One of the nutrient contents in waste that causes an increased number of *Chlorella* sp. cells is nitrate [13]. The same nutrient content is found in tempeh waste that supports the growth of *Chlorella* sp. cells. This indicates that tofu and tempeh waste have the main nutrient source for the growth of *Chlorella* sp. [14]

Based on the results, the addition of 50% of tofu waste in growth media was able to support the increase of biomass, while the addition of 30% of tempeh waste in growth media could support the increase of cells. These differences are caused by the differences in waste characteristics that can affect the growth of *Chlorella* sp. These characteristics include nutritional content, clarity, and color of waste. However, the addition of cheese whey waste inhibited the growth of *Chlorella* sp. It assumed that the high nitrate content (14.07 mg/l) in cheese whey waste have the contribution for the growth inhibition. Moreover, it is suspected that the presence of lactose content in cheese whey waste also have plays a role in inhibiting the consumption of nitrate by *Chlorella* sp. The lactose content in cheese whey waste ranges from 5-6% [15]. Moreover, Garcia *et al.* [16] confirm that the source of lactose is not able to support the growth of microalgae. The source of carbon that can be used by microalgae is glucose. Glucose is a beneficial carbon source for microalgae as well as other species of microorganisms. The increased growth and respiration rates can be affected by carbon sources or glucose compared to other carbon sources, where glucose is used by microalgae to form other organic compounds such as lipids and proteins [17].

3.3 Nitrate utilization of Chlorella sp.

Based on the data, the nitrate content in the raw materials varies. The following is a table of nitrate content of culture media at the beginning and end of cell growth (Table 4).

Table 4. The Decrease in the Nitrate Content of the Growth Media.

Medium	Initial (ppm)	Final (ppm)	Consumption (%)
Control	1.9	0.47	75.26 ± 1.05
TH50	4.61	0.31	93.27 ± 0.78
TM30	3.99	0.48	87.97 ± 0.50
KJ10	1.40	0.52	62.62 ± 1.09

During the growth period, the nitrate content decreases. This is because nitrate is used by *Chlorella* sp. in the metabolic pathway. Based on the table, the nitrate on TH50 medium has been consumed more than the nitrate on TM30 and KJ10 media, with a percentage of 93.27%. This is because the growth of biomass in the TH50 medium is higher than in TM30 and KJ10 media. This is equivalent to the amount of protein produced, which is higher than the production of protein on TM30 and KJ10 media.

4. Conclusion

Tofu waste with a concentration of 50% produces 42.5 x 10^6 cells/m, with a total protein of 52.32%. Tempeh waste with a concentration of 30% produces 37.1 x 10^6 cells/ml, with a total protein of 52%. Cheese waste with a concentration of 10% produces 8.5 x 10^6 cells/ml, with a total protein of 15.43%. This study indicated that tofu waste has potential to be used as cultivation medium for SCP production.

References

[1] Nasseri A T, Rasoul-Amini S, Morowvat M H, Ghasemi Y 2011 Single cell protein: production and process *American Journal of Food Technology* ISSN 1557-4571
[2] Becker E W 2007 Microalgae as a source of protein *Biotechnology Advances* Series **25** 207-210.
[3] De Morais, Michele. Bruna da Silva. Etiele de Morais. Jorge, Vieira 2015 *Biologically active metabolites synthesized by microalgae. Biomed Research International.* Article ID 835761.
[4] Raja R. Hemaiswarya S. Venkatesan G, Isabel C 2014 Biomass from Microalgae: an overview. *Oceanography* **2** 1
[5] Nugroho W. 2016. Promoting the growth of *Chlorella vulgaris* in Secondary Waste Treatment Effluent of Tofu Industry using *Azospirillum* sp. *International Journal on Advanced Science engineering Informastion Technology* **6** 3
[6] Fadilla Z 2010 The Effect of Tofu Waste on Microalgae *Scenedemus* sp Growth Universitas Islam Negeri Syarif Hidayatullah Jakarta (in Indonesian)
[7] Dianursantia, Rizkytata B T, Gumelara M T, Abdullah T H 2014 Industrial Tofu Wastewater as a Cultivation Medium of Microalgae Chlorella vulgaris. Conference and Exhibition Indonesia Renewacle Energy & Energy Conservation [Indonesia EBTKE CONEX 2013]. *Energy Procedia* **47** 56-61
[8] Choi B, Jun-Heok, Lim J L, Taeyoon L 2103 Optimum condition for cultivation of Chlorella sp FC-21 using Light Emiting Diodes *Korean J. Chem. Eng* **30** 8 1614-1619.
[9] Danang A 2009 Optimation of *Chlorella* sp Growth Medium in Laboratory Scale Institut Pertanian Bogor (in Indonesian)
[10] Harun R 2009 Bioprocess Enginering of Microalgar to Produce a Variety of Consumer Products.
[11] Ravindra P 2014 Biotechnology and Molecular Biology Reviews *Academic Journals* **9** 2
[12] Ali M 2013 Nitrat Degradation of Waste by using Algae *Chlorella sp*. UPN Veteran JawaTimur. Surabaya (in Indonesian)
[13] Fatha A 2007 The use of activated Zeolite to reduce BOD and COD of tofu waste *Journal Bioprocess* (in Indonesian)
[14] Taziki M, Hossein A, Marcia M, Stephen L 2015 Nitrate and nitrite removal from wastewater using algae *Current Biotechnology* **4** 3
[15] Guimaraes P M R, Texeira J A, Domingues L 2010 Lactose fermentation to bioethanol by yeasts as part of integrated solutions for the valorisation of cheese whey *Research Review Paper JBA* 06293 **1** 1-10
[16] Garcia C, Garcia C, Sanchez M 2006 Mixotrophic production of marine microalga phaeodactylumtricornutum on various carbon sources. *J MIcrobiol. Biotechnol* ***16*** *5 689-694*
[17] Perez O, Froylan E, Luz E 2011 Heterotrophic cultures of microalgae: metabolism and potential products *Water Research* 11-36.

IOP Conf. Series: Earth and Environmental Science **131** (2018) 012053 doi:10.1088/1755-1315/131/1/012053

UV-Ilmenite based photo-catalysis in lignin based black liquor

F Amriani[1], H Abimanyu[1], M Natsir[2], L Sutrizal[2] and A Nursin[2]

[1] Research Center for Chemistry, Indonesian Institute of Sciences, South Tangerang, Indonesia
[2] Chemistry Department, University of Haluoleo, Kendari, Indonesia

E-mail: feni.amriani@lipi.go.id

Abstract. Ilmenite can be found abundantly in iron sand from sea shore along Wolowo beach in Button district, Southeast Sulawesi, Indonesia. The ability of ilmenite in degrading lignin in black liquor has been investigated. The results of lignin degradation process in black liquor are supposed to be the potential resources for fungicide such as coniferyl, sinapyl, and p-coumaryl alcohol. The process was conducted in 10 watt ultraviolet (UV) light chamber with two parameters applying include exposure time and ilmenite composition. Two scheme of process are used by differentiating the feed, raw black liquor (scheme 1) and the liquor after adding of 1% sodium hydroxide into lignin-based sludge (scheme 2). Decolourization and lignin degradation analysis after the process were conducted by using UV-Vis spectrophotometer and LCMS, respectively. The results showed that the treatment from the scheme 1 was better than the scheme 2. Both lignin degradation and decolourization can effectively result in more than 31% by using 0.3 g ilmenite for 10 minutes UV exposure. The interim analysis by liquid chromatography-mass spectrophotometer (LCMS) exhibits the suspected target in range 309.4 to 311.39 g/mol as p-coumaryl alcohol while two other targets did not found in chromatogram. Thus, this research requires further evaluation and development to maximise the degradation result so the final goal can be achieved successfully.

1. Introduction

Iron sand based ilmenite has already been investigated regarding its content of TiO_2 which is well known as a potentially used catalyst [1-2]. As a facile material which has valuable content, the investigations are announced relating its preparation and application. The preparation of ilmenite is related with the characteristic of material that influences its performance [3]. The calcination by applying high temperature in a furnace is one of potential methods to be conducted as it may reduce the use of chemicals. At high temperature (800 °C and above), the TiO_2 is separated from Fe_2O_3. $2TiO_2$ that composed during the increasing of temperature up to 800 °C [4]. TiO_2 is the most used catalyst material that works well under light exposure. Ultraviolet (UV) and other kinds of light can be used to enhance the TiO_2 work in ilmenite. This method is considered as the safest way process since it is easy to tackle eventhough electricity based light is not available at a place [5]. In addition, TiO_2 is the most stable chemical compound but has high reactivity under UV or other kinds of light exposure so it opens a wide application [5]. One example of its application refers to wastewater treatment that mostly can degrade the recalcitrant dissolving compounds up to 90% [3, 6].

Another commonly found wastewater problem in pulp and paper industry or lignocellulosic biomass-based bioethanol is lignin-based black liquor (LBBL). A high amount of LBBL is usually obtained during the pretreatment process as a delignification method of lignocellulosic biomass. LBBL can be further utilized at a time of its treatment process. Lignin in LBBL can be acquired by neutralization process that result in lignin based sludge [7]. Lignin is valuable compound that can be converted into another high-added product which potentially developed for industry needs [8, 9]. Lignin is structured by several aromatic hydrocarbon compounds, such as *p*-coumaryl alcohol, sinaphyl alcohol and coniferyl alcohol in which they are potentially utilized as fungicide [10]. However, to result in the monomer of lignin, materials are needed in decomposing the lignin into simple monomer. Natural-based materials are favorable for establishing green environmental research.

This work aimed to investigate local ilmenite capability as potential-natural resources for catalyst in wastewater treatment and also evaluate the resource recovery of lignin degradation result in black liquor which can be used as biopesticide namely coniferyl, sinapyl, and *p*-coumaryl alcohol.

2. Materials and Method

2.1 Black liquor and ilmenite

The black liquor as lignin containing wastewater was collected from bioethanol second generation pilot plant at Research Center for Chemistry, Indonesian Institute of Sciences (LIPI). The sample was then used as the lignin source in this experimental study.

Ilmenite containing TiO_2 was given by Department of Chemistry, University of Haluoleo Kendari. It was procured from the beach shore in Wolowa, Button District, Southeast Sulawesi, Indonesia. The enhancement of TiO_2 performance in ilmenite was conducted by employing a mechanism of ilmenite preparation, i.e calcination at 800 °C for 5 hours to reduce Fe content. The calcination was performed twice and the magnet was used to eliminate Fe content from the calcinated ilmenite.

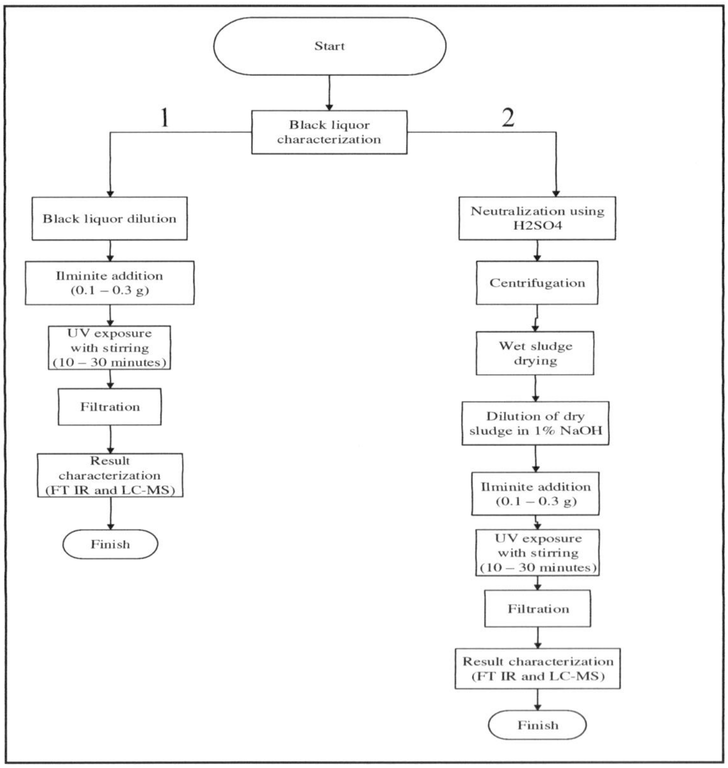

Figure 1. The procedures of lignin based liquor degradation and decolorization

2.2 Experimental set up
Two schemes were carried out (Figure 1) to extract the lignin based pesticide compound from the samples. The scheme was conducted to investigate alkaline condition of black liquor that may influence catalyst work in lignin degradation as well as black liquor decolorization. Black liquor from the first scheme was conditioned at the pH up to 10, meanwhile black liquor pH from the second scheme was set below 10.

The photocatalytic process was conducted following the chart with UV exposure process in a hand-made chamber as shown in Figure 2. All experiments were performed at room temperature.

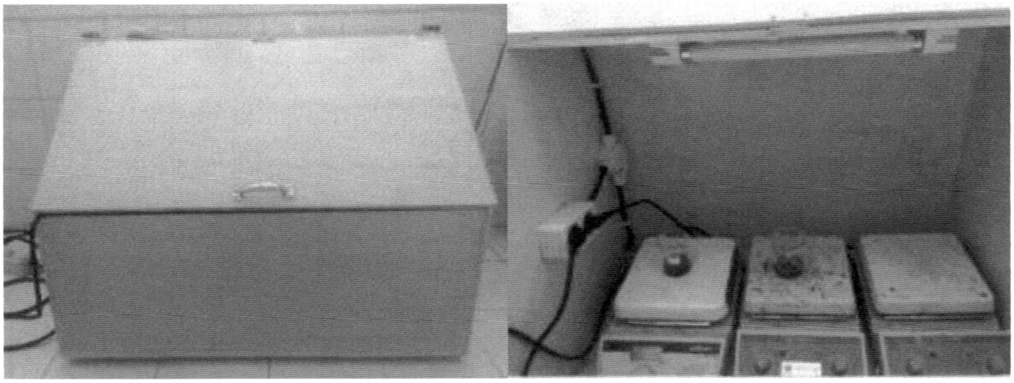

Figure 2. UV chamber for lignin based black liquor processs

2.3 Analysis
The ilmenite content was characterized by using X-Ray Fluorescence (XRF) (Torontech Tt-Edxprt, Canada) and X-Ray Diffraction (XRD) (Rigaku, Japan). Meanwhile, black liquor decolourization results were measured by spectrophotometer UV-Vis at 280 nm (maximum wave length) to measure concentration of lignin dissolved in black liquor by assistance of standard curve from commercial lignin (Sigma Aldrich, Singapore), then the decolorization percentage is calculated by following the equation [11] :

$$\% \text{ decolourization} = 100 \times (C_0 - C_t) / C_0 \quad (1)$$

Liquid chromatograph mass spectrophotometer (LCMS) was used to confirm and find the molecular weight of the targeted samples. Lignin can be degraded into its three main compound i.e synaphyl alcohol (molecular weight/MW : 210.23 $g.mol^{-1}$), *p*-coumaryl alcohol 4-O-Glucoside (molecular weight/MW : 311.32 $g.mol^{-1}$), and coniferyl alcohol (molecular weight/MW : 180.2 $g.mol^{-1}$).

3. Results and Discussion

3.1 Ilmenite characterization
The ilmenite results before and after calcination processes are shown in Table 1. It can be seen that the increasing of Ti content occurs though it still has lower content than Fe content in ilmenite before and after the calcination process.

Although the ilmenite still has low Ti content, but it cannot be assumed affecting the performance of ilmenite. Presenting high content of iron in ilmenite, the ilmenite work is considered getting a positive support from the iron in ilmenite which roles doping catalyst that is able to enhance the TiO_2 work. However, the number of doping catalyst shall be estimated in an adequate amount, therefore, the effective performance of ilmenite can be well achieved [12]. The right combination between metal based doping catalyst and TiO_2 allows reduction-oxidation mechanism and improves the performance of photocatalysis process that can degrade the recalcitrant compond such as dyes in textile industry wastewater [13, 14].

Table 1. Quantitative content of ilmenite before and after calcination process.

Ilmenite content before calcination process (% w/w)		Ilmenite content after calcination process (% w/w)	
Fe	62.17 ± 0.22	Fe:	52.1 ± 0.4
Cr	26.69 ± 0.21	Cr:	34.37 ± 0.21
Ti	3.64 ± 0.10	Ti:	4.66 ± 0.16
Si	4.02 ± 0.12	Si:	4.6 ± 0.3
Al	2.14 ± 0.13	Al:	2.6 ± 0.5
Mn	0.76 ± 0.04	Mn:	0.74 ± 0.03
Ni	0.43 ± 0.03	Zn:	0.42 ± 0.02
V	0.14 ± 0.03	Ni:	0.28 ± 0.03
		Sb:	0.14 ± 0.24
		Sn:	0.09 ± 0.11
		Zr:	0.03 ± 0.01

The XRD result translates the form of structure of ilmenite as illustrated in Figure 3. The figure shows the XRD result depicted the distribution peaks at 27°, 36°, and 55° [3]. These indicated that the structure of ilmenite after the calcination was rutile with micro powder. In addition, the pattern is quite similar to standard as explained in elsewhere report. Rutile is one of the structure of catalyst that has stable performance at different temperature and pressure [15]. This property is important in deciding a suitable process by considering time and also pressure to achieve the best result.

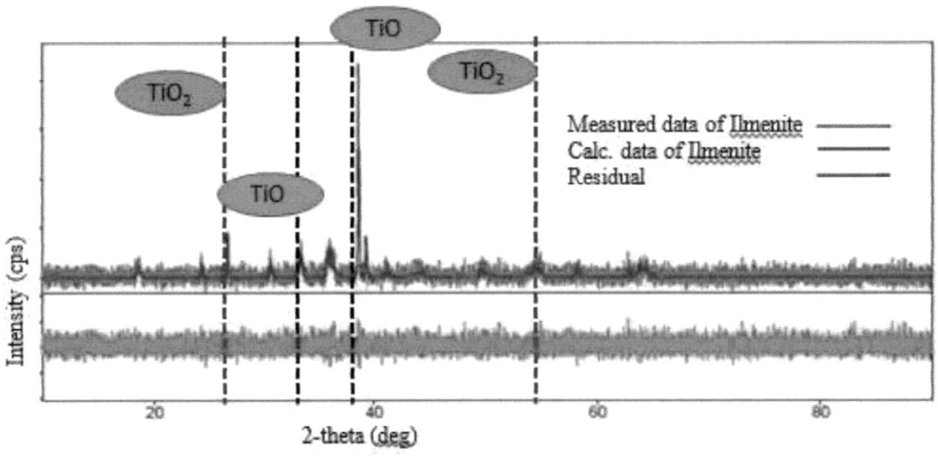

Figure 3. XRD result of ilmenite after the calcination process

3.2 Decolorization and Degradation Process

Investigating the condition of sample which may influence the process is important work to do. Nevertheless, the process was not only relying on the sample's condition but also the catalyst ability that was assisted by photo or light. This is a basic reason for present experiment where ilmenite performance as TiO_2 sources can be increased by the assistance of light either UV or visible light (VL).

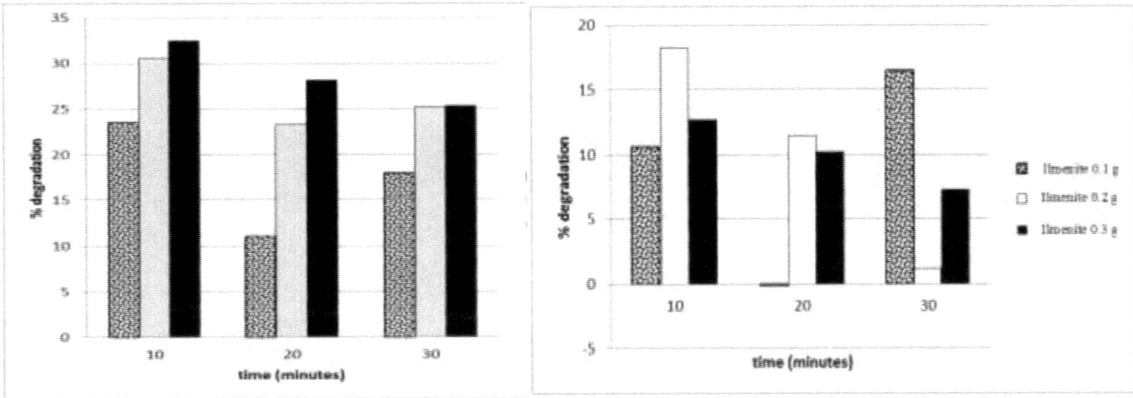

(a) (b)

Figure 4. The first scheme process, decolorization and degradation process of lignin based black liquor: a) 3000 ppm and b) 5000 ppm.

Figure 4 illustrates that the decolorization result was linear to the ilmenite addition. The more ilmenite added in lignin based black liquor, the bigger result of decolorization. It may be caused by the light aid that has greater energy than the catalyst material, then induce the creation of electron-hole pairs which may form/degrade a compound due to the reduction-oxidation method [3]. Although, fluctuated result is shown in the process which use higher concentration (5000 ppm) of raw black liquor. At a fixed time of 30 minutes, the addition of 0.2 and 0.3 g ilmenite resulted in a dramatically decrease if compared to the 0.1 g ilmenite addition result. The overloading of ilmenite affects the degradation where the reduction of surface area happened due to forming clump of ilmenite in black liquor [16].

The findings also confirmed that the UV exposure time of 10 minutes was effective, indicated by less lignin was degraded in decolorization process. The selected best treatment was the addition of ilmenite 0.3 g, 10 minutes with 3000 ppm of lignin based black liquor concentration.

The trends in the second scheme process was similar to that of the first scheme (Figure 5). It can be concluded that the condition of samples does not influence the ilmenite performance. This study is in agreement with the previous report which found that the performance of catalyst in degrading lignin or other organic-dissolved compounds shows better performance at the range of pH up to 10 [17]. These can be explained that the number of ilmenite added provides more active surface that can work effectively in decolorization, therefore the higher concentration will cause an inhibitory effect for catalyst/ilmenite work because it will reduce the irradiation field[18]. In addition, the higher pH support catalyst work as the hydroxyl groups induced the enhancement of reduction-oxidation mechanism that will lead to an effective decolorization as well as degradation result [19].

Figure 6 shows that both schemes have almost similar peaks as the results of degradation process. It is found that p-coumaryl alcohol molecular weight (MW) emerge in both LCMS result, meanwhile other two compounds (sinapyl and coniferyl alcohol) MW are not found in the chromatograph. Previous work also has revealed the lignin depolymerization that occurs as the increasing of time. The longer time of process, the more the obtained MW that will boost the reaction of involving-radical intermediate and polymerize the dissolving lignin in black liquor [20].

IOP Conf. Series: Earth and Environmental Science **131** (2018) 012053 doi:10.1088/1755-1315/131/1/012053

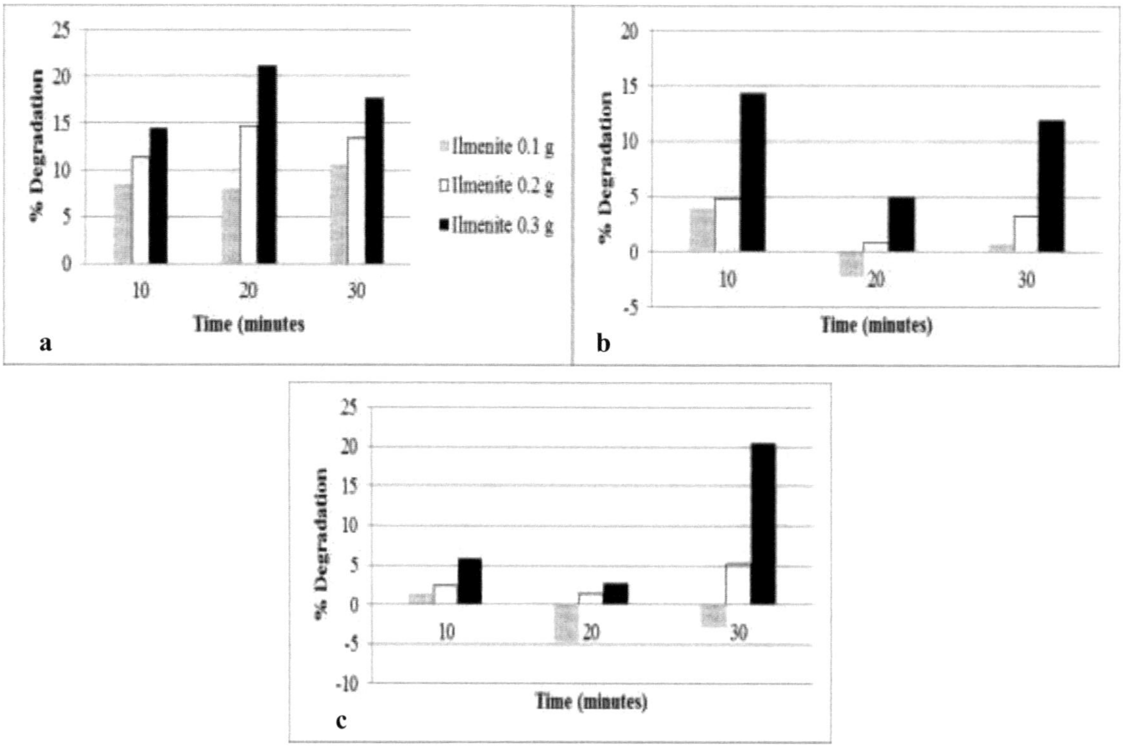

Figure 5. The second scheme process, decolorization and degradation process of lignin based black liquor: a) 300 ppm, b) 500 ppm, and c) 700 ppm.

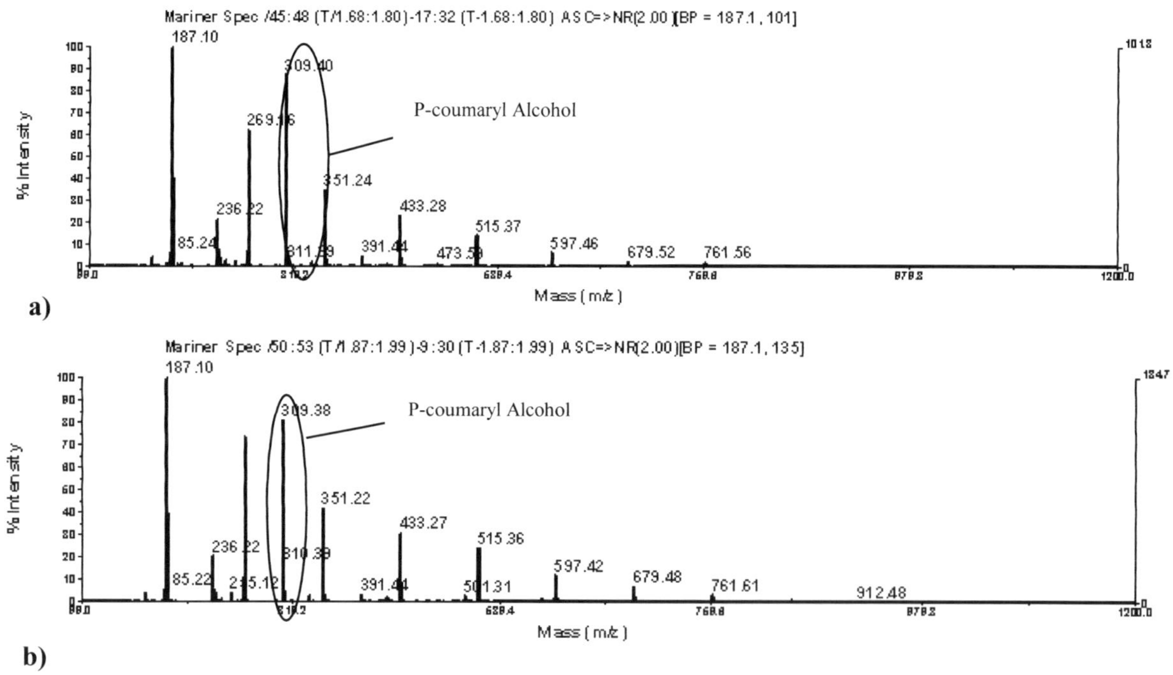

Figure 6. the LC MS results of a) first scheme, b) second scheme

4. Conclusions

The photocatalysis has not succeeded enough in decolorizing lignin-based black liquor since the ilmenite can only decolorize about 31% by using 0.3 g ilmenite for 10 minutes, but it can decompose lignin into its monomer compound, *p*-coumaryl alcohol. Therefore, further research is suggested, particularly on the process and evaluation, to enhance the interim analysis and also the isolation of the degrading compound.

Acknowledgement

The authors would like to thank to LIPI in providing the instrument for this research from Unggulan DIPA Fiscal Year 2017 and also INSINAS project 2017 for supporting our research work.

References

[1] Mulyono J E, Soepriyanto, S 2017 Synthesis and characterization of TiO_2 from Ilmenite by caustic fusion process for photocatalytic application *AIP Conf. Proc.* **30010**

[2] Bailey S W, Cameron E N, Spedden, H, Weege R J 1956 Synthesis and characterization of TiO 2 from ilmenite by caustic fusion process for photocatalytic application *Soc. Econ. Geol.* **51** 263–279

[3] Yuan P, Fan C, Ding G, Wang Y, Zhang X 2012 Preparation and photocatalytic properties of ilmenite NiTiO 3 powders for degradation of humic acid in water *Int. J. Miner. Metall. Mater.* **19** 4 372–376

[4] Zhan G, Ostrovski O 2002 Effect of preoxidation and sintering on properties of ilmenite concentrates *Int. J. Miner. Process.* **64** 201–218

[5] Schneider J, Matsuoka M, Takeuchi M, Zhang J, Horiuchi Y, Anpo M, Bahnemann D W 2014 Understanding TiO 2 photocatalysis: mechanisms and materials *Chem. Rev.* **114** 9919–9986

[6] Pataquiva-Mateus A Y, Zea H R, Ramirez J H 2017 Degradation of orange II by Fenton reaction using ilmenite as catalyst *Environ. Sci. Pollut. Res.* **24** 7 6187–6194

[7] Sari A A, Amriani F, Muryanto M, Triwulandari E, Sudiyani Y, Barlianti V, Lotulung P D N, Hadibarata T 2017 Mechanism, adsorption kinetics and applications of carbonaceous adsorbents derived from black liquor sludge *J. Taiwan Inst. Chem. Eng.* **77** 236–243

[8] Norgren M, Edlund H 2014 Lignin : Recent advances and emerging applications *Curr. Opin. Colloid Interface Sci.* **19** 5 409–416

[9] Lora J H, Glasser W G 2002 Recent Industrial applications of lignin : a sustainable alternative to nonrenewable materials *J. Poly. and the Env.* **10** 1 39-48

[10] Mohammed-Ziegler I, Holmgren A, Forsling W, Lindberg M, Ranheimer M 2004 Mechanism of the adsorption process of pinosylvin and some polyhydroxybenzenes onto the structure of lignin *Vibrational Spectro.* **36** 1 65-72

[11] Amriani F, Sari A A, Anggraini R I F, Abimanyu H, Tachibana S 2016 Evaluation of lignin-based black liquor decolorization by *Trametes versicolor* U 80 *AIP Conference Proceedings* **1803** 1–8

[12] Daghrir R, Drogui P, Robert, D 2013 Modified TiO_2 for environmental photocatalytic applications : a review *Ind. Eng. Chem. Res.* **52** 3581–3599

[13] Khairy M, Zakaria W 2014 Effect of metal-doping of TiO_2 nanoparticles on their photocatalytic activities toward removal of organic dyes *Egypt. J. Pet.* **23** 419–426

[14] Carp O, Huisman C L, Reller A 2004 Photoinduced reactivity of titanium dioxide *Prog. Solid State Chem.* **32** 33–177

[15] Sakurai K, Mizusawa M 2010 X-ray diffraction imaging of anatase and rutile *Anal. Chem.* **82** 3519–3522

[16] Yunus N N, Hamzah F, So'aib M S, Krishnan, J 2016 Effect of catalyst loading on photocatalytic degradation of phenol by using N, S co-doped TiO_2 *IOP Conf. Proceed. Mat. Sci. Eng.* **206** 012092

[17] Kansal S K, Singh M, Sud D 2008 Studies on TiO_2 / ZnO photocatalysed degradation of lignin

J. Hazard. Mater. **153** 412–417
[18] Doong R, Chen C, Maithreepala R A 2001 The influence of pH and cadmium sulfide on the photocatalytic degradation of 2-chlorophenol in titanium dioxide suspensions *Water Res.* **35** 12 2873–2880
[19] Hoffmann M R, Martin S T, Choi W, Bahnemannt D W 1995 Environmental applications of semiconductor photocatalysis *Chem. Rev.* **95** 69–96
[20] Prado R, Erdocia X, Labidi J 2013 Effect of the photocatalytic activity of TiO_2 on lignin depolymerization *Chemosphere* **91** 1355–1361

IOP Conf. Series: Earth and Environmental Science **131** (2018) 012054 doi:10.1088/1755-1315/131/1/012054

Extraction of essential oil from baby Java orange (*Citrus sinensis*) solid waste using water and steam distillation

I A Dewi, A M Prastyo and S Wijana

Department of Agro-industrial Technology, Faculty of Agricultural Technology, Universitas Brawijaya, Malang, Indonesia

E-mail: ikaatsaridewi@ub.ac.id

Abstract. Baby java orange (*Citrus sinensis*) is commonly consumed as juice. Processing of baby java orange leaves organic waste which consist of the mesocarp, exocarp, seed, and wall of the orange. Therefore, it is necessary to process baby java orange waste to be valuable products. The purpose of this study was to provide added value to unutilized baby java orange waste, and to find out the pretreatment of time-delay process that maximize the yield of essential oil produced. Essential oil processing can be done by water and steam distillation. The study used randomized block design with one factor namely distillation time-delay process by air drying consisted of 4 levels i.e. the distillation delay for 2, 4, 6, and 8 days. The best treatment was determined based on the yield. The best essential oil from baby java orange waste was obtained from the treatment of distillation delay-process of 8 days. This pretreatment generated yield value of 0.63% with moisture content of 24.21%. By estimating the price of essential oil showed that this effort not only reduced the bulky organic waste but also provided potential economical value.

1. Introduction

Orange is one of Indonesian widely cultivated commodity. Malang Regency has productively cultivated baby java orange. Baby java orange belongs to sweet oranges due to its dominant distinctive sweetness so it is commonly processed as fruit juice [1]. However, baby java orange contains vitamin, nutrient, minerals, and volatile compound which could be extracted as essential oil [2].

The high productivity of baby java orange in Malang is not accompanied by sufficient post-harvest process. As a result, plenty of unprocessed matter such as mesocarp, exocarp, seed, and wall of the orange that have potential economic value remained wasted. Un-processed parts of oranges contain high amount of aromatic volatile compound which can be processed into essential oil [3]. Volatile compound is commonly found in epidermis, flavedo, albedo, and oil glands [4]. Processing of these untreated parts of baby java orange may become the first step to produce value added by-product considering the fact that orange's essential oil are sold IDR 600,000 to IDR 700,000 each kilogram [5]. The final product may undergo another process to become industrial derivative products such as perfume, soap, air freshener, health products and many more [6].

Essential oil extraction can be done through distillation. There are some techniques of distillation which can be used, they are: steam distillation, water distillation, and water and steam distillation which was used in this research. Water and steam distillation works by separating or clarifying water-insoluble compound. The first step that needs to be done is air drying of baby java orange organic solid waste in order to disintegrate the orange cells and to facilitate releasing the oil [7].

One example of material handling process as pretreatment of distillation is air drying. Selected material is put in a room avoided by direct sun exposure but still aerated. This pretreatment is chosen because it doesn't cause excessive heat which can lead to water and volatile substances evaporation. Another advantage of air drying is the low required cost [8]. Sipahelut [9] stated air drying executed before distillation to extract essential oils could reduce moisture content and resulted in higher yields.

The study aims to discover the pretreatment of delay process time that maximize the yield of essential oil produced. The estimation of the gross revenue from this product is presented.

2. Materials and Methods

2.1. Instruments and materials

The instruments used in this study were water and steam distillation unit. Baby java orange organic solid waste was obtained from street juice vendor nearby the orange plantation in Dau Subdistrict, Malang District, East Java Province. As additional material, water was used in distillation process [10].

2.2. Research design

Research design had been carried out to determine the best treatment that produced the best essential oil out of baby java orange waste. Several analysis including yield and moisture content were done. This study used Randomized Block Design accomplished by using 1 factor, which was air drying duration: 2; 4; 6; and 8 days. Each of these 4 levels has been repeated 5 times.

2.3. Essential oil production

The production of essential oil was started by preparing organic solid waste from baby java orange. As many of 2.5 kg of baby java orange waste were air-dried for 2, 4, 6 and 8 days. After this time-delay process pretreatment, the waste were chopped into smaller pieces with maximum area of 4 cm^2. Before the process of distillation, water content of each treatment was measured.

Distillation process was started by filling 2.5 L of water into the kettle (distillatory cylinder), then heated up to 100-115°C. The material was distillated for 4 hours. Remaining water and waste after distillation were measured. From this process, baby java orange essential oil was obtained. The yield was determined by following formula [11].

$$\text{Yield} = \frac{\text{Oil volume x Essential oil specific gravity}}{\text{Total distilled material}} \text{ x } 100\% \quad (1)$$

3. Results and Discussion

3.1. Moisture content and yield

The air drying of baby java orange organic solid waste was done by spreading material in a room that was not exposed directly to the sun. The delay process was carried out at a temperature range of 24 ^{0}C to 26.9 ^{0}C and room humidity ranged from 60% to 83%. The time-delay process resulted in reduced moisture content in the material due to water evaporation. The decrease in moisture content is parallel with an increase in time-delay process, as shown in Figure 1.

Average yield of baby java orange waste essential oil was rising as the increased distillation delay time of 2 days, 4 days, 6 days and 8 days. Correlation between average yield value and time-delay process is presented on Table 1.

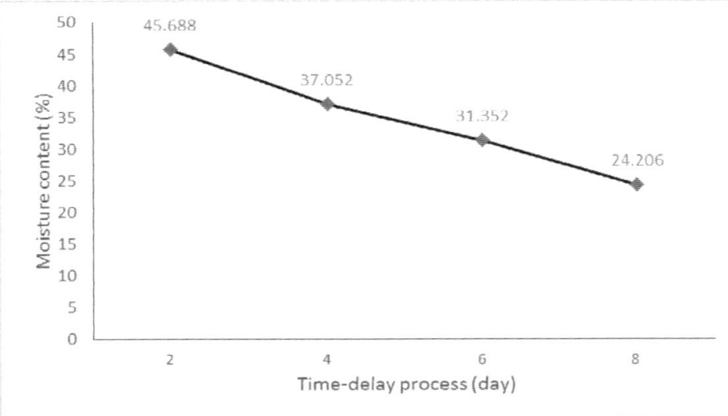

Figure 1. The effect of time-delay process on moisture content of baby java orange organic waste prior to distillation

Essential oil yield is affected by its moisture content since the less water in material, the more essential oil yield [9]. According to [12][13], water evaporation caused oil cell to rupture which allows the vapor to enter the cell and pulls out the oil. Damage to the pores could affect the damage of oil cell walls so that the oil and the content of volatile compounds in it would be more easily taken. Treatment of distillation delay process for 8 days is the best treatment because it affected the decrease in moisture content and damage to the texture of the material distilled. According to [14], the longer the delay process, the more essential oil produced.

Table 1. The Effect of Time-Delay Process of Distillation on Essential Oil Yield (α=5%)

Time-Delay Process (Days)	Average (%)	Least significant difference	Notation
2	0.18 ± 0.018	0.28	a
4	0.41 ± 0.038	0.51	b
6	0.48 ± 0.047	0.58	b
8	0.63 ± 0.095	-	c

Note: different notation indicates considerable difference among treatments

3.2. Gross revenue estimation

Opportunity of utilization of baby java orange waste to be essential oil is very high. The production figures of baby java orange in Dau Subdistrict in 2011 reached 24.5 million kg and continues to increase every year [15]. From the whole fruit, the waste of baby orange processed into juice reached 54.9%. Suppose that 50% of total waste in one year can be utilized to become essential oil. By considering the average yield of essential oil of about 0.63%, the amount of essential oil obtained can reach approximately 42,369 kg per year. Given the high price of citrus essential oil can attain IDR 700,000/kg [5], the gross revenue from the production of essential oils of baby java orange waste if accumulated in one year can reach IDR 29,658,300,000.

Citrus waste contains many kinds of bioactive compounds which can be used as anti-cholesterol, healthy digestion, improve respiratory conditions, as peptin flour, and utilization as essential oil [3]. From essential oil, many benefits can be obtained.

4. Conclusion

Time-delay process of the production of baby java orange juice waste into essential oil showed significant effect to the yield. The best treatment was obtained from distillation delay process of air drying for 8 days. This pretreatment generated yield value of 0.63% with moisture content of 24.206%.

References

[1] Kurniawan E 2015 Design of fruit sorting system based on skin colour difference *Teknik Industri Heuristic* **12** 2 93-99

[2] Bermejo A, Cano, A 2012 Analysis of nutritional constituents in twenty citrus cultivars from the Mediterranean Area at different stages of ripening *Food and Nutrition Sciences* **1** 3 639-650

[3] Yana Y (online) 9 Benefits of Citrus for Skin Health. Retrieved from http://manfaat.co.id/manfaat-kulit-jeruk. [Accessed on 11 September 2016]

[4] Widowati E, Utami R, Nurhatadi E, Andriani M A M, Wigati, W A 2014 Production and characterization of pectinase enzymes by pectinolytic bacteria in orange juice (*Citrus cinensis*) clarification *Aplikasi Teknologi Pangan* **3** 1 16-20

[5] Armando R 2009 Producing Qualified Essential Oil Penebar Swadaya Jakarta

[6] Rusli M S 2010 Producing Volatile Oil Successfully PT Agromedia Pustaka Jakarta

[7] Wartini N M, Harijanto, Susanto S, Retnowati R, Yunianti 2007 Effect of curing process on composition of Indonesian bay leaf (*Eugenia polyantha* Wight.): components' profile and preference of flavor extracted by simultaneous distillation-extraction method *Teknologi Pertanian* **8** 1 10-18

[8] Pramono S 2006 Post-harvest handling and its influence on natural drug therapy effect. Proceeding of the XXVIII Indonesian Medicinal Plants National Seminar, Bogor, 2005, pp. 1-6

[9] Sipahelut S G 2012 Chemical characteristic of nutmeg (*Myristica fragrans* Houtt) flesh oil by some drying and distillation methods *Agroforestri* **7** 1 59-64

[10] Lempang I R, Fatimawali, Pelealu N C 2016 Quality test of bulky and packaged vegetable oil in Manado Pharmacon Ilmiah Farmasi **5** 4 155-161

[11] Rahayoe S, Suharyo Y, Tetuko, Mega T 2007 Kinetic study of water content and chopping effect to distillation rate of essential oils. Proceeding of Agricultural Engineering National Seminar, Agricultural Technology Faculty, Gadjah Mada University, Yogyakarta, pp. 393-406

[12] Tanasale M L P 2012 The basic commodity treatment and refining water-steam application towards the rendement and organoleptik characteristic of sssential oil *Ekosains* **1**1 35-40

[13] Kanduser M, Miclavic D 2008 Electroporation in Cell and Tissue an Overview. In: Vorobiev E, Lebovk Elwcrotecmologies For Extraction From Food Plant Material Springer New York

[14] Sumarsono 2005 Change in water content of patchouli leaves during drying processwith different air stream velocities and tray dryer rotation *Ilmu-ilmu Pertanian Indonesia* **7** 1 59-67

[15] Kompas (online) Farmers in Malang Enjoy Citrus Harvest. Retrieved from http://regional.kompas.com/read/2011/09/26/02231820/Petani.di.Malang.Nikmati.Panen.Jeruk. [Accessed on 11 September 2016]

IOP Conf. Series: Earth and Environmental Science **131** (2018) 012055 doi:10.1088/1755-1315/131/1/012055

Characterization the potential of biochar from cow and pig manure for geoecology application

I M Gunamantha and G A B Widana

Chemical Analysis Department, Faculty of Mathematics and Natural Science, Universitas Pendidikan Ganesha, Singaraja, Indonesia

E-mail: md_gunamantha@yahoo.com

Abstract. Biochar is a solid product generated from the carbonization of biomass with various potential benefits. The utilisation of biochar should be adapted to its characteristic which is mainly influenced by its feedstock. In this study, cow and pig manure biochar generated by a conventional process, were characterized by its physical and chemical analysis and its potential to be used as soil amendment. For this purpose, several main parameters were analyzed: organic carbon, Nutrient (total-N, available P and K) status, Cation Exchange Capacity (CEC), proximate data analysis (moisture content, ash, volatile matter and fixed carbon) and its ash composition. The comparison between biochar and feedstock will be based on these parameters. The results of this study show that the organic carbon, available P, ash, and fixed carbon content of pig-manure biochar is higher than cow manure-derived biochar; while total-N, available K, CEC and volatile matter is lower. On its ash composition, the pig manure-derived biochar is dominated by SiO_2, Al_2O_3, Fe_2O_3, P_2O_5, and CaO while the cow manure-derived biochar is dominated by SiO_2, CaO, Al_2O_3, K_2O, and P_2O_5. However, both biochar show potential for improving soil quality and reducing carbon emission from animal manure.

1. Introduction

Biochar, in scientific literature, is known by many names such as black carbon, pyrogenic carbon, and charcoal [1]. Biochar is a co-product of the bio-energy pyrolysis (carbonization) process during which biomass is heated in a low-oxygen to zero-oxygen state which results in carbon-rich and recalcitrant carbon-burning ingredients [1-3]. In other words, biochar is the result of combustion or carbonization process from carbonized lignocellullosic material at 350-700^oC [4-5] and its activity is still low and contains solid and porous carbon [6-7]. Therefore, the biochar is not produced with the same characteristics and depends on the pyrolysis condition and the source of the feedstock [8-10].

Prior studies indicate that the greatest importance of biochar properties is the feedstock used for biochar production [11-13], which has a substantial impact on the compositional constituents of the biochar [10, 14, 15]. According to Shalini et al. [13], before and after production, biochar differed in physical, chemical, and nutrient characteristics. Feedstock was a better predictor of biochar ash content and C/N ratio [15]. In Zhao et al. [10], the biochar parameters most affected by feedstock properties were for e.g. total organic carbon, fixed carbon, and mineral elements of biochar. Thus, biochar is not pure carbon but a mixture of carbon (C), hydrogen (H), oxygen (O), nitrogen (N), sulfur (S), and ash in various proportions [16]. This properties are required in the soil amendment materials.

IOP Conf. Series: Earth and Environmental Science **131** (2018) 012055 doi:10.1088/1755-1315/131/1/012055

In previous studies, biochar's feedstock usually consisted of material from plants and very limited animal manure. In this present study, the cow and pig manure feedstock and their derived biochar were characterized by organic C, total-N, K and P available, CEC, and ash mineral composition. The abundant availability of this type of manure, especially in Bali, makes them good feedstock for biochar production and an excellent resource for waste management.

2. Materials and Methods

2.1. Biochar preparation

Cow and pig manure were used and converted into biochar. These materials were obtained from individual farms in Buleleng, Bali. For this biochar research we used a carbonizer made from a clay pot and its lid (2.5 L traditional beverage container), obtained from the local market. As for the pyrolysis process we used an LPG stove. During the carbonization process, the temperature was not controlled. The manure was dried in the sun for two weeks and then placed in the carbonizer. Once the fire started burning, the lid was put on the stove. Twenty minutes later, the smoke exuding from the small crevices of the lid was somewhat black and was subsequently replaced by white smoke, which acted as a sign that the manure was burnt. After 4 hours, the manure was completely combust indicated by the depletion of smoke. Prior to testing, the resulting biochar was then crushed and sieved into a 2 mm mesh sieve to acquire a more homogenous substrate [17].

2.2. Biochar chemical analysis

Feedstock and the biochar were analyzed for organic carbon, available P and K, N-Total, Cation Exchange Capacity (CEC), proximate analysis data, and the ash composition. The organic carbon was determined by Walkley-Black procedure, available P by Bray I method, available K by titrisol method, N-total by Kjeldahl method, and CEC using the percolation method with ammonium acetate [18]. The proximate analysis and ash composite determination were conducted at the Centre for Mineral and Coal Technology (tekMIRA), Bandung. A typical proximate analysis includes the moisture, volatile matter, ash content, and fixed carbon [19, 20]. ASTM D.3173 was used to determine the moisture content, ISO 562 to determine volatile matter (VM), ASTM D.3174 to determine ash content, and ASTM D.3172 to determine fixed carbon. The moisture content was determined by the weight loss after the sample was heated at 105°C for an hour. The volatile matter was thus determined by measuring the weight loss following the combustion of approximately 1 g sample in a crucible at 900°C. Following the same procedure, the ash content was determined at 750°C. The fixed carbon (FC) derived by substracting from 100 the percentage value of moisture content, VM and ash. The ash compositions were determined by the gravimetric method for SiO_2, LOI, and SO_3; spectrofotometry for TiO_2 and P_2O_5; and atomic absorption spectroscopy (AAS) for Al_2O_3, Fe_2O_3, K_2O, Na_2O, CaO, MgO, and MnO. Hence, all acquired elements were translated into oxides.

3. Results and Discussion

3.1. Organic carbon of biochar

As is the raw material, organic carbon in pig manure-derived biochar (18.88%) was lower than in cow manure-derived biochar (27.31%). These results were much lower than in wood-derived biochar (71.47%), coconut shell-derived biochar (60.07%) and in organic waste-derived biochar (31.41%); as reported by Widowati et al. [21] and Widowati and Asnah [22]. According to Verheijen et al. [3], in addition to the operating conditions of pyrolysis, the composition and chemical structure of biomass feedstocks is closely related to the composition and chemical structure of the biochar produced. Sellulose and lignin, for example, had thermal degradation in a range between 240-350°C and 280-500°C [23]. Therefore, the relative proportions of these components will affect the carbon content of biomass where higher lignin content biomass will provide higher carbon content in biochar compared with that produced from biomass with lower lignin content [3]. In Winter et al. [24], the lignin content is 16 - 30% in cow manure and <5% in pig manure. However, the results taken from this study show that organic carbon content in cow manure-derived biochar is lower than in pig manure-derived biochar. This is possible because the content of protein and fat in pig manure is higher than that of in cow manure. According to Winter et al. [24], protein content in pig manure ranges from 16 - 40% and fat 10% while in cow manure the protein content ranges from 17-20% and fat between 4 - 7%. As discussed above, organic-C in cow and pig manure-derived biochar are relatively high when measuring against International Biochar Initiative (IBI) Standard which required 10% minimum organic-C in biochar [25].

3.2. Nutrient of biochar

According to Piash et al. [26], the content of N, P and K is generally not determined by its concentration in the feedstock. The result observed in this study suggested that 0.5% nitrogen concentration in cow manure and 0.75% in its biochar, while 1.075% in pig manure and 0.74% in its biochar. Thus, the nitrogen conversion efficiency for biochar obtained from cow and pig manure is 150% and 70.48%, respectively. This data indicates that converting livestock manure into biochar does not reduce its role as a nitrogen provider. Although the nitrogen content of pig manure is higher, the conversion efficiency of N in pig manure is less than half the efficiency of N conversion in cow manure. The loss of N on the carbonization of pig manure may be caused by organic compounds in pig manure dominated by more labile organic compounds than those available in cow manure. This is understandable because pig feed is much softer than cow feed.

Unlike the conversion efficiency of nitrogen, the conversion efficiency of P on the carbonization of pig manure (104%) is greater than that of cow manure (71%). However, the conversion is relatively small. This is because the fraction P in feedstock is converted into stable P during biomass carbonization thereby reducing P mobility [27]. The results observed from this study show that the content of P available in cow manure and its biochar are 0.746g/kg and 0.529g/kg while for pig manure and its biochar are 0.620g/kg and 0.646g/kg. The P available content of both biochars are higher than in wheat straw-derived biochar (0.21 g/kg) and corn cobs-derived biochar (0.45 g/kg), but lower than in corn stalk-derived biochar (2.10 g/kg) as reported by Chan and Xu [28]. However, consistent with Dai et al. [27], the results observed in this study indicate that P in livestock manure can be recycled into biochar as source P to reduce the need for chemical fertilizers.

As is the case with total-N and available P, the data found in this study also showed that the content of available K in the bioachar was not consistent with the concentration in its feedstock. The content of available K in cow manure is lower than in pig manure, but in their biochar available K content in cow manure-derived biochar is much higher than in pig manure-derived biochar. The results observed in this study show available K in cow manure and its biochar are 15.96g/kg and 21.26g/kg while for pig manure and its biochar are 16.22g/kg and 15.28g/kg. These are higher than in biochar which are reported by Chan and Xu [28], where the available K for wheat straw-derived biochar is 2.90 g/kg, corn cobs-derived biochar is 9.40 g/kg, and corn stalks-derived biochar is 0.03 g/kg. The data also indicated that the K conversion efficiency of cow and pig manure was 133.22% and 94.19%, respectively. This indicates that the volatility of these elements during the carbonization process is very low. In other words, there was 33.22% increase of available K in cow manure-derived biochar and a slight decrease of 5.81% in the biochar obtained from pig manure.

3.3. Cation exchange capacity

Another factor that determines the quality of biochar is the CEC values. CEC is the measure of the number of cations (positively charged ions) that can be retained by biochar. The results from this research show that CEC decreased in both biochars when compared to its feedstock. However, when comparing results between the two biochars and feedstock, CEC in cow manure-derived biochar (2.11 me/100g) is slightly higher than in pig manure-derived biochar (2.06 me/100g) even though CEC in cow manure (5.37 me/100g) is much higher than in pig manure (3.19 me/100g). According to Montecillo [29], the presence of organic matter contributes to CEC. However, when viewed from the content of organic-C, both cow manure (30.32%) and its biochar (27.31%) are much higher than pig manure (22.51%) and its biochar (18.88%). As mentioned in the previous section, the content of lignin in cow manure is higher than that of pig manure, in contrast to fat and protein. In comparison to Zhao et al. [10], CEC biochar produced in this study are much less. This is consistent with relatively low readings in organic-C content and volatile matter.

3.4. Proximate analysis

Except for moisture content, all values of proximate analysis were reported on a dry weight basis. VM is a measure of gas, oil, and tar that is released when the biomass is heated above 300°C. The major constituents of feedstock (C, H, and O) evaporate during dehydration and pyrolysis with H and O are lost in much greater quantities than C. O and H were initially lost as water and then as hydrocarbons, slowly evaporating agents, H_2, CO, and CO_2 [30]. Thus, the biochar contains a lower VM when compared with its feedstock. Based on the dry weight it was founded that VM of both feedstock was 51.54% and 39.96% respectively for cow and pig manure. The VM decreased to 51.16% in cow manure and 38.94% in pig manure. VM of cow manure-derived biochar (19.36%) was higher than that of pig manure-derived biochar (13.62%). The results of this study are very close to those reported by Zhao et al. [10] but lower than VM of biochar obtained from various types of grass [31]. A study conducted by Spokas [32] showed that VM in biochar may affect overall microbial activity when added to the soil and in some case result in increased mineralization due to stimuli in microbial respiration. Theoretically, biochar with high VM content is less stable and has a higher proportion of labile carbon that provides energy for microbial growth and limits the nitrogen availability necessary for plant growth so it will suppress plant growth. A study by Deenik et al. [33] consider that VM materials above 35% are high (causing nitrogen deficiency) and VM below 10% is low, therefore, both biochars in this study are not expected to limit plant growth.

Organic material in the feedstock is not released as the remaining VM as Fixed carbon (FC). FC in cow and pig manure were founded in this study to be 14.05% and 6.46%, respectively. Generally, the fixed carbon in biochar increases relative to the FC content of the raw material with increasing pyrolysis temperature even up to 500% [18]. However, in this study the increase occurred only by 53.74 % for the carbonization of cow manure and 120.43% for pig manure. The FC for each cow and pig manure-derived biochar is 21.60% and 14.24% and in their feedstock it is 14.05% and 6.46% respectively.

According to Dume et al. [18] the fixed carbon content of biochar will be lower than 30% if the ash content is more than 35%. Ash is the remaining mineral material after complete combustion has occurred in feedstock. The results observed in this present study showed that the ash content of cow manure-derived biochar reached 59.04% and pig manure-derived biochar reached 72.14%, this is consistent with its feedstock that ash content of pig manure (53.58%) is higher than cow manure (34.41%). They are determined by the ash content of its feedstock. In this present research, for both biochars, the ash composition was rich in Si oxide (SiO_2) and to a lesser extent in Mn oxide (MnO). The ash of cow manure-derived biochar showed considerably higher concentration of CaO than the ash in pig manure-derived biochar. It is important to note that the percentage of CaO in the former was about 13.73%, while that of Al_2O_3 in the ash of pig manure-derived biochar was 13.65%. In Brewer et al. [34] it is also reported that SiO_2 was predominate in the ash composition of switchgrass (61.86%) and cast stover-derived biochar (47.06%), but not so in the composition of hardwood-derived biochar which CaO (22.37%) is the oxide predominant. In more detail, the order of metal oxides in ash composition of cow manure-derived biochar from the highest to the lowest are SiO_2 (50.10%), CaO (13.73%), P_2O_5 (7.75%), Al_2O_3 (7.28%), K_2O (5.54%), Fe_2O_3 (4.71%), MgO (4.09%), Na_2O (1.95%), LOI (1.32%), SO_3 (1.05%), TiO_2 (0.39%), and MnO (0.15%). A slightly different order on ash composition of pig manure-derived biochar; SiO_2 (48.09%) followed by Al_2O_3 (13.65%), Fe_2O_3 (9.54%), P_2O_5 (7.67%), CaO (7.22%), K_2O (5.59%), MgO (4.17%), Na_2O (2.48%), SO_3 (1.05%), LOI (0.92%), TiO_2 (0.87%), and MnO (0.26%). The findings further confirm that the application of both biochar into agricultural soil, in addition to reducing carbon emissions by increasing the fixed carbon content from their initial conditions, may have a vital effect due to their mineral content.

4. Conclusion

Biochar are chemically different in cow and pig manure feedstocks. Biochar from cow manure has a higher concentration of C-organic, P-available, ash, and fixed carbon content when compared with pig manure. While, on the other hand, biochar from pig manure has a higher concentration of N, K, CEC, and volatile matter than cow manure. But all of these were not consistent with their feedstock. Si, Al, and Ca are the most widely available minerals in both biochars, whereas Mn is the least available component.

Acknowledgments

Financial support for this research was provided by Universitas Pendidikan Ganesha (Undiksha). The authors thank the following for their assistance on various other aspects of the completion of this research: Undiksha Institute for Research and Community Service, Undiksha Chemical Analysis Laboratory, Udayana Soil Laboratory, and Center for Mineral and Coal Technology (tekMIRA), Bandung.

References

[1] Lehmann J, Gaunt J, Rondon M 2006 Biochar sequestration in terrestrial ecosystems: a review. *Mitig. Adapt. Strategy Global Change* **11** 403–27
[2] Glaser B, Lehmann J, Zech,W 2002 Ameliorating physical and chemical properties of highly weathered soils in the tropics with charcoal – a review Biol. *Fertil. Soils* **35** 219–230
[3] Verheijen, F, Jeffery, S, Bastos, A C, van der Velde, M , Diafas, I 2010 Biochar Application to Soils: A Critical Scientific Review of Effects on Soil Properties, Processes and Functions European Commission, Joint Research Centre Institute for Environment and Sustainability
[4] Domingues R R, Trugilho P F, Silva C A, de Melo I C N A, Melo L A C A, Magriotis Z M, SaÂnchez-Monedero M A 2017 Properties of biochar derived from wood and high-nutrient biomasses with the aim of agronomic and environmental benefits. *PLoS ONE* **12** 5 1-19
[5] Denyes M J, Matovic, D, Zeeba B A, Rutter, A 2012 The use of biochar to reduce soil PCB bioavailability to *Cucurbita pepo* and *Eisenia fetida Sci. Tot. Environ.* **437** 76–82
[6] Sohi S, Lopez-Cape l E, Krull E, Bol R 2009 Biochar, climate change and soil: A review to guide future research. CSIRO Glen Osmond Australia
[7] Atkinson C J, Fitzgerald J D, Hipps N A 2010 Potential mechanisms for achieving agricultural benefits from biochar application to temperate soils: a review *Plant and Soil* **337** 1-2 1-18
[8] Downie A, Monroe P, Crosky A 2009 Physical properties of biochar In Biochar for Environmental Management: Science and Technology (Eds. Lehmann, J. & Joseph, S.) Earthscan London UK 13-29
[9] Lehmann J 2007 A Handful of Carbon *Nature* **447** 7141 143–144
[10] Zhao L, Cao X, Masek O, Zimmerman A 2013 Heterogeneity of biochar properties as a function of feedstock sources and production temperatures *J. Hazard. Mater.* **256-257** 1-9
[11] Kannan, P, Arunachalam, P, Prabukumar, G, Govindaraj, M 2013 Biochar an alternate option for crop residues and solid waste disposal and climate change mitigation *Afr. J. Agric. Res.* **8** 21 2403-2412
[12] Maftu'ah, E, Nursyamsi, D 2015 Potensi berbagai bahan organik rawa sebagai sumber biochar. Proseeding Seminar Nasional Masyarakat *Biodiversiti Indonesia* pp. 776-781
[13] Shalini R, Pugalendhi S, Subramanian P, Gopal N O 2017 Characteristic study on biochar production from biological substrates by slow pyrolysis for carbon sequestration *Int. J. Curr. Microbiol. App. Sci.* **6** 4 314-323
[14] Jindo K, Mizumoto H, Sawada Y, Sanchez-Monedero M A, Sonoki, T 2014 Physical and chemical characterization of biochars derived from different agricultural residues *Biogeosciences* **11** 6613–6621
[15] Mukome F N D, Zhang,X, Silva L C R, Six, J, Parikh S J 2013 Use of chemical and physical

characteristics to investigate trends in biochar feedstocks *J. Agric. Food Chem.* **61** 2196–2204

[16] Pituello C, Francioso O, Simonetti G, Pisi A, Torreggiani A, Berti A, Morari F 2015 Characterization of chemical–physical, structural and morphological properties of biochars from biowastes produced at different temperatures *J. Soils Sediments* **15** 792–804

[17] Bouqbis L, Daoud S, Koyro H W, Kammann C I, Ainlhout L F Z, Harrouni M C 2016 Biochar from argan shells: production and characterization *Int. J Recycl. Org. Waste Agric.* **5** 361–365

[18] Dume B, Berecha G, Tulu S 2015 Characterization of biochar produced at different temperatures and its effect on acidic nitosol of jimma, Southwest Ethiopia *Int. J. Soil Sci.* **10** 2 63-73

[19] Enders A, Hanley K, Whitman T, Joseph J, Lehmann J 2012 Characterization of biochars to evaluate recalcitrance and agronomic performance *Bioresour. Technol.* **114** 644–653

[20] Mary G S, Sugumaran P, Niveditha S, Ramalakshmi B, Ravichandran P, Seshadri S 2016 Production, characterization and evaluation of biochar from pod (*Pisum sativum*), leaf (*Brassica oleracea*) and peel (*Citrus sinensis*) wastes *Int. J Recycl. Org. Waste Agric.* **5** 43–53

[21] Widowati W, Asnah A, Utomo W H 2014 The use of biochar to reduce nitrogen and potassium leaching from soil cultivated with maize *J. Degraded Mining Lands Manage.* **2** 1 211-218

[22] Widowati W, Asnah A 2014 Biochars effect on potassium fertilizer and leaching potassium dosage for two corn-planting seasons *Agrivita* **36** 1 65-71

[23] Demirbas A 2004 Combustion characteristics of different biomass fuels *Prog. Energy Combus. Sci.* **30** 219–230

[24] Winter J, Hilpert R, Schmitz H 1992 Treatment of animal manure and wastes for ultimate disposal – review *Asian-Austr. J. Animal Sci.* **5** 2 199-215

[25] International Biochar Inititive (IBI) 2013 Standardized Product Definition and Product Testing Guidelines for Biochar that is Used in Soi. IBI-STD-1.1

[26] Piash M I, Hossain Md F, Parveen Z 2016 Physico-chemical properties and nutrient content of some slow pyrolysis biochars produced from different feedstocks Bangladesh *J. Sci. Res* **29** 2 111-122

[27] Dai L, Li H, Tan F, Zhu N, He M, Hu G 2016 Biochar: a potential route for recycling of phosphorus in agricultural residues GCB *Bioenergy* **8** 852–858

[28] Chan K Y, Xu Z 2009 Biochar: Nutrient Properties and Their Enhancement Biochar for Environmental Management: Science and Technology (Eds. Lehmann, J. & Joseph, S.) Earthscan London UK 67-84

[29] Montecillo L C 1983 Total clay and organic matter in relation to soil cation exchange capacity *Philipp J. Crop Sci.* **8** 1 41-44

[30] Antal M J, Grønli M 2003 The art, science, and technology of charcoal production. *Ind. Eng. Chem. Res.* **42** 1619–1640

[31] Oram N J, Tess F J, van de Voordeb, Ouwehanda G J, Bezemerc T M, Mommer L, Jeffery S, van Groenigena J W 2014 Soil amendment with biochar increases the competitive ability of legumes via increased potassium availability *Agric. Ecosystems Environ.* **191** 92–98

[32] Spokas K A 2014 Review of the stability of biochar in soils predictability of O/C molar ratios *Carbon Manage.* **1** 2 289–303

[33] Deenik J L, McClellan T, Uehara G, Antal M J, Campbell S 2010 Charcoal volatile matter content influences plant growth and soil nitrogen transformations *Soil Sci. Soc. Am. J.* **7** 1259-1270

[34] Brewer C E, Schmidt-Rohr K, Satrio J A, Brown R C 2009 Characterization of biochar from fast pyrolysis and gasification systems *Environ. Prog. Sustain. Energ.* **28** 3 386-396

IOP Conf. Series: Earth and Environmental Science **131** (2018) 012056 doi:10.1088/1755-1315/131/1/012056

Impact of thermal pretreatment and MSW origin on composition and hydrolysability in a sugar platform biorefinery

L P Vaurs, S Heaven and C J Banks

Faculty of Engineering and the Environment, University of Southampton, Southampton, UK

E-mail: Leopaul.vaurs@gmail.com

Abstract. Municipal solid waste (MSW) is a widely available large volume source of lignocellulosic material containing a waste paper/cardboard mixture which can be converted into fermentable sugars via cellulolytic enzyme hydrolysis in a sugar platform biorefinery. Thermal pre-treatments are generally applied to MSW to facilitate the extraction of the lignocellulosic material from recyclable materials (plastics, metals etc.) and improve the paper pulp conversion to sugars. Applying high temperature might enhance food waste solubilisation but may collapse cellulose fibre decreasing its hydrolysability. Low temperature pre-treatment will reduce the energy demand but might result in highly contaminated pulp. Preliminary results showed that the enzymatic hydrolysis performances were dependent on the MSW origins. Using 8 different samples, the impact of thermal pretreatment and MSW origin on pulp composition and hydrolysability was assessed in this work. Low pre-treatment temperature produced pulp which contained less lignocellulosic material but which hydrolysed to a higher degree than MSW treated at high temperatures. High temperature pre-treatment could have exposed more of the inhibiting lignin to cellulase. This information would have a significant economic impact on a commercial plant as expensive autoclave could be advantageously replaced by a cheaper process. Glucan conversions were also found to vary depending on the region, the recycling rate possibly because of the lower recycling rate resulting in the use of less paper additive in the material or the difference in paper production technology (chemical VS mechanical pulping). This could also be explained by the differences in paper composition.

1. Introduction

Many efforts are currently made to reduce fossil fuel dependence while ensuring worldwide food supply to cope with increasing population. In the meantime, waste generation inevitably associated to society development, is raising as a priority concern for municipalities [1].

Municipal solid waste (MSW) represents a large amount of lignocellulosic material which can be enzymatically converted into monosaccharides (mostly glucose from cellulose) in a platform biorefinery. This raw material has the advantages of not being in competition with crops for arable land, it does not require additional energy supplementation to be grown, is widely available and poorly exploited.

To extract the lignocellulosic fraction, mainly constituted of wastepaper and cardboard, it is commonly accepted that a thermal pre-treatment must be applied [2-5]. Preliminary results also indicated that the MSW origin impacted upon these performances. These issues were addressed in this work. These results are confidential and were therefore expressed as percentages of the maximum conversions or autoclave temperatures.

2. Materials and Methods

2.1. Material collection and pre-treatment applied

Eight different MSW samples were collected (Table 1) and thermally pre-treated at different temperatures and a constant material/water ratio (confidential). The material was then washed and the paper pulp was extracted using a patented process (Fiberight). Office papers, newspapers, magazines and cardboards were collected from local recycled paper collections. They were shredded or manually cut into small pieces. A mix stream was also created based on UK waste paper composition [6-8].

Table 1. Summary table of all sample characteristics and pre-treatment conditions.

Designation	Region	Paper recycling rate	Pulping process	Pre-treatment temperature (% of the maximum applied)
Sample 1	Region 1	Low	Chemical	70
Sample 2-4	Region 2	Medium	Mechanical	75-100
Sample 5-6	Region 2	Medium	Mechanical	60-75
Sample 7-8	Region 3	South Wales	Mechanical	50-75

2.2. Composition analysis

All compositional analyses were carried out on dried and ground material. Structural carbohydrate and Acid Insoluble Material (AIM) were analysed based on the NREL method [9]. Pseudo lignin, which is a non-lignin acid insoluble organic fraction [10], was determined by applying an Acid Detergent Fibre (ADF) followed by an Acid Detergent Lignin (ADL) on AIM (see [11]). Food waste content was evaluated by measuring lipid [12], protein by TKN determination [13, 14] and by measuring carbohydrate from lignocellulose using the NREL method after washing the pulp with Neutral Detergent Fibre (NDF) [15], and applying the following equation:

$$\%\text{Food waste (TS)} = \%\text{Lipid} + \%\text{TKN} \times 6.25 + (\%\text{sugar before NDF} - \%\text{NDF} \times \%\text{sugar after NDF}) \quad (1)$$

2.3. Enzymatic hydrolysis

All materials were enzymatically hydrolysed at 5 % total solids (TS) using 2 % commercial cellulase enzyme on a TS basis for 72 h in a 50°C tumbler mixer. Sugars were measured by HPLC at the end of the reaction. To assess the impact of recycling on pulp hydrolysis, filter papers were soaked and mixed for 4 h at 60% in presence of 2 different paper additives.

3. Results and discussions

3.1. Impact of thermal pre-treatment temperature on composition

High pre-treatment temperature seemed to decrease the glucan content and increase the AIM percent in pulp (Figure 1a). This had a negligible effect on the content of food waste and pseudo lignin associated with the pulp.

The lignocellulosic content of the material increases when using higher temperature treatments, most probably due to a higher degree of material solubilisation during the pre-treatment, enhancing pulp separation (lab scale observation).

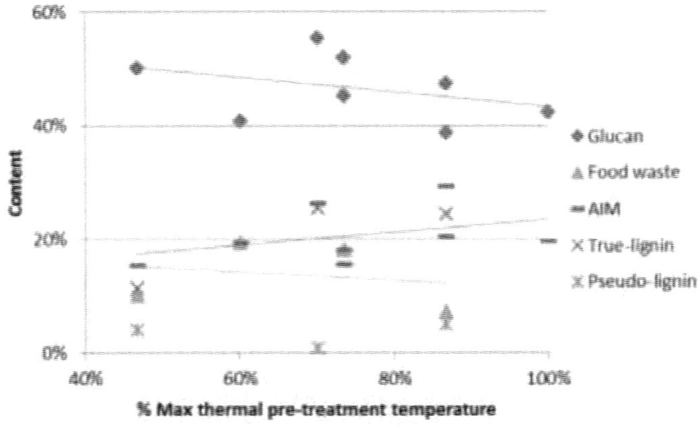

a)

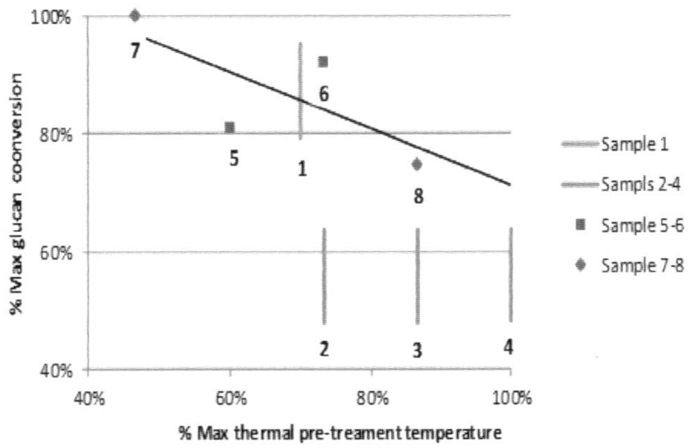

b)

Figure 1. Impact of thermal pre-treatment temperature on a) pulp composition and b) glucan conversion. Vertical bars correspond to ranges of conversion. For linear regression, averaged values were taken.

3.2. Comparison between different paper streams

Results from compositional analysis of the different paper streams are presented in Table 2. As can be seen, the composition varies quite a lot depending on the type of paper, especially the glucan content, AIM, and ash (which includes $CaCO_3$) content. Missing mass could be ink or any acid soluble lignin or additives. Note that the high missing mass in magazines could be caused by the application of organic filler used in coated paper products to provide high gloss and brightness [16].

Table 2. Composition of different paper streams used in this experiment.

All %TS	Glucan	Xylan	Mannan	Galactan	Arabinan	AIM	Ash	$CaCO_3$	Missing mass
Newspaper	47.1 (0.9)	5.6 (1.0)	5.2 (0.2)	0.7 (0.1)	0.4 (0.1)	17.3 (1.0)	17.3 (0.1)	9.6 (0.5)	6.5
Cardboard	46.2 (0.7)	7.7 (0.2)	4.2 (0.0)	0.6 (0.0)	0.4 (0.0)	15.6 (0.0)	11.4 (0.1)	3.6 (0.1)	13.9
Magazine	42.4 (0.8)	7.6 (0.3)	0.7 (0.0)	0.2 (0.0)	0.1 (0.0)	5.9 (1.8)	17.2 (0.2)	10.7 (0.1)	26.1
Office paper	66.8 (0.8)	9.00 (0.3)	1.2 (0.0)	0.0 (0.0)	0.1 (0.0)	1.2 (0.0)	20.5l (0.2)	15.1 (0.1)	0.2
Mixed paper	46.4 (2.9)	7.2 (0.3)	3.7 (0.5)	0.6 (0.1)	0.3 (0.1)	10.6 (0.2)	23.9 (0.8)	19.9 (0.7)	7.4

Results from enzymatic hydrolyses of the different paper streams can be found in Figure 2a. Glucan conversions differ depending on the type of paper. Office paper and cardboard show similar hydrolysis performances with virtually the same glucan conversions. The glucan conversion of newspaper was approximately 30 % lower, while that of magazines was only 25.0 ± 3.1 %. Surprisingly, the paper mixture showed very good conversion, close to that of office paper and cardboard, while it should theoretically be equal to the weighted average of 68.3 %. This may indicate the difficulty of obtaining representative samples at laboratory scale.

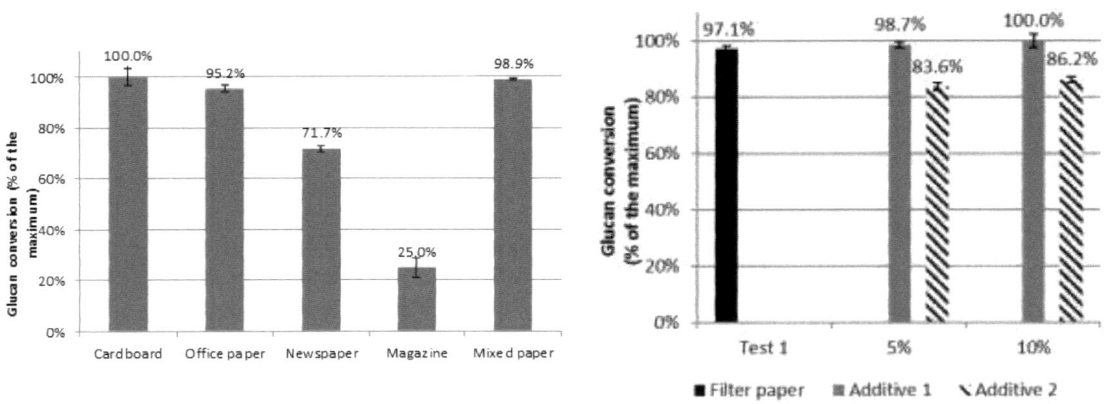

a) b)

Figure 2. Impact of a) paper stream and b) different paper additive on enzymatic hydrolysis

Waste paper found in MSW from region 1 contains a much higher proportion of easily hydrolysable office paper compared to MSW wastepaper from region 2&3 which contain approximately 40% poorly hydrolysable newspaper and magazine [6-8, 17]. This could partially explain differences in pulp glucan conversions.

3.3. Impact of paper recycling and paper pulping process on enzymatic hydrolysis

The impact of 2 paper additives on enzymatic hydrolysis is shown Figure 2b. Additive 1 appeared not to have any impact on the glucan conversion of filter paper while additive 2 decreased the conversion by around 20%. Paper additives generally used in the paper recycling industry to cope with short fibres might coat the cellulose fibrils and therefore limit the access of cellulolytic enzymes to the fibres. SEM comparison of virgin softwood fibres and recycled newsprint (not shown) seemed to confirm this hypothesis.

The different paper recycling rate within the different regions might therefore contribute to differences in hydrolysability. Paper from region 1 is mostly made from chemical pulping processes which target lignin removal while only mechanical processes are currently operating in region 2&3. Lignin is known to negatively influence enzymatic hydrolysis by physically limiting the access of cellulose to cellulase and by unproductively binding to cellulolytic enzymes [18]. Differences in pulping process can therefore impact the hydrolysability of the various pulps tested.

4. Conclusion

This preliminary study shows that low pre-treatment temperatures produce pulp which contains less lignocellulosic material but which hydrolyses to a higher degree than MSW treated at high temperatures. The composition of the wastepaper stream in MSW has also an impact on the enzymatic hydrolysis: Office paper and cardboard give the highest conversion while newspaper and magazine give the lowest.

Some paper additives typically used in the paper recycling industry was found to reduce enzymatic conversion of paper to sugars possibly due to fibre coating. The different pulping processes are also likely to influence MSW pulp hydrolysability. All these aspects could contribute to the large differences in glucan conversion reported in this work.

References

[1] Barbalace K, The History of Waste: Do you want to be a garbologist? Retrieved from http://environmentalchemistry.com/yogi/environmental/wastehistory.html. [Accessed on 2002]
[2] Li S, Zhang X, Andresen J M 2012 Production of fermentable sugars from enzymatic hydrolysis of pretreated municipal solid waste after autoclave process *Fuel* **92** 1 84-8
[3] Ballesteros M, Sáez F, Ballesteros I, Manzanares P, Negro M J, Martínez J M, Castañeda R, Oliva Dominguez J M 2009 Ethanol production from the organic fraction obtained after thermal pretreatment of municipal solid waste *Appl. Biochem. Biotechnol.* **161** 1 423-31
[4] Zheng Y, Pan Z, Zhang R, Labavitch J M, Wang D, Teter S A, Jenkins B M 2007 Evaluation of different biomass materials as feedstock for fermentable sugar production *Appl. Biochem. Biotechnol.* **137** 1 423-35
[5] Eley M H, Guinn G R, Bagchi J 1995 Cellulosic materials recovered from steam classified municipal solid wastes as feedstocks for conversion to fuels and chemicals *Appl. Biochem. Biotechnol.* **51** 1 387-97
[6] AEA Wastes Work 2010 The composition of municipal solid waste in Wales Welsh Assembly Government 1-60
[7] Zero Waste Scoland 2010 The composition of municipal solid waste in Scotland Scottish Government 1-54
[8] DEFRA 2009 Municipal Waste Composition: A Review of Municipal Waste Component Analyses. Retrieved from http://randd.defra.gov.uk/Default.aspx?Module=More&Location=None&ProjectID=15133 [Accessed on 9 October 2017]

[9] Sluiter J B, Ruiz R O, Scarlata C J, Sluiter A D, Templeton D W 2010 Compositional analysis of lignocellulosic feedstocks 1: review and description of methods, *J. Agric, Food Chem.* **58** 9043-53

[10] Hu F, Jung S, Ragauskas 2012 Pseudo-lignin formation and its impact on enzymatic hydrolysis *Bioresour. Technol.* **117** 7-12

[11] Goering H K, Van Soest P J 1970 Forage fibre and Fibercap methodologies Agricultural Handbook US Department of Agriculture **379** 1-20

[12] US Environmental Protection Agency 1998 Test Methods for Evaluating Solid Waste: Physical/Chemical Methods / SW-846, Method 9071B: N-Hexane Extractable Material (Hem) For Sludge, Sediment, and Solid Samples USA

[13] APHA 2005 Standard Methods for the Examination of Water and Wastewater American Public Health Association American Water Works Association Water Environment Federation Washington DC USA

[14] Agriculture and Consumer Protection, Method of food analysis Retrieved from http://www.fao.org/docrep/006/y5022e/y5022e03.htm. [Accessed on 2002]

[15] Mertens D R 2002 Gravimetric determination of amylase-treated neutral detergent fiber in feeds with refluxing in beakers or crucibles: collaborative study *J. AOAC Int.* **85** 6 1217-1240

[16] Mollaahmad M A 2008 Sustainable Fillers for Paper Master in Chemical and Biochemical Engineering **(**Department of Chemical Engineering and Geosicences, Division of Biochemical and Chemical Engineering. Lulea, Sweden, Lulea University of Technology) 1-33

[17] US Environmental Protection Agency 2013 Data Gov All Tables 1960 to 2013

[18] Vermaas J V, Petridis L, Qi X, Schulz R, Lindner B, Smith J C 2015 Mechanism of lignin inhibition of enzymatic biomass deconstruction *Biotechnol Biofuels* **8** 217 1-16

IOP Conf. Series: Earth and Environmental Science **131** (2018) 012057 doi:10.1088/1755-1315/131/1/012057

Municipal solid waste transportation optimisation with vehicle routing approach: case study of Pontianak City, West Kalimantan

M A Kamal[1] and D Youlla[2]

[1]Department of Agro-Industrial Technology, Faculty of Agricultural Technology, Universitas Brawijaya, Malang, Indonesia
[2]Doctoral Student of Natural Resource and Environment Management, Faculty of Agriculture, Universitas Brawijaya, Malang, Indonesia

E-mail: m.arif.kamal@ub.ac.id

Abstract. Municipal solid waste (MSW) transportation in Pontianak City becomes an issue that need to be tackled by the relevant agencies. The MSW transportation service in Pontianak City currently requires very high resources especially in vehicle usage. Increasing the number of fleets has not been able to increase service levels while garbage volume is growing every year along with population growth. In this research, vehicle routing optimization approach was used to find optimal and efficient routes of vehicle cost in transporting garbage from several Temporary Garbage Dump (TGD) to Final Garbage Dump (FGD). One of the problems of MSW transportation is that there is a TGD which exceed the the vehicle capacity and must be visited more than once. The optimal computation results suggest that the municipal authorities only use 3 vehicles from 5 vehicles provided with the total minimum cost of IDR. 778,870. The computation time to search optimal route and minimal cost is very time consuming. This problem is influenced by the number of constraints and decision variables that have are integer value.

1. Introduction

The volume of waste generated from household waste, offices, markets and public facilities in Pontianak reaches 1,807 m^3 of waste per day [1]. Pontianak City has 114 temporary garbage dumps (TGD) located at 58 points and spread in 6 Sub-districts [1]. The volume of waste that can be accommodated at all pooling stations in Pontianak city is 655,525 m^3 of waste consisting of 493.03 m^3 at market GD, 122 m^3 at office GD, 26 m^3 of garbage at public facility and 14.50 m^3 at industrial area GD. These conditions forced Pontianak City Sanitation Office to impose three categories of transportation, namely regular transportation, overtime transportation and transportation by involving the cooperation with the third party. On regular garbage transportation, the total volume of waste that can be transported is 972 m^3 per day. The garbage transportation with the total amount of waste that can be transported is 394 m^3 of garbage every day. Transportation of waste with the cooperation of the third party is able to transport the waste as much as 120 m^3.

High waste volume requires vehicle and greater freight frequency. These resulted in high costs and vehicle operating expenses, particulary if there are damaged or repaired vehicles. As a result, the level of waste transportation services in Pontianak city can only reach its maximum service level is 85% and tends to decrease along with the increase of population every year resulting in an increase in the volume of waste. Pontianak City Government still has to work hard to improve the service and of course it needs to spend very big cost for garbage transportation. The cost of MSW transportation in this city is still very high. Thus, the MSW transportation study by considering vehicle routing is necessary. Therefore, this study is intended to minimise the number of vehicles, lower operational costs and vehicle maintenance, maximise vehicle utility, and maximise waste hauling at TGD.

2. Materials and Methods

The method of MSW transport optimisation in Pontianak City in this study uses the vehicle routing approach or often called as vehicle routing problem (VRP) method. In this method, it is assumed that the vehicle capacity is the same and each node is visited only once. The demand for each node is assumed not to exceed the vehicle's capacity. The vehicle journey begins from the vehicle depot then proceeds at each point to meet the demand per vehicle load limit and ends back at the vehicle depot [2]. The assumption used in the method makes the VRP approach still very common to be applied in the case of MSW transportion. Therefore it is necessary to add decision variables and constraint to see the actual conditions closer.

Transportation of MSW from one or more TGD to final garbage dump (FGD) needs to take into consideration the following aspect: the number and capacity of the fleet used, the operational cost of each vehicle, the volume of waste from each TGD, the distance from the TGD to the FGD and the distance between one TGD and another TGD. Each assigned vehicle transports the garbage from the Depo of the vehicle to a TGD or more to the FGD and return to the Depo vehicle when there is no garbage in the TGD to be transported. One vehicle can transport more garbage than one TGD if the garbage volume at the previous TGD is less than the capacity of the vehicle [3,4].

Vehicles that have reached maximum capacity move towards the TGD to clear the load. The vehicle returns to the Depo vehicle if there is no garbage in the TGD to be transported. If there is still waste left in GD that cannot be transported in advance because it exceeds the maximum capacity of the vehicle, the vehicle must return to the TGD to pick the remaining waste or be transported by another vehicle. Therefore, a TGD can be visited more than once by the same carrier or other vehicles. The approach used for transporting waste is commonly known the vehicle routing with split-deliveries [3]. Another combination approach is called the mix-vehicle capacities, used when the vehicle capacity for transporting the garbage is not identical [5].

The main purpose of the VRP method is to investigate the cost-efective routes for transporting the MSW [2]. The vehicle routing case, especially garbage hauling, is a linear integer programming problem, where the resulting decision variable must be in a rounded form. Therefore, the problem needs to be solved by integer programming. One of the reliable integer programming algorithms for the case is the Branch and Bound method. The Branch and Bound method divides each solution in linear programming into sub-solutions for rounding each sub-solution, the value of the continuous optimum solution of the purpose function is set as the upper limit. The best round solution becomes the lower limit. Sub-solution with a lower limit than the existing lower border is not included in the next analysis. A feasible integer solution is either as good as or better than the upper limit for each sub-solution sought. If such a solution occurs, a sub-issue with the best upper limit is selected for branching [6].

3. Modelling

3.1. Model restriction and assumption

The model of MSW transport in this study is formed using basic model of VRP by imposed some constraints. These constraints are used in order to relax the limitation of vehicle trip to the TGD. The TGD can be visited by more than one vehicle with the same vehicle or different vehicle. Vehicle operating costs are assumed to be part of transportation costs per vehicle for every kilometre travelled. The MSW volume from each TGD is assumed to be fixed in one day and does not increase during the transportation time. All types of vehicles are assumed to pass the route from the TGD to the FGD without any obstacles. All vehicles are in good operation and are capable of transporting waste in accordance with their maximum capacity.

3.2. Model formulation
3.2.1. Sets

Notation	Definition
N	The set of TGD having a volume of less than or the same as the vehicle maximum capacity $= (2,3,...,n)$,
M	The set of TGD having a volume exceeding the vehicle maximum capacity $= (n+1,n+2,...,n+m)$,
S	The set of all *node* $= (1,2,...,n,n+1,n+2,...n+m,n+m+1)$, as 1 states the depo and n+m+1 states the TGD
V	The set of all vehicles $= (1,2,...,v)$.

3.2.2. Indexes

Notation	Definition
i, j, p	Indexes to state the node,
k	Indexes to state the vehicle

3.2.3. Parameters

Notation	Definition
S_i	The garbage volume on *node i*,
w_k	Vehicle capacity *k*,
c_k	The cost of transportation per km on the transportation of *k*,
d_{ij}	the distance of *node i* and *node j*
$BigM$	Positive constants having a relatively bigger value

3.2.4. Decision Variables

Notation	Definition
f_{ik}	Part of garbage volume on *node i* transported by vehicle *k*,
Q_{ik}	The garbage cumulative volume on vehicle *k* after leaving node *i*,
N_k	The number of vehicle retaking to *k*,
x_{ijk}	It scores 1 if vehicle K visits *node j* after *node i* and it scores 0 if it is on the contrary,
y_k	It scores 1 if vehicle k is used to transport garbage and it scores 0 if it is on the contrary

IOP Conf. Series: Earth and Environmental Science **131** (2018) 012057 doi:10.1088/1755-1315/131/1/012057

3.2.5. Objective Function

Notation	Definition
$\sum_{k\in V}\sum_{j\in S}\sum_{i\in S} c_k d_{ij} x_{ijk}$	The objective function in this model aims to determine the garbage transportation route for the purpose of minimising the total transportation cost

3.2.6. Constraints

Definition

Not every vehicle leaves the depo

$$\sum_{j\in N\cup M} x_{1jk} \le 1, \forall k \in V$$

Vehicles leaving the depo will be used to transport garbage,

$$\sum_{j\in N\cup M} x_{1jk} = y_k, \forall k \in V$$

No node visited by unused vehicle,

$$x_{ijk} \le y_k, \forall_i, j \in S, \forall k \in V$$

Vehicles visiting the node must leave the node

$$\sum_{\substack{i\in S\\ i\ne p}} x_{ipk} - \sum_{\substack{j\in S\\ p\ne j}} x_{pjk} = 0, \forall p \in S, \forall k \in V$$

TGD with garbage volume of less than or equal to the maximum capacity of vehicles is visited exactly once,

$$\sum_{k\in V}\sum_{\substack{i\in S\\ i\ne j}} x_{ijk} = 1, \forall j \in N$$

TGD with waste volume exceeding the maximum capacity of vehicles are visited more than once,

$$\sum_{k\in V} f_{ik} = 1, \forall i \in M,$$

$$\sum_{j\in S} x_{jik} \ge f_{ik}, \forall i \in M, \forall k \in V.$$

Cumulative waste volume on vehicles leaving TGD will increase

$$Q_{ik} + s_j - Q_{jk} \le (1 - x_{ijk})BigM,$$

$$\forall j \in N, \forall i \in S, \forall k \in V,$$

$$Q_{ik} + f_{jk}s_j - Q_{jk} \le (1 - x_{ijk})BigM,$$

$$\forall j \in M, \forall i \in S, \forall k \in V.$$

Definition

The number of retake of vehicle is the frequency of each vehicle visiting the TGDs

$$\sum_{i\in N} x_{i(n+m+1)k} = N_k, \forall k \in V.$$

The used vehicle must return to the depo

$$x_{(n+m+1)1k} = y_k, \forall k \in V.$$

Cumulative waste volume should not exceed the capacity of each vehicle

$$Q_{ik} \le w_k, \forall k \in V, \forall i \in S$$

The cumulative waste volume of each vehicle leaving the depo and TGD is 0.

$$Q_{1k} = 0, \forall k \in V$$

$$Q_{(n+m+1)k} = 0, \forall k \in V$$

No trips to the same node,

$$x_{iik} = 0, \forall i \in S, \forall k \in V$$

Constraints of non-negativity,

$$Q_{ik} \ge 0, \forall i \in S, \forall k \in V$$

$$0 \le f_{ik} \le 1, \forall i \in M, \forall k \in V$$

Binary constraint

$$x_{ijk} = \{0,1\}, \forall k \in V, \forall i, j \in S$$

$$y_k = \{0,1\}, \forall k \in V$$

IOP Conf. Series: Earth and Environmental Science **131** (2018) 012057 doi:10.1088/1755-1315/131/1/012057

The vehicles used should visit the landfill

$$\sum_{i \in N \cup M} x_{i(n+m+1)k} \leq y_{k,} \forall k \in V.$$

3.3. *Model implementation*

The model implementation of garbage transport can be seen in Figure 1. The transport vehicles move starting from the Depo to the TGDs to pick up the scattered waste in several villages. The vehicles continue to haul the garbage up to the maximum capacity of the vehicle. After reaching maximum capacity, the vehicles move to the TGDs to dispose the garbage and empty its load. Then, the vehicle has two options, namely moving to the vehicle Depo or taking the remaining garbage that cannot be transported at the previous TGDs.

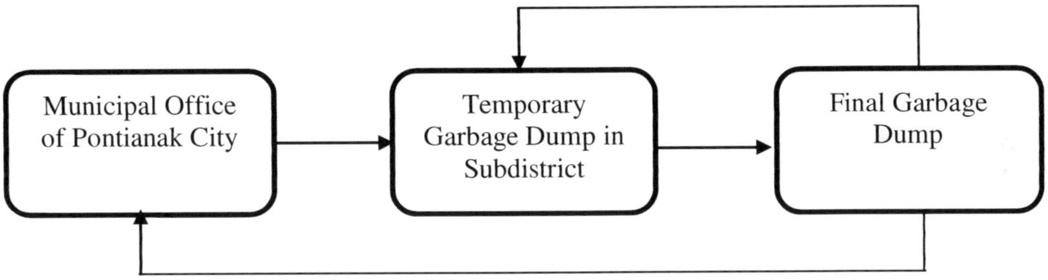

Figure 1. Waste transportation mechanism

The optimisation research on garbage transportation was focused on Darat Sekip Village of Pontianak Kota sub-district. In this area, there are 6 TGDs with total capacity equal to 63.62 m^3. There is a polling station with a volume of garbage 40.38 m^3 located on Jalan Matahari Timur. The location and volume of waste of each TGD can be seen in Table 1 below. This study is limited to six polling stations in Pontianak sub-district because the area is located in the middle of the city. In addition, the area has a high waste load that will disturb the cleanliness and comfort of the city therefore it needs more attention.

Table 1. The location and volume of TGD waste.

No	Location	Subdistrict	Volume (m^3)
1	Jl Serayu	Darat Sekip	3.99
2	Jl Matahari Timur	Darat Sekip	40.38
3	Jl Tanjungpura Indah	Darat Sekip	5.03
4	Jl KH. Ahmad Dahlan	Darat Sekip	5.88
5	Jl HOS Cokroaminoto	Darat Sekip	5.88
6	Jl Wolter Monginsidi	Darat Sekip	2.46
	Total		63.62

The garbage transport vehicles owned by the Pontianak City Government consisted of several types and each vehicle has different transport capacity variations. In this research, the vehicles used were limited to the types of large vehicles (i.e. big trucks and *Armroll* Trucks). Both types of trucks are the vehicles that are mostly used and considered reliable in carrying out waste transport activities. Five unit vehicles were allocated for garbage transportation at Pontianak Kota Sub-district. The volume of vehicle transport and travel expenses for each vehicle can be seen in Table 2.

Table 2. Characteristics of waste transportation vehicles.

Transportation Type	Volume of Vehicles (m^3)	Travel expense (IDR per km)	Unit Amount
Dump Truck	16.98	IDR 30,000	3
Armroll Truck	11.32	IDR 25,000	2

Based on the data from TGD, vehicle depo, and FGD, the nodes or points that must be visited in the routing of garbage transport vehicles can be determined. Distance of each node is used to calculate the cost of the vehicle from each transport route. Node 1 states the vehicle depo is located in the Sanitation Service of Pontianak City, while Node 8 states the FGD. Both nodes do not have capacity because they serve as the starting point and dumping point. Nodes 2 up to 6 are TGD with variations in the volume of waste that must be transported as shown in Table 3.

Table 3. Point (*node*) used in the model.

Node	Description	Location	Capacity (m^3)
1	Vehicles Depot	Sanitary agency of Pontianak city	-
2	TGD 001	Jl Serayu	3.99
3	TGD 002	Jl Matahari Timur	40.38
4	TGD 003	Jl Tanjungpura Indah	5.03
5	TGD 004	Jl KH. Ahmad Dahlan	5.88
6	TGD 005	Jl HOS Cokroaminoto	5.88
7	TGD 006	Jl Wolter Monginsidi	2.46
8	FGD	Kecamatan Batulayang	-

Based on the data, nodes are made the distance matrix that connects each node. The matrix represents the mileage each vehicle must pass when transporting garbage from the vehicle depots to the TGD. Table 4 shows the distance between nodes from Depo to TGD.

Table 4. Distance between nodes (km).

Node	1	2	3	4	5	6	7	8
1	0	4.8	4.2	5.1	5.3	5.4	5.2	14.6
2	5.3	0	0.95	2.4	2.6	2	1.8	8
3	4.4	0.95	0	2.4	2.6	2	1.8	8
4	6.6	2.4	1.9	0	0.7	0.8	0.5	7
5	7.3	3.6	2.5	1.2	0	0.079	1.4	7.1
6	7	2.9	2.4	1.2	0.079	0	1	7.2
7	6.6	2.4	1.3	0.19	0.23	0.3	0	7.2
8	14	6.5	5.8	6.3	6.7	6.7	6.5	0

4. Results and Discussion

Based on the calculation obtained, it was found that 3 vehicle with 3 routes are able to transport all the volume of waste in Pontianak Kota District optimally with the lowest cost. The route of the 3 vehicles is the best route that will provide the least cost. The total cost generated by the 3 routes through which 3 vehicles are IDR 778,870. The calculation results show optimum global solution parameters which means the solution is the best solution in terms of the number of vehicles used, the route travelled and the cost incurred. The global optimal solution is the best search result above the near optimal and satisfied solution criteria [7]. Computational results can be seen in Figure 2. Vehicles used for garbage transport at 6 TGD in Pontianak Kota sub-district are as many as 3 vehicles out of 5 vehicles. The 3rd vehicle must visit node 3 twice as the volume of garbage if the node exceeds the maximum capacity of the vehicle. The vehicle route and the volume of waste that can be transported by each vehicle can be seen in Table 5.

Table 5. Route of garbage transport.

Code Transport	Type of Transport	Trip	Transportation route
TRCK001	Big Truck	1	1 → 2 → 3 → 8 → 1 [0] [3,99] [11,46] [0] [0]
TRCK002	Big Truck	1	1 → 3 → 4 → 8 → 1 [0] [11,94] [5,03] [0] [0]
TRCK003	Big Truck	2	1 → 3 → 8 → 7 → 6 → 5 → 8 → 1 [0] [16,95] [0] [0] [0] [0] [0] [0]
ARM001	*Armroll*	0	-
ARM002	*Armroll*	0	-

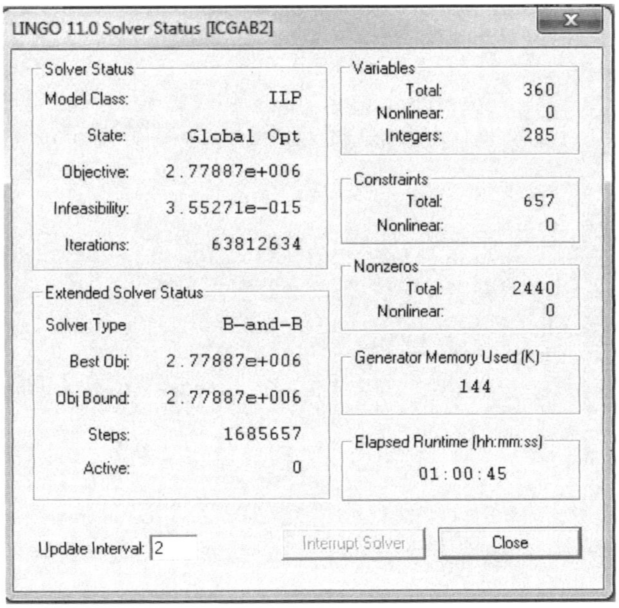

Figure 2. Minimum cost of transport routing with Branch and Bound algorithms

The model built on the waste transport research in Pontianak City consist of 360 decision variables which 285 of them are in the form of integer variables. Meanwhile, the constraints in the created mathematical model consisted of 657 constraints. The number of variables and constraints raised generated the solution space in the search for optimal value becomes very large. In addition, most variables are integer constraints. The amount of search space and enumeration of candidate solutions into variables make the complexity of the problem higher [8]. The consequence is that the time spent on finding the optimal solution becomes very long. In this research, it took a computation time of

1:00:45, to find the minimal cost of the vehicle routing with 8 points (node) of TGD and 5 vehicles. The computation time of optimal solution could affect the application of vehicle route optimisation with the number of points and larger vehicles. The computation time is affected by the type of variables and constraints.

5. Conclusion

The findings confirmed that the obtained minimum cost of MSW transportation in Pontianak City was equal to IDR 778,870, - for one day. The computation results suggest that the most optimal solution is to use 3 travel routes for MSW transport with only 3 vehicles from the total of 5 vehicles provided. Particularly on the third vehicle, it must carry two trips to transport garbage at the third TGD because it exceeds the carrying capacity. The computation time to find the optimal solution is still very high because of the number constraints and integer decision variables. Therefore, the complexity to find the optimal solution is needed to be further studied by simplifying the model equations and using better searching algorithms.

References

[1] Lestari S, Nurlaili S, Fitrianingsih Y 2016 Evaluasi pengangkutan sampah kota Pontianak (Evaluation on the municipal solid waste transportation in Pontianak City) *J. Mahasiswa Teknik Lingkungan UNTAN* **1** 1–10 [In Indonesian]

[2] Vigo D, Toth P 2014 Vehicle Routing: Probles, methods, and Applications 2nd Edition mathematical Optimization Society Bologna Italy 467

[3] Sahoo S, Kim S, Kim B-I, Kraas B, Popov A 2005 Routing optimization for waste management *Interfaces* **35** 24–36

[4] Son L H, Louati A 2016 Modeling municipal solid waste collection: A generalized vehicle routing model with multiple transfer stations, gather sites and inhomogeneous vehicles in time windows *Waste Manag.* **52** 34–49

[5] Golden B, Assad A, Levy L, Gheysens F 1984 The fleet size and mix vehicle routing problem *Comput. Oper. Res.* **11** 49–66

[6] Archetti C, Bianchessi N, Speranza M G 2014 Branch-and-cut algorithms for the split delivery vehicle routing problem *Eur. J. Oper. Res.* **238** 685–98

[7] Fischetti M, Toth P, Vigo D 1994 A Branch-and-bound algorithm for the capacitated vehicle routing problem on directed graphs *Oper. Res.* **42** 846–859

[8] Klein C M, Morin T L 1991 Conjugate duality and the curse of dimensionality *Eur. J. Oper. Res.* **50** 220–8

Quantifying and analysing food waste generated by Indonesian undergraduate students

P Mandasari

Agribusiness Study Program, Faculty of Agriculture, Sebelas Maret University, Indonesia

E-mail: putriesti.mandasari.uns@gmail.com

Abstract. Despite the fact that environmental consequences derived from food waste have been widely known, studies on the amount of food waste and its influencing factors have relatively been paid little attention. Addressing this shortage, this paper aimed to quantify monthly avoidable food waste generated by Indonesian undergraduate students and analyse factors influencing the occurrence of avoidable food waste. Based on data from 106 undergraduate students, descriptive statistics and logistic regression were applied in this study. The results indicated that 4,987.5 g of food waste was generated in a month (equal to 59,850 g yearly); or 47.05 g per person monthly (equal to 564.62 g per person per a year). Meanwhile, eating out frequency and gender were found to be significant predictors of food waste occurrence.

1. Introduction

Studies on food waste amount and its influencing factors have relatively been received little attention compared to environmental consequences derived from food waste. Some researchers, such as Morgan [1] and Young et al. [2], examined how a firm influences its consumer to reduce environmental impacts of the product usage in their household. More specifically, Young et al. [2] applied the behavioral change approaches on environmental issue in hoped that the change in attitude and behaviour would be able to shift consumers' lifestyle into more environmental friendly. On the other hand, some previous studies by Koivupuro et al. [3], Bortoleto et al. [4], and Cox et al. [5] have been focused on calculating the amount of wasted food, finding the influence of behavioral factors towards food waste, and analyzing waste prevention behavior in household scope.

Schneider and Obersteiner [6] and European Commission [7] stated that food waste was categorized into avoidable and unavoidable based on its edibility. In this study, only the avoidable food is referred to as food waste, which defined by Koivupuro [3] as any kinds of food that could have been consumed prior to disposal.

As the number of avoidable food waste tends to escalate each year all over the world, conducting research in this area is emerged. United States, one of the biggest food wasters, disposing nearly 277 kg of food per person each year [8]. Meanwhile, United Kingdom threw away edible food equal to £13bn per year [9]. Surprisingly, not only the developed countries but also developing country have to deal with this issue. Indonesia, for example, has becoming the world's second largest food waster, disposing almost 300 kg of food per person each year [8]. This is an ironic fact since millions in the country suffer from malnutrition [8], while at the same time, the amount of avoidable food waste is huge.

Addressing those gaps, this research purposed to quantify monthly avoidable food waste generated by Indonesian undergraduate students and analyse factors influencing the occurrence of avoidable food waste. This study focused on specific group (youngsters), as Wassermann and Schneidar [6] mentioned that young people waste more food than the elders.

2. Materials and Methods

2.1. Sample

Undergraduate students from four faculties (Faculty of Agriculture, Faculty of Economic and Business, Faculty of Teacher Training and Education Science, and Faculty of Engineering) in Sebelas Maret University (Surakarta City) were conveniently chosen as respondents. A total of 110 questionnaires were distributed to respondents for each faculty sequentially started from September 1st up to 9th 2017. Among those number, 106 (96.36%) questionnaires were used due to its completeness.

The selected respondents consisted of 90 females (84.91%) and 16 males (15.09%); both groups came from various semesters (1st, 3rd, 5th, and 9th) with age category ranged from 17 to 24 years old. The majority of respondents originated from the 5th semesters (60%), and most of them were 20 years old (52%).

2.2. Instruments measurement

2.2.1. Food waste

Different from household unit analysis used by Koivupuro et al. [3], this study applied an individual unit as quantification basis. More specifically, the quantification of avoidable food waste was based on each meal taken by respondents. Meanwhile in household unit, food waste was measured not only from individual meal but also from all kinds of food sourced from household. As a consequence, it is estimated that the amount of food waste measured in individual unit is fewer than the household unit.

Some questions were depicted in the questionnaires in order to measure the amount of avoidable food waste. Since the questionnaires was only once given, the respondents were expected to remember their previous eating habits, specifically on the intensity of wasting the edible food in each meal they consumed (breakfast, lunch, dinner, or any meals taken in between). This intensity was measured in a day and in a week. In addition, respondents were also expected to measure how many spoons they used to waste the edible food in each of their meal.

In regards to measure the amount of avoidable food waste, data generated by 106 respondents were processed by multiplying the intensity of weekly avoidable food waste and the amount of food being wasted in terms of tablespoons. Then, the result was converted into gram and month. A tablespoon of food waste was equivalent to 15 g.

2.2.2. Factors influencing the avoidable food waste occurrence

In this research, variables of individual characteristic (gender), financial profile (monthly living cost), and behavior (food shopping frequency and eating out frequency) were selectively chosen from Kuivopuro et al. [3]. Then, those factors were modified into more suitable context for undergraduate students.

Gender was considered as one of the most common category or predicted factor used in a scientific study. In this research, male and female were predicted to produce different amount of food waste since both groups tended to have different role and eating habits. Most woman, for example, take in charge for their family meals, for that reason, decisions regarding food menu and quantity were often lead to the level of food waste generated in her household. Additionally, some woman might consume less amount of meals than men, therefore, finishing one food portion could be a burden for them. This would lead to higher avoidable food waste. In order to measure whether gender was a predictor of food waste occurrence, the respondents were divided using nominal scale: 1 (woman) and 2 (man).

IOP Conf. Series: Earth and Environmental Science **131** (2018) 012058 doi:10.1088/1755-1315/131/1/012058

This study also measured respondent's monthly living cost (excluded housing rent). The housing rent was not included due to the variety of student's accommodation. Many of them stayed in boarding houses, while some of them lived with their parents or other family members. Therefore, in order to eliminate bias, monthly payment on housing was excluded in measuring living cost.

It is expected that the different amount of living cost would affect one's buying power for monthly needs, including for food. Higher living cost could lead into higher chance of uncontrolled grocery (or meal) buying that eventually triggers more avoidable food waste. The monthly living cost was measured by 7 scale: 1 (< IDR 500,000); 2 (IDR 501,000 - IDR 1,000,000); 3 (IDR 1,001,000 - IDR 1,500,000); 4 (IDR 1,501,000 - IDR 2,000,000); 5 (IDR 2,001,000 - IDR 2,500,000); 6 (IDR 2,501,000 - IDR 3,000,000); 7 (>IDR 3,000,000).

Furthermore, the food shopping frequency was predicted as factors influencing food waste occurrence. The higher frequency of food shopping was expected to lead into higher food being wasted due to its durability limit and too much cook servings [3]. The expired date might be forgotten as many foods were stacked even before consumed. Therefore, this research measured the food shopping frequency by asking the respondents on how often they buy "raw" food.

Additionally, the eating out frequency was also expected to be a predictor of food waste occurrence. As someone has less control over the amount and taste from the food they bought, the higher eating out frequency was expected to trigger more waste from edible food. This factor was measured by asking the respondents on how often they buy meal outside, such as in food stalls or restaurants. Then, those numbers were estimated into monthly frequencies.

H_1: Gender is predictor of the avoidable food waste occurrence.

H_2: Monthly living cost is predictor of the avoidable food waste occurrence.

H_3: Food shopping frequency is predictor of the avoidable food waste occurrence.

H_4: Eating out frequency is predictor of the avoidable food waste occurrence.

2.3. Statistical analysis

The data collected through questionnaire were analysed using Microsoft excel and SPSS software. Firstly, the food waste was quantified with Excel. Subsequently, using SPSS, logistic regression model was applied to determine factors influencing the avoidable food waste occurrence. Finally, descriptive statistics from SPSS was used to classify the food wasters.

3. Results and Discussion

3.1. Food waste quantification

The descriptive statistics show that the range of wasted food resulted from the respondents was 0 up to 420 g monthly, or 47.05 g per person on average. In other words, the sample provided evidence that the average of avoidable food waste resulted by Indonesian youngsters was 564.62 g per person yearly.

Table 1. Quantification of avoidable food waste per month.

Food Waste Quantity per Month	Number of persons	%
0 g	49	46.20
1 – 100 g	34	32.00
101 – 200 g	18	17.00
201 – 300 g	3	2.80
301 – 400 g	2	1.89

The data depicts that among 106 respondents, the number of non food waster and food waster was almost equal. In particular, almost a half of Indonesian undergraduate students did not waste the edible food since they fully consumed their meals. Perhaps this was caused by the cost living limitation and the need to fulfil higher energy from the food. For instance, the university students usually have certain amount of cost living, especially if they live separately from their family. In this case, they would use the money allowance wisely for monthly living cost, including when spending it for food. Having this limitation, they tend to have more intention to finish their meals. Moreover, the youngsters are considered to be more active than the older ages, therefore, they have a higher need to fulfil energy required to do those activities.

On the other hand, there were 53.80% respondents produced up to 400 grams of food waste each month. The fact that more than a half of students were food waster could be a concern since any kinds of edible food waste are dissipating the farmer's effort in producing raw food materials. Koivupuro et al. [10] and Gustavsson et al. [11] highlighted that when edible food is wasted, not only the energy and resources to produce the food were useless, but also the environment effects from food production and transformation were neglected.

3.2. Factors influencing the occurrence of avoidable food waste

The result of Hosmer and Lameshow test indicates that the logistic regression model fits the data (sig. 0.215). In addition, the Nagelkerke R Square shows that living cost, food shopping frequency, eating out frequency, and gender contributed as many as 18.7% in influencing the occurrence of avoidable food waste among Indonesian youngsters. Table 2 depicts that, among four predicted factors influencing the avoidable food waste existence, only two were found to be statistically significant. The two factors namely eating out frequency and gender, resulting 0.006 and 0.035 significant values, respectively. Meanwhile, the two other factors (living cost and food shopping frequency) were not predictors of avoidable food waste occurrence among the youngsters.

Table 2. Factors influencing the occurrence of avoidable food waste.

Chi-square	df	Sig.
10.765	8	0.215

Table 3. Factors influencing the occurrence of avoidable food waste.

-2 Log likelihood	Cox and Snell R Square	Negelkerke R Square
130.011	0.140	0.187

Table 4. Factors influencing the occurrence of avoidable food waste.

							95.0% C.I.for EXP(B)	
	B	S.E.	Wald	df	Sig.	Exp(B)	Lower	Upper
LivingCost	.396	.269	2.162	1	.141	1.486	.876	2.521
FoodShopFreq	.061	.077	.632	1	.427	1.063	.914	1.236
EatingOutFreq	.022	.008	7.425	1	.006*	1.023	1.006	1.039
Gender	-1.325	.629	4.435	1	.035*	.266	.077	.912
Constant	-.163	.939	.030	1	.862	.850		

Based on the statistical result, eating out frequency was strongly significant in influencing the occurrence of edible food being wasted. The higher eating out frequency, the more edible food was wasted by the youngsters. This was probably because most of respondents (85%) eating out on a regular bases, especially when they had activities on campus. When that happened, they could not freely control the food amount and taste from the food they bought. The data from questionnaire revealed that the main reasons why the youngsters wasted edible food were 38.60% due to the dissatisfaction with the food taste (although the meal was in the right portion) and 36.84% due to the excessive food portion given by the seller (or waiter/ waitress). Accordingly, the more avoidable food being wasted occurred.

The result also mentioned that gender was moderately significant as predictor of food waste occurrence. In line with this result, Kuivopuro et al. [3] stated that single woman was the one who wasted more edible food compared to other groups (single man, young households, and adult households). Perhaps one of the main reasons was related to the high intensity of eating out, where woman often given the excessive food portion. Since the portion was usually much more than they could consume, the woman ended up wasting the avoidable food.

4. Conclusion

To underline, this research found that Indonesian youngsters produced 47.05 g of avoidable food waste per person each month (564.62 g each year). Moreover, eating out frequency and gender were proven to be predictors of food waste occurrence.

References

[1] Morgan E 2015 Plan A: Analysing business model innovation for sustainable consumption in mass-market clothes retailing *Journal of Corporate Citizsh* 73-98

[2] Young W, Russel S V, Robinson C A, Barkemeyer R 2016 Can social media be a tool for reducing consumers' food waste? A behaviour experiment by a UK retailer *Resources, Conservation and Recycling* 3401

[3] Koivupuro H K, Hartikainen H, Silvennoinen K, Katajajuuri J M, Heikintalo N, Reinikainen A, Jalkanen L 2012 Influence of socio-demographical, behavioural and attitudinal factors on the amount of avoidable food waste generated in Finnish households *International Journal of Consumer Studies* **36** 183-191

[4] Bortoleto A P, Kurisu K H, Hanaki K 2012 Model development for household waste prevention behaviour *Waste Management* **32** 2195-2207

[5] Cox J, Giorgi S, Sharp V, Strange K, Wilson D C, Blakey N 2010 Household waste prevention-a review of evidence *Waste Management and Research* **28** 193-219

[6] Schneider F, Obersteiner G 2007 Food waste in residual waste of households- regional and socio-economic differences. Proceedings of the International Waste Management and Landfill Symposium, Sardinia, Italy, 2007, pp. 469-470

[7] Uropean Commision (online) 2010 Preparatory study on food waste across EU 27. Retrieved from http://ec.europa.eu/environment/eussd/pdf/bio_foodwaste_report.pdf. [Accessed on 20 August 2017]

[8] The Jakarta Globe (online) 2017 Indonesia second largest food waster. Retrieved from http://thejakartaglobe.com. [Accessed on 07 December 2017]

[9] The Guardian (online) 2017 UK Throwing away 13bn pounds of food each year, latest figures show. Retrieved from http://theguardian.com. [Accessed on 07 December 2017]

[10] Koivupuro H K, Jalkanen L, Katajajuuri J M, Reinikainen, A, Silvennoinen K (online) 2010 Food waste in the supply chain. Retrieved from http://www.mtt.fi/mttraportti/pdf/mttraportti12.pdf. [Accessed on 22 August 2017]

[11] Gustavsson J, Cederberg C, Sonesson U, van Otterdijk R, Meybeck A 2011 Global food losses and food waste. Rome: Food and Agriculture Organization of the United Nations, 2011

Institute of Physics
Dirac House, Temple Back
Bristol BS1 6BE UK

ISSN: 1755-1307
ISBN 978-1-5108-6010-0

9 781510 860100

CURRAN ASSOCIATES INC.
proceedings.com